PROCEEDINGS OF
SYMPOSIA IN PURE MATHEMATICS

VOLUME XX

1969 NUMBER THEORY INSTITUTE

AMERICAN MATHEMATICAL SOCIETY
PROVIDENCE, RHODE ISLAND
1971

Proceedings of the 1969 Summer Institute on Number Theory: Analytic Number Theory, Diophantine Problems, and Algebraic Number Theory
Held at the State University of New York at Stony Brook
Stony Brook, Long Island, New York
July 7–August 1, 1969

Prepared by the American Mathematical Society
under the National Science Foundation Grant GP-9551

DONALD J. LEWIS
Editor

International Standard Book Number 0-8218-1420-6
Library of Congress Catalog Number 76-125938

AMS 1970 Subject Classifications. Primary 10XX, 12XX.
Printed in the United States of America

CANDID CAMERA—
1969 NUMBER THEORY INSTITUTE

J. AX AND A. PFISTER

P. X. GALLAGHER AND W. SCHMIDT

T. KUBOTA

B. DWORK AND B. J. BIRCH

K. IWASAWA

A. SELBERG

PAUL TURÁN

A. BAKER

H. M. STARK

H. P. F. SWINNERTON-DYER

KURT MAHLER

P. T. BATEMAN AND E. GROSSWALD

D. J. LEWIS

Y. KAWADA

H. E. RICHERT AND E. WIRSING

A. SCHINZEL

E. BOMBIERI

CONTENTS

PREFACE

This book is an outgrowth of the American Mathematical Society's Sixteenth Summer Research Institute, which had as its topics algebraic number theory, diophantine problems, and analytic number theory.

The Organizing Committee for the institute consisted of James Ax, Paul T. Bateman, K. Iwasawa, D. J. Lewis (Chairman), and Atle Selberg. The institute was held at the State University of New York at Stony Brook from July 7 to August 1, 1969, and was financed by grants from the National Science Foundation and the New York State Science and Technology Foundation.

During the 1960's a large number of old problems in number theory were solved: some by refinements of known methods, others by the introduction of entirely new methods. One of the purposes of the institute was to acquaint the participants from the various areas of number theory with the important results and methods developed recently, especially in areas other than their own. It is impossible to cover all areas of number theory in a single institute; many of the areas not emphasized at this institute were the subject of other institutes and conferences held here and abroad this past year. In order to survey the achievements of the decade, the Organizing Committee invited sixteen speakers to each give a series of lectures. In addition to the lecture program, there was a seminar program. The list of seminars with speakers and titles are given below; for the most part, the results announced in the seminars will appear elsewhere. This volume consists of the sixteen invited lecture series, plus nine seminar talks which were felt to have been particularly effective surveys. The papers are addressed to a general number theory audience rather than to a group of specialists and are meant to enable a number theorist to become acquainted with important innovations in areas outside his own specialty. It is hoped that this collection of papers will facilitate access to various parts of number theory and foster further development.

In this book the papers are arranged so that those treating related topics or using related techniques appear together. The first few papers treat the role of algebraic geometry in number theory. The highlight of the institute was the series of fourteen lectures by H. P. F. Swinnerton-Dyer on this topic. His paper is an excellent introduction to possible uses of algebraic geometry. The paper by W. Waterhouse and J. Milne treats abelian varieties over finite fields. The paper by N. Katz on p-adic cycles covers the same material as that presented by B. Dwork at the institute, but the presentation is different. Dwork's proof will appear elsewhere.

These papers are followed by a sequence of papers by O. T. O'Meara on automorphisms of the orthogonal group; K. Iwasawa on Jacobians for number fields; B. J. Birch on K_2-theory; Y. Kawada on class formations; J. A. Shalika on nonabelian class field theory; T. Storer on cyclotomy; A. Schinzel on reducibility of polynomials; and A. Pfister on the quantitative form of Hilbert's seventeenth problem. These papers treat questions in algebraic number theory or make use of algebraic techniques. Several of these papers serve as an introduction to difficult and sophisticated theory, while others are thorough surveys of a subject.

The first paper by J. Ax demonstrates the relevancy of logic as a tool in number theory. The paper by Julia Robinson is a revision of her lectures which incorporates the recent proof of Ju. V. Matijasevič of Hilbert's tenth problem. This is followed by a report of A. Baker on his effective methods for solving binary equations, methods which at first sight might be judged to be effective from a theoretical point of view but not from a computational one; however, Baker, Davenport, Ellison, and others have demonstrated that with skill these methods can be used very satisfactorily to find all solutions.

The next set of papers deals with transcendental numbers and diophantine approximations. There is a discussion by J. Ax of Schanuel's all-encompassing conjectures, and there is the long-awaited paper by E. Wirsing on approximation of algebraic numbers by algebraic numbers, including some refinements and extensions of those ideas. The paper by K. Mahler is an extensive survey of the theory of transcendental numbers including that by Shidlovsky, and the paper by W. Schmidt discusses his recent work on the existence of Mahler's T-numbers

Next, there are two extensive papers by H.-E. Richert and by Atle Selberg on sieve methods. The paper by Selberg contains proofs of results obtained over several decades but not previously published.

These are followed by papers by E. Bombieri on density theorems for the zeta function; P. Turán on recent results in analytic number theory; and E. Wirsing on characterizing the logarithm as an additive function. The paper by T. Kubota treats the reciprocity law and automorphic functions, and a second paper by Birch treats elliptic curves and modular forms. Finally, there is a paper by H. Stark surveying the class number problem for complex quadratic fields, and there is a paper by D. Shanks on class number and genera.

The photographic insert consists of photos taken of the participants at work and at play by the institute's roving photographer, Carolyn Dana Lewis.

It is an immutable fact of mathematical publishing that there is a substantial period of time between the completion of a manuscript and its appearance in printed form. In a field such as number theory, peopled with energetic and imaginative researchers, it is a foregone conclusion that during this interval a number of important results will be discovered, including answers to problems raised at the institute and in the published proceedings. We note in passing that since these manuscripts were submitted, W. Schmidt has proved the n-dimensional Thue-Siegel-Roth theorem; A. Baker and H. Stark have determined the complex quadradic fields with class number two; and E. Bombieri (with P. X. Gallagher and H. Montgomery) has given a simple version of the large sieve.

SEMINARS

Algebraic Number Theory

Olga Taussky, Hilbert's Theorem 94.
Richard B. Lakein, Euclid's algorithm in imaginary quartic fields.
H. Heilbronn, Density theorems for cubic fields.
S. Ullom, Groups, rings and cyclotomic fields.
B. Dwork, p-adic cycles.
William C. Waterhouse, Abelian varieties over finite fields, I: Classification up to Isogeny.
William C. Waterhouse, Abelian varieties over finite fields, II: Endomorphism rings and isomorphism classes.
J. S. Milne, Relations to the conjectures of Birch and Swinnerton-Dyer.
J. A. Shalika, Representations of p-adic groups.
J. A. Shalika, Some conjectures in class field theory.
J. Fresnel, A new definition of p-adic L-functions.
Koji Doi, On a problem in the theory of automorphic forms.
B. J. Birch, On elliptic curves and modular functions.
Koji Katayama, On some new zeta-functions.

Armand Brumer, Chairman

Analytic Number Theory

H. L. Montgomery, Mean and large values of Dirichlet polynomials and zeros of L-functions.
Lowell Schoenfeld, An improved estimate for the summatory function of the Möbius function.
Karl K. Norton, The distribution of power residues and non-residues.
C. Ryavec, The variation of an additive function.
P. X. Gallagher, A larger sieve.
M. Goldfeld, An application of the large sieve to the Goldbach problem.
L. Ehrenpreis, The zeros of the zeta function on the critical line.
Bruce C. Berndt, On the average order of a class of arithmetical functions.
Wolfgang Schwarz, Weak asymptotic and asymptotic properties of partitions.
S. L. Segal, Tauberian theorems of Landau-Ingham type.
H. M. Stark, An all purpose Tauberian theorem.
E. Bombieri, Density theorems for the zeta function.
P. T. Bateman, Linear relations connecting the imaginary parts of the zeros of the zeta function.
S. Chowla, Some remarks on number theory.

P. T. Bateman, Chairman

Cyclotomy, Combinatorics, and Addition Theorems

Albert Leon Whiteman, Residue Difference Sets.
Thomas Storer, Extensions of Cyclotomic theory.
Joseph B. Muskat, Cyclotomy and evaluation of character sums.
George E. Andrews, Partition identities.
Kenneth B. Stolarsky, Is partition theory inherently quadratic?
Lorne Houten, Plane partitions.

L. Carlitz, Eulerian numbers.
Henry Mann, Addition Theorems.
C. Ryavec, Addition of residue classes modulo n.
Robert A. Lee, On the e-transform.
E. G. Straus, Some problems concerning sum free and average-free sets.
Calvin T. Long, Factorization of sets of integers.
Emanuel Vegh, Arithmetic progressions of primitive roots of a prime.

A. L. Whiteman and T. Storer, Chairmen

Diophantine Approximation and Probabilitsic Number Theory

William W. Adams, Diophantine approximation in a cubic field.
E. G. Straus, Entire Functions.
W. Philipp, 1. An attempt to unify probabilistic number theory, 2. On limit theorems for additive functions.
P. Szüsz, On the metrical theory of continued fractions.
T. W. Cusick, Diophantine approximation for ternary linear forms.
Wolfgang Schwarz, On the congruence behavior of equivalent power series (a problem of Turán's).
Wolfgang Schwarz, A remark on an asymptotic formula of Renyi.
E. Wirsing, On approximation of algebraic numbers by algebraic numbers of fixed degree.
P. D. T. A. Elliott, The distribution of the values of Dirichlet L-series on, and to the left, of the line $\sigma = 1$.
R. T. Bumby, How to double a continued fraction.

D. Cantor and W. Philipp, Chairmen

Diophantine Equations

J. Ax, Schanuel's conjecture and decomposable forms by Skolem's methods.
M. Fried, Diophantine equations related to $h(x) - y = a$.
A. Schinzel, An improvement of Runge's Theorem.
L. J. Gerstein, Decompositions of Hermitian forms.
J. M. Ghandi, Fermat's last theorem and generalizations.
O. T. O'Meara, The automorphisms of the orthogonal groups and their congruence subgroups over arithmetic domains.
R. Bumby, On solutions of $ax^4 + by^2 = 4$.
S. Abyankar, Some remarks on the Fermat problem.
W. J. Ellison, An easy proof of Waring's problem.
Gordon Pall, Factorization of representations by binary quadratic forms.
L. Carlitz, Gaussian Sums.
John W. Schuck, Counting zeros of polynomials modulo p^k via integration on p-adic manifolds.
David Burgess, On the multiplicative group generated by the values of a polynomial.
J. H. H. Chalk and R. A. Smith, Exponential sums and a distribution problem of Mordell.
P. A. Leonard, Polynomial factorization over $GF(p)$.
D. J. Lewis, Simultaneous equations of additive type.

N. C. Ankeny, Chairman

Imaginary Quadratic Fields with Small Class Numbers

H. M. Stark, Some historical remarks on class numbers of complex quadratic fields.

P. Weinberger, Complex quadratic fields with class number two and even discriminant.

Carlos J. Moreno, Class number two and related problems.

Larry J. Goldstein, Imaginary quadratic fields with small class numbers.

B. J. Birch, The appropriate field for class invariants.

Daniel Shanks, Class number, a theory of factorization, and genera.

H. M. Stark, Chairman

D. J. Lewis
Ann Arbor, Michigan

APPLICATIONS OF ALGEBRAIC GEOMETRY TO NUMBER THEORY

H. P. F. SWINNERTON-DYER

1. Introduction

The object of these lectures is to show that a knowledge of algebraic geometry can be useful for number theorists, and to give an indication of the geometric ideas and methods involved and the kind of number-theoretic problems to which they have been successfully applied. I have tried to make the definitions and theorems involved intelligible to someone with very little background in geometry. Fortunately, in order to apply results from algebraic geometry it is in most cases necessary to know only the statements of the theorems and not also their proofs; for the way in which a theorem is applied and the way in which it is proved usually have nothing in common. Geometric theorems will therefore be quoted without much indication of the proof. Also, in the interests of simplicity some of the definitions below (for example, those of a *complete* variety and of a *polarized* Abelian variety) will be only approximations to the correct ones—though it will be made clear which of the definitions are only approximate. Exact definitions may be found for example in Baldassarri [2], which is an excellent survey of what is known in algebraic geometry and is copiously equipped with references. (Baldassarri's book, and these lectures, are both concerned with geometry as it was before the introduction of *schemes*; but though schemes will undoubtedly be necessary for some applications of geometry to number theory, in most present applications they only complicate matters.)

In effect, these lectures are a survey of parts of the long frontier between two neighbouring subjects; and so they must be concerned with a sequence of isolated topics rather than having a coherent theme. On the geometric side they could, with a little distortion, have been represented as various applications of the theory of Abelian varieties; for much of §§7, 8 can be rewritten in terms of Abelian varieties—see for example Lang [14]. But on the number-theoretic side there is no possible common link. Four topics will be considered:

(i) Diophantine equations corresponding to surfaces, in §§3–5.

(ii) Zeta-functions over finite fields, in §§6–9.

(iii) Diophantine equations corresponding to curves, in §§10–12.

(iv) Complex multiplication and class field theory, in §§13–15.

The geometric background which is common to more than one topic is described in §2; this is a prerequisite for everything else. Given this, the accounts of the four topics are essentially independent and can be read in any order; each account contains a summary of the additional geometry needed for that topic. The order in which the topics are presented corresponds to the increasing sophistication of the methods involved. Diophantine equations have been treated as two topics because there are two very different bodies of geometric knowledge that can be applied to them; one can hope that the methods that have been applied to surfaces will in due course also be applied to higher dimensional varieties, but at the moment the geometric tools are not sufficiently developed to make this possible.

The obvious omission from this list of topics is the behaviour of global zeta-functions and the ideas associated with the Birch-Swinnerton-Dyer and Tate conjectures. This seems to be an area in which number theory may be able to contribute something towards the solution of problems which are of interest to geometers (even of the most classical sort), whereas in the topics listed above it is the number theorists who benefit from the work of geometers. I have recently given a survey of the state of this subject, in my Driebergen talk [22]; and I have nothing to add to what is contained there.

Everyone who works in this subject knows how much it has been created and shaped by André Weil, both by his own work and by his influence on others. Like many others, I owe him a great debt of gratitude.

2. The Riemann-Roch Theorem and abelian varieties

Let Γ be a complete nonsingular curve, not necessarily embedded in a plane. (*Nonsingular* means that at each point of Γ there is a unique tangent line; in other words Γ has no multiple points or cusps. *Complete* can be taken to mean that Γ lies in projective space of some dimension and has no points missing—an affine curve cannot be complete because of the "points at infinity".) If Γ is defined over the complex numbers, then it can also be regarded as a compact Riemann surface without a boundary; and conversely any such Riemann surface represents a curve Γ. A *function* on Γ is a rational function of the coordinates of the underlying space, which must be homogeneous and of degree zero since the coordinates are projective; because the coordinates are related by the equations for Γ, a given function on Γ has more than one representation as a rational function of the coordinates. A function on Γ corresponds precisely to a meromorphic function on the associated Riemann surface, in the complex variable case.

The principal concepts of elementary complex variable theory can now be lifted back to Γ, and can there be defined in purely geometric terms so that they have a meaning whatever is the field of definition of Γ. Thus a *divisor* on Γ is a formal finite sum $\sum n_i P_i$ where the n_i are rational integers and the P_i are points on Γ; and the divisors form an additive group G. The *degree* of the divisor $\sum n_i P_i$ is the rational integer $\sum n_i$; and the divisors of degree 0 form a subgroup G_a, the group of divisors *algebraically equivalent to zero.* Two divisors on Γ are said to be *algebraically equivalent* if their difference is in G_a; clearly this is an equivalence relation. A function f, not identically zero, has *zeros* and *poles* with associated multiplicities; and the *divisor of* f, written (f), is defined as the sum of its zeros minus the sum of its poles, each taken with the correct multiplicity. A divisor is called a *principal divisor*, and is said to be *linearly equivalent to zero*, if it can be written as the divisor of some function; such divisors form a group G_l, which is a subgroup of G_a. Two divisors are said to be *linearly equivalent* if their difference is in G_l. We write $\mathfrak{A}_1 \sim \mathfrak{A}_2$ to denote that the divisors $\mathfrak{A}_1, \mathfrak{A}_2$ are linearly equivalent, and $\mathfrak{A} \sim 0$ to denote that the divisor $\mathfrak{A}$ is linearly equivalent to zero.

We can also define a *differential* $\omega = u\,dv$ and its divisor (ω), where u and v are any functions on Γ; and differentials satisfy the usual formal rules of the differential calculus. To find whether ω has a pole or zero at a given point P, we choose any uniformizing variable v_1 at P and write $\omega = u_1\,dv_1$ for some u_1 which is in fact $u_1 = u(dv/dv_1)$; then ω is said to have a pole or zero at P if u_1 has one there. Divisors are partially ordered in the obvious way, so that one can talk of *positive divisors*—that is, divisors $\sum n_i P_i$ with each $n_i \geq 0$. Note that 0 is a positive divisor; it would be more logical to speak of nonnegative divisors, but the word "positive" is too firmly established to be changed. The differentials whose divisors are positive are called *differentials of the first kind*; together with 0 they form a vector space, whose dimension is finite. The divisors of all differentials precisely fill a coset of G_l in G; this coset is called the *canonical class*, and any divisor in it is called a *canonical divisor.*

For a variety V of arbitrary dimension n some of these definitions need to be modified, and the analogy of complex variable theory is less helpful. V contains subvarieties of each dimension r with $0 \leq r \leq n$; and a formal finite sum of subvarieties of dimension r with rational integer coefficients is called an *r-cycle.* A *divisor* is now simply an $(n - 1)$-cycle. The *degree* of a cycle has a natural meaning only for 0-cycles, when it is defined exactly as before. *Linear equivalence* is also defined as before, in terms of the divisor of a function; it is therefore only defined for divisors. There are generalizations to other values of r, but they are not very helpful.

Algebraic equivalence, however, can be defined for every r, though the definition now has to be much more elaborate. A variety W is said to be *absolutely irreducible* if it cannot be written as the union of two strictly smaller varieties. Let T be a cycle on the product variety $V \times W$, let pr_1 denote the projection map from $V \times W$ to V and let a dot denote intersection on $V \times W$, all these operations being only defined when they are sufficiently well-behaved. To each point P on W corresponds the cycle $pr_1\{T \cdot (V \times P)\}$ on V, which we denote by $T'(P)$; its dimension is $\dim T - \dim W$. As P varies on W, $T'(P)$ varies in an *algebraic family* of cycles on V. Two cycles $\mathfrak{A}_1$ and $\mathfrak{A}_2$ on V are said to be *algebraically equivalent* if there is a family of cycles on V to which they both belong. It can be shown that this is an equivalence relation, and is compatible with addition and with forming intersections; hence "algebraically equivalent to zero" is a well-defined phrase. For each r the r-cycles algebraically equivalent to zero form an additive group G_a^r; and G_a will always denote G_a^{n-1}. It will be seen that algebraic equivalence for cycles in geometry is analogous to homology for cycles in topology; and indeed they have very similar properties.

A much more detailed account of all this may be found in Samuel [19], primarily in Chapter I. He also gives a comparative table of the terminologies used by different writers.

Now return to the case of a complete nonsingular curve Γ. To classify all divisors, we note that $G/G_a = Z$, the homomorphism being given by the degree of a divisor; it is therefore natural to ask about the structures of G_a/G_l and of G_l. It turns out that one has to consider G_l first, and that the right question to ask about it is: Given a divisor $\mathfrak{A}$, what is the set of positive divisors linearly equivalent to $\mathfrak{A}$? Let $\mathfrak{b}$ be any such divisor; then $\mathfrak{A} \sim \mathfrak{b}$ is the same as $\mathfrak{A} + (f) = \mathfrak{b} \geq 0$ for some function f, so this question is equivalent to asking for the set of all f such that $\mathfrak{A} + (f) \geq 0$. These f, together with the constant function 0, form a vector space $L(\mathfrak{A})$ of dimension $l(\mathfrak{A})$ say; thus $l(\mathfrak{A}) > 0$ if and only if there exists a positive divisor linearly equivalent to $\mathfrak{A}$. The value of $l(\mathfrak{A})$ is given by the Riemann-Roch theorem for curves:

THEOREM 1. *For any divisor $\mathfrak{A}$ on a complete nonsingular curve Γ,*

$$l(\mathfrak{A}) = \deg(\mathfrak{A}) + 1 - g + l(\mathfrak{k} - \mathfrak{A})$$

where $g \geq 0$ is an integer depending only on Γ, and $\mathfrak{k}$ is the divisor of any differential on Γ.

This is not a formula which gives $l(\mathfrak{A})$ in terms of simpler objects, because of the presence of the last term on the right-hand side; rather it is a duality theorem. There are also Riemann-Roch theorems for varieties of higher dimension, but they involve terms which cannot be expressed in the language of classical algebraic geometry.

The integer g is called the *genus* of Γ. It can be defined in many ways—for example as the number of "handles" on the Riemann surface for Γ, or as the dimension of the vector space of differentials of the first kind on Γ. Indeed since $l(0) = 1$, the associated functions being just the constants, it follows at once from Theorem 1 that

$$l(\mathfrak{k}) = g \quad \text{and} \quad \deg(\mathfrak{k}) = 2g - 2.$$

Moreover $l(\mathfrak{A}) = 0$ whenever $\deg(\mathfrak{A}) < 0$; thus Theorem 1 implies

$$(1) \qquad l(\mathfrak{A}) = \deg(\mathfrak{A}) + 1 - g \quad \text{whenever } \deg(\mathfrak{A}) > 2g - 2.$$

This is the form in which it is most often used.

If Γ is being regarded as a curve defined over a given field k, which is not necessarily algebraically closed, then $L(\mathfrak{A})$ is defined to be the k-vector-space of functions f which are defined over k and satisfy $\mathfrak{A} + (f) \geq 0$. Increasing k to a field $K \supset k$ has no effect on g, $\mathfrak{k}$ or $l(\mathfrak{A})$; and it replaces $L(\mathfrak{A})$ by its tensor product with K. Hence this machinery is essentially independent of the chosen field of definition —a fact which is important for applications to number theory.

With the help of the Riemann-Roch theorem we can study G_a/G_l on any given curve Γ and also find canonical models for Γ, whether or not we allow an extension of the field of definition. The case $g = 0$ is wholly exceptional, and the case $g = 1$ is not typical, so these have to be considered separately.

Consider first the case when Γ has genus 0 and contains a point P_0 which is defined over the given ground field k. Since $l(P_0) = 2$ by (1), the vector space $L(P_0)$ contains a nonconstant function defined over k. Let x be such a function; then x has a simple pole at P_0 and no other pole. Thus for any c in k the function $x - c$ has just one pole and therefore just one zero on Γ; so x takes each value (including infinity) exactly once, and therefore establishes a one-one correspondence defined over k between Γ and the projective straight line. This result is still valid if we know merely that there is a divisor $\mathfrak{A}$ on Γ which is defined over k and has odd degree; for since $\deg(\mathfrak{k}) = -2$ we can find an integer n such that $\deg(\mathfrak{A} + n\mathfrak{k}) = 1$, and then since $l(\mathfrak{A} + n\mathfrak{k}) = 2$ there is a point P on Γ which is linearly equivalent to $\mathfrak{A} + n\mathfrak{k}$ and is defined over k.

One consequence is that in this case Γ can be parametrized by x—that is, the coordinates of a general point of Γ can be written as rational functions of x with coefficients in k in such a way that to each point of Γ corresponds just one value of x. In these circumstances the point of Γ is defined over k if and only if the value of x is defined over k, so that this parametrization gives a complete description of the points of Γ defined over k. Now suppose only the converse, that the coordinates of a general point of Γ can be written as rational functions of some parameter x with coefficients in k—in other words there is a map from the projective line to Γ which is defined over k but is not necessarily one-one. Then it can be shown that Γ has genus 0 and can therefore be put into one-one correspondence over k with the projective line, since it obviously contains points defined over k. This result is known as Lüroth's Theorem. The analogous result for surfaces is false in general (as can be deduced from Theorem 6 below), but is true when k is algebraically closed. In higher dimensions the analogous result is generally supposed to be false even when k is the complex numbers, but no wholly convincing counter-example has yet been given.

If we assume only that Γ has genus 0, but not that it has a point defined over k, we cannot obtain quite so much. If $\mathfrak{k}$ is any canonical divisor, $\deg(-\mathfrak{k}) = 2$ whence $l(-k) = 3$; replacing $\mathfrak{k}$ by a linearly equivalent divisor if necessary, we can assume that $\mathfrak{k}$ is defined over k and that $-\mathfrak{k} > 0$. Let $1, x, y$ be a base for $L(-\mathfrak{k})$; then $1, x, y, x^2, xy, y^2$ all lie in $L(-2\mathfrak{k})$, and since $l(-2\mathfrak{k}) = 5$ these six functions must be linearly dependent. The resulting equation is that of a conic, which must be nondegenerate since $1, x, y$ are known to be linearly independent even over the algebraic closure of k. Moreover since y has two poles and therefore takes each value twice, there is a one-one correspondence defined over k between Γ and the conic. If k is an algebraic number field, it is known that the "local-to-global" principle holds for conics—that is, if a conic contains points defined over each $\wp$-adic extension of k (where $\wp$ runs through all finite and infinite primes of k), then the conic contains points defined over k. By what we have just proved, such a principle must hold for any nonsingular curve of genus 0.

The results for $g = 0$ can be summarized as follows:

THEOREM 2. *Let Γ be a nonsingular curve of genus 0 defined over a field k; then Γ is birationally equivalent over k to a line if and only if there is a divisor on Γ defined over k and of odd degree. If k is an algebraic number field, a local-to-global principle holds for Γ.*

In view of (1), algebraic and linear equivalence are the same on a curve of genus 0, and so there is nothing more to be said.

Next, consider the case when Γ has genus 1 and contains a point P_0 which is defined over the given ground field k. Since $l(P_0) = 1$, there is no nonconstant function which has a simple pole at P_0 and no other pole. Since $l(2P_0) = 2$ we can choose a function x defined over k so that $1, x$ is a base for $L(2P_0)$; and because of the last sentence x has a genuine double pole at P_0. Similarly since $l(3P_0) = 3$ we can find y such that $1, x, y$ is a base for $L(3P_0)$; and y has a genuine triple pole at P_0. Now the functions $1, x, y, x^2, xy, x^3, y^2$ all lie in $L(6P_0)$; and since $l(6P_0) = 6$ these seven functions must be linearly dependent. They have poles at P_0 of orders respectively 0, 2, 3, 4, 5, 6 and 6; and in the linear dependence relation there must be at least two terms which have a pole at P_0 whose order is the highest that occurs, in order to have a cancellation. Thus y^2 and x^3 both appear in the relation. Assuming that the characteristic of k is not 2 or 3, we can replace y by $y + ax + b$ so as to remove the terms in xy and y, and then replace x by $x + c$ to remove the term in x^2, and then replace x and y by λx and λy respectively, and thereby reduce the relation to the form

$$y^2 = 4x^3 - g_2x - g_3 \tag{2}$$

where g_2, g_3 are in k; the point P_0 on Γ corresponds to the point at infinity on (2), and there is a one-one correspondence between (2) and Γ because x takes each value twice on Γ. There are similar but more complicated formulae when char $(k) = 2$ or 3. The coefficients g_2 and g_3 in (2) are not uniquely determined by Γ and P_0 because of the possibility of a transformation $x \to \mu^2x$, $y \to \mu^3y$; but

$$j = 1728g_2^3/(g_2^3 - 27g_3^2) \tag{3}$$

is uniquely determined. The reason for taking j in this form, rather than using the apparently simpler g_2^3/g_3^2, is that for (2) to be a curve of genus 1 the right-hand side must not have a repeated factor as a polynomial in x—that is to say, the curve must not have a cusp; and the condition for this is just $g_2^3 - 27g_3^2 \neq 0$. Thus j is finite; and conversely it is easy to show that to any finite value of j in k there corresponds at least one curve (2) defined over k.

The most important fact about this reduction is that j does not depend on P_0 but only on k. To see this, let P_1 be another point on Γ defined over k, and let P be a general point on Γ. Since $l(P + P_0 - P_1) = 1$, there is just one point Q on Γ such that $P + P_0 - P_1 \sim Q$; and Q is defined over $k(P)$ because it is the unique zero other than P_1 of a function which is in $L(P + P_0 - P_1)$ and therefore defined over $k(P)$. Hence there is a well-defined map $P \to Q$ of Γ to itself; and this map is one-one and defined over k, and takes P_1 into P_0. Thus to reduce Γ to the form (2) using P_1 is the same as to apply this map and then reduce Γ to the form (2) using P_0; so the value of j is the same in both cases.

For fixed P_0, any linear equivalence class in G_a contains just one divisor of the form $P - P_0$; for if deg $(\mathfrak{A}) = 0$ then $l(\mathfrak{A} + P_0) = 1$. Thus there is a natural one-one correspondence between the points of Γ, or of (2), and the elements of G_a/G_l; and in this correspondence P_0, or the point at infinity on (2), corresponds to G_l itself. Now by the results of the last paragraph, the correspondence between

(2) and G_a/G_l does not depend on the choice of P_0. Thus we can identify the curve (2), or to be more exact its image

$$(4) \qquad y^2z = 4x^3 - g_2xz^2 - g_3z^3$$

in projective space, with G_a/G_l; and this induces a natural commutative group structure on the points of (4).

Now suppose only that Γ has genus 1, but not that it necessarily has a point defined over k. However, by extending k to a suitable larger field we can ensure that there is a point P_0 on Γ defined over $k(P_0)$, and we can then repeat the construction above. The resulting value of j lies in $k(P_0)$ and does not depend on the choice of P_0; thus it lies in the intersection of all possible $k(P_0)$, which is just k. A similar but more complicated argument shows that the curve (2) can be chosen to be defined over k. Thus there is a natural identification of G_a/G_l with a curve defined over k. On the other hand, it is known that there is no canonical model for Γ which is birationally equivalent to it over k; in particular, no variant of the device used for the case $g = 0$ will work, since now deg $(\mathfrak{k}) = 0$.

The results for $g = 1$ can be summarized as follows:

THEOREM 3. *Let Γ be a nonsingular curve of genus 1 defined over a field k; then G_a/G_l can be realized as a curve defined over k, called the Jacobian of Γ, and this curve can be written in the form* (4) *provided that* char $(k) \neq 2$ *or* 3. *The value of j, given by* (3), *is finite. If there is a point on Γ defined over k, then Γ is birationally equivalent over k to its Jacobian.*

There is no local-to-global principle for curves of genus 1, as is shown by the example

$$3x^3 + 4y^3 + 5z^3 = 0$$

discovered by Selmer; this has points defined over every p-adic field but not over the rationals. More complicated examples may be obtained by taking any plane section of the surface (6) in §3 below.

Finally, consider the case when Γ has genus $g > 1$. It is possible to put Γ into a canonical form by means of arguments analogous to those used in the case $g = 0$, considering $L(n\mathfrak{k})$ for various $n > 0$; but the results are of little value. More important is the problem of G_a/G_l, where the results are similar to those for $g = 1$ but their proof involves overcoming substantial technical difficulties. After making a field extension if necessary, we can assume that Γ contains a divisor $\mathfrak{A}_0$ of degree g; thus for any $\mathfrak{A}$ in G_a we have $l(\mathfrak{A} + \mathfrak{A}_0) \geq 1$. If there were equality here for all $\mathfrak{A}$ (as there is when $g = 1$), then to each $\mathfrak{A}$ there would correspond a unique positive divisor $\mathfrak{b} > 0$ on Γ such that $\mathfrak{b} \sim \mathfrak{A} + \mathfrak{A}_0$; and this would induce a one-one correspondence between G_a/G_l and S, the g-fold symmetric product of Γ—that is, the variety which parametrizes g-tuples of points of Γ. For general $\mathfrak{A}$ we do have equality and therefore a unique corresponding $\mathfrak{b}$ and point of S; but there are $\mathfrak{A}$ for which $l(\mathfrak{A} + \mathfrak{A}_0) > 1$, and each of these corresponds to a subvariety of positive dimension on S. To obtain a variety whose points are without exception in one-one correspondence with the elements of G_a/G_l, it is necessary to collapse each of these subvarieties to a point. This is what Weil did in his original proof of the existence of the Jacobian variety of a curve; and it was to make the proof work that he was forced to introduce the concept of an *abstract variety*. A more elegant but less direct technique was later produced by Chow. The final result is as follows:

THEOREM 4. *Let Γ be a complete nonsingular curve of genus $g > 0$ defined over a field k; then G_a/G_l can be realized as a complete nonsingular g-dimensional variety J, called the Jacobian of Γ, defined over k. If Γ contains a point defined over k, then Γ can be embedded in J by means of a map defined over k.*

Now forget about Γ and consider only the intrinsic properties of J; these are that it is a complete nonsingular variety whose points form a commutative group under a law of composition that can be expressed geometrically—that is, the map $J \to J$ which takes an element of the group into its inverse, and the map $J \times J \to J$ which takes an ordered pair of elements into their product, are defined by means of rational functions of the coordinates. It turns out that the commutativity of the group can be deduced from the other properties. We shall therefore say that a variety A is an *Abelian variety* if it is complete and nonsingular, and if its points form a group under a law of composition that can be expressed geometrically. Not every Abelian variety is a Jacobian, but most of the properties of Jacobians can be extended to arbitrary Abelian varieties. The most complete account of the theory in book form is in Lang [14]; the two most elegant results are:

THEOREM 5. (i) *The group law on an Abelian variety is commutative.* (ii) *Let A be an Abelian variety, V a group variety not necessarily complete, and ϕ: $V \to A$ a geometric map which takes the identity element of V onto the identity element of A. Then ϕ is also a group homomorphism.*

Another formulation of (i) is that any Abelian variety is abelian; the name however comes from the connection with Abel's theorem in complex variable theory. Both results depend fundamentally on A being complete. An example of a group variety which is neither complete nor commutative is that given by the equation

$$X(X_{11}X_{22} - X_{12}X_{21}) = 1$$

in affine 5-dimensional space, with the group law induced by the obvious one-one correspondence between the points of this and the elements of $GL(2)$. The map ϕ is called an *isogeny* if V is also an Abelian variety and ϕ is onto and has finite kernel; and we say that A_1 is *isogenous* to A_2 if there exists an isogeny $A_1 \to A_2$. It can be shown that this is an equivalence relation, though this is not trivial.

The geometric theory of Abelian varieties was created by Weil in the last twenty-five years; but in the case when the ground field is the complex numbers there is a much older analytic theory. This is not a prerequisite for any of what follows, but it represents useful background knowledge. For a more complete account than is given here see Conforto [6], Shimura and Taniyama [21] or Weil [23].

Let Γ be a complete nonsingular curve of genus $g > 0$ defined over the complex numbers, and let $\mathcal{S}$ be the corresponding Riemann surface. Let $\omega_1, \ldots, \omega_g$ be a base for the vector space of differentials of the first kind on $\mathcal{S}$, and let Λ be the lattice in C^g consisting of all the points

$$\left(\oint \omega_1, \oint \omega_2, \ldots, \oint \omega_g\right)$$

where the integrals are taken over any closed path (the same for every integral) on $\mathcal{S}$. By the known topology of $\mathcal{S}$, Λ is a free group on $2g$ generators and C^g/Λ is compact, so that Λ is discrete as a subspace of C^g. The map

$$P \to \left(\int^P \omega_1, \ldots, \int^P \omega_g\right)$$

maps Γ into C^g/Λ, and can be extended to a homomorphism $G_a \to C^g/\Lambda$. Abel's theorem states that this map is onto and its kernel is precisely G_l. In the classical case this substantially proves Theorem 4, and identifies the Jacobian of Γ with C^g/Λ. The functions on J correspond to the multiply periodic meromorphic functions on C^g which have each element of Λ for a period.

The obvious generalization is to consider any lattice Λ in C^g such that Λ is discrete and C^g/Λ is compact. However, for a general lattice Λ there are no non-trivial meromorphic functions on C^g which admit every element of Λ as a period. The condition for such functions to exist is essentially number-theoretic. Let Ω be a $g \times 2g$ matrix whose columns give a set of generators of Λ; then a necessary and sufficient condition for the existence of such functions is that there exists a $2g \times 2g$ skew-symmetric matrix P such that

(i) the elements of P are rational integers not all zero,

(ii) $\Omega P \Omega' = 0$,

(iii) $i\Omega P\bar{\Omega}'$, which is automatically Hermitian, is at least positive semidefinite.

Such a matrix P is known as a *principal matrix* associated with Ω. Assuming that $i\Omega P\bar{\Omega}'$ is actually positive definite (which rules out certain degenerate cases), one can then show that the meromorphic functions on C^g which admit every element of Λ as a period form a finitely generated field of transcendence degree g over C; and by means of these functions one can canonically construct a projective variety whose points are in one-one correspondence with those of C^g/Λ. With the group law induced from that on C^g/Λ, this is an Abelian variety.

An alternative version of the same necessary and sufficient condition is that there exists a real-valued R-bilinear form $E(x, y)$ on $C^g \times C^g$ such that

(i) the value of $E(x, y)$ is an integer whenever x and y are both in Λ,

(ii) $E(x, y) = -E(y, x)$,

(iii) $E(x, ix)$ is a positive semidefinite quadratic form.

The R-bilinear form $E(x, y)$ is called a *Riemann form* associated with Λ. It is easy to show, by elementary algebra, that Λ admits a Riemann form if and only if it admits a principal matrix.

3. Diophantine equations corresponding to surfaces

The basic problem in Diophantine equations is to find the solutions of a given set of polynomial equations in a given field k, or in a given commutative ring R. Those parts of the problem which depend on the detailed nature of k or R essentially belong to number theory, but those results which hold for all k have close connections with geometry. For the equations define a variety V, and the geometry of V will be relevant to the problem—for example, in virtue of Lüroth's theorem a one-parameter solution of the equations corresponds to a curve of genus 0 on V. Unfortunately the most natural problems for a number theorist are not always those on which the geometry sheds any light. Moreover, as the dimension of V increases the known geometric theory of V becomes less complete; so these methods have yielded a lot for curves, a certain amount for surfaces, and very little in higher dimensions.

The geometry of V becomes richer as its field of definition is increased; thus we normally study the geometry of V over $\bar{k}$, the algebraic closure of k, or even over C when k is an algebraic number field. The geometric results over $\bar{k}$ must then be lifted back to k to provide number-theoretic results. One direction in which this leads (which will not be followed in these lectures) is towards the use of Galois cohomology.

Even if V was originally given in affine space, a geometer will normally wish to embed it in projective space; thus the most natural applications of the geometry are to problems of rational rather than integral solutions—though as will be seen below some results about integral solutions can be obtained in this way. Again, most of the interesting questions are unaffected by a birational transformation of V defined over k; thus we can usually assume that V is nonsingular as well as complete.

We therefore assume henceforth that V is a complete nonsingular surface, not necessarily embedded in three-dimensional space. The most fruitful way of exploiting the geometry seems to be to consider the set of curves on V; this approach was originated by B. Segre and has been extensively used by Châtelet and Manin among others. It automatically restricts one to looking at special classes of surface, for the general surface has too few curves on it—in fact only complete intersections with hypersurfaces.

In the notation of §2, it is known that G/G_a is a finitely generated free abelian group, which is called the Néron-Severi group. It has properties very similar to those of the homology group of a topological manifold (and indeed when the ground field is the complex numbers it can be regarded as a subgroup of a homology group). However, while it is quite straightforward to determine the homology group of a given topological manifold, the determination of the Néron-Severi group of a given variety is a difficult problem which seems to require number-theoretic methods. (For example, the Néron-Severi group of the Abelian variety C^g/Λ is naturally isomorphic to the additive group generated by the principal matrices associated with Λ, in the notation of the end of §2; so its determination involves investigating linear dependence of given complex numbers over the rationals.)

Now let $\Gamma_1, \ldots, \Gamma_r$ be divisors on V which are representatives of a complete set of generators of the Néron-Severi group of V. Then to any curve Γ on V there corresponds a uniquely determined set of integers $n_1, \ldots, n_r$ such that

$$\Gamma \text{ is algebraically equivalent to } n_1\Gamma_1 + \cdots + n_r\Gamma_r.$$

The most interesting facts about Γ—such as its degree, its genus and its intersection number with any given curve—depend only on the n_i; but in respect of the genus this statement needs some qualification. The *arithmetic genus* of Γ is defined as the genus of a general curve of the algebraic equivalence system to which Γ belongs, and this only depends on the n_i. The *geometric genus* of Γ is the genus in the sense of §2 of any nonsingular curve birationally equivalent to Γ. We have

$$\text{(5)} \qquad \text{geometric genus of } \Gamma \leq \text{arithmetic genus of } \Gamma,$$

the difference arising from the singularities of Γ if any. (An explicit example is given in §4 below.) Thus all members of an algebraic equivalence class of curves will have the same arithmetic genus; almost all of them will have this also as their geometric genus, but a subset will have singularities and will therefore have a smaller geometric genus. What one usually looks for are curves of geometric genus 0, because these can be parametrized; and now the inequality (5) is the convenient way round, because it shows that an irreducible curve which has arithmetic genus 0 must also have geometric genus 0. (A reducible curve—that is, one which can be written as a union of several curves—has well-defined arithmetic and geometric genera satisfying (5); but these may now be negative.)

Given a class of algebraically equivalent divisors, does it contain a positive divisor? For the surfaces that occur in practical problems, G_a/G_l is trivial; and indeed when G_a/G_l is not trivial one has better ways of investigating the number-theoretical properties of V, using the theory of Abelian varieties. There is a Riemann-Roch inequality for surfaces, analogous to Theorem 1; and in the corresponding notation this gives a lower bound for $l(\mathfrak{A})$. Thus it provides a technique for showing that a given linear equivalence class contains a positive divisor, if in fact it does so. (In the problems discussed in these lectures, simpler arguments will be used.) However, in practice the difficulty is to show that the positive divisor is an irreducible curve; and two of the methods for doing this will be illustrated below.

The most interesting surfaces for this purpose are the Del Pezzo surfaces (which include quadrics, cubic surfaces and the intersection of two quadrics in four dimensions); these are surfaces which over $\bar{k}$ are birationally equivalent to a plane. Among their advantages is the fact that it is easy to find the Néron-Severi group of any given Del Pezzo surface. If k is an algebraic number field, there is a local-to-global theorem for quadric surfaces; but there is no local-to-global theorem for cubic surfaces, as is shown by the example

$$\text{(6)} \qquad 5x^3 + 9y^3 + 10z^3 + 12w^3 = 0.$$

This has solutions in every p-adic field but not in the rationals, as was proved by Cassels and Guy [5]. It is not known whether a local-to-global principle holds for the intersection of two quadrics in four dimensions.

Taking cubic surfaces as the first interesting case, two obvious problems arise:

(i) What is the obstruction to a local-to-global principle, and under what extra conditions does it hold?

(ii) In what circumstances can we exhibit all points defined over k on the surface?

These are sensible questions for other Del Pezzo surfaces also, but probably only for them. If the cubic surface is in an affine space, one can reasonably ask also for a description of all or some of the integer points on it. For slightly more complicated surfaces—quartics for example—we may reasonably ask about one-parameter solutions defined over k.

Partial answers to these questions are given in the next three sections. As yet, these are the only types of problem about surfaces for which a knowledge of the geometry has proven valuable.

4. Cubic surfaces

Throughout this section, V will denote a complete nonsingular cubic surface. The geometry of V has been known for about a century; it contains exactly twenty-seven straight lines, which have known incidence properties, and its Néron-Severi group G/G_a has rank 7 and is generated by the classes of these lines. On V, linear equivalence and algebraic equivalence are the same. Among the birational correspondences between V and a plane, there is one which gives a convenient model and which will be used throughout this section; it is one-one except that there are six skew lines $l_1, \ldots, l_6$ on V each of which is mapped onto a single point on the plane. Call the points $P_1, \ldots, P_6$; the only constraints on them are that no three lie on a straight line and that there is no conic through all six. The plane sections of V correspond to cubic curves through $P_1, \ldots, P_6$; and the other 21 lines on V correspond to the 15 lines P_iP_j and the 6 conics each through five of the P_i. The intersections of two curves on V correspond to the intersections other than the P_i of their images on the plane. The Néron-Severi group on V is generated by the l_i and any one of the twisted cubics on V which meet none of the l_i; the latter correspond to the straight lines in the plane. Let Γ be any curve of V other than an l_i, and let Γ' be its image in the plane; let n be the degree of Γ' and n_i the multiplicity with which it passes through P_i. Then the class of Γ in the Néron-Severi group is determined by n and the n_i; and the arithmetic genus and freedom of Γ are given by the formulae

$$\begin{aligned} \text{arithmetic genus} &= \tfrac{1}{2}(n-1)(n-2) - \textstyle\sum \tfrac{1}{2}n_i(n_i - 1), \\ \text{freedom} &= \tfrac{1}{2}n(n+3) - \textstyle\sum \tfrac{1}{2}n_i(n_i + 1). \end{aligned}$$

(The *freedom* of Γ is the dimension of the vector space of functions f satisfying $(f) + \Gamma \geq 0$; it is also the number of independent constraints that can be imposed on a curve linearly equivalent to Γ.) These notations can also be given a meaning for the l_i; for example l_1 has $n = 0$, $n_1 = -1$ and $n_2 = \cdots = n_6 = 0$ as can be seen by considering a plane section of V consisting of l_1 and two other lines. The notation of this paragraph will be used throughout this section.

Of course, not all of this apparatus will be defined over k, though it is defined over $\bar{k}$. We usually specify a curve Γ by imposing enough conditions on Γ'; to show that the resulting Γ is defined over k, we have to lift the conditions back to conditions on Γ and show that in that form they are invariant under the Galois group of $\bar{k}/k$. In particular, the class of Γ in the Néron-Severi group must be invariant under the Galois group. Such classes form a subgroup of the full Néron-Severi group, which is generated by the class of plane sections of V and the classes generated by the union of any one of the 27 lines and its conjugates over k. Finding all the possibilities for this subgroup is therefore a finite combinatorial exercise.

The first natural question about V is under what conditions there is a parametric representation of all the points of V defined over k. The obvious way to achieve this is to have a birational map from V to a plane, which is defined over k. For this it is clearly necessary that V should contain points defined over k; but this is not sufficient.

THEOREM 6. *A necessary and sufficient condition for V to be birationally equivalent to a plane over k is that V should contain points defined over k and that there should be a set of 2, 3 or 6 skew lines on V whose union is a divisor on V defined over k.*

Let P be a point on V defined over k; then the tangent plane to V at P meets V in a cubic with a singularity at P, and in general this is a curve of genus 0 and so contains an infinity of points defined over k. It follows by repeated use of this argument that we can find points defined over k which lie on V and lie off any preassigned subvariety—that is, we can find points where ever we need them. (The case in which the curve of intersection degenerates into three lines through P is more complicated, but the result is still true.)

If we have a birational map from V to a plane π, then the straight lines on π correspond to a doubly infinite linear system of curves of geometric genus 0 on V, any two curves of the system having just one free intersection. Conversely, suppose we have such a system of curves on V and suppose that the curves are irreducible. We can choose a curve Γ_0 of the system which is defined over k, by constraining it to go through two points of V defined over k. The functions f such that $(f) = \Gamma - \Gamma_0$ for some curve Γ of the system form a vector space of dimension 2. Let $1, f_1, f_2$, all be defined over k, be a base for this vector space; then it is easy to see that $P \to (f_1(P), f_2(P))$ is a birational map from V to an affine plane, defined over k.

Thus we have only to find such a doubly infinite system. In general this will consist of curves with given n and n_i, which are further constrained to pass through certain points Q_j with multiplicities m_j. The equations for the genus and the freedom now give

$$\tfrac{1}{2}(n-1)(n-2) - \sum \tfrac{1}{2}n_i(n_i - 1) - \sum \tfrac{1}{2}m_j(m_j - 1) = 0,$$
$$\tfrac{1}{2}n(n+3) - \sum \tfrac{1}{2}n_i(n_i + 1) - \sum \tfrac{1}{2}m_j(m_j + 1) = 2;$$

and these are equivalent to

$$(7) \qquad n^2 - 1 = \sum n_i^2 + \sum m_j^2, \qquad 3n - 3 = \sum n_i + \sum m_j.$$

To prove the theorem, we have to determine when there exists a solution of (7) which corresponds to a system of irreducible curves in a class of the Néron-Severi group which is defined over k.

If V has a set of 6 skew lines, the set being defined over k, then we can use the construction in the first paragraph of this section; this corresponds to $n = 1$, each $n_i = 0$ and no m_j occur, which is a solution of (7). The curves Γ are irreducible since the Γ' are irreducible, being straight lines.

Next suppose that V has a set of 2 skew lines, the set being defined over k. The correspondence needed is easy to define geometrically. Take a fixed plane π in the space in which V is embedded, not containing either of the given skew lines; to a general point P on V there corresponds a unique line through P which meets the two given lines, and this unique line meets π in a point Q. If P is defined over k, so is the transversal and hence so is Q. Conversely Q uniquely defines the transversal, and hence uniquely defines P as the only intersection of the transversal with V which does not lie on one of the given lines. This gives a one-one correspondence $P \to Q$ from V to π defined over k. If the given lines are chosen to be l_1 and l_2 then any linear equivalence class with

$$n = 3\lambda, \qquad n_1 = n_2 = \lambda - \mu, \qquad n_3 = n_4 = n_5 = n_6 = \lambda$$

is defined over k; and the corresponding solution of (7) is given by $\lambda = 2$, $\mu = -1$, $m_1 = 1$.

If V has a set of 3 skew lines, the corresponding map $V \to \pi$ seems to have no convenient geometrical description. If the lines are taken to be l_1, l_2 and l_3, then any linear equivalence class with

$$n = 3\lambda, \qquad n_1 = n_2 = n_3 = \lambda - \mu, \qquad n_4 = n_5 = n_6 = \lambda$$

is defined over k; and a possible solution of (7) is given by $\lambda = 1$, $\mu = 1$, $m_1 = 2$, $m_2 = 1$. The corresponding curves Γ' are cubics constrained to have a fixed double point and to go through four other given points; and such a curve is in general not degenerate. (We know there are a double infinity of such Γ', but the degenerate ones form finitely many sets of singly infinite families.)

For the negative part of the theorem, one starts with a tedious combinatorial argument which shows that the worst that can happen (that is, the most linear equivalence classes defined over k without the condition in the theorem being satisfied) is when there is a plane section of V consisting of three lines each of which is defined over k. Choose a representation of the cubic surface which is such that these three lines map into P_1P_2, P_3P_4 and P_5P_6; then the linear equivalence classes defined over k are just those given by

$$n = \lambda + \mu + \nu, \qquad n_1 = n_2 = \lambda, \qquad n_3 = n_4 = \mu, \qquad n_5 = n_6 = \nu$$

for some λ, μ, ν. The conditions (7) that have to be satisfied become

$$\text{(8)} \qquad \begin{aligned} \sum m_j^2 &= 2\lambda\mu + 2\mu\nu + 2\nu\lambda - \lambda^2 - \mu^2 - \nu^2 - 1, \\ \sum m_j &= \lambda + \mu + \nu - 3. \end{aligned}$$

These equations have a large number of integral solutions; and to complete the proof of the theorem we have to show that the curves corresponding to any solution are reducible. The algebraic details of such a proof are lengthy, and are omitted here; but the idea is straightforward. For each solution of (8) we find integers n', n_i' and m_j' satisfying

$$\text{(9)} \qquad 0 < n' < n, \qquad 0 \le n_i', \qquad 0 \le m_j',$$

$$\text{(10)} \qquad \tfrac{1}{2}n'(n' + 3) - \sum \tfrac{1}{2}n_i'(n_i' + 1) - \sum \tfrac{1}{2}m_j'(m_j' + 1) \ge 0,$$

$$\text{(11)} \qquad nn' < \sum n_i n_i' + \sum m_j m_j'.$$

Conditions (9) and (10), the latter being that the freedom of the system of curves specified by the n', n_i' and m_j' is nonnegative, ensure that there is at least one curve with the assigned values of the parameters. Let C' be such a curve in the representing plane; we do not know that C' is irreducible, but this does not matter. If Γ' is any curve of the doubly infinite system specified by the undashed letters, then C' and Γ' must have a component in common; for otherwise they would have altogether nn' intersections, and (11) shows that they have more intersections than this at the points P_i and Q_j alone. This common component cannot be the whole of Γ', for C' has lower degree than Γ' by (9); hence it is a proper component of Γ', and Γ' is not irreducible. This is what we needed to prove, and it completes the proof of the theorem.

This is in principle not the only way in which one could describe all the points of V which are defined over k. Manin has shown that given any finite set of many-to-one maps $\phi_i: \pi \to V$ there are points on V defined over k which are not the image of a point of π defined over k under any of the ϕ_i. This blocks one possible approach; but nothing is known about the existence of complete solutions in more than two parameters—that is, about maps from n-dimensional projective space (where $n > 2$) to V such that each point on V defined over k arises from at least one point defined over k in the projective space.

The methods used in the proof of Theorem 6 can also be used to attack the more general question: given two cubic surfaces defined over k, is there a birational map from one to the other defined over k? This question is not yet completely solved; the most elegant known result is that if V contains no set of skew lines defined over k, and if V is birationally equivalent to another cubic surface W over k, then there is a *linear* change of variables which takes V into W.

The next natural question is whether there exist local-to-global theorems for cubic surfaces, where now k is assumed to be an algebraic number field. The example quoted in §3 shows that such a theorem cannot hold without some further restriction on V; and the most general result known is as follows:

THEOREM 7. *Let V be a nonsingular cubic surface defined over an algebraic number field k, and suppose*

(i) *that V contains points defined over each $\wp$-adic completion of k; and*

(ii) *that V contains a set of either* 3 *or* 6 *skew lines defined over k.*

Then V contains points defined over k.

This result was apparently first proved by Châtelet, and at least three other people have subsequently obtained it independently; but as far as I know it has never been published. It contains as a special case Selmer's theorem on diagonal cubic surfaces. The proof given here is due to Cassels.

Assume first that we have such a set of 6 lines, which we take as $l_1, \ldots, l_6$ in our plane model for V; then the straight lines in the plane model correspond to those twisted cubics on V which meet no l_i, and the system of twisted cubics is defined over k even though we do not yet know that any of the individual twisted cubics is. Let C be a curve of this system defined over an algebraic extension of k; let K be a finite normal extension of k over which C is defined, and let G be the Galois group of K/k. For any σ in G, σC is in the same system of curves as C since the system is defined over k; hence $C \sim \sigma C$ and there is a function f_σ on V defined over K such that $(f_\sigma) = C - \sigma C$. Now consider

$$a_{\sigma,\tau} = (\tau f_{\tau^{-1}\sigma}) f_\sigma^{-1} f_\tau. \tag{12}$$

On the one hand, as a function on V this has been defined as a multiplicative coboundary and it therefore satisfies the equation for a cocycle; on the other hand its divisor is trivial and it is therefore in K^*. Moreover, its cohomology class does not depend on the choice of C and the f_σ. For if we multiply the f_σ by arbitrary nonzero b_σ in K^*, this alters $a_{\sigma,\tau}$ by an arbitrary coboundary in K^*; and to replace C by another curve C_1 of the same system, also defined over K, corresponds to replacing each f_σ by $f_\sigma\phi(\sigma\phi)^{-1}$ where $(\phi) = C_1 - C$, and this does not alter the $a_{\sigma,\tau}$. Thus (12) defines a 2-cohomology class in K^* which depends only on the set of 6 lines.

This class splits over every $\wp$-adic completion $\hat{k}$ of k; for we can choose points P, P' on V defined over $\hat{k}$ and determine C in its linear equivalence class by requiring it to pass through P and P'. The C thus chosen is defined over $\hat{k}$ and hence is invariant under G; thus we can take every $f_\sigma = 1$ and $a_{\sigma,\tau} = 1$. But it is known that a 2-cohomology class which is trivial in every $\wp$-adic completion is trivial; thus $a_{\sigma,\tau}$ is a coboundary and by multiplying the f_σ by suitable elements of K^* we can assume that $a_{\sigma,\tau} = 1$. Now (12) shows that f_σ is a 1-cocycle, in the multiplicative group of the function field of V over K. By Hilbert's Theorem 90, every 1-cocycle in the multiplicative group of a field is a 1-coboundary; thus $f_\sigma = g^{1-\sigma}$ for some g in the function field. It is easily verified that $C - (g) = D$, say, is invariant under G and is therefore defined over k. The vector space of functions f such that $D + (f) > 0$ is defined over k, and its tensor product with K is nontrivial since it contains g; hence it contains a nonzero function f^0 defined over k. Now $C^0 = D + (f^0)$ is a twisted cubic on V defined over k; and so by Theorem 2 it contains points defined over k. This proves the theorem in the case of a set of 6 skew lines.

Suppose instead that V contains a set of 3 skew lines defined over k. There are just two ways in which we can choose 3 further lines to make a set of 6 skew lines; so each of these new sets is defined over at worst a quadratic extension k_1 of k. By the part of the theorem already proved, there are points on V defined over k_1; let Q be one such. If Q is not defined over k, let Q' be its conjugate over k; then the line QQ' is defined over k and hence so is its one remaining intersection with V. This completes the proof of Theorem 7.

The other question about cubic surfaces to which geometric methods of this sort can be applied is the determination of an infinity of integral points on V, in the case when V is affine and k is an algebraic number field. One way of doing this is to exhibit one-parameter *polynomial* solutions of the equation of V. The classical example of this is the equation

$$x^3 + y^3 + z^3 = 1; \tag{13}$$

Euler showed that this had an infinity of integer solutions, by means of the identity

$$(-9t^4 - 3t)^3 + (9t^4)^3 + (9t^3 + 1)^3 = 1. \tag{14}$$

In general, let Γ_∞ be the curve at infinity on V; then a one-parameter polynomial solution of the equation of V defines a curve Γ on V which is of geometric genus 0 and which meets Γ_∞ in a single point P corresponding to $t = \infty$; P must not be a multiple point of Γ, but in principle it could be a cusp. The multiplicity of the intersection will be the degree of Γ, which is $3n - \sum n_i$ in the present notation.

Conversely, any such curve (defined over k) and point P give rise to a parametric polynomial solution; for P must be defined over k and so we can parametrize Γ over k in such a way that P corresponds to $t = \infty$. The expressions for the coordinates do not become infinite for any finite t, and hence they must be polynomials. (Of course their coefficients may not be integers; this seems to be a matter of luck.)

Assume first that Γ_∞ is nonsingular and therefore of genus 1. Writing $r = 3n - \sum n_i$, and letting $\mathfrak{A}$ denote the intersection of Γ'_∞ with some line in the representing plane, we have

$$n\mathfrak{A} \sim \Gamma' \cdot \Gamma'_\infty = rP + \sum n_i P_i \text{ on } \Gamma'_\infty, \tag{15}$$

so that P has to be one of the finitely many solutions of $rP' \sim n\mathfrak{A} - \sum n_i P_i$. In compensation, (15) shows that the condition of r-fold tangency to Γ'_∞ at P' imposes only $r - 1$ independent conditions on Γ'; for if $r - 1$ intersections are at P' the rth one must also be there. Hence the residual freedom of Γ' is just

$$\tfrac{1}{2}n(n + 3) - \sum \tfrac{1}{2}n_i(n_i + 1) - (r - 1) = \tfrac{1}{2}(n - 1)(n - 2) - \sum \tfrac{1}{2}n_i(n_i - 1)$$

which is also the arithmetic genus of Γ. Denote this by g; then to reduce the geometric genus to 0 involves imposing g further nonlinear constraints, which will completely fix Γ. As these constraints are nonlinear, one would expect that Γ will not be defined over k when $g > 0$, and thus will not meet our needs. This is not altogether true—indeed there exist V with infinitely many such Γ defined over k; but there seems no general method of determining when a field extension is unnecessary, and so we confine ourselves to the case $g = 0$. Note that, even without this hypothesis, there are only a finite number of possible P defined over k regardless of the values of n and the n_i. For by considering the number of intersections of Γ' with a conic through any five of the n_i we obtain (unless $n = 2$ and $r = 1$)

$$2n \geq \text{sum of any five of the } n_i,$$

whence $\sum n_i \leq 12n/5$ and $r \geq 3n/5$. Now the Mordell-Weil theorem (see §10) and (15) show that there are only boundedly many possible P defined over k; and by (15) again, the choice of one of these P will usually impose several linear relations between n and the n_i.

If Γ_∞ is singular, and therefore of genus 0, much of this survives. There is a natural group law on Γ_∞ with the singularity deleted, the group being isomorphic to the additive group of k if the singularity is a cusp, or to the multiplicative group of k if the singularity is a double point. Linearly equivalent curves on V meet Γ_∞ in divisors which are equivalent under the group law; thus there is still a relation having the form of (15), and the residual freedom of Γ' is still equal to its arithmetic genus. There is still only a finite choice for P in the case of a double point, but there is an infinite choice in the case of a cusp.

To find all the acceptable curves with $g = 0$ and given P, one has therefore to examine the solutions of $g = 0$ and the equations arising from (15). Usually, almost all of these can be shown to be degenerate, by methods analogous to those used in the proof of Theorem 6; thus for example, the only parametric polynomial solutions of (13) with $g = 0$ are those arising from (14) and the obvious straight lines. However there do exist V with an infinity of essentially distinct parametric polynomial solutions.

5. A special quartic surface

If we try to extend these methods to more general surfaces, we immediately encounter two difficulties. For most surfaces, the Néron-Severi group $G/G_a = Z$ and there are not enough curves on the surface to give anything of interest; thus we have to confine ourselves to special surfaces within any family. Moreover, even if there are plenty of curves on the surface, the formulae for the freedom and the arithmetic genus of a curve on the surface will usually imply that there are only finitely many families of curves of genus 0 on the surface. This happens, for example, for all nonsingular surfaces of degree greater than 4 in projective three-dimensional space. Thus there are rather few other kinds of surface worth considering from this point of view; the simplest are certain special quartic surfaces, of which a good example is

$$V\colon A^4 + B^4 = C^4 + D^4. \tag{16}$$

This has a large Néron-Severi group G/G_a, as can already be seen by considering the straight lines on it. Euler found one nontrivial parametric solution of (16), given by polynomials of degree 7; can we find all parametric solutions defined over Q, or at any rate all those of arithmetic genus 0?

In considering V, the first step is to find that part of the Néron-Severi group which is defined over Q. The line $A = C, B = D$ lies on V, and the planes through it defined by $A - C = \lambda(B - D)$ cut out on V cubic curves Γ_λ; moreover Γ_λ has on it a point defined over $Q(\lambda)$, for example the point P_λ where it meets the line $A = -C, B = -D$ which lies in V and does not meet the original line. The general Γ_λ is nonsingular, and is therefore a curve of genus 1 with a natural group law on it; for special values of λ, Γ_λ can acquire singularities or even split. Each point of V lies on exactly one Γ_λ.

Now let Γ be any curve on V; it meets Γ_λ in a divisor on Γ_λ, which can be reduced to a divisor of degree 0 by subtracting an appropriate multiple of P_λ. The linear equivalence class of this Γ_λ-divisor is not altered if Γ is replaced by another curve on V linearly equivalent to it. (Note that on V linear and algebraic equivalence are the same.) Thus to any element of the Néron-Severi group on V we have associated an element of G_a/G_l on Γ_λ, and hence a point of Γ_λ since Γ_λ is its own Jacobian. If the element of the Néron-Severi group is defined over Q, then the resulting point on Γ_λ is defined over $Q(\lambda)$. This process works backwards also; given a point P'_λ on Γ_λ defined over $Q(\lambda)$, its locus as λ varies is a curve Γ on V which is defined over Q and which gives rise to P'_λ under the process described. What is the most general divisor on V defined over Q which gives rise to P'_λ? It must meet Γ_λ in a divisor

$$(\phi_\lambda) + nP_\lambda + P'_\lambda$$

where n is a fixed integer and ϕ_λ is a function on Γ_λ defined over $Q(\lambda)$. Because each point of V is on just one Γ_λ, ϕ_λ is induced by a function ϕ on V defined over Q; and the most general divisor on V which gives rise to P'_λ must have the form

$$(\phi) + n(\text{locus of } P_\lambda) + (\text{locus of } P'_\lambda) + (\text{components of various } \Gamma_\mu).$$

The possible P'_λ form a finitely generated group, by the Mordell-Weil theorem; thus the part of the Néron-Severi group defined over Q is generated by

(i) the loci of the generators of the group of P'_λ,

(ii) the line $A = -C$, $B = -D$ which is the locus of P_λ,

(iii) any Γ_λ, and the components of those Γ_μ which split.

To find the group of P'_λ involves a descent argument of standard type on an elliptic curve; this is straightforward and the group turns out to have only one generator. The rest of the calculation is trivial; and we find that the relevant part of the Néron-Severi group has rank 9. Thus there are explicitly determined curves $\Gamma_1, \ldots, \Gamma_9$ such that any Γ on V defined over Q satisfies

$$\Gamma \sim n_1\Gamma_1 + \cdots + n_9\Gamma_9$$

for just one set of integers $n_1, \ldots, n_9$. The arithmetic genus of Γ is an indefinite quadratic form in the n_i, and it only remains to determine which of the zeros of this quadratic form correspond to *irreducible* curves Γ; luckily all such zeros will turn out to correspond to curves of odd degree, so that the problem of whether the resulting Γ can be parametrized over Q is trivial. (The *degree* of Γ is the number of its intersections with an arbitrary plane, and is a linear function of the n_i.)

It is worth remarking that if we had wished to find the full Néron-Severi group of V we would have proceeded in the same way with Q replaced by C, the complex numbers. This would have led us to the problem of finding all the points on Γ_λ defined over $C(\lambda)$, which is still a number-theoretic problem since $C(\lambda)$ is not algebraically closed. This is further confirmation of the statement in §3 that the determination of the Néron-Severi group is a problem for which algebraic geometers need the help of number theory.

Now return to the problem of the Γ corresponding to solutions $n_1, \ldots, n_9$ of the equation

$$\text{arithmetic genus of } \Gamma = 0.$$

We do not as yet know that such solutions correspond to positive divisors at all. However, the Riemann-Roch inequality for surfaces gives a lower bound for the freedom of any divisor, which in the case of a divisor of strictly positive degree on the quartic surface V reduces to

$$\text{freedom of } \Gamma \geq \text{arithmetic genus of } \Gamma;$$

thus any relevant set $n_1, \ldots, n_9$ can be realized as a positive divisor. We shall not need this in what follows, since we construct all the irreducible Γ explicitly. It turns out that nearly all positive divisors Γ of arithmetic genus 0 are reducible, but infinitely many are irreducible. Moreover there seems to be no simple direct condition for irreducibility, such as a set of inequalities on the n_i. How then do we specify the irreducible Γ?

What we have to do is to consider birational maps $\phi\colon V \to V$ defined over Q. (For geometric reasons, all such maps are actually biregular, in contrast with the cubic surface case.) The symmetries of V, including sign changes of the coordinates, give some such maps. Less trivial ones can be obtained from any pencil of elliptic curves on V, such as the Γ_λ; for consider the map defined by $P \to -P$, where the minus denotes the operation of the group law on that particular Γ_λ on which P

lies. Clearly the image of an irreducible curve Γ of arithmetic genus 0 under any birational map ϕ is another such curve; but we have also a converse result:

THEOREM 8. *There exist two biregular maps ϕ_1, ϕ_2 from V to V defined over Q such that if Γ is any irreducible curve on V of arithmetic genus 0 defined over Q then Γ can be generated from the line $A = C$, $B = D$ by repeated applications of ϕ_1, ϕ_2 and the symmetries of V in such a way that each application of ϕ_1 or ϕ_2 strictly increases the degree of the curve being produced.*

Because of the last clause, all such Γ of degree lower than any preassigned bound can be found constructively. The maps ϕ_1 and ϕ_2 come from pencils of elliptic curves in the way described above, and they can be written down explicitly.

The idea of the proof of Theorem 8 is simple, though the details are tedious. One produces an explicit list of positive divisors $C_1, \ldots, C_n$ defined over Q but not necessarily irreducible. If Γ is to be irreducible and not one of the C_i (which have to be discussed individually) then its intersection number with each C_i must be nonnegative, and this is equivalent to a linear inequality in the n_i. Again, if the degree of Γ cannot be reduced by a symmetry of V followed by the operation of ϕ_1^{-1} or ϕ_2^{-1}, this implies a linear inequality in the n_i for each symmetry and choice of ϕ^{-1} or ϕ_2^{-1}; for ϕ_1 and ϕ_2 induce automorphisms of the Néron-Severi group and thus linear transformations on the n_i. If we consider the Γ of minimal degree which cannot be generated in the way described in the theorem, its n_i satisfy a large number of homogeneous linear inequalities; and the corresponding point $(n_1, \ldots, n_9)$ therefore lies in a convex polyhedral cone. If $A(n_1, \ldots, n_9)$ is the formula for the arithmetic genus of Γ, to prove the theorem it is enough to prove that $A > 0$ for all points of the cone; and because of the particular form of A it turns out to be enough to prove this for the extremal rays of the cone—that is, the lines joining the vertex of the cone to the vertices of the convex polyhedron in which any hyperplane meets the cone. The proof is thereby reduced to a finite computation, which has been carried out on a computer.

This theorem leaves two questions open, both of which seem to be very difficult. First, it says nothing about irreducible curves defined over Q which are parametrizable but because of accidental double points have arithmetic genus greater than 0. Such curves should be rare, since they correspond to solutions of a nonlinear equation and this should need a field extension; but in fact there are infinitely many of them. For take any curve Γ of the type described in Theorem 8 which meets each Γ_λ in more than one point; and let ψ be the 4-to-1 map $V \to V$ defined by $P \to 2P$, where the addition is given by the group law on that Γ_λ on which P lies. The image of Γ under ψ will inherit a parametrization from Γ, but it will have a double point on each Γ_λ whose intersection with Γ contains two points whose difference under the group law of Γ_λ is a 2-division point. In this way we can even get an example with as many double points as we choose.

Again, searches by means of computers have shown that (16) has a very large number of rational solutions—probably more than can be accounted for by the parametric solutions. Do there exist rational solutions of (16) which are not special cases of one-parameter solutions; and if so, can a specific example be given?

6. Zeta functions over finite fields, and the Weil conjectures

Let V be a complete nonsingular variety defined over $k = GF(q)$, the finite field of q elements. How many points defined over k lie on V? In the special case when q is a prime p and there are no restrictions on V, this problem has a much older formulation: if $f(x_1, \ldots, x_n)$ is a polynomial which is irreducible mod p, how many solutions are there of the congruence $f \equiv 0 \bmod p$? In this formulation the problem arises naturally in the use of the Hardy-Littlewood method in analytic number theory, and techniques were developed which gave the leading term p^{n-1} of the answer and a good enough error term. However the problem is of interest in its own right, and considerable efforts were made (by nongeometric methods) to improve the error term.

A major step forward was to recognize that this was essentially a problem of algebra or of algebraic geometry, and to generalize the problem by defining and studying the local zeta-function of V. For this, write $k_\nu = GF(q^\nu)$, the unique extension of degree ν of k, and define M_ν and N_ν by

$$M_\nu = \text{number of positive 0-cycles of degree } \nu \text{ on } V \text{ defined over } k,$$
$$N_\nu = \text{number of points on } V \text{ defined over } k_\nu.$$

Define the *zeta function* of V with respect to k to be

$$Z(u) = \sum_{\nu=0}^{\infty} M_\nu u^\nu = \sum_{\mathfrak{A}} u^{\deg(\mathfrak{A})} = \prod_{\wp} (1 - u^{\deg(\wp)})^{-1} \tag{17}$$

where the sum over $\mathfrak{A}$ is taken over all positive 0-cycles defined over k, and the product over $\wp$ is taken over all irreducible 0-cycles defined over k, that is, positive nonzero cycles which cannot be written as a sum of two positive nonzero cycles defined over k. Every positive cycle can be written as a sum of irreducible cycles in essentially one way, so the last equality in (17) follows in the same way as the analogous identity for the classical zeta-function. Now a point on V counts towards N_ν if and only if it is a member of an irreducible cycle whose degree divides ν; thus N_ν is the sum of deg $(\wp)$ taken over all $\wp$ for which deg $(\wp)$ divides ν, and we obtain (at least formally)

$$u\frac{Z'(u)}{Z(u)} = \sum \frac{u^{\deg(\wp)} \deg(\wp)}{1 - u^{\deg(\wp)}} = \sum_{\nu=1}^{\infty} \sum_{\wp} \deg(\wp) u^{\nu \deg(\wp)} = \sum_{\nu=1}^{\infty} N_\nu u^\nu.$$

If V is embedded in R-dimensional space then $N_\nu = O(q^{R\nu})$ because there cannot be more points defined over k_ν on V than in the space; thus all the series converge and all the manipulations are justified provided u is small enough.

If V is a curve of genus g, it follows from the Riemann-Roch theorem that

$$Z(u) = \left\{\prod_{i=1}^{2g} (1 - \alpha_i u)\right\} \Big/ (1 - u)(1 - qu) \tag{18}$$

for some complex numbers α_i; and $Z(u)$ satisfies the functional equation

$$Z(1/qu) = (qu^2)^{1-g} Z(u), \tag{19}$$

which can be more sensibly expressed by saying that the q/α_i are just the α_i in a different order. The *Riemann hypothesis for function fields* is the statement

$$|\alpha_i| = q^{1/2}; \tag{20}$$

this is essentially equivalent to the estimate

$$|N_\nu - q^\nu - 1| \leq 2gq^{\nu/2} \tag{21}$$

which is best possible. The estimate (21) was first proved by Hasse in the case $g = 1$, and by Weil in the general case, and (20) was deduced from it. No direct proof of (20) which does not involve cohomology theory is known; but a proof by means of cohomology is implicit in Kleiman [13].

In general, if V is a complete nonsingular variety of dimension d, Weil made three conjectures about its zeta-function. The first is that it is a rational function of the form

$$Z(u) = \frac{Z_1(u) \ldots Z_{2d-1}(u)}{Z_0(u)Z_2(u) \ldots Z_{2d}(u)} \quad \text{where} \quad Z_n(u) = \sum_{i=1}^{B_n} (1 - \alpha_{ni}). \tag{22}$$

From this it follows in particular that

$$N_\nu = \sum_n (-1)^n \sum_i \alpha_{ni}^\nu. \tag{23}$$

The second conjecture is that there is a functional equation connecting $Z(1/q^d u)$ and $Z(u)$, which can most easily be expressed by saying that for each n the q^d/α_{ni} are just the $\alpha_{2d-n,i}$ in a different order. The third conjecture is the Riemann hypothesis

$$|\alpha_{ni}| = q^{n/2}. \tag{24}$$

Weil gave reasons for making these conjectures, which amounted to a program for proving the first two conjectures and possibly the third; and he gave many examples of varieties V for which the zeta-function could be calculated and obeyed the conjectures. In fact the rationality of the zeta function was first proved by Dwork [7] by quite different methods (which are now understood as involving a rudimentary p-adic cohomology), and he also proved the functional equation in the case when V is a hypersurface. Weil's program has been carried out, by means of l-adic cohomology, largely by Grothendieck, M. Artin and Verdier, so far as the rationality and the functional equation are concerned; see [11] and [13]. However it does not at the moment seem likely that these methods will suffice for the Riemann hypothesis. Moreover, the proof only shows that the individual $Z_n(u)$ have l-adic coefficients, though $Z(u)$ is obviously defined over Q and it is conjectured that the coefficients of the $Z_n(u)$ are rational integers.

To see why these conjectures are plausible, we must start from the Lefschetz fixed point theorem in topology. Let K be a topological complex, and let ϕ be an acceptable map of K into itself. K has rational homology groups in each dimension, which are finite dimensional Q-vector-spaces; and ϕ induces an endomorphism ϕ_n on the nth homology group. The Lefschetz fixed point formula is

$$\text{Weighted number of fixed points of } \phi = \sum (-1)^n \operatorname{Tr}(\phi_n), \tag{25}$$

where the weights are integers which are not necessarily positive. The formula works for real topological manifolds under suitable restrictions. Now let V be a d-dimensional variety defined over the complex numbers, and let ϕ be a geometric map of V to itself; V can be regarded as a $2d$-dimensional real topological manifold, and ϕ as a topological map, and they satisfy the conditions under which (25) holds. Moreover, the weight of a fixed point can be calculated in a purely geometric way and is necessarily positive; in fact it is an intersection number. The homology groups can now be replaced by their duals, the cohomology groups, and the ϕ_n can be replaced by the dual maps $\phi^{(n)}$ induced by ϕ on the nth cohomology group. (The elements of homology groups are topological cycles, and the elements of cohomology groups are differentials with suitable conditions about poles; the pairing called for by the duality is just the operation of integrating a differential round a cycle.) The formula (25) now becomes

$$\text{(26)} \qquad \text{Number of fixed points of } \phi = \sum(-1)^n \operatorname{Tr}(\phi^{(n)}).$$

Now suppose that V is a variety defined over an arbitrary field. Formally, everything in (26) should still be definable—this is why we went from homology to cohomology. It is necessary to develop a cohomology theory in which the cohomology groups are finite dimensional vector spaces (over the field of coefficients for the cochains) and in which the Lefschetz formula (26) can still be proved. That being done, suppose that V is defined over $GF(q)$ and let ϕ be the Frobenius map π which replaces every coordinate by its qth power. The fixed points of π are precisely the points of V defined over $GF(q)$, and each of them has multiplicity 1. Let the α_{ni} be the characteristic roots of the endomorphism induced by π on the nth cohomology group; then (26) becomes

$$N_1 = \sum(-1)^n \sum \alpha_{ni},$$

and working with the νth power of π instead of π it becomes just (23). This gives (22) immediately, and the functional equation is just a duality statement.

The reasons for believing the Riemann hypothesis (24) lie very much deeper, and it is not possible to explain them here. However, it is known that (24) can be deduced from certain conjectures of Lefschetz and of Hodge; for if these conjectures hold then the $q^{-n/2}\alpha_{ni}$ can be identified with the eigenvalues of an automorphism of a certain positive definite quadratic form. For a full account see Kleiman [13].

7. The inequality for N_ν, for curves

The proof of the rationality and the functional equation of the zeta function for a curve presupposes some estimate, though a very weak one, for N_ν. To avoid quoting results obtained in other ways, it is therefore desirable to prove (21) before (18) and (19). The proof that will be outlined below is substantially Weil's original proof, which is closely related to the heuristic arguments above and which does not explicitly involve Abelian varieties. Another proof of the key Theorem 9, depending on the Riemann-Roch theorem for surfaces, has been given by Mattuck and Tate [16]; see also Grothendieck [10].

If f is any map $\Gamma \to \Gamma$, then f can be represented by its graph Γ_f; this is a curve on $\Gamma \times \Gamma$, with the property that it meets each $P \times \Gamma$ in the single point $P \times f(P)$.

For geometrical purposes this last restriction is unnatural and inconvenient; we therefore define a *correspondence* on Γ to be any divisor on $\Gamma \times \Gamma$. If G is the group of divisors on Γ, then a correspondence on Γ induces a homomorphism $G \to G$; and if two correspondences induce the same homomorphism they differ only by a component of the form $\mathfrak{A} \times \Gamma$. (By convention, a correspondence of the form $\mathfrak{A} \times \Gamma$ maps all divisors on G into 0—even those which have a component in common with $\mathfrak{A}$.) If Z is any correspondence, we can define its transpose Z' as the image of Z under the map of $\Gamma \times \Gamma$ to itself which takes any point $P \times Q$ into $Q \times P$. The correspondence Z takes any point P into the divisor $Z(P)$, whose degree does not depend on P; we write this degree as $d(Z)$ and for convenience define $d'(Z)$ to be $d(Z')$, which is also the degree of the intersection $(\Gamma \times P) \cdot Z = Z'(P) \times P$. With the help of the transpose, we can define a satisfactory ring structure on the correspondences compatible with that on the homomorphisms of G:

LEMMA 1. *Let X, Y be any correspondences on Γ; then there is a unique correspondence $Z = X \circ Y$ such that $Z(P) = X(Y(P))$ and $Z'(P) = Y'(X'(P))$ for all P on Γ. Moreover $d(Z) = d(X)d(Y)$ and $d'(Z) = d'(X)d'(Y)$.*

The proof is a matter of straightforward technique, z being given by

$$Z = pr_{13}\{(Y \times \Gamma) \cdot (\Gamma \times X)\}$$

whenever the intersection has the correct dimension. Here the intersection is taken on $\Gamma \times \Gamma \times \Gamma$ and the projection is onto the product of the first and third factors.

A correspondence maps an algebraic system of divisors into an algebraic system of divisors; since a divisor is in G_a if and only if it is the difference of two divisors in the same algebraic system, a correspondence maps G_a into G_a. A similar argument, with "linear" for "algebraic" throughout, shows that a correspondence maps G_l into C_l. Hence it induces a map of the Jacobian variety of G_a/G_l into itself; and every geometric endomorphism of G_a/G_l can be obtained in this way. There is therefore a ring epimorphism

$$\text{(27)} \qquad \text{Ring of correspondences on } \Gamma \to \text{End}\,(G_a/G_l).$$

What is the ideal which is its kernel?

LEMMA 2. *A correspondence Z maps G_a into G_l if and only if it can be written in the form $Z = \mathfrak{A}_1 \times \Gamma + \Gamma \times \mathfrak{A}_2 + (\phi)$ where $\mathfrak{A}_1$, $\mathfrak{A}_2$ are divisors on Γ and ϕ is a function on $\Gamma \times \Gamma$.*

"If" is easy, since the three components of Z map a point P into respectively 0, $\mathfrak{A}_2$ and (ϕ_P), where ϕ_P is the restriction of ϕ to $P \times \Gamma$. For "only if", let A be a fixed point on Γ, and let k be a common field of definition for Γ, Z and A. If P is a general point of Γ, $P - A$ is in G_a and hence $Z(P) - Z(A)$ is in G_l; as it is defined over $k(P)$ it has the form (ϕ_P) for some function ϕ_P on Γ defined over $k(P)$. Now ϕ_P induces a function ϕ on $\Gamma \times \Gamma$ defined over k and given by the formula $\phi(P \times Q) = \phi_P(Q)$; and the divisor of ϕ meets $P \times \Gamma$ in $Z(P) - Z(A)$. It follows that $(\phi) - Z + \Gamma \times Z(A)$, which is defined over k, has zero intersection with $P \times \Gamma$; so it must be of the form $\mathfrak{A}_1 \times \Gamma$. This proves the lemma.

This technique of lifting from ϕ_P to ϕ is a major tool of this kind of algebraic geometry; and it is largely because of it that fields of definition are important to the geometer.

In view of (27) we shall describe the elements of End (G_a/G_l) as *correspondence classes*, and shall write $Z_1 \equiv Z_2$ if the images of Z_1 and Z_2 are in the same class. Correspondence classes will be denoted by Greek letters. In particular, δ will denote the identity class which is the image of the *diagonal* Δ, the locus of $P \times P$ on $\Gamma \times \Gamma$; and when k is $GF(q)$, Π will denote the divisor corresponding to the Frobenius map $\Gamma \to \Gamma$ which raises every coordinate to its qth power, and π will denote the class of Π.

If X_1, X_2 are correspondences, then the intersection $X_1 \cdot X_2$ is undefined if X_1 and X_2 have a common component; but we can always choose functions f_1, f_2 on $\Gamma \times \Gamma$ such that $(X_1 + (f_1)) \cdot (X_2 + (f_2))$ is defined. Define $I(X_1, X_2)$, the intersection number of X_1 and X_2, to be the degree of this intersection for some f_1, f_2 for which it is defined; it does not depend on the choice of f_1 and f_2. In the language of the Lefschetz fixed point theorem, $I(X, \Delta)$ is the number of fixed points of X; and the 0- and 2-cohomology groups of Γ are one-dimensional vector spaces on which X induces multiplication by $d(X)$ and $d'(X)$ respectively. Thus the trace of the map induced by X on the 1-cohomology should be $S(X) = d(X) + d'(X) - I(X, \Delta)$. We take this as a definition of $S(X)$ and prove that it has the usual properties of a trace. $S(X)$ depends only on the class of X, for it is additive and if $X \equiv 0$ then X is given by Lemma 2 and it is easy to verify that $S(X) = 0$. Thus $S(\xi)$ is well defined, where ξ is any correspondence class. Clearly $S(X)$ is linear in X and $S(X') = S(X)$; and $S(X \circ Y) = S(Y \circ X)$ in view of Lemma 1 and

$$I(X \circ Y, \Delta) = I(X, Y') = I(Y, X') = I(Y \circ X, \Delta).$$

Moreover, $S(\delta) = 2g$; for $d(\Delta) = d'(\Delta) = 1$, and if f is any function on $\Gamma \times \Gamma$ which has Δ as a component with multiplicity $+1$ then $I(\Delta, \Delta) = I(\Delta - (f), \Delta)$ is minus the degree of the divisor of the differential induced by df on Δ.

The further property of the trace which we shall need is contained in the following theorem, due in the classical case to Castelnuovo:

THEOREM 9. *If* $\xi \neq 0$ *then* $S(\xi\xi') > 0$.

We can assume that $g > 0$, for if $g = 0$ then $G_a = G_l$ and there are no non-trivial correspondence classes. The first step is to show that we can choose a representative X for ξ such that $X > 0$, $d(X) = g$ and for general P the image $X(P)$ is the sum of g distinct points. Let X_0 be any representative of ξ, and $\mathfrak{A}$ a sufficiently general divisor on Γ of degree $d(X_0) - g$; and let K be a common field of definition for Γ, X_0 and $\mathfrak{A}$. For generic P, the divisor $X_0(P) - \mathfrak{A}$ has degree g, and hence by the Riemann-Roch theorem there exist a function ϕ_P on Γ and a divisor $\mathfrak{b}_P > 0$ on Γ, both defined over $K(P)$, such that $X_0(P) - \mathfrak{A} + (\phi_P) = \mathfrak{b}_P$; and by suitable choice of $\mathfrak{A}$ we can assume that $\mathfrak{b}_P$ consists of g distinct points. Now lift ϕ_P to a function ϕ on $\Gamma \times \Gamma$ defined over K by means of the formula $\phi(P \times Q) = \phi_P(Q)$. The divisor $X_0 - \Gamma \times \mathfrak{A} + (\phi)$ is in the class of X_0, is defined over K, and meets $P \times \Gamma$ in g distinct points each with multiplicity $+1$; after removing any components of the form $\mathfrak{A}_1 \times \Gamma$ it is the X we are looking for.

If $g = 1$ then for this X we have $X \circ X' = d'(X)\Delta$ and so $S(\xi\xi') = 2d'(X)$, which we show below is strictly positive; thus we may assume for the time being that $g > 1$. Write $X(P) = Q_1 + \cdots + Q_g$; then the correspondence $X \circ X'$ is just the locus of $\sum\sum Q_i \times Q_j$ over $K(P)$. The terms with $i = j$ contribute $d'(X)\Delta$, for each Q arises from $d'(X)$ points P; suppose that the rest of the locus is Y, so that $X \circ X' = d'(X)\Delta + Y$ and

$$S(X \circ X') = 2gd'(X) + (2g - 2)d'(X) - I(\Delta, Y),$$

the first term being $d(X \circ X') + d'(X \circ X')$ and the second being $-I(d'(X)\Delta, \Delta)$. It remains to evaluate $I(\Delta, Y)$, which is simply the number of P for which two of the Q_i coincide.

Now let $\mathfrak{k}$ be a positive canonical divisor defined over K, and let $f_1, \ldots, f_g$ defined over K be a base for the vector space $L(\mathfrak{k})$ whose dimension is known to be g. Consider

$$F(P) = \{\det f_i(Q_j)\}^2;$$

the right-hand side is a symmetric function of the Q_j and so $F(P)$ is a single-valued function on Γ defined over K. The poles of F are double poles at those points P for which any Q_j is a component of $\mathfrak{k}$; since $\deg \mathfrak{k} = 2g - 2$, there are just $(2g - 2)d'(X)$ of these double poles. But $F(P)$ vanishes whenever two of the Q_j coincide; and since F has as many poles as zeros, this can happen for at most $4(g - 1)d'(X)$ points P. Thus $I(\Delta, Y) \leq 4(g - 1)d'(X)$ and so $S(X \circ X') \geq 2d'(X)$, a result which we have already proved in the case $g = 1$. Since X is positive, $d'(X) \geq 0$; and if we have equality then $X = \Gamma \times \mathfrak{A}_2$ for some $\mathfrak{A}_2$ and so ξ is trivial. This completes the proof of the theorem.

COROLLARY. $|N_\nu - q^\nu - 1| \leq 2gq^{\nu/2}$ *for all* $\nu > 0$.

We have $\pi^\nu \pi'^\nu = q^\nu \delta$, the multiplicity arising from having to take a q^νth root. Let $\xi = u\delta + v\pi^\nu$ for integers u, v; then by the theorem

$$0 \leq S(\xi\xi') = 2gu^2 + 2uvS(\pi^\nu) + 2gq^\nu v^2.$$

Since this holds for all integers u, v, it follows that $|S(\pi^\nu)| \leq 2gq^{\nu/2}$. But

$$S(\pi^\nu) = S(\Pi^\nu) = 1 + q^\nu - N_\nu$$

since $d(\Pi) = 1$, $d'(\Pi) = q$ and $I(\Delta, \Pi^\nu)$ is the number of points which are fixed under Π^ν, which are just the points defined over k_ν. This proves the corollary.

8. Proofs of (18), (19) and (20) for a curve

If $N_\nu > 0$ then $M_\nu > 0$; for if P is defined over k_ν then a suitable multiple of the sum of P and its conjugates over k is a positive divisor of degree ν defined over k. By the estimate above, $N_\nu > 0$ for all large ν; thus we can find divisors of degrees $\nu + 1$ and ν defined over k, and their difference is a (not necessarily positive) divisor $\mathfrak{b}$ defined over k and of degree 1. Let h be the number of linear equivalence classes of degree 0 which contain divisors defined over k, and let $\mathfrak{A}_1, \ldots, \mathfrak{A}_h$ be representatives of these classes. We know that h is finite because, by adding a suitable multiple of $\mathfrak{b}$ to each of a complete set of representatives, h is the number of such linear equivalence classes of any assigned degree; thus $h \leq M_g$ since each class of degree g contains positive divisors.

Let $\mathfrak{A}$ be any positive divisor of degree ν defined over $k = GF(q)$; then $\mathfrak{A} - \nu\mathfrak{b}$ has degree 0 and is therefore linearly equivalent to some $\mathfrak{A}_i$. Thus $\mathfrak{A} = \nu\mathfrak{b} + \mathfrak{A}_i + (\phi)$ for some i and some nonzero ϕ in $L(\nu\mathfrak{b} + \mathfrak{A}_i)$; and conversely any such i and ϕ define a positive $\mathfrak{A}$ of degree ν and defined over k. If $\mathfrak{A}$ is fixed, then i is unique and ϕ is determined up to multiplication by an arbitrary nonzero element of k. Thus to given ν and i there correspond $\{q^{l(\mathfrak{A}_i+\nu\mathfrak{b})} - 1\}/\{q - 1\}$ suitable $\mathfrak{A}$, and we have for $g > 0$

$$Z(u) = \sum_{\nu=0}^{\infty} \sum_{i=1}^{h} \frac{u^\nu\{q^{l(\mathfrak{A}_i+\nu\mathfrak{b})} - 1\}}{q - 1}$$

(28)

$$= \sum_{\nu=0}^{2g-2} \sum_{i=1}^{h} \frac{u^\nu q^{l(\mathfrak{A}_i+\nu\mathfrak{b})}}{q - 1} + \frac{h}{q-1}\left\{\frac{q^g u^{2g-1}}{1 - qu} - \frac{1}{1 - u}\right\}$$

the second expression following from the first because $l(\mathfrak{A}_i + \nu\mathfrak{b}) = \nu + 1 - g$ if $\nu > 2g - 2$. (If $g = 0$ then $h = 1$ and $l(\mathfrak{A}_i + \nu\mathfrak{b}) = \nu + 1$ always; thus

$$Z(u) = \frac{1}{q-1}\left\{\frac{q}{1 - qu} - \frac{1}{1 - u}\right\} = \frac{1}{(1 - qu)(1 - u)}$$

which satisfies (18) and (19). Moreover, (20) is now trivial since there are no α_i.) Now (28) shows immediately that $(1 - qu)(1 - u)Z(u)$ is a polynomial in u of degree at most $2g$; and the polynomial has constant term 1 since $M_0 = 1$. This proves (18). To prove (19) we write (28) as $Z(u) = Z_1(u) + Z_2(u)$. It is easy to check that $(qu^2)^{g-1}Z_2(1/qu) = Z_2(u)$, and we have only to prove the corresponding result for Z_1. Let $\mathfrak{A}_j$ be the representative of the class of $\mathfrak{k} - \mathfrak{A}_1 - (2g - 2)\mathfrak{b}$, where $\mathfrak{k}$ is a fixed canonical divisor defined over k; then as i runs through the set $1, 2, \ldots, h$ so does j. We have, writing $\mu = 2g - 2 - \nu$,

$$l(\mathfrak{A}_i + \nu\mathfrak{b}) = \nu + 1 - g + l(\mathfrak{k} - \mathfrak{A}_i - \nu\mathfrak{b}) = \nu + 1 - g + l(\mathfrak{A}_j + \mu\mathfrak{b});$$

thus

$$(qu^2)^{g-1}Z_1(1/qu) = \sum_{\nu=0}^{2g-2} \sum_{i=1}^{h} \frac{q^{g-1-\nu}u^{2g-2-\nu}q^{l(\mathfrak{A}_i+\nu\mathfrak{b})}}{q - 1}$$
$$= \sum_{\mu=0}^{2g-2} \sum_{j=1}^{h} \frac{u^\mu q^{l(\mathfrak{A}_j+\mu\mathfrak{b})}}{q - 1} = Z_1(u)$$

as required. This proves (19).

Finally we prove (20). By the corollary to Theorem 9,

(29) $$\left|\sum \alpha_i^\nu\right| = |N_\nu - q^\nu - 1| \leq 2gq^{\nu/2}.$$

By simultaneous approximation to the $\arg(\alpha_i/2\pi)$, we can choose ν so that $|\arg \alpha_i^\nu| < \pi/4$ for all i; and it then follows from (29) that $|\alpha_i^\nu| \leq 2gq^{\nu/2}$. Since this holds for some arbitrarily large ν, it follows that $|\alpha_i| \leq q^{1/2}$. But by the functional equation q/α_i is an α_j, and so $|q/\alpha_i| \leq q^{1/2}$. These two statements together give (20).

9. Zeta-functions of some special varieties

Most of the explicit determinations of zeta-functions, and in particular most of the verifications of the Riemann hypothesis for particular varieties, are due to Weil and are obtained by nongeometric methods. The most important of these are in [24] and [25], in which he determines the zeta functions of all varieties of the form

$$a_0x_0^{n_0} + a_1x_1^{n_1} + \cdots + a_rx_r^{n_r} = b;$$

in particular he shows that the zeta function of the complete nonsingular variety

$$a_0x_0^n + a_1x_1^n + \cdots + a_rx_r^n = 0 \tag{30}$$

satisfies the conjectures. He proves the Riemann hypothesis for this variety by showing that the characteristic roots $\alpha_{\nu i}$ of the zeta function are products of Gauss sums; and since all Gauss sums are characteristic roots of zeta functions of curves admitting complex multiplication, one can say that the Riemann hypothesis for (30) is a consequence of the known Riemann hypothesis for curves. Certainly, what makes it easier to prove results for (30) than for a general hypersurface is that there is a natural fibring of (30) in which the base has an equation of the same type with $r - 1$ for r, and all the fibres are birationally equivalent to each other over an extension field.

Other cases are known in which the zeta function of a variety V can be expressed in terms of the zeta functions of curves. For example, suppose that V is a nonsingular cubic threefold and L is a line on V, both defined over $k = GF(q)$; given V, such an L certainly exists if q is large enough. There are $q^{2\nu} + q^\nu + 1$ planes through L defined over k_ν; and a general plane through L meets V residually in a nonsingular conic, which contains $q^\nu + 1$ points defined over k_ν. Let Γ' parametrize the planes for which this residual intersection breaks up into two lines, and let Γ'' parametrize the resulting lines (which are just those lines on V that meet L). Γ'' is a double covering of Γ'; and if P' is a point of Γ' defined over k_ν, and P_1'', P_2'' the corresponding points on Γ'', then the lines are individually defined over k_ν if and only if the P_i'' are defined over k_ν. Thus the number of points defined over k_ν on the residual intersection corresponding to P' is $2q^\nu + 1$ if the P_i'' are defined over k_ν, and 1 otherwise. Suppose that the numbers of points defined over k_ν on V, Γ', Γ'' are respectively N_ν, N_ν', N_ν''. Counting the numbers of points of V on each residual intersection, and noting that in this way each point of L gets counted $q^\nu + 1$ times, we obtain

$$\begin{aligned} N_\nu &= (q^{2\nu} + q^\nu + 1 - N_\nu')(q^\nu + 1) + \tfrac{1}{2}N_\nu''(2q^\nu + 1) \\ &\quad + (N_\nu' - \tfrac{1}{2}N_\nu'') - q^\nu(q^\nu + 1) \\ &= q^{3\nu} + q^{2\nu} + q^\nu + 1 - q^\nu(N_\nu' - N_\nu''). \end{aligned}$$

Since Γ'' is a covering of Γ', every characteristic root of the zeta function of Γ'' is a characteristic root of the zeta function of Γ'. Suppose that $\alpha_1, \alpha_2, \ldots, \alpha_n$ are the remaining characteristic roots of the zeta function of Γ'; then $N_\nu' - N_\nu'' = \sum \alpha_i^\nu$, which determines N_ν and the zeta function of V. For more details, see [3].

Similarly let F, G be homogeneous quadratic forms in $x_0, \ldots, x_r$ such that their intersection V: $F = G = 0$ is a nonsingular variety. The condition for this is that $\phi(u, v) = \det(uF + vG)$ has no repeated factor. Let W run through all those distinct quadrics of the pencil $uF + vG = 0$ which are defined over k_ν; the points of V lie on each of the $q^\nu + 1$ quadrics W, and any other point of the space lies on just one W. Thus if there are N_ν points on V defined over k_ν,

$$q^\nu N_\nu + (q^{\nu(r+1)} - 1)/(q^\nu - 1) = \sum(\text{Number of points on } W \text{ defined over } k_\nu).$$

The singular W in the pencil correspond to the zeros of ϕ, and they are cones with a point vertex. Suppose first that r is even; then whenever W is nonsingular it has $(q^{\nu r} - 1)/(q^\nu - 1)$ points defined over k_ν on it, and the number of points on a singular W depends only on whether the derivative of ϕ at the corresponding value of u/v is a square or not. Thus N_ν can be explicitly determined. Similarly if r is odd the number of points on a nonsingular W defined over k_ν is

$$(q^{\nu r} - 1)/(q - 1) + q^{\nu(r-1)/2}\chi_\nu((-1)^{(r+1)/2}\phi(u, v))$$

where $\chi_\nu(y) = 1$ if y is a nonzero square in k_ν, $\chi_\nu(0) = 0$ and $\chi_\nu(y) = -1$ otherwise. This formula happens to work also when W is singular. Thus in this case N_ν can be expressed in terms of the characteristic roots of the zeta function of the curve $y^2 = \phi(x, 1)$. A similar argument works for the intersection of more than two quadrics. For more details see [26].

If V is a nonsingular surface which is birationally equivalent to a plane over an extension field, then its zeta-function has the form

$$Z(u) = \{(1 - u)(1 - q^2u)\prod(1 - \eta_i qu)\}^{-1}$$

where the η_i, which are roots of unity, are the characteristic roots of the Frobenius element of the Galois group of $\bar{k}$ over k, acting on the Néron-Severi group. This was first proved directly by Manin, but it follows also from the cohomological proof of the rationality of the zeta-function, because in this case the cohomology can be expressed in terms of the Néron-Severi group.

10. The weak Mordell-Weil theorem

Let K be an algebraic number field and $m > 1$ a positive integer. For an Abelian variety A defined over K, denote by A_K the group of points on A defined over K, and by A_m the group of m-division points on A—that is, the points P such that mP is the identity element of A. The Mordell-Weil theorem states that A_K is finitely generated; and in any particular case its proof gives a sensible upper bound for the number of generators. The weak Mordell-Weil theorem, some version of which is an essential step in the proof of the full theorem, states that A_K/mA_K is finite. Replacing K by a larger field can only increase A_K, and hence to prove the full Mordell-Weil theorem we can restrict ourselves to large enough fields K; this would not be true of the weak theorem if it were being regarded as an end in itself. Throughout this section we shall therefore assume that the points of A_m are defined over K, and (which in fact follows from this) that K contains the mth roots of unity.

So as not to break up the main line of the argument, we state two preliminary lemmas:

LEMMA 3. *Let K be an algebraic number field containing the* mth *roots of unity, and let $\mathcal{S}$ consist of all but finitely many primes of K. If K_1 is an abelian extension of K of exponent m (that is, if the* mth *power of every element of the Galois group of K_1 over K is the identity), and if K_1 is unramified over K at each prime in $\mathcal{S}$, then the degree $[K_1\colon K]$ is bounded, the bound depending only on K, m, and $\mathcal{S}$.*

The Galois group of K_1/K can be written as the direct product of cyclic groups whose orders divide m, and hence K_1 can be written as the compositum of finitely many cyclic extensions L of K whose degrees divide m. Since K contains the mth roots of unity, $L = K(\alpha^{1/m})$ for some α in K^*; and two α in the same coset of K^{*m} give rise to the same L. But for any $\wp$ in $\mathcal{S}$, L is to be unramified over K at $\wp$; thus the exponent to which $\wp$ divides α must be a multiple of m. Denote by M the group of those α in K^* which satisfy this condition for each $\wp$ in $\mathcal{S}$; what we have shown so far is that $K_1 = K(\alpha_1^{1/m}, \ldots, \alpha_r^{1/m})$ where the α_i lie in distinct cosets of K^{*m} in M. To prove the lemma, it is enough to show that these conditions impose an upper bound on r—that is, that M/K^{*m} is finite. But the $\mathcal{S}$-divisor of α (that is, the part of the divisor composed of primes in $\mathcal{S}$) is an mth power, say $\mathfrak{A}^m$; and the map which takes α into $\mathfrak{A}$ induces a homomorphism from M/K^{*m} to the $\mathcal{S}$-ideal class group of K. Indeed there is an exact sequence.

$$0 \to E/E^m \to M/K^{*m} \to C_m \to 0,$$

where E is the group of $\mathcal{S}$-units of K (that is, the elements of K^* whose $\mathcal{S}$-divisor is trivial) and C_m is the group of those $\mathcal{S}$-ideal classes whose mth power is the principal class. Now E/E^m is finite because E is finitely generated, and C_m is finite because the $\mathcal{S}$-ideal class group is finite. Hence M/K^{*m} is finite, and the lemma is proved.

LEMMA 4. *Let B be a subgroup of A_K which contains mA_K, and suppose that B/mA_K is finite. Then $K(m^{-1}B)$ is a finite abelian extension of K, and its Galois group G has exponent m. Moreover, there is a bilinear map $G \times (B/mA_K) \to A_m$ whose kernel in each factor is trivial.*

Here $m^{-1}B$ denotes the set of points P on A such that mP is in B. Suppose that b is in B and that β on A satisfies $m\beta = b$. The other solutions of $mx = b$ are obtained by adding any element of A_m to β; since A_m consists of points defined over K, the field $K(\beta)$ is a normal extension of K and its Galois group over K is isomorphic to a subgroup of A_m. Since $K(m^{-1}B)$ is the compositum of such fields $K(\beta)$ as b runs through a complete set of representatives of the cosets of mA_K in B, this proves the statements about G in the lemma.

Now let σ be in G and consider the map that takes σ, β into $\sigma\beta - \beta$, which is in A_m. The image depends only on b and not on the choice of β for given b; thus it defines a map $G \times B \to A_m$, which is easily seen to be bilinear. If σ is in the kernel of the first factor, then $\sigma\beta = \beta$ for all β and so σ is the identity since the β generate $K(m^{-1}B)$ over K; if b is in the kernel of the second factor, then $\sigma\beta = \beta$ for all σ, whence β is defined over K and b is in mA_K. This completes the proof of the lemma.

THEOREM 10. *Let $m > 1$ be an integer, and let A be an Abelian variety defined over an algebraic number field K. Suppose that K contains A_m and the* mth *roots of unity. Then A_K/mA_K is finite.*

In fact we prove that if B is as in Lemma 4 then the order of B/mA_K is bounded; this is a technical device, to avoid any possibility of infinite field extensions. Because A_K/mA_K has exponent m, this result is equivalent to the theorem.

Now let $\wp$ be a finite prime of K. The variety A is embedded in projective space and is therefore given by a finite set of equations, and the group law on A is defined by a further finite set of equations. The proof that A is irreducible, the calculation of its dimension, and the proof that the law of composition on A is everywhere defined and gives A the structure of a group, can all be obtained by straightforward manipulation of these equations together with the remark that certain polynomials or elements of K which emerge from these manipulations are nonzero. Suppose that the equations for A and for the law of composition on A contain only coefficients which are integers at $\wp$; then we can map these coefficients into the residue field mod $\wp$, thus obtaining a variety $A^{\wp}$ and a law of composition on it. Suppose also that the coefficients in all the further equations obtained in the course of the manipulations described above are integers at $\wp$, that there has been no division by an element of $\wp$, and that the polynomials and elements of K which had to be nonzero are not divisible by $\wp$; then $A^{\wp}$ has the same dimension as A, and is an Abelian variety under the law of composition obtained on it. In such a case we say that *A has a good reduction* mod $\wp$; clearly A has a good reduction at all but finitely many $\wp$. (This is a naive approach to the problem of reduction mod $\wp$, in that it causes us to throw away a needlessly large number of primes $\wp$; for a more sophisticated approach see [21, Chapter III], or [17].) Let $\mathcal{S}$ be the set of those finite $\wp$ such that A has a good reduction mod $\wp$ and all the points of A_m are incongruent mod $\wp$.

Now let B be as in Lemma 4, let $\wp$ be in $\mathcal{S}$ and let b be in B. If β is any solution of $mx = b$, then the other solutions are the $\beta + \alpha$ where α runs through A_m; and the degree of the equation $mx = b$ is equal to the number of elements of A_m. Thus if $\tilde{\beta}$ is any root of the equation which is the reduction mod $\wp$ of $mx = b$, then $\tilde{\beta} + \tilde{\alpha}$ is also a root where $\tilde{\alpha}$ is the reduction mod $\wp$ of any α. These roots are all distinct because the $\tilde{\alpha}$ are all distinct, and the number of them is equal to the degree of the equation; hence each of them has multiplicity one and there are no other roots. It follows that $K(\beta)$ is unramified over K at $\wp$; and hence $K(m^{-1}B)$, which is a compositum of several $K(\beta)$, is unramified over K at $\wp$. If G is the Galois group of $K(m^{-1}B)$ over K, it now follows from Lemma 3 that the order of G is bounded, the bound depending only on m, K and A. But by Lemma 4, G and B/mA_K have the same order; so the order of B/mA_K is bounded and this proves the theorem.

Given an elliptic curve defined over the rational number field Q, one determines the group of rational points on it by the process of "infinite descent", which goes back to Fermat. The fact that at each stage of the descent one has only finitely many possibilities to consider is essentially just the weak Mordell-Weil theorem, even though the proof above seems to have little resemblance to a descent argument as normally formulated. It is therefore desirable to give a formulation of the descent algorithm which makes clear its connection with the arguments above.

For this purpose we assume that Γ is an elliptic curve defined over an algebraic number field K, and that $m > 1$ is an integer such that the m-division points of Γ are defined over K; thus Γ corresponds to A in the arguments above. Only minor modification would be needed to cope with the case where m is replaced by an arbitrary endomorphism of Γ; but the hypothesis that the points of Γ_m are defined over K is essential.

An m-covering of Γ is defined to be a curve Δ defined over K and a commutative diagram of maps

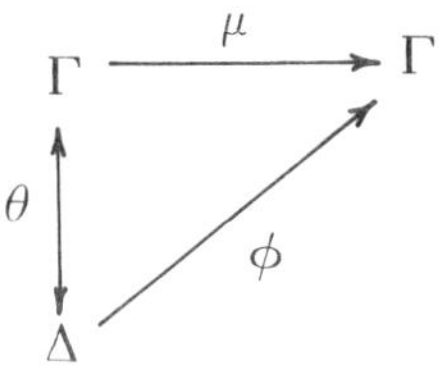

where θ is a birational map not necessarily defined over K, μ is the map $P \to mP$ and ϕ is defined over K. In virtue of θ, Δ is a curve of genus 1 whose Jacobian can be identified with Γ; and the operation of the Jacobian on Δ can be described by means of θ. There is an obvious definition of equivalence of m-coverings, and the classes of m-coverings form a commutative group with a geometrically defined group law; but we need not pursue this topic here. (For more details see [4] and the references contained therein.) To perform an *m-descent* on Γ is to give a set Σ of m-coverings Δ_i for which one can show (without necessarily knowing Γ_K) that each point of Γ_K is for some i the image under ϕ_i of a point on Δ_i defined over K. It will usually be evident that each point of Γ_K comes from only one Δ_i, though from m^2 points on that Δ_i differing by m-division points on the Jacobian. If a point P of Γ_K comes from P_i on Δ_i, then $P + mP'$ on Γ comes from $P'(P_i)$ on Δ_i, where the notation corresponds to the operation of the Jacobian. Thus if Δ_i contains points defined over K, the images of all such points make up a complete coset of $m\Gamma_K$ in Γ_K; and so the number of Δ_i in Σ which contain points defined over K is precisely the order of $\Gamma_K/m\Gamma_K$.

Before constructing Δ we need a digression. In it we shall use one trivial argument repeatedly, and so we formulate it in general terms:

LEMMA 5. *Let $\psi(P)$ be a function on Γ defined over K. Then $\psi(P) = \Psi(mP)$ for some function Ψ on Γ if and only if the divisor of ψ has the form $m^{-1}\mathfrak{A}$ for some $\mathfrak{A} \sim 0$; and in this case Ψ is defined over K.*

If Ψ exists let its divisor be $\mathfrak{A}$; then $(\psi) = m^{-1}\mathfrak{A}$ whence $m^{-1}\mathfrak{A}$ and so also $\mathfrak{A}$ is defined over K, and so Ψ is defined over K. Conversely suppose that $(\psi) = m^{-1}\mathfrak{A}$ with $\mathfrak{A} \sim 0$, and let $\Psi(P)$ be the function with divisor $\mathfrak{A}$; then $\psi(P)$ and $\Psi(mP)$ have the same divisor and are therefore equal up to multiplication by a constant. This proves the lemma.

Now let I be the identity element of Γ considered as a group, and let Q be any m-division point on Γ. Since $m^{-1}(Q - I) \sim 0$ and is defined over K, there is a function $\phi_Q(P)$ on Γ defined over K whose divisor is $m^{-1}(Q - I)$. Moreover $(\phi_Q(P))^m$ has divisor $m^{-1}(mQ - mI)$ and hence by Lemma 5 there is an F such that $(\phi_Q(P))^m = F(mP) = (\phi_Q(P + Q'))^m$ for any m-division point Q'. It follows that

$$\phi_Q(P + Q') = \chi(Q, Q')\phi_Q(P) \tag{31}$$

where $\chi(Q, Q')$ is an mth root of unity which does not depend on P.

LEMMA 6. *χ defines a bilinear map $\Gamma_m \times \Gamma_m \to$ (mth roots of unity) whose kernel in each factor is trivial. Hence every character on Γ_m is a $\chi(\cdot, Q')$ for some Q'.*

The function $\phi_{Q_1+Q_2}/\phi_{Q_1}\phi_{Q_2}$ satisfies the conditions of Lemma 5 and is therefore an $F(mP)$; so it is unchanged by writing $P + Q'$ for P. After (31), this implies that

$$\chi(Q_1 + Q_2, Q') = \chi(Q_1, Q')\chi(Q_2, Q')$$

which is the linearity in the first factor. The linearity in the second factor is equivalent to

$$\phi(P + Q_1' + Q_2')\phi(P) = \phi(P + Q_1')\phi(P + Q_2')$$

which follows from (31) in its present form and with $P + Q_2'$ for P. Now suppose that for some Q and all Q' we have $\chi(Q, Q') = 1$; then $\phi_Q(P)$ is unaltered by replacing P by $P + Q'$ for any Q', and hence it depends only on mP. After Lemma 5 this implies that Q is I; thus the kernel of χ in the first factor is trivial. The rest of the lemma now follows from the duality theory of finite abelian groups.

Now let λ be any element of Hom $(\Gamma_m, \bar{K}^*/K^*)$ where $\bar{K}$ is the algebraic closure of K; and for each Q in Γ_m let λ_Q be a representative in $\bar{K}^*$ of $\lambda(Q)$, which is a coset of K^*. Consider Δ, the locus as y varies on Γ of $z = (x, t_{Q_1}, t_{Q_2}, \ldots)$ where x is on Γ, the t_Q are scalars and there is one for each of the m^2 points of Γ_m, and the values of these are given by

$$x = my, \qquad \phi_Q(y) = \lambda_Q t_Q. \tag{32}$$

This locus is a curve which is evidently defined over $\bar{K}$. In fact it is defined over K; for let σ be any element of the Galois group of $\bar{K}$ over K. We have $\sigma\lambda_Q = \mu_Q\lambda_Q$ for some μ_Q which depends only on λ and not on the choice of the λ_Q; and evidently μ_Q is a multiplicative character on Γ_m. By Lemma 6 there is a Q' such that $\chi(Q, Q') = \mu_Q$ for all Q; and so the effect of σ is precisely balanced by writing $y + Q'$ for y. Thus Δ is invariant under every σ and hence is defined over K. Moreover a different choice of the λ_Q merely involves multiplying the t_Q by elements of K^*.

Clearly Δ gives rise to an m-covering of Γ, in virtue of the maps $z \to y$ and $z \to x$; and up to equivalence this covering depends only on λ. In fact we obtain in this way an isomorphism between the group of λ and the group of all m-coverings of Γ with its geometrically defined law of composition. In particular, let P be any point of Γ_K and define P_1 on Γ by $P = mP_1$. We have

$$\phi_{Q_1+Q_2}(y)/\phi_{Q_1}(y)\phi_{Q_2}(y) = F(my)$$

for some F, by Lemma 5; writing P_1 for y we see that $Q \to \phi_Q(P_1)$ induces a homomorphism $\Gamma_m \to \bar{K}^*/K^*$. If this homomorphism is λ, then the $\phi_Q(P_1)$ are proper choices for the λ_Q. It is easy to see that the resulting covering Δ depends only on P and not on the choice of P_1, that P is the image of a point of Δ defined over K, and that up to equivalence Δ is the only m-covering of this type for which this is so. Moreover the resulting map

$$\Gamma_K/m\Gamma_K \to \mathrm{Hom}\,(\Gamma_m, \bar{K}^*/K^*)$$

is a group monomorphism, for by an analogue of Lemma 5

$$\phi_Q(y_1 + y_2)/\phi_Q(y_1)\phi_Q(y_2) = F(my_1, my_2)$$

for some F.

Thus we obtain an m-descent by taking the set of all those m-coverings obtained in this way for which Δ has points defined over every $\wp$-adic completion of K. Suppose first that $\wp$ is such that Γ has a good reduction at $\wp$ and the points of Γ_m are all incongruent mod $\wp$. If x in (32) is defined over $K_\wp$ then y must be defined over an unramified extension, by an argument like that in the proof of Theorem 10, and hence so must the λ_Q. The λ_Q^m are in $K_\wp$, and by the last result the exponent to which $\wp$ divides any of them must be a multiple of m. As in the proof of Lemma 3, this already shows that there are only finitely many λ which we need to consider; and in any particular case consideration of the remaining $\wp$ may lead to a further reduction of the list.

In the case where A is an elliptical curve Γ, we have thus obtained as a by-product a second proof of Theorem 10; and it is obvious that this is closely related to the original proof. The descent process described here may be generalized to an arbitrary Abelian variety A, provided that one considers simultaneously A and its dual $\hat{A}$. But so far as I know, no one has ever carried out a descent argument in detail on an Abelian variety of dimension greater than 1; and it would be extremely laborious to do so.

11. Heights and the strong Mordell-Weil theorem

To conclude a classical "infinite descent" argument, it is necessary to show that the image of a point of Γ_K under an m-descent is in some sense smaller than the original point. To do this, we introduce the idea of *height*. Let V be any variety defined over K; then a height on V is any function h defined on V_K and satisfying $h(P) \geq 1$ for all P in V_K. However, we shall only be concerned with a very special class of heights, the simplest of which is defined as follows. Assume that V is embedded in n-dimensional projective space and that P can be written as $(x_0, x_1, \ldots, x_n)$; then write

$$h(P) = \prod \left\{\operatorname*{Max}_i \|x_i\|_\wp\right\} \tag{33}$$

where $\| \ \|_\wp$ denotes the usual $\wp$-adic norm. The product converges since only finitely many factors are not equal to 1. Replacing the x_i by λx_i only multiplies the right-hand side by $\prod\|\lambda\|_\wp = 1$, so $h(P)$ is well defined; and if we choose a representation of P for which some $x_i = 1$, then each factor on the right-hand side is at least 1, so that $h(P) \geq 1$.

More generally, let D be a divisor on V and let $f_0, \ldots, f_r$ be functions on V (D and the f_i being defined over K) such that all the divisors $(f_i) + D$ are strictly positive and their intersection is empty. We shall call D *adequate* if there is such a system of functions associated with it; the term is chosen as being slightly weaker than "ample", which is a standard geometric concept. For any given P in V_K, there is an i such that P is not on the divisor $(f_i) + D$; and since for each j

$$(f_j/f_i) = \{(f_j) + D\} - \{(f_i) + D\},$$

P is not a pole of any f_j/f_i. Thus we can define a height on V by the formula

$$h(P) = \prod_{\wp} \left\{\operatorname*{Max}_j \|f_j(P)/f_i(P)\|_{\wp}\right\}; \tag{34}$$

and $h(P)$ is independent of the choice of i, by an argument similar to that used for (33). Formula (33) is a special case of (34), given by choosing D to be $x_0 = 0$ and the f_i to be $1, x_1/x_0, \ldots, x_n/x_0$; conversely, if D is ample and the functions f_i induce a projective embedding of V in r-dimensional space, then formula (34) is just (33) for this embedding. We shall call (34) a height *associated with D.*

Two heights h_1 and h_2 on V are called *equivalent* if there is a constant C such that

$$h_1(P) < Ch_2(P) \quad \text{and} \quad h_2(P) < Ch_1(P)$$

for all P in V_K. They are called *quasi-equivalent* if for each $\epsilon > 0$ there is a C_ϵ such that

$$h_1(P) < C_\epsilon(h_2(P))^{1+\epsilon} \quad \text{and} \quad h_2(P) < C_\epsilon(h_1(P))^{1+\epsilon}.$$

Clearly each of these is an equivalence relation.

We have now to prove a series of purely geometrical results about heights, culminating in Theorem 11. The reader who is interested only in the number theory should go straight to the statement of Theorem 11, followed by the statement and proof of Theorem 12. When a descent argument is applied to a specific elliptic curve, the general geometric argument is unnecessary, since it can be replaced by an ad hoc argument based on the specific formulae of the descent.

LEMMA 7. *Let V be a complete variety defined over K, let D_1 and D_2 be adequate divisors on V such that $D_1 \sim D_2$, and let h_1 and h_2 be heights associated with D_1 and D_2 respectively. Then h_1 and h_2 are equivalent.*

Let the functions used to define h_1, h_2 be the f_i and the g_i respectively. If $D_2 - D_1 = (\phi)$ then D_1 and the ϕg_i also define h_2; thus without loss of generality we can assume $D_2 = D_1 = D$ say. Now fix j and write $\phi_i = f_i/g_j$; since $(\phi_i) = \{(f_i) + D\} - \{(g_j) + D\}$ the ϕ_i cannot all vanish simultaneously. This means that there is no ring homomorphism $K[\phi_0, \ldots, \phi_r] \to K$ which takes each ϕ_i to 0; for since V is complete such a homomorphism could be extended so as to map P, the argument of the ϕ_i, onto a point defined over $\bar{K}$, and we have just seen that there is no point on V for which all the ϕ_i vanish. But the only possible obstruction to such a homomorphism is the existence of a relation of the form

$$\sum c_\alpha \phi^\alpha = 1, \tag{35}$$

where the α are $(r + 1)$-vectors in a standard notation, the c_α are in K and there is no constant term on the left-hand side. Let $\wp$ be a finite prime, and consider the relation (35) at any point P in V_K. At least one of the terms on the left-hand side must have $\wp$-norm at least 1 and hence there is an inequality of the form

$$\{\text{Max } \|f_i\|_\wp\}/\|g_j\|_\wp = \text{Max } \|\phi_i\|_\wp \geq \text{Min } \|b_\alpha\|_\wp \tag{36}$$

where the b_α are a finite set of elements of K^* which depend on the c_α but not on $\wp$ or P. If $\wp$ is infinite, we get a relation like (36) but with an extra factor on the right-hand side which depends only on the number of nonzero c_α. There is an equation like (36) for each j; and they can be combined into the form

$$\text{Max } \|f_i\|_\wp \geq C_\wp \text{ Max } \|g_j\|_\wp$$

where the $C_\wp$ do not depend on P and all but finitely many of them are 1. Multiplying all these together gives $h_1 \geq Ch_2$ for some nonzero constant C; and in view of the symmetry between h_1 and h_2 this proves the lemma.

If D is an adequate divisor on V, we shall write h_D for any height associated with D; by the lemma h_D is defined up to equivalence. Now let D and E be any two adequate divisors on V, with associated systems of functions f_i and g_j. It is easy to see that $D + E$ is an adequate divisor, for which the $f_i g_j$ provide a satisfactory associated system of functions; and for this system we have Max $\|f_i g_j\|_\wp = \{\text{Max } \|f_i\|_\wp\}\{\text{Max } \|g_j\|_\wp\}$ for any $\wp$, whence

$$h_{D+E}(P) = h_D(P)h_E(P).$$

We have next to prove the analogue of Lemma 7 for algebraically equivalent divisors; this will be stated as Lemma 9, since its proof needs the following purely geometrical intermediate result:

Lemma 8. *Let X be a positive divisor of degree d_0 on a complete nonsingular variety V, and let Y be a hyperplane section of V. If $d > d_0$ then $dY - X$ is adequate.*

The proof in fact gives the stronger statement that $dY - X$ is ample. Let r, n be the dimensions of V and the ambient space respectively, let L be a general linear space of dimension $n - r - 1$ and H a general hyperplane. Let C be the cone with vertex L and base X, that is, the union of all straight lines which meet both L and X; by counting degrees of freedom, C is a hypersurface. The degree of C is the number of intersections of C with a general line Λ; and this is just the number of intersections of X with the linear space generated by L and Λ, which is d_0. Thus $C \sim d_0 H$ in the ambient space, and the intersection $C \cdot V \sim d_0 H \cdot V \sim d_0 Y$ on V. It follows that

$$dY - X \sim (C \cdot V - X) + (d - d_0)Y > 0$$

on V, and since we can choose L so that $C \cdot V - X$ avoids any assigned point on V, this proves the lemma.

Lemma 9. *Let X, Y be adequate positive divisors on V which are algebraically equivalent; then h_X is quasi-equivalent to h_Y.*

It follows from the theory of the Picard variety that there exists an algebraic equivalence system $\mathcal{E}$ of positive divisors on V such that any divisor on V algebraically equivalent to zero is linearly equivalent to $E_1 - E_2$ for some E_1, E_2 in $\mathcal{E}$. For the proof of Lemma 9 we assume this statement, and apply it to $n(X - Y)$ where n is a large positive integer. Thus $n(X - Y) \sim E_1 - E_2$ for some E_1, E_2 in $\mathcal{E}$ which depend on n. But all E in $\mathcal{E}$ have the same degree; thus by Lemma 8 and the fact that (since X is adequate) a suitable multiple of X is induced by a hyperplane section of some variety biregularly equivalent to V, there is an n_1 independent of n such that $n_1X - E_1$ is adequate for each E_1. Repeating the argument, there is an n_2 independent of n such that $Z = n_2X - (n_1X - E_1) - E_2$ is adequate for each E_1, E_2. Write $n_0 = n_2 - n_1$; then $(n + n_0)X \sim nY + Z$ whence, by Lemma 7, $h_X^{n+n_0}$ is equivalent to $h_Y^n h_Z$. Here the constant implied in the equivalence depends on n, so that we have

$$h_X^{n+n_0} \geq C_n h_Y^n h_Z \geq C_n h_Y^n.$$

Taking nth roots and letting $n \to \infty$, we obtain one of the two conditions for quasi-equivalence; and by symmetry this proves the lemma.

THEOREM 11. *Let A be an Abelian variety defined over an algebraic number field K, let $m > 1$ be an integer, b a point in A_K and h the height on A defined by* (33) *for a given projective embedding of A. Then $h(mP + b)$ is quasi-equivalent to $(h(P))^{m^2}$ as P varies in A_K.*

Let μ: $A \to A$ be the map defined by $\mu x = mx + b$. If X is a hyperplane section of A, it follows from the general theory of Abelian varictics that $\mu^{-1}X$ is algebraically equivalent to m^2X; and it is obvious that both these divisors are adequate. But the height induced by $\mu^{-1}X$ is just $h(mP + b)$, and this proves the theorem.

THEOREM 12. *Let A be an Abelian variety defined over an algebraic number field K; then A_K is finitely generated.*

Choose an integer $m > 1$. After replacing K by a larger field if necessary (which can only increase A_K) we can assume that K contains A_m and the mth roots of unity. Let $b_1, \ldots, b_r$ be a set of representatives for the cosets of mA_K in A_K, which are finite in number by Theorem 10. After Theorem 11 there is a constant $C > 0$ such that $h(mP + b_i) > C(h(P))^2$ for each b_i and all P in A_K. To any P_0 in A_K there correspond P_1 in A_K and b_i so that $P_0 = mP_1 + b_i$; and $h(P_1) < \frac{1}{2}h(P_0)$ provided $h(P_0) > 4C^{-1}$. Thus starting with any P_0 in A_K we reach a P in A_K with $h(P) \leq 4C^{-1}$ after finitely many steps; and to prove the theorem it is enough to prove that the set of such P is finite. Choose a representation $(x_0, x_1, \ldots, x_n)$ of P for which one of the x_i is 1, so that $\text{Max } \|x_i\|_\wp \geq 1$ for each prime $\wp$ of K. It follows that $\|x_i\|_\wp \leq 4C^{-1}$ for all i and $\wp$, and it is well known that the set of x_i in K with this property is finite. This completes the proof of the theorem.

The procedure of the last two sections gives a sensible and constructive upper bound for the number of generators of A_K, but it does not provide a sure means of determining A_K in any particular case. It is a major unsolved problem to give a more complete description of A_K. For some of the ideas which at the moment seem most relevant, see [22].

12. Siegel's theorem

Mordell has conjectured that if Γ is a curve of genus $g > 1$ defined over an algebraic number field K, then Γ_K is finite. More generally, Lang has conjectured that if A is an Abelian variety defined over K and V is a subvariety of A which contains no translation of a nontrivial Abelian subvariety of A, then V_K is finite. These conjectures are both plausible, in the sense that A_K is only just infinite; but there is at present no known way to attack them. The most that is known about Mordell's conjecture is a theorem of Chabauty, that if J is the Jacobian of Γ and if the rank of J_K is less than g, then Γ_K is finite; but even in this case there is no known algorithm for determining the points of Γ_K. However, if instead of considering rational points we consider integral points, the problem has been completely solved by Siegel:

THEOREM 13. *Let Γ be an affine curve defined over an algebraic number field K. If Γ_K contains infinitely many points with integral coordinates, then Γ has genus $g = 0$ and has at most two distinct points at infinity.*

The proof of the theorem falls into two parts, one for $g = 0$ and one for $g > 0$, which have little in common. Here we shall only consider the case $g > 0$, which is closely connected with the ideas of §11. It can be restated in the following apparently more general form:

THEOREM 14. *Let Γ be a curve of genus $g > 0$ defined over an algebraic number field K, and let ϕ be a nonconstant function on Γ defined over K. Then the set of points of Γ_K at which the value of ϕ is an algebraic integer is finite.*

This includes the part of Theorem 13 for which $g > 0$, by taking ϕ to be a nonconstant coordinate function; and it has the advantage of being invariant under birational transformation of Γ. We may assume that Γ_K is not empty; thus by birational transformation we may suppose Γ to be embedded in its Jacobian J. The basic tools of the proof are Theorems 11 and 12 and Roth's theorem, which we restate in the more convenient form:

LEMMA 10. *Let Γ be a complete nonsingular curve defined over an algebraic number field K, let ψ be a function on Γ defined over K, none of whose poles has order greater than r, and let $\| \ \|$ be an archimedean valuation on K. Then for any fixed $c > 2$ the set of points P in Γ_K for which $\|\psi(P)\| > (h(P))^{cr}$ is finite.*

For each pole P_i of ψ let ξ_i be a quotient of linear functions of the coordinates which is a local uniformizing variable at P_i and is defined everywhere on Γ; thus $h(\xi_i(P))$, which is by definition $\prod\{\text{Max}\,(\|\xi_i(P)\|_\wp, 1)\}$, is equivalent to $h(P)$ by Lemma 7. Let N_i be a closed neighborhood of P_i which contains no pole of ψ other than P_i, and in which ξ_i is bounded. Since $\psi(P)$ is bounded outside the union of the N_i, if there are infinitely many P for which $\|\psi(P)\| > (h(P))^{cr}$, there must be infinitely many within some N_i. But $\|\psi(P)\{\xi_i(P) - \xi_i(P_i)\}^r\|$ is bounded in N_i; and Roth's theorem in its orthodox form shows that for any given $C > 0$ the number of P defined over K and in N_i for which $\|\xi_i(P) - \xi_i(P_i)\| < C(h(\xi_i(P))^{-c}$ is finite. Putting together all these inequalities gives the lemma.

We now revert to the proof of Theorem 14. Let $N = [K:Q]$, let d_1 be the degree of ϕ as a function on Γ, and d_2 the degree of Γ as a curve embedded (by means of J) in projective space, and let r be the largest of the orders of the poles of ϕ. Choose an integer $m > 1$ so that

$$m^2 > 4rNd_2/d_1. \tag{37}$$

Let the b_i be a set of representatives for the cosets of J_K in mJ_K; this set is finite by Theorem 12. (We cannot use Theorem 10 here because once m is chosen we cannot increase K without contravening (37).) For each i let C_i be the locus of those points x on J such that $mx + b_i$ is in Γ; each C_i is an unramified covering of Γ and hence the function ψ_i on C_i defined by $\psi_i(x) = \phi(mx + b_i)$ has no pole of order greater than r. Moreover, each point on Γ defined over K lifts back to a point defined over K on some C_i. By Lemma 10 applied to ψ_i on C_i, there are only finitely many x on C_i defined over K for which $\|\phi(mx + b_i)\| = \|\psi_i(x)\| > (h(x))^{3r}$; and by Theorem 11 there are only finitely many x for which $h(x) > (h(mx + b_i))^{4/3m^2}$. Thus for all but finitely many P in Γ_K, $\|\phi(P)\| \leq (h(P))^{4r/m^2}$ for each archimedean norm on K. But there are only N of these norms at most, and if $\phi(P)$ is an integer its norm is at most 1 for each nonarchimedean valuation; thus

$$h(\phi(P)) \leq (h(P))^{4rN/m^2} \tag{38}$$

for all but finitely many of the P in Γ_K for which $\phi(P)$ is integral. But the divisor of poles of ϕ is an adequate divisor on Γ, with associated height $h(\phi(P))$; and so by Lemma 9

$$h(\phi(P)) \quad \text{is quasi-equivalent to } (h(P))^{d_1/d_2}.$$

In virtue of (37), this and (38) prove that $h(P)$ is bounded, so that P must belong to a finite set. This concludes the proof of Theorem 14.

13. Complex multiplication on elliptic curves and its relation to class field theory

Let k be a given algebraic number field. The aim of classical class field theory, which was largely achieved, was to give a complete description of the fields K abelian over k in terms of the structure of k alone. Let $\mathfrak{m}^*$ be an extended ideal in k, that is, the formal product of an ideal $\mathfrak{m}$ in k with possibly some of the real archimedean primes of k; then $S(\mathfrak{m}^*)$, the *primitive ideal group* mod $\mathfrak{m}^*$, consists of those fractional principal ideals of k which can be written in the form (ξ) with $\xi \equiv 1 \bmod \mathfrak{m}^*$, where this is to be understood as $\xi \equiv 1 \bmod \mathfrak{m}$ and $\xi > 0$ for each real archimedean prime in $\mathfrak{m}^*$. Let $I(\mathfrak{m}^*)$ denote the group of all fractional ideals of k prime to $\mathfrak{m}$; then $S(\mathfrak{m}^*)$ has finite index in $I(\mathfrak{m}^*)$ and any group H with $I(\mathfrak{m}^*) \supseteq H \supseteq S(\mathfrak{m}^*)$ is called an *ideal group* mod $\mathfrak{m}^*$. If $\mathfrak{m}_1^*$ is a multiple of $\mathfrak{m}^*$, then $H \cap I(\mathfrak{m}_1^*)$ is an ideal group mod $\mathfrak{m}_1^*$ which is almost the same as H so far as class field theory is concerned. In this way we define an equivalence relation between ideal groups to different moduli; the least among the moduli to which ideal groups in a given equivalence class are defined is called the *conductor* of that equivalence class, and of any ideal class in it. From now on we only consider ideal classes which are defined modulo their conductor. Now let K be abelian over k with Galois group G, let $\wp$ be a finite prime ideal in k which does not ramify in K and let $\mathfrak{P}$ be a prime ideal in K which divides $\wp$. If $N\wp$ denotes the absolute norm of $\wp$ (that is, the number of integral residue classes mod $\wp$) then there is a unique element σ of G, called the *Frobenius automorphism* of K, such that $\sigma\alpha \equiv \alpha^{N\wp} \bmod \mathfrak{P}$ for all α in K prime to $\mathfrak{P}$; and because G is abelian, σ depends only on $\wp$ and not on the choice of a factor $\mathfrak{P}$. A knowledge of σ determines the factorization of $\wp$ in K; in particular $\wp$ splits completely (that is, $\wp$ can be written as a product of $[K: k]$ distinct prime ideals) if and only if σ is the identity element of G.

The fundamental theorems of classical class field theory, so far as we shall need them, can be stated as follows:

THEOREM 15. *There is a one-one inclusion-reversing correspondence between equivalence classes of ideal groups in k and finite abelian extensions of k. Let H be an ideal group defined modulo its conductor $\mathfrak{f}^*$ and let K be the corresponding abelian extension of k with Galois group G; then there is a natural isomorphism between $I(\mathfrak{f}^*)/H$ and G which maps each prime ideal $\wp$ in k prime to $\mathfrak{f}^*$ into its Frobenius automorphism. The primes of k which ramify in K are precisely the primes which divide $\mathfrak{f}^*$, including the infinite ones.*

K is called the *class field* associated with H, and the prime ideals in H are precisely those prime ideals of k which split completely in K. K is unramified over k if and only if H contains all principal ideals; in particular the class field associated with the group of all principal ideals is the maximal unramified abelian extension of k, and is called the *absolute class field* of k.

To prove that a given field $K \supset k$ is the class field corresponding to a given ideal group H, it is not necessary to prove all the properties enumerated in Theorem 15. Note that if K is any algebraic extension of k, not necessarily even normal, then $\wp$ splits completely in K if and only if it is unramified in K and $\alpha^{N\wp} \equiv \alpha \bmod \wp$ for all α in K prime to $\wp$.

LEMMA 11. *Let H be an ideal group* mod $\mathfrak{m}^*$ *in k, and let K be a finite algebraic extension of k. A sufficient condition for K to be the class field associated with H is that either*

(i) *with at most finitely many exceptions, the first degree primes $\wp$ in k which split completely in K are just those in H; or*

(ii) $[K: k] \geq [I(\mathfrak{m}^*)/H]$ *and with at most finitely many exceptions every first degree prime $\wp$ in H splits completely in K.*

Here $\wp$ is a *first degree* prime if and only if $N\wp$ is a rational prime. Condition (i) is in fact Weber's original definition of a class field.

The main problem which the classical theory left unsolved was that of constructing the class field K corresponding to a given ideal group H, and hence of constructing all abelian extensions of k. Because the relation between H and K is inclusion-reversing, it is enough to do this for the special case where H is $S(\mathfrak{m}^*)$, a primitive ideal group. When $k = Q$, this problem was solved by Kronecker: if $\mathfrak{m}^* = (m)\wp_\infty$, where $m > 0$ and $\wp_\infty$ is the unique infinite prime, then the class field corresponding to $S(\mathfrak{m}^*)$ is just the field of mth roots of unity. Hence every abelian extension of Q is cyclotomic; but the "hence" here involves the use of class field theory, and Kronecker's proof of this result without class field theory is very lengthy. Kronecker conjectured, and he and Weber proved, that if k is a complex quadratic field all its abelian extensions could be generated by values of elliptic and elliptic modular functions. The precise form of Weber's result is as follows:

THEOREM 16. *Let k be a complex quadratic field, and let ω_1, ω_2 be a base for some nonzero ideal $\mathfrak{A}$ of k regarded as a Z-module. Assume that ω_1, ω_2 are so ordered that* $\mathrm{Im}(\omega_2/\omega_1) > 0$, *and consider the elements of $\mathfrak{A}$ as a lattice. If j is the fundamental elliptic modular function then the value of $j(\omega_2/\omega_1)$ depends only on the ideal class to which $\mathfrak{A}$ belongs, the h values thus obtained are a complete set of conjugates over k and each one of them generates over k the absolute class field of k. Moreover, let $\mathfrak{m}$ be any integral ideal in k, let z_0 be a primitive $\mathfrak{m}$-division point of the lattice defined by ω_1 and ω_2, and denote by $\wp(z)$ the Weierstrass function with respect to that lattice. Then the class field corresponding to* $S(\mathfrak{m})$ *is generated over k by $j(\omega_2/\omega_1)$ together with $g_2g_3\wp(z_0)/\Delta$ if $g_2g_3 \neq 0$, $g_3\wp^3(z_0)/\Delta$ if $g_2 = 0$ or $g_2^2\wp^2(z_0)/\Delta$ if $g_3 = 0$.*

Kronecker and Weber also proved that the $j(\omega_2/\omega_1)$ are algebraic integers, and are already a complete set of conjugates over Q; but these results are less relevant to what follows. A primitive $\mathfrak{m}$-division point of the lattice is a point z_0 such that, for α an integer in k, αz_0 lies in the lattice if and only if α is in $\mathfrak{m}$; such points exist for every $\mathfrak{m}$. The condition $g_2 = 0$ corresponds to $k = Q(\sqrt{-3})$, and $g_3 = 0$ to $k = Q(\sqrt{-1})$; and the functions of z_0 involved in the statement of the theorem are in each case homogeneous of degree zero in the lattice. Theorem 16 can be stated and proved in a purely geometric way, without any mention of analytic functions, as was first recognized by Deuring. In its original form it led naturally to the question: for a given type of field k, what analytic functions are needed to generate the class field over k? Until recently, the only further result known was that of Hecke, who generated the absolute class fields of certain biquadratic fields by means of values of Hilbert modular functions of two variables. However, in the last fifteen years Deuring's reformulation of Theorem 16, and the work of Weil on Abelian varieties, have suggested a different approach to the problem; and great advances have been made first by Shimura and Taniyama and then by Shimura alone. The remainder of this section consists of a proof of Theorem 16 in geometric language; the two following sections contain a brief account of subsequent developments.

Let Γ be an elliptic curve defined over a field of characteristic zero; what possibilities are there for its ring of endomorphisms? We can take Γ to be defined over C, so that an analytic model for it is C/Λ for some lattice Λ; let ω_1, ω_2 be a base of Λ. If τ is in End (Γ) it induces a linear transformation on C which takes Λ into itself; thus τ can be regarded as a complex number and

$$\tau\omega_1 = a_{11}\omega_1 + a_{12}\omega_2, \qquad \tau\omega_2 = a_{21}\omega_1 + a_{22}\omega_2$$

for some rational integers a_{ij}. It follows that

$$\tau^2 - (a_{11} + a_{22})\tau + a_{11}a_{22} - a_{21}a_{12} = 0,$$

so that τ is an integer in some quadratic field; and since $\omega_2/\omega_1 = a_{12}^{-1}(\tau - a_{11})$

if τ is not in Z then ω_2/ω_1 must lie in the quadratic field, which must therefore be complex rather than real. Since End $(\Gamma) \supset Z$, End (Γ) must be a unitary subring of a complex quadratic field. Conversely, let R be such a subring; then End $(\Gamma) = R$ if and only if R is the precise ring of endomorphisms of Λ, which is to say that Λ is a projective R-module of rank 1. Thus to any R there are finitely many equivalence classes of Λ, and these form a commutative group under the operation $(\Lambda_1, \Lambda_2) \rightarrow \Lambda_1 \otimes_R \Lambda_2$. In particular, when R is the ring of integers of a complex quadratic field k, then this group is just the ideal class group of k.

The situation is made somewhat untidy by the fact that there is no natural way to normalize Λ and Γ; indeed one believes that if the elements of Λ are in k then the least field of definition for Γ is transcendental over Q. There are two sorts of maps between two curves C/Λ with the same endomorphism ring R which will be relevant. The first is the scaling map $C/\Lambda \rightarrow C/\alpha\Lambda$ induced by $z \rightarrow \alpha z$ for any nonzero α in C; this sends (ξ, η) on $y^2 = 4x^3 - g_2x - g_3$ into $(\alpha^{-2}\xi, \alpha^{-3}\eta)$ on $y^2 = 4x^3 - \alpha^{-4}g_2x - \alpha^{-6}g_3$ and is one-one. The second and more interesting one is induced by

$$\text{Hom}\,(\Lambda_2, C/\Lambda_1) = C/\Lambda_2^{-1}\Lambda_1$$

where $\Lambda_2^{-1}\Lambda_1$ is the R-module consisting of all those α such that $\alpha\Lambda_2 \subset \Lambda_1$; this satisfies $\Lambda_2 \otimes \Lambda_2^{-1}\Lambda_1 = \Lambda_1$. If $\Lambda_2 \subset R$ the inclusion $\Lambda_2^{-1}\Lambda_1 \supset \Lambda_1$ induces a $[R/\Lambda_2]$-to-one map

$$C/\Lambda_1 \rightarrow C/\Lambda_2^{-1}\Lambda_1 \tag{39}$$

which corresponds to $z \rightarrow z$. In the special case $\Lambda_2 = \alpha R$ where α is in R, we get a combined map

$$C/\Lambda \rightarrow C/\alpha^{-1}\Lambda \xrightarrow{\alpha} C/\Lambda$$

which is just α considered as an element of End (C/Λ). Now the value of j is unaffected by the scaling map; hence it depends only on the ideal class group to which Λ belongs, and there are only finitely many values of j corresponding to a given R. Given any such j, we can choose a Γ which is defined over $k(j)$, and then the elements of End (Γ) are also defined over $k(j)$. Thus if j' is such that $k(j')$ is isomorphic to $k(j)$ as an extension of k, then this isomorphism can be extended to give a suitable Γ' and elements of End (Γ'), and so End (Γ') is isomorphic to R. We have already seen that there are only finitely many j with this property; thus each of them is algebraic over k, and with any j we get all its conjugates over k. Again, if j_1 and j_3 are values of j associated with R and Γ_1 is a curve with $j = j_1$, then (39) gives an isogeny $\Gamma_1 \rightarrow \Gamma_3$ where Γ_3 is some curve with $j = j_3$. Thus j_3 is in $k(j_1)$. It follows that $k(j)$ is a normal extension of k, whose degree is at most that of the ideal class group of R.

For simplicity, we now confine ourselves to the case when R is the full ring of integers in k. Theorem 16 can be restated in purely geometric language as follows:

THEOREM 17. *Let k be a complex quadratic field, of class number h, and let R be the ring of integers of k. There exist h complex numbers $j_1, \ldots, j_h$ such that an elliptic curve Γ defined over C has* End $(\Gamma) = R$ *if and only if $j = j(\Gamma)$ is some j_i; the j_i form a complete set of conjugates over k and each one of them generates* *over k the absolute class field of k. Moreover let Γ be any one of these curves, normalized so as to be defined over $k(j)$, let E be the group of units of R and let $\Delta = \Gamma/E$ be the associated Kummer variety (which is birationally equivalent to the projective line). If $\mathfrak{m}$ is any integral ideal in R and P_0 is a primitive $\mathfrak{m}$-division point on Γ, then the class field corresponding to $S(\mathfrak{m})$ is generated over $k(j)$ by the coordinates of the image of P_0 in Δ.*

In each ideal class choose an integral ideal $\mathfrak{A}_i$, with $\mathfrak{A}_1 = (1)$. Let Γ_1 be an elliptic curve with $j = j_1$, normalized so as to be defined over $K = k(j_1)$; let Γ_i be derived from Γ_1 by (39) with $\Lambda_2 = \mathfrak{A}_i$, and let $j_i = j(\Gamma_i)$. Let $\wp$ be a first degree prime in k which is unramified in K and prime to each $\mathfrak{A}_i$, and write $N\wp = p$; and let $\mathfrak{P}$ be a prime factor of $\wp$ in K. Assume that $\wp$, $\mathfrak{P}$ are such that each Γ_i has a good reduction mod $\mathfrak{P}$, that the j_i are all integers at $\mathfrak{P}$ and that no two of them are congruent mod $\mathfrak{P}$. These conditions only exclude finitely many first degree primes $\wp$. Write $j(\mathfrak{A})$ for the value of j corresponding to the ideal class of $\mathfrak{A}$, and let $\mathfrak{A}_j$ be the representative of the class of $\mathfrak{A}_i\wp^{-1}$; then $\wp$ induces an isogeny $\Gamma_i \to \Gamma_j$ by means of (39) with $\Lambda_2 = \wp$, followed by scaling by $\wp\mathfrak{A}_j\mathfrak{A}_i^{-1} = (\alpha)$ say, and this induces $z \to \alpha z$ in the model C/Λ. But this also involves $dz \to \alpha dz$, where dz is the normalized differential of the first kind on Γ, which can also be defined algebraically as $y^{-1}\,dx$. Now reduce everything mod $\mathfrak{P}$ and denote the reduction by a tilde; the isogeny $\Gamma_i \to \Gamma_j$ obtained above induces an isogeny $\tilde{\Gamma}_i \to \tilde{\Gamma}_j$ which is of degree $p = N\wp$ and which is inseparable because it annihilates the differential of the first kind. This can only be the pth power map on $\tilde{\Gamma}_i$, and hence

$$(40) \qquad j(\mathfrak{A}\wp^{-1}) \equiv \{j(\mathfrak{A})\}^p \bmod \mathfrak{P}.$$

Since this congruence holds modulo each prime factor of $\wp$ it holds mod $\wp$; and since every conjugate of $j(\mathfrak{A})$ over k is a $j(\mathfrak{A}_i)$, the Frobenius automorphism of K/k associated with $\wp$ must take $j(\mathfrak{A})$ to $j(\mathfrak{A}\wp^{-1})$. Now since there are infinitely many $\wp$ in each ideal class of k, all the $j(\mathfrak{A})$ must be conjugates over k; hence $[K\colon k] = h$, and K is the absolute class field of k by Lemma 11. This proves the first half of the theorem.

The second half of the proof works similarly. We need to discover the effect of the Frobenius automorphism on $\tilde{P}_0$ only in the case when $\wp$ is a principal ideal, because after (40) the Frobenius automorphism certainly does not act like the identity on the field we are concerned with in any other case. Now the map (39) is itself canonical in the analytic theory; but in the algebraic theory it is merely the canonical map of Γ to its quotient by the group of $\wp$-division points. Because $\wp$ is principal, this quotient can be identified with Γ; but it is not identified canonically and thus the induced map $\Gamma \to \Gamma$ is only determined up to an automorphism of Γ, that is, up to an element of E. What is uniquely determined is the associated map $\Delta \to \Delta$ of Kummer varieties. Now let $\wp = (\pi)$ and, in addition to the conditions already imposed on $\wp$, assume that $\wp$ is prime to $\mathfrak{m}$. From the analytic model, the isogeny $\Gamma \to \Gamma$ induced by $\wp$ is just $z \to \pi z$; note that π also is only determined up to an element of E. This map sends P_0 to itself if and only if $\pi \equiv 1 \bmod \mathfrak{m}$; thus it sends the image of P_0 in Δ into itself if and only if $\wp$ is in $S(\mathfrak{m})$. But we have already seen that $\wp$ induces an inseparable map of degree p

on $\tilde{\Gamma}$; thus it must be the pth power map on $\tilde{\Gamma}$, up to an element of E, and hence it induces the pth power map on $\tilde{\Delta}$. Hence the Frobenius automorphism acts like the identity on the image of $\tilde{P}_0$ in $\tilde{\Delta}$ if and only if $\wp$ is in $S(\mathfrak{m})$. In view of Lemma 11 this completes the proof of the theorem.

14. Abelian varieties determined by their rings of endomorphisms

Once it was realized that the theorems of Kronecker and Weber described in the previous section were best expressed geometrically, in terms of the elliptic curves Γ with a given ring of endomorphisms, it was natural to look for generalizations to Abelian varieties A of higher dimension. For any given dim $(A) = g$ the possible End $(A) \otimes_Z Q$ had already been determined by Albert [1], whose methods depended on the analytic representation of A as C^g/Λ described at the end of §2; a corresponding investigation for nonzero characteristic is still lacking. By analogy with §13, one is interested in those A which are essentially determined by a knowledge of g and End (A). For given g there is just one family of rings End (A) with this property; these are orders in totally complex algebraic number fields k such that $[k: Q] = 2g$ and k contains a totally real subfield k_0 with $[k_0: Q] = g$. The investigation of such A was begun simultaneously by Weil, Shimura and Taniyama (see their talks in [18]), and a comprehensive theory is given in [21]. (An alternative proof of one of the key formulae can be found in Giraud [9].) The proofs are much more intricate than in the classical case, but the ideas are basically the same except for the complications arising from polarization and the CM-types. These complications also affect the statements of the results, which are therefore less similar to those of the classical theory and also less satisfactory as a solution to the original problem of constructing class fields than one might have hoped.

In order to define the *field of moduli*, which is the generalization of $k(j)$, we have to introduce the idea of a *polarization*. Previously, all the varieties which have occurred have been defined or constructed explicitly, and it has been natural to think of them as embedded ab initio in projective space, and hence as having a least field of definition and so on. But A is only given up to biregular transformation, because of the way it has been specified; and there is no canonical way of embedding it in projective space. (There is such a canonical embedding if A is the Jacobian of a curve; this is one reason why this difficulty did not arise in the previous section.) To choose a projective embedding is the same as to choose an ample linear equivalence class of positive divisors on A. More generally, to choose an algebraic equivalence class of ample divisors is to choose a family of projective embeddings—one for each linear equivalence class within the algebraic equivalence class. We therefore define a *polarization* of A to be the choice of an algebraic equivalence class of divisors of A containing at least one divisor which induces a projective embedding; but for most purposes one should think of a polarization simply as a projective embedding. If A is represented as C^n/Λ, then the polarizations correspond to the Riemann forms $E(x, y)$ of §2. When necessary, we denote a polarized Abelian variety by (A, X) where X is any divisor in the algebraic equivalence class which defines the polarization.

We have already defined isogenies of Abelian varieties. Now let (A, X), (A', X') be two polarized Abelian varieties, and let $\lambda: A \to A'$ be an isogeny of the underlying varieties; then λ is said to be an isogeny $(A, X) \to (A', X')$ of the polarized

Abelian varieties if $\lambda^{-1}X'$ is algebraically equivalent to a multiple of X as divisors on A, and similarly for endomorphisms and isomorphisms. In what follows, we shall need to distinguish between isogenies of abstract and of polarized Abelian varieties.

For a general polarized Abelian variety one does not know in detail how to define analogues of the invariant $j = j(\Gamma)$; but (at least in characteristic 0) one can define an analogue of the field $Q(j)$. This is the *field of moduli*, which is the field K_0 of the following theorem; moduli are the modern equivalents of invariants.

THEOREM 18. *Let (A, X) be a polarized Abelian variety defined over a field K; then there exists a field $K_0 \subset K$ with the following property; Let σ be any isomorphism of K; then (A^σ, X^σ) is isomorphic to (A, X) as a polarized Abelian variety if and only if σ is the identity on K_0.*

In characteristic 0 this property defines K_0 uniquely, and K_0 is the least field of definition for A with the given polarization; in characteristic p, K_0 is only defined up to a possible totally inseparable extension. This does not however mean that K_0 is a possible field of definition for A with the given projective embedding, but only for some projective embedding of A in the same equivalence class. For example, the least field of definition of the elliptic curve

$$\Gamma: y^2 = 4x^3 - g_2x - g_3$$

is $Q(g_2, g_3)$; but its field of moduli is $Q(j)$ and over a large enough field it is biregularly equivalent to the curve

$$y^2 = 4x^3 - 3\lambda x - \lambda \quad \text{where} \quad \lambda = j/(j - 1728).$$

This last curve is defined over $Q(j)$.

In the rest of this section we sketch the methods and results of Shimura and Taniyama [21], regarding them as a generalization of the classical case described in §13 and emphasizing the additional complications that occur.

In the classical case, the procedure for constructing the elliptic curves Γ with a given endomorphism ring R (which is an order in a complex quadratic field k) is as follows: choose one of the finitely many isomorphism classes of one-dimensional projective R-modules, realize it as a lattice Λ in C, and identify C/Λ as an elliptic curve. In the present case, R is an order in a field k of the kind described above. When R is the full ring of integers of the field k, the associated classes of projective R-modules correspond to the ideal classes in the field. There are now several essentially different ways of embedding any one of these modules as a lattice Λ in C^g, and one has to show that Λ satisfies certain conditions in order to be able to identify C^g/Λ as an Abelian variety A. To explain all this, it is convenient to start with the natural matrix representations of End (A).

Since we are only concerned with the case when the characteristic is 0, we can take A to be defined over C and can identify it with C^g/Λ. Any element of End (A) induces an endomorphism of C^g, so that there is a natural complex g-dimensional representation of End (A); similarly it induces an endomorphism of Λ, so that there is a rational $2g$-dimensional representation of End (A), actually even integral but the induced representation of End $(A) \otimes Q$ is only rational. The rational representation is equivalent to the sum of the complex representation and its complex conjugate. By means of the rational representation we obtain the *characteristic polynomial* of an endomorphism, and in particular its *trace*; this is the trace that has already appeared in §7 for Jacobians, without the restriction to characteristic zero. Both the complex and the rational representation are faithful. To any polarization of A there corresponds (in any characteristic) an anti-automorphism $\xi \longrightarrow \xi'$ of End (A) which has the property that Tr ($\xi\xi'$) > 0 for all $\xi \neq 0$ in End (A); this too has already appeared in §7 in the case when A is a Jacobian and therefore has a canonical polarization. In characteristic 0, let ξ_0, ξ_0' denote the endomorphisms of C^g induced by ξ, ξ' respectively; then ξ_0' is just the adjoint of ξ_0 with respect to the Riemann form E described in §2.

All this restricts the possibilities for End (A). Suppose henceforth that A is simple, so that End $(A) \otimes_Z Q$ is a skew field with centre k say; then $\xi \longrightarrow \xi'$ induces an automorphism ρ on k such that $\rho^2 = 1$ and Tr ($\xi\xi^\rho$) > 0 for $\xi \neq 0$. Let k_0 be the fixed field of ρ; then it follows that k_0 is totally real, ρ is complex conjugacy, and either $k = k_0$ or $[k: k_0] = 2$ and k is totally complex. Moreover $[k: Q]$ divides $2g$, because the rational representation is by nonsingular matrices; and $[k_0: Q]$ divides g because the rational representation is equivalent to the sum of two complex conjugate representations. Thus k is at most a field of degree $2g$ over Q. For the rest of this section we confine ourselves to this case; thus k is a totally complex quadratic extension of a totally real field k_0 with $[k_0: Q] = g$. Moreover, since the restriction of the rational representation to k must be equivalent to the sum of the one-dimensional representations of k and since there are no nondiagonal matrices which commute with all diagonal matrices, k is the whole of End $(A) \otimes Q$. So far k has been an abstract field. But because the rational representation is equivalent to the sum of the complex representation and its complex conjugate, the complex representation of k is equivalent to the sum of g one-dimensional representations, one from each pair of complex conjugate one-dimensional representations. Thus to realize a suitable End (A)-module as a lattice Λ in C^g, it is necessary to choose g embeddings ϕ_i: $k \longrightarrow C$, one from each complex conjugate pair. Shimura and Taniyama call any such specification $(k, \{\phi_i\})$ a *CM-type*; but their definition is more general than the one above.

THEOREM 19. *Let $(k, \{\phi_i\})$ be any CM-type. There exist Abelian varieties of dimension g having this CM-type and any two such varieties are isogenous. Moreover, if $R \supset Z$ is any finitely generated subring of k such that $R \otimes Q = k$, then there are such Abelian varieties for which* End $(A) = R$.

To construct a C^g/Λ whose ring of endomorphisms is precisely R, we choose any one-dimensional projective R-module $M \subset k$, for example R itself; then we can take Λ to be the set of $(\phi_1\alpha, \phi_2\alpha, \ldots, \phi_g\alpha)$ as α runs through the elements of M. Conversely, suppose that C^g/Λ has CM-type $(k, \{\phi_i\})$. By suitable choice of coordinates in C^g we can assume that the complex representation of R is diagonal, and we know that it is the sum of the one-dimensional representations ϕ_i; thus the corresponding lattice Λ is the set of $(\phi_i\alpha, \ldots, \phi_g\alpha)$ where α runs through the elements of some Z-submodule of k of rank $2g$. If M_1, M_2 are any two such submodules there is a rational integer $n \neq 0$ such that $M_1 \supset nM_2$; this proves the statement about isogeny. For the rest, we need to know when C^g/Λ can be identified with an Abelian variety. A necessary and sufficient condition for this is the existence of a *Riemann form* as described in §2; that is, a function E: $C^g \times C^g \to$ reals such that

(i) E is a real bilinear form and $E(x, y) = -E(y, x)$,

(ii) $E(x, iy)$ is a symmetric positive definite bilinear form,

(iii) $E(x, y)$ is in Z whenever x and y are both in Λ.

For the Λ constructed above, a suitable E can be written down directly. For let η be any element of k such that $-\eta^2$ is a totally positive element of k_0 and $\mathrm{Im}(\phi_j\eta) > 0$ for each j; it is easy to show that such η exist. Then for a suitably chosen positive integer n,

$$E(x, y) = n\sum\{(\phi_j\eta)(x_j\bar{y}_j - \bar{x}_jy_j)\} \tag{41}$$

is a Riemann form. This completes the proof of the theorem. Conversely, it may be shown that every Riemann form associated with Λ is given by (41) for suitable η and n, though this fact will not be needed below.

The *CM*-type is preserved under isogeny; thus two Abelian varieties with the same k and essentially dissimilar sets of ϕ_i are not isogenous. As in the case $g = 1$, given an R of the type described above there are only finitely many biregularly inequivalent Abelian varieties A of dimension g with End $(A) = R$, and all such A are simple because End (A) is an integral domain. However, each of them now has infinitely many essentially different polarizations; but in what follows it does not matter which of these polarizations is chosen.

In the classical case the next step is to take any "good" first degree prime $\wp$ in k. Write $N\wp = p$ and let K be a field of definition for Γ and $\mathfrak{P}$ a prime factor of $\wp$ in K. Denoting reduction mod $\mathfrak{P}$ by a tilde, the pth power homomorphism on $\tilde{\Gamma}$ can be identified with the reduction mod $\mathfrak{P}$ of the isogeny denoted by $\wp$, because this reduced map has degree p and annihilates the unique differential of the first kind on Γ. This was proved by looking at the complex representation of End (Γ), which is simply one of its embeddings into C. In the present case the situation is more complicated, largely because even when k is normal over Q it may not be the field most naturally associated with the complex representation. Let C, C^* be two copies of the complex numbers, and regard k as a subfield of C and the maps ϕ_i as maps $k \to C^*$. Let $L \subset C$ be a finite normal algebraic extension of Q containing k, and let L^* be the corresponding subfield of C^*. The complex representation of End (A) is a representation in C^*, because the images of the ϕ_i are in C^*; let $k^* \subset L^*$ be the field generated by the traces of the complex representations of the elements of End (A). Each ϕ_i can be extended, in finitely many ways, to an isomorphism ϕ: $L \to L^*$. Let the ψ_j be the restrictions to k^* of all the possible ϕ^{-1}; thus the ψ_j are maps $k^* \to C$ and are to be regarded as a list without repetitions. Then $(k^*, \{\psi_j\})$ is also a *CM*-type, though in general with a different value of g; $(k^*, \{\psi_j\})$ is called the *dual* of $(k, \{\phi_i\})$, and it turns out that this relation is symmetric. The abstract field naturally associated with the complex representation of End (A) is k^*.

Now let L^*, as well as having the properties above, be a field of definition for A and for all the endomorphisms of A; this can be achieved by choosing L large enough. Let $\wp$ be a "good" first degree prime in k^*, with $N\wp = p$, and let $\mathfrak{P}$ be any prime factor of $\wp$ in L^*. Denote reductions mod $\mathfrak{P}$ by a tilde. It can be shown that $\mathfrak{A} = \prod\psi_j(\wp)$ is an ideal in k, and therefore induces a homomorphism of A; and moreover $\mathrm{Norm}_{k/Q}\mathfrak{A} = (p^g)$. Since $\mathfrak{P}$ divides $\phi_i(\mathfrak{A})$ for each i, the reduction mod $\mathfrak{P}$ of the homomorphism of A induced by $\mathfrak{A}$ annihilates every differential of the first kind on $\tilde{A}$. It follows that $\tilde{\mathfrak{A}}$ is just the Frobenius map on $\tilde{A}$. Taking for simplicity the case when R is the full ring of integers in k, it follows that $\tilde{A}^p$ is birationally equivalent to $\tilde{A}$ if and only if $\mathfrak{A}$ is a *principal* ideal in k.

However, we have so far been considering A as an abstract variety, ignoring the existence of a polarization on it. The choice of a polarization corresponds to the choice of a Riemann form on C^g/Λ, up to multiplication by a constant; thus if the Frobenius map on $\tilde{A}$ is to preserve $\tilde{A}$ as a polarized Abelian variety, multiplication by the principal ideal $\mathfrak{A}$ must induce multiplication of the Riemann form by a rational integer. Examination of the Riemann form produced in the proof of Theorem 19 shows that this happens if and only if there is a number π in k such that $\mathfrak{A} = (\pi)$ and $\pi\bar{\pi} = p$. Now let H_0^* be the group of those fractional ideals $\mathfrak{b}$ in k^* for which there exists μ in k such that $\prod\psi_j(\mathfrak{b}) = (\mu)$ and $N\mathfrak{b} = \mu\bar{\mu}$. If $\mathfrak{b} = (\beta)$ is a principal ideal in k^*, then $\gamma = \prod\psi_j(\beta)$ is in k, and $\gamma\bar{\gamma} = N_{k^*/Q}\beta$; since $\gamma\bar{\gamma} > 0$ it follows that $N\mathfrak{b} = \gamma\bar{\gamma}$. Thus H_0^* is an ideal group with trivial conductor, and if σ is the Frobenius automorphism of L^*/k^* then A^σ is equivalent to A as a polarized Abelian variety if and only if $\wp$ is in H_0^*. Now let K_0^* be the field of moduli of A, and let $K^* = K_0^*k^*$; then σ preserves K^* elementwise if and only if A^σ is equivalent to A as a polarized Abelian variety, by definition of the field of moduli. Thus by Lemma 11, K^* is an unramified abelian extension of k^*, and is the class field corresponding to H_0^*; but in general H_0^* will contain non-principal ideals and so K^* will not be the absolute class field of k^*.

As in the classical case, we can also consider the effect of adjoining the coordinates of division points on A. The group of units of R is infinite when $g > 1$, but only finitely many of them preserve the polarization. Thus we can form the Kummer variety of A, which is the quotient of A by the group of its automorphisms as a polarized Abelian variety. If P is a point on the Kummer variety which is the image of a division-point on A, then we can define the "true" field of definition for P in the same way as we described the field of moduli as the "true" field of definition for A. It can then be shown that the compositum of K^* with the "true" field of definition for P is a class field over k^*; and if P is the image of a proper $\mathfrak{b}$-division point, for some integral ideal $\mathfrak{b}$ of k, then the corresponding ideal group consists of all ideals A in k^* for which there exists $\mu \equiv 1 \bmod \mathfrak{b}$ such that $\prod\psi_i(\mathfrak{A}) = (\mu)$ and $\mu\bar{\mu} = N\mathfrak{A}$.

15. Families of abelian varieties

The results of the previous section give a complete answer to the problem of generalizing Theorem 17 to arbitrary Abelian varieties with enough complex multiplications. Unfortunately they do not provide a description of the absolute class field of a field k^*, even when k^* is a field of the special type which appears in that theory. (Moreover, the theory does not provide one with a means of calculating explicitly those abelian extensions which it does describe. The value of $j(\omega)$, where ω is a complex quadratic irrational, can be obtained by means of the modular equations for j. It is believed that the field of moduli of an Abelian variety is generated over Q by the values of certain Siegel modular functions of n variables, see [8], exposés 18–20; but even this does not provide a means of calculating them.)

The case described in the previous section is the only one in which a knowledge of End (A) determines A, up to isomorphism, to within a finite set. From this point of view, the next line of attack is to choose a ring R and a dimension $g > 0$ such that R is capable of being the ring of endomorphisms of an Abelian variety A of dimension g. One then asks what are the families of polarized Abelian varieties A of dimension g such that End $(A) \supset R$, and more particularly what are the varieties V which parametrize the fields of moduli of the members of such a family. This problem has been attacked by Shimura in an impressive sequence of papers in Annals of Mathematics during the last ten years; he has also given a simplified account of part of the theory in [20], which contains references to all the papers. As well as the problem of determining the natural field of definition of V, and of particularly interesting points on V, Shimura has investigated the zeta-function of V; but that aspect of his work will not be described here. Shimura's proofs depend largely on the theory of automorphic functions, and the geometry plays only a minor role.

Write $B = \text{End}(A) \otimes_Z Q$ and $B_R = \text{End}(A) \otimes_Z \mathfrak{R}$ where $\mathfrak{R}$ denotes the reals; and suppose for simplicity that B is a division algebra, which is equivalent to supposing that A is simple as an Abelian variety. Let B_0 consist of those elements of B which are fixed under the canonical antiautomorphism, and as in §14 write k for the centre of B and k_0 for $k \cap B_0$; then k_0 is a totally real algebraic number field, and either $k = k_0$ or k is a totally complex quadratic extension of k_0. Because of the properties of the canonical antiautomorphism, B must be of one of the following four types:

(I) $B = k = k_0$ is totally real;

(II) $k = k_0$ is totally real, and $B = B_0$ is a totally indefinite quaternion algebra over k, that is, one which splits at every infinite prime of k;

(III) $B_0 = k = k_0$ is totally real, and B is a totally definite quaternion algebra over k, that is, one which ramifies at every infinite prime of k;

(IV) k is a totally complex quadratic extension of a totally real field k_0, and B is a central simple algebra over k.

There are further conditions on type IV, which need not be described here, to ensure the existence of the canonical antiautomorphism. Conversely, let B^* be a ring of one of these types equipped with a suitable antiautomorphism such that Tr $(\xi\xi') > 0$ for all $\xi \neq 0$, and let g be such that B^* has the necessary complex representation of dimension g compatible with the antiautomorphism; then except in certain trivial cases there is an Abelian variety $C^g/\Lambda = A$ of dimension g such that $B^* = \text{End}(A) \otimes Q$. In the exceptional cases $B^* \subset \text{End}(A) \otimes Q$ but A is not simple. Moreover, if B is given then $R = \text{End}(A)$ can be any order in B which contains Z.

Now write $m = 2g/[B\colon Q]$; then B acts on C^g and so there is a natural isomorphism of real linear spaces $\phi\colon B_R^m \to C^g$. Let M be the R-lattice in B_R^m which is the inverse image of Λ under ϕ. By standard theorems of algebra there is a unique form $T = T(x, y)$ from $B^m \times B^m$ to B which is B-skew-Hermitian with respect to the antiautomorphism—that is,

$$T(x, y) = -T(y', x'), \qquad T(\beta x, y) = \beta T(x, y), \qquad T(x, \beta y) = \beta' T(x, y)$$

for any β in B, and is such that $E(\phi x, \phi y) = \text{Tr}\,(T(x, y))$ where E is the Riemann form associated with the given polarization of A and Tr is the reduced trace from B to Q. Moreover Tr $(T(x, y))$ is in Z for x, y in M. The choice of such a T determines a family of polarized Abelian varieties. Moreover let

$$\mathfrak{H} = \{\alpha \mid \alpha \text{ in } GL(m, B), T(x\alpha, y\alpha) \equiv T(x, y)\}$$

and

$$\Gamma = \{\alpha \mid \alpha \text{ in } \mathfrak{H}, M\alpha = M\};$$

then $\mathfrak{H}$ has an induced complex structure and the Abelian varieties associated with T are in natural one-one correspondence with the points of the quotient space $\mathfrak{H}/\Gamma$. Shimura's first major result is the following:

THEOREM 20. *There exist a nonsingular projective variety V, an analytic monomorphism ψ: $\mathfrak{H}/\Gamma \to V$, an algebraic number field K and automorphic functions $f_1, \ldots, f_r$ such that*

(i) *$C(f_1, \ldots, f_r)$ is the full field of automorphic functions on $\mathfrak{H}/\Gamma$;*

(ii) *V is defined over K (though K need not be the least field of definition for V);*

(iii) *the image of ψ is an open subset of V;*

(iv) *for any z in $\mathfrak{H}/\Gamma$, the field of moduli of the Abelian variety associated with z is precisely $K(f_1(z), \ldots, f_r(z))$.*

The field K is uniquely determined, and is in an obvious sense the "natural" field of definition for V. If B is of type I or II then K is Q. In principle a similar type of argument should determine K when B is of type III, but there are two differences from the previous case. First, a strong approximation theorem is needed in the course of the argument, and the one applicable to type III is as yet unproved; second, the fact that $K = Q$ is a consequence of a "one class in each genus" theorem similar to Meyer's theorem on quadratic forms, and such a theorem is probably false in this case because the quadratic form over k_0 associated with B is definite. For these reasons very little is known about type III.

For type IV, K has been determined only when $B = k$. Just as in §14, k and its complex representation determine a dual field k^* of similar type; and K is now a well-determined totally unramified abelian extension of k^*—but once again it need not be the absolute class field of k^*.

The proof of Theorem 20 is predominantly concerned with automorphic functions, and Abelian varieties scarcely appear except in connection with (iv). In particular the field K arises naturally in connection with automorphic functions. Moreover types II and III are the two extreme cases of a quaternion algebra B over a totally real field k; so it is natural to put aside the Abelian varieties and see whether the rest of the theory will generalize to an arbitrary quaternion algebra over k. Set $[k: Q] = d$ and take the simplest case $g = 2d$, so that $m = 1$. If B splits at precisely t of the infinite primes of k, and ramifies at the other $d - t$ infinite primes, then B_R is the direct sum of t copies of $GL(2, \mathfrak{R})$ and $d - t$ copies of the classical quaternions. If we confine ourselves to B^+, the set of elements of B whose images in all of the t copies of $GL(2, \mathfrak{R})$ have positive determinants, then we can take $\mathfrak{H} = H^t$ where H denotes the upper half plane $\operatorname{Im} z > 0$; and $GL^+(2, \mathfrak{R})$ acts on H in the traditional manner

$$\begin{pmatrix} a & b \\ c & d \end{pmatrix} z = \frac{az + b}{cz + d}.$$

If also R is a maximal order in B then we can take Γ to be

$$\Gamma_1(R) = \{\alpha \mid \alpha \text{ in } R, \alpha\alpha' = 1\}$$

where the dash is the canonical antiautomorphism of B. Once again the field of automorphic functions on $H^t/\Gamma_1(R)$ has a geometric model V and a natural field of definition K associated with it. The cases $t = 0$ and $t = d$ have already been described. If $t = 1$ then V is a curve and Kk is an abelian extension of k which is unramified except perhaps at some infinite primes and which can be described explicitly as a class field. For $1 < t < d$, there are phenomena similar to those of §14; there is a field k^* associated with k, and Kk^* is a class field over k^*.

However, the disappearance of the Abelian varieties is an illusion, brought about by looking at the situation too superficially. There are infinitely many families Σ of Abelian varieties which are parametrized by a given $H^t/\Gamma_1(R)$, and to each such Σ there is a field K_Σ and a map ψ_Σ: $H^t/\Gamma_1(R) \to V_\Sigma$ where V_Σ is a variety defined over K_Σ. Moreover, if z is a point of $H^t/\Gamma_1(R)$ then the field of moduli of the associated Abelian variety is just $K_\Sigma(\psi_\Sigma(z))$. To obtain such a family Σ for $0 < t < d$ it is enough to choose any totally imaginary quadratic extension k_1 of k_0 and consider the suitably polarized Abelian varieties A with dim $(A) = 2td$ and End $(A) \otimes Q \supset B \otimes_k k_1$. It is not true that, in the notation above, V parametrizes a family of Abelian varieties. On the other hand, it can be shown that K is the intersection of all the K_Σ and every map ψ_Σ can be factored through ψ.

Now let L be any quadratic extension of k contained in B, and assume that all the integers of L are in R. If L^+ is regarded as operating on H^t then all the elements of L^+ not in k have a common fixed point, z_L say. Let $\hat{K}$ be the maximal abelian extension of k unramified except at infinity; then Shimura's final theorem is that in the case $t = 1$, the field $\hat{K}L(\psi(z_L))$ is the Hilbert class field of L.

REFERENCES

1. A. A. Albert, *On the construction of Riemann matrices.* I, II, Ann. of Math. (2) **35** (1934), 1–28; ibid., **36** (1935), 376–394.

2. M. Baldassarri, *Algebraic varieties*, Ergebnisse der Mathematik und ihrer Grenzgebiete, Neue Folge, Heft 12, Springer-Verlag, Berlin, 1956. MR **18,** 508.

3. E. Bombieri and H. P. F. Swinnerton-Dyer, *On the local zeta function of a cubic threefold*, Ann. Scuola Norm. Sup. Pisa (3) **21** (1967), 1–29. MR **35** #2894.

4. J. W. S. Cassels, *Diophantine equations with special reference to elliptic curves*, J. London Math. Soc. **41** (1966), 193–291; Corrigenda, ibid., **42** (1967), 183. MR **33** #7299; MR **34** #2523.

5. J. W. S. Cassels and M. J. T. Guy, *On the Hasse principle for cubic surfaces*, Mathematika **13** (1966), 111–120. MR **35** #2841.

6. F. Conforto, *Abelsche Funktionen und Algebraische Geometrie*, Springer-Verlag, Berlin, 1956. MR **18,** 68.

7. B. M. Dwork, *On the rationality of the zeta function of an algebraic variety*, Amer. J. Math. **82** (1960), 631–648. MR **25** #3914.

8. *Séminaire H. Cartan* 1957/58. *Fonctions automorphes*, Secrétariat Mathématique, Paris, 1958. MR **21** #2750.

9. J. Giraud, *Remarque sur une formule de Shimura-Taniyama*, Invent. Math. **5** (1968), 231–236. MR **37** #2757.

10. A. Grothendieck, *Sur une note de Mattuck-Tate*, J. Reine Angew. Math. **200** (1958), 208–215. MR **25** #75.

11. ———, *Formule de Lefschetz et rationalité des fonctions L*, Séminaire Bourbaki 279 (1964/65); reprinted as pp. 31–45 of [12].

12. A. Grothendieck and N. H. Kuiper (editors), *Dix exposés sur la cohomologie des schémas*, Amsterdam, 1968.

13. S. L. Kleiman, *Algebraic cycles and the Weil conjectures*, printed as pp. 359–386 of [12].

14. S. Lang, *Abelian varieties*, Interscience Tracts in Pure and Appl. Math., no. 7, Interscience, New York, 1959. MR **21** #4959.

15. Ju. I. Manin, *Correspondences, motives and monoidal transformations*, Mat. Sb. **77 (119)** (1968), 475–507 = Math. USSR Sb. **6** (1968), 439–470.

16. A. Mattuck and J. Tate, *On the inequality of Castelnuovo-Severi*, Abh. Math. Sem. Univ. Hamburg **22** (1958), 295–299. MR **20** #5202.

17. A. Néron, *Modèles minimaux des variétés abéliennes sur les corps locaux et globaux*, Inst. Hautes Études Sci. Publ. Math. No. 21 (1964). MR **31** #3423.

18. Proceedings of the International Symposium on Algebraic Number Theory (Tokyo-Nikko, 1955) Tokyo, 1956.

19. P. Samuel, *Méthodes d'algèbre abstraite en géométrie algébrique*, Ergebnisse der Mathematik und ihrer Grenzgebiete, Neue Folge, Heft 4, Springer-Verlag, Berlin, 1955. MR **17,** 300.

20. G. Shimura, *Automorphic functions and number theory*, Lecture Notes in Math., no. 54, Springer-Verlag, Berlin, 1968. MR **38** #3229.

21. G. Shimura and Y. Taniyama, *Complex multiplication of abelian varieties and its applications to number theory*, Publ. Math. Soc. Japan, 6, The Math. Soc. of Japan, Tokyo, 1961. MR **23** #A2419.

22. H. P. F. Swinnerton-Dyer, *The conjectures of Birch and Swinnerton-Dyer, and of Tate*, Proc. Conf. Local Fields (Driebergen, 1966) Springer, Berlin, 1967, pp. 132–157. MR **37** #6287.

23. A. Weil, *Introduction à l'étude des variétés kählériennes*, Actualités Sci. Indust., no. 1267, Hermann, Paris, 1958. MR **22** #1921.

24. ———, *Numbers of solutions of equations in finite fields*, Bull. Amer. Math. Soc. **55** (1949), 497–508. MR **10,** 592.

25. ———, *Jacobi sums as "Grössencharaktere"*, Trans. Amer. Math. Soc. **73** (1952), 487–495. MR **14,** 452.

26. ———, *Footnote to a recent paper*, Amer. J. Math. **76** (1954), 347–350. MR **15,** 778.

TRINITY COLLEGE,
CAMBRIDGE, ENGLAND.

ABELIAN VARIETIES OVER FINITE FIELDS[1]

W. C. WATERHOUSE and J. S. MILNE

I. Classification up to isogeny

1. We begin by fixing a field k, which will eventually be finite but need only be perfect until further notice. An *Abelian variety* is a subset of some projective n-space which

(i) is defined by polynomial equations on the coordinates (with coefficients in k),

(ii) is connected, and

(iii) has a group law which is algebraic (in the sense that the coordinates of the product of two points are rational functions of the coordinates of the factors).

The first theorem one proves is that Abelian varieties are commutative; this shows why we insist on connectedness, since otherwise we would be allowing all finite groups and the geometry would be of no help.

In the classical case $k = \mathbf{C}$ the structure of Abelian varieties is fairly well understood: they are all of the form $\mathbf{C}^g/\Lambda$ where Λ is a certain kind of lattice in $\mathbf{C}^g$. These lattices (sometimes disguised as homology groups) are basic to the classical treatment of the subject. They unfortunately disappear in characteristic $p > 0$, and part of our job will be developing substitutes for them. Nevertheless, $\mathbf{C}^g/\Lambda$ is a useful model to keep in mind when considering properties of Abelian varieties.

Let A be an Abelian variety of dimension g, for example, and n a nonzero integer. Multiplication by n is a homomorphism of the group A into itself, and in the classical case it obviously is surjective with finite kernel of cardinality n^{2g}. In general what one proves is that it is surjective (in a reasonable sense, though it need not be surjective on points with coordinates in k), and that its degree is n^{2g}. Here the degree of a map is a number, definable algebraically, which in good cases counts

[1]Preparation of the part by Waterhouse was supported by the contract NSF GP-9395.

the number of points with the same image; thus the degree of a homomorphism is the size of its kernel.

A homomorphism is called an *isogeny* if it is surjective with finite kernel; multiplication by n is a leading example. We say A and B are *isogenous* if there is an isogeny φ from A to B; this is reasonable terminology because there is then also an isogeny from B to A. Indeed, let $G = \ker \varphi$. If n is the degree of φ, then $G \subseteq A_n = \ker (A \xrightarrow{n} A)$. Hence $A \xrightarrow{n} A$ factors through $A/G = B$. The resulting map $\psi\colon B \to A$ is surjective (since $A \xrightarrow{n} A$ is) and has finite kernel (since φ is surjective and $\psi \circ \varphi$ has finite kernel).

Honesty compels me to confess that this proof is not so simple as it looks. It is not obvious that B has the reasonable properties of a quotient A/G; as we will see later, it is not even true without a rather fancy definition of $\ker \varphi$. But now that due warning has been given, I intend to ignore most such problems, and I urge you to do likewise. We therefore know that isogeny is an equivalence relation, and our main concern will be with the structure of Abelian varieties up to isogeny. This means formally that we are making multiplication by n invertible; in practice it means that instead of looking at objects like End A, we look mainly at objects like End $A \otimes_{\mathbf{Z}} \mathbf{Q}$.

THEOREM 1. *Let A be an Abelian variety. Then* End A *is finitely generated and torsion-free, and* End $A \otimes_{\mathbf{Z}} \mathbf{Q}$ *is a semisimple* $\mathbf{Q}$*-algebra.*

Freedom from torsion is simple: if $n\varphi = 0$, then $\varphi(A)$ is connected and lies in the finite set A_n, and so $\varphi(A) = 0$. The rest is not too hard, but we won't linger over it, because it is just the first indication of a better result ahead. Semisimple algebras have an attractive structure: the center is a product of fields, around each field is a division algebra, and around each division algebra is a matrix algebra. The hope is to find this reflected in the structure of A. Note that if $\varphi\colon A \to B$ is an isogeny, and $\psi\colon B \to A$ is the map constructed earlier, then $\alpha \mapsto \varphi \circ \alpha \circ \psi$ is an isomorphism of End $A \otimes \mathbf{Q}$ onto End $B \otimes \mathbf{Q}$; thus the structure of the algebra can at best give structure on A up to isogeny.

DEFINITION. An Abelian variety is *elementary* (or *simple*) if it has no nontrivial Abelian subvarieties. (It of course always has finite subgroups.)

If A and B are elementary, then clearly any nonzero homomorphism from A to B is surjective and has a finite kernel, i.e. is an isogeny. In particular, End $A \otimes \mathbf{Q}$ is a division algebra (Schur's lemma, as usual). We automatically have End $(A^m) = M_m$ (End A), the $m \times m$ matrices. Thus if $A_1, \ldots, A_n$ are nonisogenous elementary Abelian varieties, End $(\prod A_i^{m_i}) \otimes \mathbf{Q}$ is the semisimple algebra $\prod M_{m_i}$ (End $A_i \otimes \mathbf{Q}$). Our hope is then fulfilled by

THEOREM 2 (POINCARÉ-WEIL). *Every Abelian variety is isogenous to a product of powers of nonisogenous elementary Abelian varieties.*

2. We next turn to finding a replacement for the lattices; the idea is to grab hold of the only small, manageable things in sight. Let A be an Abelian variety of dimension g, and let l be a prime different from char (k). We know that

$$A_{l^n} = \ker (A \xrightarrow{l^n} A)$$

is a finite group of order $(l^n)^{2g}$; since it contains exactly l^{2g} elements killed by l, we must have $A_{l^n} \simeq (\mathbf{Z}/l^n\mathbf{Z})^{2g}$. The A_{l^n} form an inverse system under the maps $l^m: A_{l^n} \to A_{l^{n-m}}$, and we fit them together to form the *Tate module*

$$T_lA = \varprojlim A_{l^n}.$$

From what we have said about A_{l^n} it is clear that $T_lA \simeq (\mathbf{Z}_l)^{2g}$, where $\mathbf{Z}_l =$ proj lim $\mathbf{Z}/l^n\mathbf{Z}$ is the l-adic integers. In the classical case there is a canonical isomorphism $\Lambda \otimes_{\mathbf{Z}} \mathbf{Z}_l \simeq T_lA$, and you should think of T_lA as capturing the nature of the lattice "locally at the prime l." If $\varphi: A \to B$ is a homomorphism, then clearly φ takes A_{l^n} to B_{l^n} and thus defines a map $T_l\varphi: T_lA \to T_lB$. This has the obvious reasonable properties (i.e., T_l is an additive functor).

It is time now to remember that k may not be algebraically closed. If for instance k is finite, then there are only finitely many points on A with coordinates in k, so there aren't enough points to make A_{l^n} as large as it should be. What we do, of course, is to take A_{l^n} consisting of points from the algebraic closure $\bar{k}$ of k; and then the statements are correct.

Once we notice this, we also pick up some additional structure. Let σ be an element of $\mathcal{G} = \mathrm{Gal}(\bar{k}/k)$. If $x \in A_{l^n}$, then $\sigma x \in A_{l^n}$, because being in A_{l^n} is a condition defined by polynomials over k. Thus T_lA is actually a $\mathcal{G}$-module. Furthermore, since by "homomorphisms" we mean group homomorphisms given by rational functions over k, the maps $T_l\varphi$ all commute with the $\mathcal{G}$-action. Finally, as evidence that T_lA captures the local structure at l we have

THEOREM 3 (WEIL). *The map*

$$\mathrm{Hom}(A, B) \otimes_{\mathbf{Z}} \mathbf{Z}_l \to \mathrm{Hom}_{\mathcal{G}}(T_lA, T_lB)$$

is injective.

PROOF. Let $\{\varphi_i\}$ be a $\mathbf{Z}$-basis of Hom (A, B), and suppose $\sum \varphi_i \otimes \lambda_i$ goes to 0. Given n, choose integers b_i with $b_i \equiv \lambda_i \pmod{l^n}$. Then $\sum b_i\varphi_i: A \to B$ has image in l^n Hom (T_lA, T_lB) and so vanishes on $A_{l^n} = \ker(l^n)$. This implies that there is a $\psi: A \to B$ such that $\sum b_i\varphi_i = \psi \circ l^n = l^n\psi$, and hence all $b_i \equiv 0 \pmod{l^n}$. Thus the λ_i are in $\bigcap l^n\mathbf{Z}_l = \{0\}$.

We have here assumed that Hom (A, B) is finitely generated; with a little further argument one can use this approach to prove it.

3. Our control over the local structure is now good except at $p = \mathrm{char}\, k$ when this is nonzero. We can still define A_p to be the kernel of multiplication by p, but here there aren't enough points even in $\bar{k}$ (there are at most p^g, and perhaps only one). To understand what is happening, look at a simpler situation. The multiplicative group of $\bar{k}$ has of course an algebraic group law, and $\varphi: x \mapsto x^p$ is a homomorphism of the group onto itself. Obviously the polynomial map φ has degree p, but you can't tell this by looking at its kernel in $\bar{k}$, since that kernel would be the pth roots of unity and there aren't any except unity itself.

To provide sufficiently large kernels for maps like φ, then, we are forced to introduce objects that can look like "pth roots of unity in characteristic p". These are furnished by the theory of schemes; in a sense it allows us to look for pth roots of unity in rings that aren't fields, and there we can find them. After developing the technique, one can then prove that A_{p^n} is a finite commutative group scheme of the right rank $(p^n)^{2g}$. The A_{p^n} fit together again in a reasonable way, forming what is called a *p-divisible group* (scheme).

At first sight it is probably not clear what has been gained by introducing these abstract-seeming objects. This will be clarified by the next theorem, for which we need to introduce a certain ring. Let W be the ring of Witt vectors over k. (If k is finite with p^a elements, then W is the ring of integers in the unramified extension of $\mathbf{Q}_p$ with degree a.) Let σ be the unique automorphism of W which reduces to the map $x \mapsto x^p$ on the residue field k. Let $\mathcal{A} = W[F, V]$ where F and V are indeterminates subjected to the relations

(1) $FV = VF = p$, and
(2) $F\alpha = \alpha^\sigma F$ and $\alpha V = V\alpha^\sigma$ for $\alpha \in W$.

THEOREM 4 (DIEUDONNÉ-CARTIER-BARSOTTI-ODA). *There is a functor from*

$$\{\textit{finite commutative group schemes over } k \textit{ of } p\textit{-power rank}\}$$

to

$$\{\mathcal{A}\textit{-modules of finite length}\};$$

it is an anti-equivalence of categories. If a group scheme G has rank p^s, then its Dieudonné module DG has length s.

It follows readily that p-divisible groups of co-rank $2g$ (like that coming from A) correspond to $\mathcal{A}$-modules free of rank $2g$ over W. We write T_pA for the module thus associated with A. Any homomorphism $\varphi\colon A \to B$ induces an $\mathcal{A}$-module map $T_p\varphi\colon T_pB \to T_pA$, and the same proof as before yields

THEOREM 5. *The map*

$$\operatorname{Hom}(A, B) \otimes \mathbf{Z}_p \to \operatorname{Hom}_{\mathcal{A}}(T_pB, T_pA)$$

is injective.

4. In this section I will try to explain some of the ideas that go into the proof of the basic

THEOREM 6 (TATE). *Suppose that k is finite. Then the maps in Theorems* 3 *and* 5 *are bijective.*

From now on we assume that k is finite with $q = p^a$ elements. There is a natural decomposition

$$\operatorname{End}(A \times B) = \operatorname{End}(A) \times \operatorname{Hom}(A, B) \times \operatorname{Hom}(B, A) \times \operatorname{End}(B)$$

and a corresponding decomposition of $\operatorname{End} T_l(A \times B) = \operatorname{End}(T_lA \times T_lB)$; hence if we have an isomorphism on $\operatorname{End}(A \times B)$ we have one on $\operatorname{Hom}(A, B)$. Thus we may restrict ourselves to endomorphism rings.

An argument like that in Theorem 3 shows that if φ lands in $l(\operatorname{End} T_lA)$ then

$\varphi = l\psi$ for some ψ; and thus the quotient of End (T_lA) by the image is torsion-free. Hence it will be enough to prove that

$$(*) \qquad E \otimes_{\mathbf{Q}} \mathbf{Q}_l \to \mathrm{End}_{\mathcal{G}}(V_lA)$$

is bijective, where $V_lA = T_lA \otimes_{\mathbf{Z}_l} \mathbf{Q}_l$ and E is the endomorphism algebra End $(A) \otimes \mathbf{Q}$.

Now the left-hand side of (*) has the same dimension for all l, and we know by Theorem 3 that the map is always injective. We will show that the right-hand sides all have the same dimension, so that it will then be enough to prove the isomorphism for a single l. At this point we need the fact that if $\varphi \in$ End A, then the function $n \mapsto$ [degree of $(\varphi - n)$] is a monic polynomial in n of degree 2 dim A with integer coefficients. One can prove that it equals the characteristic polynomial of $T_l\varphi$ on T_lA, and also equals the characteristic polynomial of $T_p\varphi$ acting W-linearly on T_pA.

We also need to recall that Gal $(\bar{k}/k)$ is generated (topologically) by $x \mapsto x^q$. This automorphism of course maps A (which is defined over k) into itself. But this map, since it is given simply by polynomials in the coordinates, actually corresponds to an element π_A in End A. We call π_A the *Frobenius endomorphism* of A, and write f_A for its characteristic polynomial. The elements of $\mathrm{End}_{\mathcal{G}}\, V_lA$ are now simply the $\mathbf{Q}_l$-linear maps on V_lA which commute with the specific map $T_l(\pi_A)$.

Since π_A is in the center of the semisimple algebra $E \otimes \mathbf{Q}_l$, it acts semisimply on V_l. Suppose its characteristic polynomial $f_A(X)$ factors as $\prod(X - \alpha_i)$ over some splitting field. Then by standard algebra the dimension of the commutant of $T_l(\pi_A)$ is the number of ordered pairs $\langle i, j\rangle$ (including $\langle i, i\rangle$) with $\alpha_i = \alpha_j$. This is obviously independent of l. A similar argument (complicated by the presence of the noncommutative ring $\mathcal{Q}$) works at $l = p$; see Part II.

It still must be shown that the map is bijective for some l. For this purpose one takes l to be a prime which splits completely in the algebra $\mathbf{Q}(\pi_A)$; this condition means that f_A splits into linear factors over $\mathbf{Q}_l$, and so the action of $T_l(\pi_A)$ is quite simple. The proof requires a clever use of a finiteness condition, however, and we will omit it.

I will instead end this section with an unsolved problem: does Theorem 6 hold when k is a number field? It has been proved (except for some cases) by Serre when A and B are elliptic curves. Here of course the Galois group $\mathcal{G}$ is much more complicated, and the representations of $\mathcal{G}$ on the T_lA seem to be quite interesting.

5. As a first consequence of Tate's Theorem we get

THEOREM 7. *Let A and B be Abelian varieties over a finite field k. The following are equivalent*:

(1) *A and B are isogenous.*

(2) *V_lA and V_lB are $\mathcal{G}$-isomorphic for some l.*

(3) $f_A = f_B$.

(4) *The zeta functions of A and B are the same.*

(5) *For each finite extension k' of k, the varieties A and B have the same number of points over k'.*

PROOF. We obviously have (1) $\Rightarrow$ (2) $\Rightarrow$ (3), and (3) $\Rightarrow$ (2) because a semisimple representation is determined by its characteristic polynomial. Suppose now that $\operatorname{Hom}_{\mathcal{G}}(V_l A, V_l B) \simeq \operatorname{Hom}(A, B) \otimes \mathbf{Q}_l$ contains an isomorphism. We can approximate it by elements of $\operatorname{Hom}(A, B) \otimes \mathbf{Q}$ which, if close enough, will also be isomorphisms. Multiplying by an integer we obtain a $\varphi \in \operatorname{Hom}(A, B)$ inducing an isomorphism $V_l A \to V_l B$. If $\ker \varphi$ contained any positive-dimensional Abelian variety C, then φ would annihilate the subspace $V_l C$ of $V_l A$; hence $\ker \varphi$ is finite. Similarly $\varphi(A)$ cannot lie in any lower-dimensional subgroup of B, and φ is an isogeny.

By the definition of the zeta function, (4) $\Leftrightarrow$ (5). To connect these with (3), consider the map $\pi_A - 1\colon A \to A$. The points of its kernel are those fixed by $x \mapsto x^q$, i.e. the points with coordinates in k. After checking that the zeros are all separated (e.g. by looking on the tangent space), we can conclude that the number of points of A in k is

$$\text{degree}\,(\pi_A - 1) = f_A(1) = \prod (1 - \alpha_i),$$

where the α_i are the roots of f_A. Similarly, degree $(\pi_A^s - 1) = \prod(1 - \alpha_i^s)$ gives the number of points in the extension of degree s, and thus f_A determines the zeta function. To get the converse, you simply check that the values $\prod(1 - \alpha_i^s)$ (in fact, a finite number of them) are enough to determine the α_i and hence f_A.

6. We are now ready for the classification up to isogeny. We know from §1 that every Abelian variety is isogenous to a product of elementary ones, so those are all we need to discuss.

THEOREM 8. *Let A be an elementary Abelian variety over the field k with q elements. Then*

1. $f_A = m_A^e$ *for some integer e and some irreducible monic polynomial m_A with integer coefficients.*

2. $E = \operatorname{End}(A) \otimes \mathbf{Q}$ *is a division algebra whose center is* $\Phi = \mathbf{Q}(\pi_A)$.

3. $|E\colon \mathbf{Q}| = e^2|\Phi\colon \mathbf{Q}|$, *and* $2 \dim A = e|\Phi\colon \mathbf{Q}|$.

4. *Let v be a prime of Φ, and $\| \ \|_v$ the normalized absolute value. If $\|\pi_A\|_v = q^{-i}$, then i is the invariant of E at v. Explicitly, this is*

$$\begin{aligned} &\tfrac{1}{2} \quad \textit{if } v \textit{ is real},\\ &0 \quad \textit{if } v \textit{ lies over a prime } l \neq p \textit{ in } \mathbf{Q},\\ &\operatorname{ord}_v(\pi_A)\cdot|\Phi_v\colon \mathbf{Q}_p|/\operatorname{ord}_v(q) \quad \textit{if } v \textit{ lies over } p. \end{aligned}$$

5. *Every embedding of Φ in $\mathbf{C}$ gives π_A the absolute value $q^{1/2}$. In other words, all roots of f_A have absolute value $q^{1/2}$.*

PROOF. We know that E is a division algebra, so its center is a field. That center is $\mathbf{Q}(\pi_A)$ because $E \otimes \mathbf{Q}_l$ is the commutant of $T_l\pi_A$ in $\operatorname{End}_{\mathbf{Q}_l}(V_l A)$. If f_A had two distinct irreducible factors, $\mathbf{Q}(\pi_A)$ could not be a field. The second statement in (3) is obvious from (1), and the first statement comes from the dimension computation in the proof of Theorem 6. The assertion in (5), which we will not prove, is an equivalent form of Weil's Riemann hypothesis for curves over finite fields. Suppose now we take a prime v. There are only a few cases where a real prime exists (cf. §7), and (4) can be verified there directly. If v lies over $l \neq p$, we look at V_lA; by (1) we see that it is a free module of rank e over $\Phi \otimes \mathbf{Q}_l$. Hence $E \otimes \mathbf{Q}_l$, the commutant, is simply the $e \times e$ matrices over $\Phi \otimes \mathbf{Q}_l$. Thus E becomes a matrix algebra, i.e. has invariant 0, at all primes of Φ lying over l. A similar but messier computation lets us deduce the invariants at p from the structure of T_pA; see Part II.

COROLLARY. *A is determined up to isogeny by π_A, that is, by m_A.*

PROOF. Given a root π of m_A, we can by the above formulas compute the invariants of E at all primes of $\mathbf{Q}(\pi)$. There is a unique division algebra with center $\mathbf{Q}(\pi)$ and these invariants; its dimension gives us e and so determines $f_A = m_A^e$. For computational purposes we may note that e is the least common denominator of the invariants.

Let us say that an algebraic integer π is a *Weil number* (for q) if it satisfies statement (5) of Theorem 8. Every Abelian variety gives us a Weil number π_A determined up to conjugacy, and π_A determines A up to isogeny. The classification theory can now be summed up in

THEOREM 9 (TATE-HONDA). *Let k be finite. Then there is a one-to-one correspondence between*

$$\{\textit{isogeny classes of elementary Abelian varieties over } k\}$$

and

$$\{\textit{conjugacy classes of Weil numbers for } q = \text{card}(k)\}.$$

To finish proving this, one just has to produce lots of Abelian varieties over k. The idea is to use the classical theory: construct Abelian varieties (with large endomorphism algebras) defined over a number field or a p-adic field, and reduce them mod p. A formula of Shimura and Taniyama describes the Weil numbers we get this way, and a few technical devices then produce enough varieties to give them all.

7. Let me conclude with a couple of remarks. First, it may not seem easy to produce algebraic integers π with all conjugates of absolute value $q^{1/2}$. In fact, however, they can be found quite simply as follows. If π is a Weil number, set $\beta = \pi + (q/\pi)$. Since $\bar{\pi} = |\pi|^2/\pi = q/\pi$ in every embedding, β is a totally real algebraic integer, and $|\beta| \leq 2q^{1/2}$ in every embedding. But now conversely, given any β satisfying these conditions, the roots of $X^2 - \beta X + q = 0$ will be Weil numbers. It is thus easy to write down a β, compute the root π, and read off properties of the corresponding isogeny class from Theorem 8. We may note that π can be real only for $\beta = \pm 2q^{1/2}$, that is, only in the special cases $\pi = \pm q^{1/2}$.

Finally, Tate's theorem is from one point of view an existence theorem for endomorphisms, and in this respect it has a converse. Let k be any field of characteristic p, let A be an elementary Abelian variety over k, let $E = \text{End}\, A \otimes \mathbf{Q}$, and let Φ be the center of E. We say that A is of *CM-type* if $2 \dim A = |E{:}\ \Phi|^{1/2}|\Phi{:}\ \mathbf{Q}|$; this holds for k finite by Theorem 8. Grothendieck has recently proved that, conversely, any Abelian variety of CM-type is isogenous to a variety defined over a finite field.

II. Two Theorems of Tate

We here present proofs of two theorems stated in Part I:

THEOREM 1. *Let A and B be Abelian varieties over a finite field k with $q = p^a$ elements. Let T_pA and T_pB be the associated Dieudonné modules. Then*

$$\mathrm{Hom}_k(A, B) \otimes Z_p \xrightarrow{\sim} \mathrm{Hom}_{\mathfrak{a}}(T_pB, T_pA).$$

THEOREM 2. *Suppose that A is elementary over k with Weil number π. Let $E = \mathrm{End}\, A \otimes \mathbf{Q}$ with center $\Phi = \mathbf{Q}(\pi)$, and let v be a prime of Φ lying over p. Then the invariant of E at v is*

$$\frac{\mathrm{ord}_v(\pi)}{\mathrm{ord}_v(q)}\,|\Phi_v : \mathbf{Q}_p|.$$

Tate announced these results in [13]; the proofs were presented to a seminar but have never been released to the public. Our proofs are basically the same, but rely more on [4].

PROOF OF THEOREM 1. We adopt the notation of Part I. In addition we write L for the fraction field $W \otimes_{\mathbf{Z}_p} \mathbf{Q}_p$, and V_pA for the L-module $T_pA \otimes_{\mathbf{Z}_p} \mathbf{Q}_p$. Then V_pA is actually a module over

$$L[F, V] = L[F, (1/p)F^{-1}] = L[F, F^{-1}],$$

or in other words an $L[F]$-module on which F is a bijection. As in Theorem 6 of Part I it is enough to prove that

$$E \otimes \mathbf{Q}_p \to \mathrm{End}_{L[F]}(V_pA)$$

is bijective for all A, and since injectivity is known it is enough to prove that the two algebras have the same dimension.

We recall from [4, Chapter 3] the structure of finite indecomposable modules V over $R = L[F]$. Such a V has the form $R/R\lambda$ for some λ in R, and there is a smallest r for which $V^r \simeq R/cR$ with c in the center of R. The center of R is $\mathbf{Q}_p[F^a]$, and in fact $c = m_V(F^a)$ with m_V a power of an irreducible polynomial; two indecomposables are isomorphic iff they have the same m_V. One can identify m_V as the minimal polynomial for F^a as a $\mathbf{Q}_p$-linear map on V. Clearly the $\mathbf{Q}_p$-characteristic polynomial for F^a on $R/m_V(F^a)R$ is $m_V^{a^2}$, and the L-characteristic polynomial is m_V^a; the L-characteristic polynomial on V is $m_V^{a/r}$.

Write V_pA now as a sum $\bigoplus V_i^{n_i}$ where the V_i are nonisomorphic indecomposables, and say $V_i^{r_i} \simeq R/m_i(F^a)R$. Let π be the Frobenius endomorphism of A, with characteristic polynomial f and minimal polynomial m. Then on V_pA we know that π acts as F^a and that f is its L-characteristic polynomial (cf. [6, p. 66]). We also know from Part I that m has no repeated roots. Since the $\mathbf{Q}_p$-minimal polynomial of F^a on $\bigoplus V_i^{n_i}$ is the least common multiple of the m_i, and this must divide m, we see that the m_i must all be without repeated roots. They are thus all irreducible, and being distinct have no roots in common. The characteristic polynomial f is $\prod m_i^{an_i/r_i}$, so the formula in Part I shows that the dimension of E is

$$\sum (\deg m_i) a^2 n_i^2 / r_i^2.$$

But the endomorphisms of $V_i^{r_i} = R/m_i(F^a)R$ are just multiplications by elements of $R/m_i(F^a)R$, which gives $\mathbf{Q}_p$-dimension $a^2 \deg m_i$. Hence End $(V_i^{n_i})$ has dimension $(n_i^2/r_i^2)a^2 \deg m_i$, and

$$\dim \operatorname{End} V_pA = \sum \dim \operatorname{End} (V_i^{n_i}) = \sum (n_i^2/r_i^2)a^2 \deg m_i.$$

PROOF OF THEOREM 2. Suppose the prime v corresponds to the irreducible factor m_i of m over $\mathbf{Q}_p$. Then Φ_v is generated over $\mathbf{Q}_p$ by a root π of m_i. We note that V_i is a simple R-module, since any indecomposable submodule would correspond to a divisor of m_i. Hence $R/m_i(F^a)R$ is a simple algebra. Clearly it can be written as $L \otimes \Phi_v[F]$ with $F^a = \pi$ in Φ_v and $F(\alpha \otimes \varphi) = (\alpha^\sigma \otimes \varphi)F$, where σ is the Frobenius automorphism of L over $\mathbf{Q}_p$; the center is Φ_v.

Let f_v be the residue degree of Φ_v, and set $g = (f_v, a)$, so that $g = |L \cap \Phi_v \colon \mathbf{Q}_p|$ and $a/g = |L\Phi_v \colon \Phi_v|$. Let D be the algebra $L\Phi_v[F']$, where $(F')^{a/g} = \pi \in \Phi_v$ and $F'(\alpha\varphi) = (\alpha^{\sigma^g}\varphi)F'$. We define a map from $L \otimes \Phi_v[F]$ into the $g \times g$ matrices over D by sending

$$\alpha \otimes \varphi \mapsto \begin{pmatrix} \alpha\varphi & & & 0 \\ & \alpha^\sigma\varphi & & \\ & & \cdots & \\ 0 & & & \alpha^{\sigma^{g-1}}\varphi \end{pmatrix}$$

$$F \mapsto \begin{pmatrix} 0 & 1 & 0 & \vdots & 0 \\ 0 & 0 & 1 & \vdots & 0 \\ & \cdots & & & \cdots \\ 0 & 0 & 0 & \vdots & 1 \\ F' & 0 & 0 & \vdots & 0 \end{pmatrix}$$

It is easy to check that this is a Φ_v-algebra homomorphism. Since $L \otimes \Phi_v[F]$ is simple, the map is injective; by dimension counting it is an isomorphism. Hence $L \otimes \Phi_v[F]$ has the same invariant as D.

Now D is in the standard form for a central simple algebra: $L\Phi_v$ is the unramified extension of degree a/g, and F' acts on it as a generator of the Galois group. Explicitly, it raises to the gth power on the residue field of Φ_v; the Frobenius of $L\Phi_v$ over Φ_v raises to the f_vth power and so is the (f_v/g)th power of our generator. Then it is well known (from explicit computation of the cocycle) that the invariant of D is $(f_v/g) \operatorname{ord}_v \pi/(a/g)$. If e_v is the ramification index of Φ_v over $\mathbf{Q}_p$, we have $e_v f_v = |\Phi_v \colon \mathbf{Q}_p|$ and $\operatorname{ord}_v q = a \cdot \operatorname{ord}_v p = ae_v$, so the number can also be written $(\operatorname{ord}_v \pi/\operatorname{ord}_v q) \cdot |\Phi_v \colon \mathbf{Q}_p|$.

This is the invariant of D, and hence of $R/m_i(F^a)R$. The ring $\operatorname{End}_R(R/m_i(F^a)R)$ is the opposite ring, and so has the negative of that invariant; since it is End $(V_i^{r_i})$, both End (V_i) and End $(V_i^{n_i})$ have this negative invariant. But End $(V_i^{n_i})$ is the part of End (V_pA) sitting over Φ_v, and so it gives the invariant at v. The map from $E \otimes \mathbf{Q}_p$ to End (V_pA) is an anti-isomorphism, however, so the sign reverses again and gives us the invariant we want for E at v.

III. Further Topics

A. ENDOMORPHISM RINGS AND ISOMORPHISM CLASSES. In studying Abelian varieties over a finite field k, what we would like best is a description of the isomorphism classes; since we know a classification up to isogeny, we can restrict to a fixed isogeny class. One invariant obviously associated with A is the ring End (A), and as a first step we can ask which orders in the algebra E occur (up to isomorphism) as endomorphism rings. For any particular E one can answer this question by computation, constructing the spaces $V_l A$ (and $V_p A$) and considering lattices in them. The problem is to formulate reasonable general theorems. One such involves a nice type of variety which we now define.

THEOREM. *Let A be an elementary Abelian variety over k. The following are equivalent:*

(1) $\pi_A + (q/\pi_A)$ *is a prime to p,*

(2) A_p *contains* $p^{\dim A}$ *points over* $\bar{k}$.

Such varieties are called *ordinary*. For them E is commutative and not changed by extending k.

THEOREM. *Let π be the Weil number of an elementary isogeny class, and E the endomorphism algebra. Then any endomorphism ring is an order containing π and q/π. The converse holds if the class is ordinary, or if $k = \mathbf{Z}/p\mathbf{Z}$ and $\pi \neq \pm p^{1/2}$. The converse does not hold in general, even when E is commutative and not changed by extending k.*

There is a reasonable classification theory for elementary Abelian varieties over $\mathbf{Z}/p\mathbf{Z}$, and a neat treatment of ordinary varieties has appeared in [1]. In general, however, the computation of the isomorphism classes seems to become quite unpleasant. Simplifications can be introduced by adding assumptions on End A, leading to results like the

THEOREM. *Let E be the algebra of an elementary isogeny class. Assume E is commutative, and let R be the ring of integers in E. Then the set of varieties in the class with* End $A = R$ *has the ideal class group of R acting freely on it. Two varieties are in the same orbit iff there is a separable isogeny between them, and one can give a formula for the number of orbits.*

The proof relies on passing from an ideal of End A to a variety isogenous to A; this process has interesting properties more generally. Details can be found in [17].

B. RELATION TO THE CONJECTURES OF BIRCH AND SWINNERTON-DYER. Tate's theorem (see I,§4) gives, in particular, the rank of the free abelian group $\mathrm{Hom}_k(A, B)$ in terms of the characteristic polynomials $f_A(X)$ and $f_B(X)$ of the Frobenius endomorphisms of A and B. It is possible to give similarly explicit descriptions of the higher extension groups of A and B, these groups being formed in the abelian category of all group schemes of finite type over k.

THEOREM 1. *Let A and B be abelian varieties over the finite field k.*

(a) $\mathrm{Hom}_k(A, B)$ *is a free abelian group of rank equal to* $r(f_A, f_B)$, *the number of ordered pairs* $\langle i, j\rangle$ *such that* $\alpha_i = \beta_j$ *where* $\alpha_1, \ldots, \alpha_{2g(A)}$ *are the roots of* $f_A(X)$ *and* $\beta_1, \ldots, \beta_{2g(B)}$ *the roots of* $f_B(X)$.

(b) $\mathrm{Ext}_k^1(A, B)$ *is finite, and its order* $[\mathrm{Ext}_k^1(A, B)]$ *is given by*

$$q^{g(A)g(B)} \prod_{\alpha_i \neq \beta_j} (1 - \alpha_i/\beta_j) = [\mathrm{Ext}_k^1(A, B)]|D|,$$

where D is the discriminant of the nondegenerate pairing

$$\mathrm{Hom}_k(A, B) \times \mathrm{Hom}_k(B, A) \to \mathbf{Z}$$

which takes two homomorphisms to the trace of their composite (as an endomorphism of A or B, indifferently).

(c) $\mathrm{Ext}_k^2(A, B)$ *is a divisible group of corank equal to* $r(f_A, f_B)$.

(d) $\mathrm{Ext}_k^i(A, B) = 0$, $i > 2$.

(a) is Tate's theorem (I, Theorem 6). (b) and (c) may be restated in terms of p-divisible group schemes and proved using their associated Galois modules of points ($p \neq \mathrm{char}(k)$) or Dieudonné modules ($p = \mathrm{char}(k)$) [6]. (d) follows from a much more general theorem on the vanishing of higher extension groups in categories of commutative group schemes over perfect fields [8].

Now let K be a function field in one variable over k. An abelian variety A over k can also be regarded as a "constant" abelian variety over K, and for any abelian variety over a global field there are the conjectures of Birch and Swinnerton-Dyer. For the constant abelian variety A, these take on an especially simple form. In fact, with the notations of [12, §1], S is empty and $\alpha_{i,v} = \alpha_i^{\deg(v)}$ where, as before, the α_i are the roots of $f_A(X)$. By comparing

$$L(s) = \prod_v \prod_i \frac{1}{1 - \alpha_i^{\deg(v)} Nv^{-s}} = \prod_i \prod_v \frac{1}{1 - (\alpha_i q^{-s})^{\deg(v)}}$$

with the known expressions for the zeta function of K

$$Z(K, T) = \frac{f_J(T)}{(1 - T)(1 - qT)} = \prod_v \frac{1}{1 - T^{\deg(v)}}$$

(J = Jacobian of the curve X associated to K/k) and with a little juggling, one reduces the conjectures to

(A) The rank of the group of K-rational points of A, $A(K)$, is $r(f_J, f_A)$.

(B) The order of the Tate-Šafarevič group Ш of A over K is given by

$$q^{g(J)g(A)} \prod_{\gamma_i \neq \alpha_j} \left(1 - \frac{\gamma_i}{\alpha_j}\right) = [\text{Ш}]|D'|$$

where the γ_i are the roots of $f_J(X)$ and D' is the discriminant of the canonical height pairing on the K-rational points of A and its dual (suitably normalized).

THEOREM 2. *In this special situation, conjectures* (A) *and* (B) *are true.*

(A) is due to Tate and follows directly from Theorem 1(a) by the obvious identification $A(K) = \mathrm{Hom}_k(J, A)$ (modulo torsion). (B) follows from Theorem 1(b) by identifying D' with D (with J for A, and A for B) and Ш with $\mathrm{Ext}_k^1(J, A)$. Ш can be interpreted as the first étale cohomology group of A over X, and the machinery of étale and flat cohomology used to reduce the identification to the statement, well known in the classical case, that a finite covering of X comes from an isogeny of its Jacobian J [7].

REFERENCES

1. P. Deligne, *Variétés abéliennes ordinaires sur un corps fini*, Invent. Math. **8** (1969), 238–243.

2. A. Grothendieck and P. Deligne, *Séminaire de Géométrie Algébrique* 7, Exposé II, Inst. Hautes Etudes Sci., Paris (to appear).

3. T. Honda, *Isogeny classes of abelian varieties over finite fields*, J. Math. Soc. Japan **20** (1968), 83–95. MR **37** #5216.

4. N. Jacobson, *The theory of rings*, Amer. Math. Soc. Math. Surveys, no. II, Amer. Math. Soc., Providence, R.I., 1943. MR **5**, 31.

5. S. Lang, *Abelian varieties*, Interscience Tracts in Pure and Appl. Math., no. 7, Interscience, New York, 1959. MR **21** #4959.

6. J. Milne, *Extensions of abelian varieties defined over a finite field*, Invent. Math. **5** (1968), 63–84. MR **37** #5226.

7. ———, *The Tate-Šafarevič group of a constant abelian variety*, Invent. Math. **6** (1968), 91–105. MR **39** #5581.

8. ———, *The homological dimension of commutative group schemes over a perfect field*, J. Algebra (to appear).

9. T. Oda, *The first deRham cohomology group and Dieudonné modules*, Ann. Sci. Ecole Norm. Sup. (4) **2** (1969), 63–135.

10. F. Oort, *Commutative group schemes*, Lecture Notes in Math., no. 15, Springer-Verlag, Berlin and New York, 1966. MR **35** #4229.

11. J.-P. Serre, *Abelian l-adic representations and elliptic curves*, Benjamin, New York, 1968.

12. J. Tate, *On the conjectures of Birch and Swinnerton-Dyer and a geometric analogue*, Séminaire Bourbaki 1965/66, Exposé 306, Benjamin, New York, 1966. MR **34** #5605.

13. ———, *Endomorphisms of abelian varieties over finite fields*, Invent. Math. **2** (1966), 134–144. MR **34** #5829.

14. ———, *Endomorphisms of abelian varieties over finite fields.* II. (This paper, though cited in the literature, does not exist.)

15. ———, *p-divisible groups*, Proc. Conference on Local Fields, NUFFIC Summer School (Driebergen, Netherlands, 1966) Springer-Verlag, Berlin and New York, 1967. MR **37** #3885.

16. ———, *Classes d'isogénie des variétés abéliennes sur un corps fini* (*d'après T. Honda*), Séminaire Bourbaki 1968/69 Exposé 352, Benjamin, New York.

17. W. Waterhouse, *Abelian varieties over finite fields*, Ann. Sci. Ecole Norm. Sup. (4) **2** (1969), 521–560.

18. A. Weil, *Variétés abéliennes et courbes algébriques*, Actualités Sci. Indust., no. 1064, Hermann, Paris, 1948. MR **10**, 621.

CORNELL UNIVERSITY
ITHACA, NEW YORK

UNIVERSITY OF MICHIGAN
ANN ARBOR, MICHIGAN 48104

INTRODUCTION AUX TRAVAUX RÉCENTS DE DWORK

NICHOLAS M. KATZ*

(0.0) Liste de donnés.

(0.1) $k = GF(q)$, le corps à q éléments.

(0.11) $\mathfrak{O} = W(k)[\zeta_p]$, $W(k)$ étant les vecteurs de Witt de k, et ζ_p une racine primitive $p$ième de l'unité.

(0.12) Ω, une clôture algébrique du corps des fractions de $\mathfrak{O}$.

(0.2) des entiers $n \geq 0$, $d \geq 1$.

(0.3) une famille $f \colon X \to \mathbf{A}^1$ d'hypersurfaces projectives de degré d dans $\mathbf{P}^{n+1}$, définie sur k (c'est-à-dire, une hypersurface de degré d dans $\mathbf{P}^{n+1}_{\mathbf{A}^1}$), à fibre générique lisse.

(0.3 bis) Souvent en pratique, une base $X_1, \ldots, X_{n+2}$ de $\mathfrak{O}_{\mathbf{P}^{n+1}}(1)$, choisie de telle façon que pour tout sous-ensemble propre A de $\{1, \ldots, n+2\}$, la variété linéaire d'équation $\{X_i = 0, \forall i \in A\}$ coupe transversalement la fibre générique de f.

(1.0) Les choix.

(1.1) Soit S un ouvert de $\mathbf{A}^1$ au-dessus duquel f soit lisse, et soit $S^\dagger$ un relèvement de Washnitzer-Monsky de S au-dessus de $\mathfrak{O}$ [6].

(1.1 bis) Explicitement, on obtient un tel $S^\dagger$ de la façon suivante. On choisit une polynôme monique $g(\lambda) \in \mathfrak{O}[\lambda]$ dont l'image $\bar{g} \in \Gamma(\mathbf{A}^1, \mathfrak{O}_{\mathbf{A}^1})$ définit le fermé $\mathbf{A}^1 - S$. On désigne par $\mathfrak{O}[\lambda, y]^\dagger$ la sous-algèbre de $\mathfrak{O}[[\lambda, y]]$ formée des séries formelles.

Je tiens à remercier P. Deligne et W. Messing de l'aide qu'ils m'ont apportés dans la préparation de ce manuscrit.

*A detailed and expanded version of the lecture presented at the Institute by B. Dwork will be published elsewhere [reference [3] below]. The present article, originally written for private distribution, covers substantially the same material but the style is different.

$\sum a_{ij}\lambda^i y^j$ qui vérifient, pour un nombre réel convenable $\alpha > 0$, une majoration

$$\text{ord}\,(a_{ij}) \geq \alpha(i + j) \quad \text{si } i, j \gg 0.$$

Alors on pose $S^\dagger = \mathfrak{O}[y, \lambda]^\dagger/(1 - yg)\mathfrak{O}[y, \lambda]^\dagger$.

(1.2) On relève la famille $f\colon X \to \mathbf{A}^1$ en une famille $\tilde{f}\colon \tilde{X} \to \text{Spec}\,(\mathfrak{O}[\lambda])$, supposée lisse au-dessus de $\tilde{S} = \text{Spec}\,(\mathfrak{O}[\lambda][1/g])$.

(2.0) Ce qu'on peut construire.

(2.1) un module $\mathfrak{H}$ sur $S^\dagger \otimes Q$, qui est localement libre de rang fini, et muni d'une connection intégrable Д. Ce module, avec sa connection, provient d'un module localement libre à connection intégrable sur $\tilde{S} \otimes Q$, à savoir, la partie primitive de la cohomologie de De Rham relative de $\tilde{f}\colon \tilde{X} \otimes Q \to \tilde{S} \otimes Q$.

(2.1 bis) le module dual $\check{\mathfrak{H}} = \text{Hom}_{S^\dagger \otimes Q}(\mathfrak{H}, S^\dagger \otimes Q)$, muni de la connection duale, donnant lieu à la formule

$$\frac{d}{d\lambda}(h^\vee, h) = \left(Д\left(\frac{d}{d\lambda}\right)h^\vee, h\right) + \left(h^\vee, Д\left(\frac{d}{d\lambda}\right)h\right)$$

où (,): $\check{\mathfrak{H}} \times \mathfrak{H} \to S^\dagger \otimes Q$ désigne l'accouplement canonique, et où $h^\vee$, h sont respectivement des sections de $\check{\mathfrak{H}}$ et $\mathfrak{H}$.

(2.2) un relèvement de l'homomorphisme de Frobenius (élévation à la $q$ième puissance) de S en une homomorphisme de $\mathfrak{O}$-algèbres $\varphi\colon S^\dagger \to S^\dagger$, qui sera noté selon le cas $h \mapsto \varphi(h)$ ou, $h(\lambda) \mapsto h(\varphi(\lambda))$. Cette dernière notation est justifiée, en qu'un relèvement "est" le choix d'un élément $\varphi(\lambda) \in S^\dagger$ qui relève l'élément $\lambda^q \in k[\lambda][1/\bar{g}]$.

(2.3) un morphisme horizontal, dépendant du choix de φ, de $S^\dagger \otimes Q$-modules $\mathfrak{F}\colon \varphi^*\check{\mathfrak{H}} \to \check{\mathfrak{H}}$.

En termes plus explicites, $\mathfrak{F}$ "est" un endomorphisme additif de $\check{\mathfrak{H}}$ qui vérifie

(2.3a) $$\mathfrak{F}(h(\lambda)\tau) = h(\varphi(\lambda))\mathfrak{F}(\tau) \quad \text{pour } \tau \in \check{\mathfrak{H}},\ h \in S^\dagger \otimes Q$$

et

(2.3b) $$Д\left(\frac{d}{d\lambda}\right)(\mathfrak{F}(\tau)) = \frac{d\varphi(\lambda)}{d\lambda}\,\mathfrak{F}\left(Д\left(\frac{d}{d\lambda}\right)(\tau)\right).$$

Si l'on considère Д comme une flèche

$$Д\colon \check{\mathfrak{H}} \to \check{\mathfrak{H}} \otimes \Omega^1_{S^\dagger \otimes Q/\mathfrak{O} \otimes Q}.$$

(2.3b) se récrit

(2.3 bis) $$Д(\mathfrak{F}(\tau)) = \mathfrak{F}(Д(\tau)).$$

(3.0) La fonction zêta et $\mathfrak{F}$.

(3.1) Soit α un point de $S^\dagger \otimes Q$ à valeurs dans Ω laissé fixe par une itération de φ: $\varphi^{(\nu)}(\alpha) = \alpha$. Le point α est donc un relèvement d'un point $\bar{\alpha}$ de S à valeurs dans $GF(q^\nu)$. De plus, d'après Monsky, les points α qui vérifient $\varphi^{(\nu)}(\alpha) = \alpha$ sont en correspondance biunivoque avec les points de S à valeurs dans $GF(q^\nu)$, car $\varphi^{(\nu)}$ agit comme une contraction dans l'espace de tous les points α qui relèvent un point $\bar{\alpha}$ donné à valeurs dans $GF(q^\nu)$.

(3.2) Soit $\check{\mathfrak{H}}_\alpha$ le Ω-vectoriel fibre de $\check{\mathfrak{H}}$ en α, de dimension égale au rang de $\mathfrak{H}$. La condition $\varphi^{(\nu)}(\alpha) = \alpha$ nous assure que $\check{\mathfrak{H}}_\alpha \simeq (\varphi^{(\nu)*}\check{\mathfrak{H}})_\alpha$, de sorte que $\mathfrak{F}^{(\nu)}$ induit un Ω-endomorphisme de $\check{\mathfrak{H}}_\alpha$

(3.21)
$$\check{\mathfrak{H}}_\alpha \simeq (\varphi^{(\nu)*}\,\check{\mathfrak{H}})_\alpha \xrightarrow{\mathfrak{F}^{(\nu)}} \check{\mathfrak{H}}_\alpha.$$

(3.3) La théorie de Dwork affirme que son polynôme caractéristique est la "partie intéressante" de la fonction zêta de la fibre $X_{\bar{\alpha}}$ de $f\colon X \to \mathbf{A}^1$ en $\bar{\alpha}$, considérée comme variété sur $GF(q^\nu)$. Explicitement, on a

(3.31)
$$Z(T, X_{\bar{\alpha}}/GF(q^\nu)) \prod_{i=0}^{n} (1 - q^i T) = \det(1 - \mathfrak{F}^{(\nu)}T \mid \check{\mathfrak{H}}_\alpha)^{(-1)^{n+1}}.$$

(4.0) La fonction zêta et les solutions locales des équations de Picard-Fuchs.

(4.1) Soit $S^\infty \otimes \Omega$ l'espace *analytique* (au sens "mou") sous-jacent de $S^\dagger \otimes Q$; au-dessus de $S^\infty \otimes \Omega$, le module à connection intégrable $\check{\mathfrak{H}}$ devient localement trivial (à savoir, est localement (*au sens analytique*) isomorphe à $\mathfrak{O}^N_{S^\infty \otimes \Omega}$ muni de la connection standard).

(4.2) Pour chaque point α de $S^\dagger \otimes Q$ à valeurs dans Ω, posons

(4.21) $\check{\mathfrak{H}}(\alpha, \text{Д}) =$ le Ω-vectoriel des germes en α de sections horizontales locales du module sur $S^\infty \otimes \Omega$ deduit de $\check{H}$ par le changement de base $S^\dagger \otimes Q \hookrightarrow S^\infty \otimes \Omega$.

Alors $\mathfrak{F}$ induit une flèche linéaire

(4.22)
$$\check{\mathfrak{H}}(\varphi(\alpha), \text{Д}) \xrightarrow{\mathfrak{F}} \check{\mathfrak{H}}(\alpha, \text{Д}).$$

Si de plus $\varphi^{(\nu)}(\alpha) = \alpha$, on obtient un Ω-endomorphisme

(4.23)
$$\check{\mathfrak{H}}(\alpha, \text{Д}) \xrightarrow{\mathfrak{F}^{(\nu)}} \check{\mathfrak{H}}(\alpha, \text{Д}).$$

(4.24) Le module à connection intégrable $\check{\mathfrak{H}}$ étant localement trivial au-dessus de $S^\infty \otimes \Omega$, la flèche "valeur en α"

$$\check{H}(\alpha, \text{Д}) \to \check{\mathfrak{H}}_\alpha$$

est un isomorphisme, et le diagramme suivant, liant (4.23) à (3.21), est commutatif.

(4.25)
$$\begin{array}{ccc} \check{\mathfrak{H}}_\alpha \simeq (\varphi^{(\nu)*}\check{\mathfrak{H}})_\alpha & \xrightarrow{\mathfrak{F}^{(\nu)}} & \check{\mathfrak{H}}_\alpha \\ \int \Big\uparrow \text{valeur en } \alpha & & \int \Big\uparrow \text{valeur en } \alpha \\ \check{\mathfrak{H}}(\alpha, \text{Д}) & \xrightarrow{\mathfrak{F}^{(\nu)}} & \check{\mathfrak{H}}(\alpha, \text{Д}) \end{array}$$

D'après (3.3), la partie intéressante de la fonction zêta de $X_{\bar{\alpha}}/GF(q^\nu)$ est le polynôme caractéristique de $\mathfrak{F}^{(\nu)}$ agissant sur $\check{\mathfrak{H}}(\alpha, \text{Д})$.

(4.3) Classiquement, on ne considère pas la connection sur $\mathfrak{H}$, mais plutôt les "équations de Picard-Fuchs", qui sont les équations différentielles définissant les sections horizontales de $\check{\mathfrak{H}}$.

(5.0) La notion de formule pour une racine de la fonction zêta.

(5.10) Soit S^{rigid} l'algèbre rigide sous-jacente à $S^\dagger \otimes Q$; S^{rigid} est par définition le complété de $S^\dagger \otimes Q$ pour la topologie de la convergence uniforme sur Spec Max $(S^\dagger \otimes Q)$. On a Spec Max $(S^{\text{rigid}}) =$ Spec Max $(S^\dagger \otimes Q)$.

(5.101) Soit d'autre part $U^\dagger$ un ouvert (†) principal de $S^\dagger$, i.e. de la forme $(S^\dagger[1/k])^\dagger$, avec k non nul dans $S^\dagger$. On désigne par U^{rigid} l'algèbre rigide sous-jacente à $U^\dagger \otimes Q$, au sens de (5.10);

$$\text{Spec Max}\,(U^{\text{rigid}}) = \{\alpha \in \text{Spec Max}\,(S^{\text{rigid}}) \mid |k(\alpha)| = 1\}.$$

On dit qu'une fonction $h \in U^{\text{rigid}}$ donne un formule au-dessus de U^{rigid} pour une des racines de zêta si pour tout entier $v \geq 1$ et tout point α de $U^\dagger \otimes Q$ à valeurs dans Ω qui vérifie $\varphi^{(\nu)}(\alpha) = \alpha$, l'endomorphisme $\mathfrak{F}^{(\nu)}$ de $\check{\mathfrak{H}}_\alpha$ admet comme valeur propre

$$h(\alpha)h(\varphi(\alpha)) \ldots h(\varphi^{(\nu-1)}(\alpha)). \tag{5.11}$$

(5.2) En prenant tous les $h \in U^{\text{rigid}}$ qui donnent une formule au-dessus de U^{rigid}, et identifiant ceux qui donnent la même formule, on obtient, par définition, l'ensemble Form(U^{rigid}) des *formules principales* au-dessus de U^{rigid}.

(5.3) Soit $\mathbf{P}(\mathfrak{H})$ le fibré des droites dans $\check{\mathfrak{H}}$. Pour chaque ouvert (†) principal $U^\dagger$, on pose, par abus de notation:

(5.31) $\mathbf{P}(\mathfrak{H})(U^{\text{rigid}}) =$ l'ensemble des orbites dans $\check{H} \otimes_{S^\dagger \otimes Q} U^{\text{rigid}} - \{0\}$ de la multiplication par $(U^{\text{rigid}})^*$.

(5.4) Désignons par $\mathbf{P}(\mathfrak{H})^{Д}$ le sous-foncteur "droites stable par Д":

(5.41) $\mathbf{P}(\mathfrak{H})^{Д}(U^{\text{rigid}}) =$ le sous-ensemble de $\mathbf{P}(\mathfrak{H})(U^{\text{rigid}})$ formé des éléments représentées par un élément $\tau \in \check{\mathfrak{H}} \otimes U^{\text{rigid}} - \{0\}$ qui vérifie $(d/d\lambda)(\tau) = k\tau$ pour un $k \in U^{\text{rigid}}$.

PROPOSITION 5.42. *Soit $\tau \in \check{\mathfrak{H}} \otimes U^{\text{rigid}}$, qui représente un élément de $\mathbf{P}(\mathfrak{H})^{Д}(U^{\text{rigid}})$. Alors τ ne s'annule en aucun point fermé de $U^{\text{rigid}} \otimes \Omega$.*

DÉMONSTRATION. Sinon, il existe un point α de U^{rigid} à valeur dans Ω et un entier $m > 0$ tel que $\tau = (\lambda - \alpha)^m\varphi$, avec $\varphi \in \check{\mathfrak{H}} \otimes U^{\text{rigid}}$ et $\varphi(\alpha) \neq 0$. Par hypothèse, il existe $k \in U^{\text{rigid}}$ avec $Д(d/d\lambda)(\tau) = k\tau$. Faisant la substitution $\tau = (\lambda - \alpha)^m\varphi$, on trouve

$$m(\lambda - \alpha)^{m-1}\varphi + (\lambda - \alpha)^m Д(d/d\lambda)(\varphi) = k(\lambda - \alpha)^m\varphi$$

d'où $m\varphi = -(\lambda - \alpha)Д(d/d\lambda)(\varphi) + (\lambda - \alpha)k\varphi$ ce qui contredit $\varphi(\alpha) \neq 0$.

(5.5) $\mathfrak{F}$ opère (φ-linéarement) sur $\check{\mathfrak{H}}$, $\check{\mathfrak{H}} \otimes U^{\text{rigid}}$, et $\check{\mathfrak{H}} \otimes U^{\text{rigid}} - \{0\}$, (car son noyau est nul) donc sur $\mathbf{P}(\mathfrak{H})(U^{\text{rigid}})$. Etant horizontal, $\mathfrak{F}$ opère sur $\mathbf{P}(\mathfrak{H})^{Д}(U^{\text{rigid}})$. On pose

(5.51) $\mathbf{P}(\mathfrak{H})^{Д,\mathfrak{F}}(U^{\text{rigid}}) =$ l'ensemble des points fixes de $\mathfrak{F}$ opèrant dans

$$\mathbf{P}(\mathfrak{H})^{Д}(U^{\text{rigid}}).$$

(5.6) Nous allons définir une application naturelle

$$\mathbf{P}(\mathfrak{H})^{Д,\mathfrak{F}}(U^{\text{rigid}}) \to \text{Form}\,(U^{\text{rigid}}).$$

Soit $\tau \in \check{\mathfrak{H}} \otimes U^{\text{rigid}}$ un représentant d'un élément de $\mathbf{P}(\mathfrak{H})^{Д,\mathfrak{F}}(U^{\text{rigid}})$. Par définition, il existe $h \in (U^{\text{rigid}})^*$ tel que

(5.61) $$\mathfrak{F}(\tau) = h\tau.$$

Alors h donne un élément de Form (U^{rigid}). Soit en effet un entier ≥ 1, et α un point de U^{rigid} à valeurs dans Ω tel que $\varphi^{(\nu)}(\alpha) = \alpha$. Appliquant $\mathfrak{F}$ aux deux membres de (5.61); on trouve par récurrence

(5.62) $$\mathfrak{F}^{(\nu)}(\tau) = h\varphi(h)\varphi^{(2)}(h)\ldots\varphi^{(\nu-1)}(h)\tau$$

et, évaluant en α (ce qui a un sens, car $\varphi^{(\nu)}(\alpha) = \alpha$)

(5.63) $$\mathfrak{F}^{(\nu)}(\tau(\alpha)) = h(\alpha)h(\varphi(\alpha))\ldots h(\varphi^{(\nu-1)}(\alpha))\tau(\alpha).$$

D'après (5.52), $\tau(\alpha) \neq 0$, de sorte que $h(\alpha)h(\varphi(\alpha))\ldots h(\varphi^{(\nu-1)}(\alpha))$ est valeur propre de l'endomorphisme $\mathfrak{F}^{(\nu)}$ de $\mathfrak{H}_\alpha$. Reste à voir que la formule donnée par h ne dépend que de l'élément de $\mathbf{P}(\mathfrak{H})^{Д,\mathfrak{F}}(U^{\text{rigid}})$ représenté par τ. Si l'on remplace τ par $u\tau$, avec $u \in (U^{\text{rigid}})^*$, alors

(5.64) $$\mathfrak{F}(u\tau) = ((\varphi(u)/u)h)(u\tau)$$

et on voit bien que h et $(\varphi(u)/u)h$ donnent la même formule.

(5.7) Réciproquement, si deux éléments h_1, $h_2 \in (U^{\text{rigid}})^*$ donnent la même formule, on peut montrer qu'il existe $u \in (U^{\text{rigid}})^*$ avec $h_2 = (\varphi(u)/u)h_1$ [3, théorèmes].

(5.8) En pratique, tout formule provient d'une section de $\mathbf{P}(\mathfrak{H})^{Д,\mathfrak{F}}$ via (5.6).

(6.0) La construction des formules: première méthode.

(6.1) Soient α un point de U^{rigid} à valeurs dans Ω qui vérifie $\varphi(\alpha) = \alpha$ et $\tau \in \check{\mathfrak{H}} \otimes U^{\text{rigid}}$, un représentant d'un élément de $\mathbf{P}(\mathfrak{H})^{Д,\mathfrak{F}}(U^{\text{rigid}})$. On a

(6.11) $$Д(d/d\lambda)(\tau) = k\tau \quad \text{avec } k \in U^{\text{rigid}},$$

(6.12) $$\mathfrak{F}(\tau) = h\tau \quad \text{avec } h \in (U^{\text{rigid}})^*.$$

PROPOSITION 6.13. *La donnée d'un τ comme ci-dessus équivaut à la donnée des objets suivants*:

1. *une boule (dans l'espace analytique "mou" $S^\infty \otimes \Omega$) Δ_α arbitrairement petite, centrée en α;*

2. *une section horizontale η de $\check{\mathfrak{H}} \mid \Delta_\alpha$, qui, en tant qu'élément de $\check{\mathfrak{H}}(\alpha, Д)$* (cf. (4.21)) *est vecteur propre de $\mathfrak{F}$;*

3. *une fonction ω holomorphe et partout non-nulle dans Δ_α, telle que $\omega(\alpha) = 1$; vérifiant*

(a) *$\omega\eta$ est la restriction à Δ_α d'un élément de $\check{\mathfrak{H}} \otimes U^{\text{rigid}}$ (à savoir, τ);*

(b) *$\varphi(\omega)/\omega$ est la restriction à Δ_α d'un élément de $(U^{\text{rigid}})^*$ (à savoir, $(1/h(\alpha))h$);*

(c) *$\omega'(\lambda)/\omega(\lambda)$ est la restriction à Δ_α d'un élément de U^{rigid} (à savoir, k).*

REMARQUE (6.131). On peut démontrer que la condition (c) est conséquence de (b), mais nous ne nous servirons pas de ce fait ici [3, Lemmes 3.1 et 3.4].

DÉMONSTRATION. Soit τ comme en (6.1). On pose

$$\omega(\lambda) = \exp\left(\int_\alpha^\lambda k(t)\,dt\right), \text{ série des puissances en } (\lambda - \alpha), \tag{6.14}$$

$$\eta = \tau/\omega(\lambda). \tag{6.15}$$

Alors les conditions (a) et (c) sont évidemment satisfaites. Quant à (b), remarquons que par hypothèse

$$\mathfrak{F}(\eta) = \epsilon\eta, \qquad \epsilon \text{ constante convenable,} \tag{6.16}$$

ce qui donne

$$\mathfrak{F}(\eta) = \mathfrak{F}\left(\frac{1}{\omega}\tau\right) = \frac{1}{\varphi(\omega)}\mathfrak{F}(\tau) = \frac{h}{\varphi(\omega)}\tau = \frac{h\omega}{\varphi(\omega)}\eta, \tag{6.17}$$

et donc

$$h \cdot \omega/\varphi(\omega) = \epsilon. \tag{6.18}$$

Prenant les valeurs en α, compte tenu de ce que $\varphi(\omega)(\alpha) = \omega(\varphi(\alpha)) = \omega(\alpha)$, on trouve que $h(\alpha) = \epsilon$, d'où (b) par (6.18).

Donnons-nous réciproquement η et ω. On pose

$$\tau = \omega\eta, \tag{6.19}$$

$$\epsilon = \text{la constante telle que } \mathfrak{F}(\eta) = \epsilon\eta, \tag{6.191}$$

$$h = \epsilon\varphi(\omega)/\omega, \tag{6.110}$$

$$k = \omega'(\lambda)/\omega(\lambda). \tag{6.111}$$

Alors

$$\text{Д}\left(\frac{d}{d\lambda}\right)(\tau) = \text{Д}\left(\frac{d}{d\lambda}\right)(\omega\eta) = \omega'(\lambda)\eta = \frac{\omega'(\lambda)}{\omega(\lambda)}\tau = k\tau \tag{6.112}$$

et

$$\mathfrak{F}(\tau) = \mathfrak{F}(\omega\eta) = \varphi(\omega)\epsilon\eta = \frac{\varphi(\omega)}{\omega}\epsilon\tau = h\tau \tag{6.113}$$

et on conclut, par "l'unicité de prolongement analytique", que les relations

$$(d/d\lambda)(\tau) = k\tau, \tag{6.112$'$}$$

$$\mathfrak{F}(\tau) = h\tau \tag{6.113$'$}$$

démontrées au-dessus de Δ_α restent valable au-dessus de U^{rigid}.

(7.0) La première méthode dans un cas particulier.

(7.1) Supposons $\check{\mathfrak{H}}$ libre, et qu'une base en ait été choisi. Ceci identifie $\check{\mathfrak{H}}$ à $(S^\dagger \otimes Q)^N$; on représentera ses sections par des "vecteurs colonnes"

$$(7.11)\qquad \begin{pmatrix} f_0 \\ \vdots \\ f_{N-1} \end{pmatrix}, \qquad \text{avec } f_i \in S^\dagger \otimes Q.$$

(7.2) Supposons de plus que la connection ∇ provienne d'une équation différentielle d'ordre N, i.e. que la connection s'écrive

$$(7.21)\qquad \nabla\left(\frac{d}{d\lambda}\right) \begin{matrix} f_0 \\ \\ \vdots \\ \\ f_{N-1} \end{matrix} = \begin{matrix} f'_0 \\ \\ \vdots \\ \\ f'_{N-1} \end{matrix} - \begin{matrix} 0 & 1 & 0 & \ldots & & 0 \\ 0 & 0 & 1 & 0 & \ldots & 0 \\ & & & & & \\ 0 & & \ldots & & 0 & 1 \\ a_0 & a_1 & & \ldots & & a_{N-1} \end{matrix} \quad \begin{matrix} f_0 \\ \\ \vdots \\ \\ f_{N-1} \end{matrix} .$$

Une section horizontale s'écrit

$$(7.22)\qquad \begin{matrix} f \\ f' \\ \vdots \\ f^{(N-1)} \end{matrix}, \qquad \text{où } f^{(\nu)} = \frac{d^\nu f}{d\lambda^\nu}$$

et où f est une solution de l'équation de Picard-Fuchs:

$$(7.23)\qquad \frac{d^N f}{d\lambda^N} = \sum_{i=0}^{N-1} a_i \frac{d^i f}{d\lambda^i}.$$

Explicitons la proposition (6.13) en termes de la fonction f, supposée être la première coordonée (7.22) d'une section horizontale η de $\check{\mathfrak{H}}$ au-dessus de Δ_α, vecteur propre de $\mathfrak{F}$:

$$(7.24)\qquad \mathfrak{F}(\eta) = \epsilon\eta.$$

Soit ω une fonction sur Δ_α qui vérifie (a), (b), et (c) de (6.13).

Ecrivons

$$(7.25)\qquad \tau = \begin{matrix} \tau_0 \\ \\ \vdots \\ \\ \tau_{N-1} \end{matrix} = \omega\eta = \begin{matrix} \omega f \\ \vdots \\ \omega f' \\ \vdots \\ \omega f^{(N-1)} \end{matrix} .$$

(7.26) On suppose que $f(\alpha) = 1$, et que $\tau_0 = \omega f \in (U^{\text{rigid}})^*$.

Alors, si l'on remplace ω *par* $1/f$, *les conditions de* (6.13) *restent vérifiées.*

En effet, par (7.25) on a

$$(7.27)\qquad \frac{1}{f}\eta = \begin{matrix} 1 \\ f'/f \\ \vdots \\ f^{(N-1)}/f \end{matrix} = \begin{matrix} 1 \\ \tau_1/\tau_0 \\ \vdots \\ \tau_{N-1}/\tau_0 \end{matrix}$$

ce qui donne (a) de (6.13). Pour (b), on note que

$$(7.28)\qquad \frac{\varphi(1/f)}{1/f} = \frac{f}{\varphi(f)} = \frac{\tau_0}{\varphi(\tau_0)} \cdot \frac{\varphi(\omega)}{\omega}$$

et, pour (c), que

$$(7.29)\qquad \frac{(1/f)'}{1/f} = -\frac{f'}{f} = \frac{\omega'}{\omega} - \frac{\tau'}{\tau}.$$

(7.210) Réciproquement, soient α un point de U^{rigid} à valeurs dans Ω, tel que $\varphi(\alpha) = \alpha$, et f une fonction holomorphe et partout non-nulle dans Δ_α. Supposons que

(0) $\eta = \begin{pmatrix} f \\ f' \\ \vdots \\ f^{(N-1)} \end{pmatrix}$ est une section horizontale de $\check{\mathfrak{H}} \mid \Delta_\alpha$,

(1) $\mathfrak{F}(\eta) = \epsilon\eta$,

(2) $\varphi(f)/f$ se prolonge en un élément de $(U^{\text{rigid}})^*$,

(3) f'/f se prolonge en un élément de U^{rigid} (cf. (6.131)).

Alors le prolongement de $\epsilon f/\varphi(f)$ donne une formule (cf. (5.0)) au-dessus de U^{rigid}.

Vérifions les conditions (a), (b), et (c) de (6.13) pour $\omega = 1/f$. Les conditions (b) et (c) resultent de (2) et (3). La condition (a), que $f^{(\nu)}/f$ se prolonge en un élément de U^{rigid} resulte de (3) et de la formule

$$(7.211)\qquad f^{(\nu+1)}/f = (f^{(\nu)}/f)' + (f^{(\nu)}/f)(f'/f).$$

(7.212) Remarque. La condition (1) de (7.210) est difficile à vérifier. Une des vertus de la deuxième méthode (ci-dessous) est d'éviter ce problème.

(8.0) La construction des formules: deuxième méthode.

(8.1) Reprenons les notations de (1.1 bis) et (1.2). Choisissons un relèvement $\varphi\colon S^\dagger \to S^\dagger$ de Frobenius qui provienne d'un relèvement $\varphi\colon \mathfrak{O}[\lambda]^\dagger \to \mathfrak{O}[\lambda]^\dagger$, de sorte qu'on puisse parler de $\varphi(\alpha)$ pour α un point de $\mathfrak{O}[\lambda]^\dagger \otimes Q$, à valeurs dans Ω même non contenu dans Spec $(S^\dagger \otimes Q)$.

(8.2) Rappelons que $S^\dagger$ avait été obtenu comme le (†) localisé de $\mathfrak{O}[\lambda]^\dagger$ en g, polynôme monique dans $\mathfrak{O}[\lambda]$. Comme $\mathfrak{F}$ est "défini sur $S^\dagger \otimes Q$", il résulte de la définition de $S^\dagger$ qu'il existe $\epsilon > 0$ tel que $\mathfrak{F}\colon \check{\mathfrak{H}} \to \check{\mathfrak{H}}$ et φ se prolongent au-dessus de

(8.21) $\mathfrak{D}_\epsilon$ = la sous-algèbre de $S^\dagger \otimes Q$ formée des fonctions "définies" dans la région $|\lambda| < 1 + \epsilon$, $|g(\lambda)| > 1 - \epsilon$.

(8.22) Reprenons les notations de (5.101), et soit $k \in S^\dagger \cap \mathfrak{D}_\epsilon$. Par "l'ouvert rigid $U_\epsilon^{\text{rigid}}$ de $\mathfrak{D}_\epsilon$ défini par $|k| = 1$", on entend la complété de $\mathfrak{D}_\epsilon[1/k]$ pour la topologie de la convergence uniforme sur $\{x \in \text{Spec Max}\,(\mathfrak{D}_\epsilon) \mid |k(x)| = 1\}$.

(8.3) L'interprétation de la formule de Picard-Lefschetz [4].

(8.31) Supposons que $g(0) = 0$, et que 0 est un point fixe de φ. Supposons que la fibre en 0 de $\tilde{f}$ (cf. (1.12)) ait un point quadratique non-dégénéré comme seule singularité. Heuristiquement, la formule de Picard-Lefschetz nous suggère ceci:

(8.311) Posons R_0 = l'anneau des séries de Laurent en λ convergentes dans la rondelle $1 - \epsilon < |\lambda| < 1$.

(8.32) *Cas* 1. Si la dimension des fibres est paire et si $\tilde{X} \otimes Q$ est régulier, il existe une section non-nulle γ de $\check{\mathfrak{H}} \otimes R_0$ telle que $\lambda^{1/2}\gamma$ soit horizontal; cette section γ est unique au produit par une constante près. $\lambda^{1/2}\gamma$ "est" le cycle évanescent en 0; il se multiplie par -1 quand on tourne autour de 0.

(8.33) *Cas* 2. Si la dimension de la fibre est impaire, il existe une section γ de $\check{\mathfrak{H}} \otimes R_0$ unique à un multiple constant près et une section β de $\check{\mathfrak{H}} \otimes R_0$ telle que $\beta + \log(\lambda)\gamma$ soit horizontal. Cette section γ est le cycle évanescent en 0, par un multiple duquel toute section horizontale augmente quand on tourne autour de 0. (N.B. Il peut arriver que γ, encore unique, soit nulle.)

(8.34) REMARQUE. On dit que $\beta + \log(\lambda)\gamma$ ou $\lambda^{1/2}\gamma$, est une section horizontale s'il devient tel lorsqu'on le restreint à une petite boule centrée en n'importe quel point fermé de R_0; dans une telle boule $\log(\lambda)$ et $\lambda^{1/2}$ se définissent comme solutions des équations différentielles

$$f'(\lambda) = 1/\lambda \qquad \text{pour } \log(\lambda),$$
$$f'(\lambda) = f/2\lambda \qquad \text{pour } \lambda^{1/2}.$$

Admettons (8.32) et (8.33), et appliquons $\mathfrak{F}$ aux cycles évanescents.

(8.35) *Cas* 1. On suppose q impaire, d'ou $\lambda^{(q-1)1/2} \in \mathfrak{D}_\epsilon$. L'élément $\varphi(\lambda)/\lambda^q$ de $\mathfrak{D}_\epsilon$, étant près de 1, admet une racine carrée dans $\mathfrak{D}_\epsilon$. La formule

$$\mathfrak{F}(\lambda^{1/2}\gamma) = (\varphi(\lambda))^{1/2}\mathfrak{F}(\gamma) = \lambda^{1/2}(\lambda^{(q-1)/2}(\varphi(\lambda)/\lambda^q)^{1/2}\mathfrak{F}(\gamma))$$

met la section horizontal $\mathfrak{F}(\lambda^{1/2}\gamma)$ sous la forme $\lambda^{1/2}$ (une section de $\check{\mathfrak{H}} \otimes R_0$), donc, par l'assertion d'unicité de (8.32), on a

(8.351) $\mathfrak{F}(\lambda^{1/2}\gamma) = \epsilon\lambda^{1/2}\gamma$ où ϵ est une constante, soit

(8.352) $$\mathfrak{F}(\gamma) = \epsilon\gamma/\lambda^{(q-1)/2}(\varphi(\lambda)/\lambda^q)^{1/2}.$$

(8.36) *Cas* 2. La fonction $\varphi(\lambda)/\lambda^q$, étant près de 1, admet une logarithme dans $\mathfrak{D}_\epsilon$. On a

$$\mathfrak{F}(\beta + \log(\lambda)\gamma) = \left(\mathfrak{F}(\beta) + \log\left(\frac{\varphi(\lambda)}{\lambda^q}\right)\mathfrak{F}(\gamma)\right) + \log(\lambda)(q\mathfrak{F}(\gamma)).$$

D'après l'assertion d'unicité de (8.33), on a

(8.361) $$\mathfrak{F}(\gamma) = \epsilon\gamma, \qquad \text{où } \epsilon \text{ est une constante.}$$

Soit $U_\epsilon^{\mathrm{rigid}}$ un ouvert rigid de $\mathfrak{D}_\epsilon$, de type considéré en (8.22). Les méthodes de (6.13) nous donnent

(8.37) CONCLUSION. Dans le cas 1 ainsi que dans le cas 2 si $\gamma \neq 0$, si l'on peut trouver une fonction ω sur R_0 telle que

(a) $\omega\gamma$ se prolonge en une section de $\check{\mathfrak{H}} \otimes U_\epsilon^{\text{rigid}}$,

(b) $\varphi(\omega)/\omega$ se prolonge en un élément de $(U_\epsilon^{\text{rigid}})^*$ (cf. (6.131))

alors le prolongement de la fonction

$$(8.371) \qquad \begin{array}{ll} \epsilon\varphi(\omega)/\omega\lambda^{(q-1)/2}(\varphi(\lambda)/\lambda^q)^{1/2} & \text{dans le cas 1,} \\ \epsilon\varphi(\omega)/\omega \quad \text{dans le cas 2} & \end{array}$$

nous fournit une formule (cf. (5.0)) au-dessus de U^{rigid}.

(8.38) REMARQUE. Pour que la conclusion (8.37) soit valable dans un cas particulier, il suffit de vérifier qu'il existe une solution γ "meilleur que les autres" dans une rondelle autour de zero, qui est un vecteur propre de $\mathfrak{F}$. Le rôle de la formule de Picard-Lefschetz est de nous signaler le cycle évanescent comme étant "meilleur que les autres".

(9.0) Un exemple.

(9.1) La famille des courbes elliptiques de Legendre, donnée en équation affine par

$$(9.11) \qquad y^2 = x(x-1)(x-\lambda)$$

est lisse au-dessus de $\tilde{S} = \mathfrak{O}[\lambda][1/\lambda(1-\lambda)]$, pour $\mathfrak{O}$ de caractéristique résiduelle autre que deux. Le module $\mathfrak{H}$ est libre de rang deux et, si ω est la classe de la différentielle invariante dx/y, il admet comme base ω et $\nabla(d/d\lambda)(\omega)$. La connection s'exprime, dans $\mathfrak{H}$, par

$$(9.12) \qquad \lambda(1-\lambda)\left(\nabla\left(\frac{d}{d\lambda}\right)\right)^2(\omega) + (1-2\lambda)\nabla\left(\frac{d}{d\lambda}\right)(\omega) - \frac{1}{4}\omega = 0.$$

Prenant la base de $\check{\mathfrak{H}}$ duale de cette base de $\mathfrak{H}$, on se trouve dans la situation de (7.1) et (7.2). L'équation (7.23) pour la première coordonnée d'une section horizontale s'écrit

$$(9.13) \qquad \lambda(1-\lambda)\frac{d^2f}{d\lambda^2} + (1-2\lambda)\frac{df}{d\lambda} - \frac{1}{4}f = 0.$$

Dans toute rondelle $1 - \epsilon < |\lambda| < 1$, le cycle évanescent au sens de (8.33), est

$$(9.14) \qquad \gamma = \begin{pmatrix} f \\ f' \end{pmatrix}$$

avec

$$(9.15) \qquad f(\lambda) = f(1/2, 1/2, 1, \lambda) = \sum_{j \geq 0} \binom{-1/2}{j}^2 \lambda^j.$$

On peut démontrer que γ/f se prolonge au-dessus de l'ouvert rigid $U_\epsilon^{\text{rigid}}$ défini par $|H(\lambda)| = 1$, où

$$H(\lambda) = \sum_{j=0}^{p-1} \binom{-1/2}{j}^2 \lambda^j$$

est le polynôme de Hasse [3], [4]. Le prolongement en $U_\epsilon^{\text{rigid}}$ de la fonction

(9.16) $$(-1)^{(p-1)/2}\frac{f(1/2, 1/2, 1, \lambda)}{f(1/2, 1/2, 1, \lambda^p)}$$

donne une formule pour l'une des racines (à savoir, celle qui est unité p-adique) de la fonction zêta des courbes (9.11) qui ne sont pas "supersingulières".

(9.17) Le cycle évanescent en $\lambda = 1$ est

(9.18) $$\eta = \begin{pmatrix} g \\ g' \end{pmatrix}$$

avec

(9.19) $$g(\lambda) = f(1/2, 1/2, 1, 1 - \lambda).$$

On démontre encore que η/g se prolonge au-dessus de $U_\epsilon^{\text{rigid}}$ et, oh miracle, on y trouve

(9.20) $$\gamma/f = -\eta/g.$$

Les cycles évanescents en $\lambda = 0$ et $\lambda = 1$ définissent donc le *même* élément de $\mathbf{P}(\mathfrak{H})^{\text{Д}, \mathfrak{F}}(U_\epsilon^{\text{rigid}})$.

(9.21) A l'infini il y a aussi un cycle "meilleurs que les autres", à savoir

(9.22) $$\mu = \begin{pmatrix} s \\ s' \end{pmatrix}$$

avec

(9.23) $$s(\lambda) = \lambda^{-1/2} f(1/2, 1/2, 1, \lambda^{-1})$$

et μ/s se prolonge au-dessus de $U_\epsilon^{\text{rigid}}$. De plus,

(9.24) $$\gamma/f = -\eta/g = \mu/s.$$

(9.25) Les cycles évanescents provenant des trois fibres singulières, en $\lambda = 0$, 1, ∞ "donnent" le *même* sous-module de $\mathfrak{H}$ au-dessus des courbes non-supersingulières. Ceci semble tout à fait différent de la théorie "classique" (sur les complexes, ou en cohomologie l-adique) de la "monodromie" agissant sur la cohomologie d'une fibre géométrique de cette famille, où les fibres supersingulières ne jouent aucun rôle exceptionel, et où les cycles évanescents engendrent toute la cohomologie d'une fibre géométrique.

REFERENCES

1. B. Dwork, *On the zeta function of a hypersurface*, Inst. Hautes Études Sci. Publ. Math. No. 12 (1962), 5–68. MR **28** #3039.

2. ———, *On the zeta function of a hypersurface.* II, Ann. of Math. (2) **80** (1964), 227–299. MR **32** #5654.

3. ———, *P-adic cycles*, Inst. Hautes Études Sci. Publ. Math. (to appear).

4. S. Lefschetz, *L'analysis situs et la géométrie algébrique*, Gauthier-Villars, Paris, 1924.

5. J. Tate, *Rigid analytic spaces*, Inst. Hautes Études Sci. (notes reproduced without the permission of the author, 1961).

6. G. Washnitzer and P. Monsky, *Formal cohomology.* I, Ann. of Math. (2) **88** (1968), 181–217.

PRINCETON UNIVERSITY
PRINCETON, NEW JERSEY 08540

THE INTEGRAL CLASSICAL GROUPS AND THEIR AUTOMORPHISMS

O. T. O'MEARA[1]

There is ample evidence to suggest that the method of involutions, which has been used so effectively for the last twenty years in the automorphism theory of the classical groups, has been pushed as far as it can go. The method works very well as long as there are many involutions in the classical group G under consideration, for instance, whenever G is the full linear group GL_n or the full symplectic group Sp_n over a field, but its efficiency diminishes as the number of involutions decreases. To take an extreme example, there are classical groups which contain no involutions and in which the method of involutions cannot therefore be used. If G is defined over an integral domain that is not a field it will, in general, contain very few involutions and the method of involutions will rarely work. When it does, it is extremely difficult to apply.

During the last two or three years methods have been developed which do not depend on the use of involutions and the main purpose of this talk is to describe one of these methods by using it to find the automorphisms of the orthogonal groups and their congruence subgroups over arithmetic domains.[2] In addition I will give a brief summary and history of the automorphism problem of the classical groups.

1. Statement of the problem

Throughout the talk V will be an n-dimensional vector space over a (commutative) field F. As usual, $GL_n(V)$ denotes the general linear group and $SL_n(V)$ the special linear group of V. If V is provided with a regular alternating bilinear form $B: V \times V \rightarrow F$, then the set of elements σ in $GL_n(V)$ which preserve B,

---

[1]Supported in part by the National Science Foundation under grants GP-8366 and GP-11 342.

[2]Complete details and proofs can be found in [17].

i.e., for which $B(\sigma x, \sigma y) = B(x, y)$ for all x, y in V, is a subgroup of $SL_n(V)$ called the symplectic group determined by B. It is written $Sp_n(V)$. If B is symmetric instead of alternating, and if the characteristic is not 2, the corresponding group is called the orthogonal group with respect to B and is written $O_n(V)$. Define

$$O_n^+(V) = O_n(V) \cap SL_n(V), \qquad \Omega_n(V) = DO_n(V),$$

where DX denotes the commutator subgroup of a group X. The orthogonal groups of characteristic 2, and the unitary groups $U_n(V)$ and $U_n^+(V)$, can be defined in somewhat similar ways [6]. So we have four families of groups and we list them in Table I.

linear	GL_n, SL_n
symplectic	Sp_n
orthogonal	O_n, O_n^+, Ω_n
unitary	U_n, U_n^+

TABLE 1

A classical group over F is any one of the groups in the table. By the projective classical groups of F we mean the groups PG where P denotes the canonical homomorphism of $GL_n(V)$ onto $GL_n(V)/\text{cen}\, GL_n(V)$. As an important side remark we mention that, under suitable assumptions, the projective classical groups

$$PSL_n, \qquad PSp_n, \qquad P\Omega_n, \qquad PU_n^+$$

are simple [6].

Now let $\mathfrak{o}$ be an integral domain, properly contained in F, with quotient field F. Let M be an $\mathfrak{o}$-module in V of the form $M = \mathfrak{o}x_1 + \cdots + \mathfrak{o}x_n$ with $x_1, \ldots, x_n$ a base for V. (Actually M can be somewhat more general, but this is of little significance to this talk.) Define the integral linear groups by the equations

$$GL_n(M) = \{\sigma \in GL_n(V) \mid \sigma M = M\}, \qquad SL_n(M) = GL_n(M) \cap SL_n(V).$$

If V has a regular symmetric bilinear form B and the characteristic is not 2, define the corresponding integral orthogonal groups by

$$O_n(M) = \{\sigma \in O_n(V) \mid \sigma M = M\}, \qquad O_n^+(M) = O_n(M) \cap O_n^+(V).$$

Similarly with the other classical groups. The groups obtained in this way are called the integral classical groups. The projective integral classical groups are the ones obtained by applying P.

For a nonzero integral ideal $\mathfrak{a}$ of $\mathfrak{o}$, the groups

$$O_n(M; \mathfrak{a}) = \{\sigma \in O_n(M) \mid (\sigma - 1_V)M \subseteq \mathfrak{a}M\},$$
$$O_n^+(M; \mathfrak{a}) = O_n(M; \mathfrak{a}) \cap O_n^+(V)$$

are called the congruence subgroups of the integral orthogonal group $O_n(M)$ determined by the ideal $\mathfrak{a}$. Note that $O_n(M)$ is itself a congruence group since it is equal to $O_n(M; \mathfrak{o})$. The same things can be done for the other classical groups and we then obtain the congruence subgroups of the integral classical groups. To define the projective congruence groups, apply P.

All these groups have simple matrix interpretations. For example, the group $O_n^+(M; \mathfrak{a})$ corresponds, via the base $x_1, \ldots, x_n$ for M, to the matrix group

$$\{X \in SL_n(\mathfrak{o}) \mid {}^tXAX = A,\ X \equiv I \bmod \mathfrak{a}\}$$

where $SL_n(\mathfrak{o})$ is the group of matrices of determinant 1 with entries in $\mathfrak{o}$, and A is the symmetric matrix $A = (B(x_i, x_j))$.

We are ready to pose the problem. Let G be a classical group, or an integral classical group, or a congruence subgroup, and let Λ be an automorphism of G. Then what does Λ look like? We will see later that we are still far from a complete solution. But when a solution is known, it almost always takes on a certain "expected form" which we now describe. In the nonlinear case we "expect" Λ to have the form

$$\Lambda = P_\chi \circ \Phi_g$$

where P_χ is an automorphism of G of the form

$$P_\chi(\sigma) = \chi(\sigma) \cdot \sigma, \qquad \forall \sigma \in G$$

for some homomorphism χ of G into its center, and Φ_g is an automorphism of G of the form

$$\Phi_g(\sigma) = g\sigma g^{-1}, \qquad \forall \sigma \in G$$

for some semilinear isomorphism g of V onto V. In the linear case we expect Λ to be either of the above form or of the form

$$\Lambda = P_\chi \circ \Psi_h$$

where P_χ is as before, and Ψ_h is an automorphism of G of the form

$$\Psi_h(\sigma) = h^{-1}\check{\sigma}h, \qquad \forall \sigma \in G$$

for some semilinear isomorphism h of V onto the dual space V', where $\check{\sigma}$ denotes the contragredient $\check{\sigma} = {}^t\sigma^{-1}$ of σ. The few known true exceptions to the expected form are for small n.

In the nonlinear case the semilinear g in the expected form will, in general, have the additional property of preserving orthogonality. The P_χ are called radial automorphisms of G. They are almost trivial since $\chi(\sigma)$ will, in general, be a scalar matrix. For example, $P_\chi(\sigma) = \pm\sigma$ in orthogonal and symplectic situations.

Similar, but not identical, things can be expected for the projective groups.

2. The automorphism theory over fields

The automorphism theory of the classical groups began with the paper of Schreier-van der Waerden in 1928 on the automorphisms of PSL_n over (commutative) fields. The matter then lay dormant until the appearance of two groundbreaking works by Dieudonné and Rickart in 1950/1951. Their methods, which depend on the use of involutions and certain double centralizer techniques of Mackey, have subsequently been exploited by several authors to the point where the automorphism question for the classical groups over (commutative) fields can be regarded as settled. Only questions in characteristic 2 and some difficulties in low dimensions remain. The status is as follows.

1. Linear and symplectic cases settled. All automorphisms of expected type, but for some exceptional behavior for Sp_4 over certain fields of characteristic 2. See Dieudonné [6] for the full story.

2. The automorphisms of O_n and U_n are of expected type when $n \geqq 3$ and the characteristic is not 2. See Dieudonné [6] again. The same for O_n^+ and U_n^+ when $n \geqq 5$. Here refer to Wonenburger [32], [33].

3. Spiegel [26] has some results for U_n in characteristic 2. There is also some information about the above low-dimensional gaps in [6].

By and large, all the results were proved using the method of involutions. The orthogonal group $\Omega_n(V)$ does not, however, succumb to this treatment. Examples are constructed in [16] of certain $\Omega_n(V)$ with "anisotropic" V which contain no involutions other than the trivial one. This means that the method of involutions cannot be applied to $\Omega_n(V)$ in general. An involution-free method was introduced in [16] to handle these groups and it subsequently developed (see §3) that this method could also be used to by-pass the shortage of involutions in the integral classical groups and their congruence subgroups. I shall refer to this involution-free approach as the method of residual spaces and I will give an illustration of its application later in the talk. Using it one can show

4. The automorphisms of Ω_n are of expected type when the characteristic is not 2 and $n \geqq 7$ with $n \neq 8$. See O'Meara [16].

Analogous, but not identical, things can be said about the projective classical groups over fields.

The automorphisms of the Chevalley groups have been studied over various fields by Steinberg [28], [29] and Humphreys [9]. These groups are related to certain of the classical groups, roughly to the linear case and the cases of "maximal Witt index". The result is that, with certain assumptions on the Chevalley group G, every automorphism of G is "the composite of inner, graph, field and diagonal automorphisms". Borel-Tits [2] are developing a homomorphism (and automorphism) theory for a much wider class of groups using the theory of algebraic groups. Their results are also related to the classical groups, roughly to the linear groups and the cases of "positive Witt index", i.e., the isotropic cases.

3. The automorphism theory over integral domains

The automorphism question over integral domains was first raised by Rickart [22] in 1951. First results were by Hua-Reiner [8] for the linear groups over **Z**. Then Landin-Reiner extended the theory to GL_n over the Gaussian integers in [10], and over any principal ideal domain of characteristic not 2 for $n \geqq 3$ in [11]. Similar things were done by Wan. Yan [34] obtained results on the isomorphisms of the group generated by the elementary matrices over an integral domain of characteristic not 2. In 1966, Cohn [3] found the isomorphisms of GL_n over certain "k-rings with a degree function" associated with a field k. Also in 1966, O'Meara [14] showed that all automorphisms of GL_n and SL_n over an integral domain of any characteristic are of expected type when $n \geqq 3$. Proofs in [14] were based on complicated applications of the method of involutions. Short conceptual proofs were subsequently given by O'Meara [18] using the method of residual spaces, and by Zassenhaus [35] using group-theoretic analogues of theorems about Lie algebras and their representations. These results were applied to the linear congruence groups over Dedekind domains by O'Meara-Zassenhaus [19]. Using residual spaces, Solazzi [24] determined the automorphisms of the projective linear congruence groups over integral domains for $n \geqq 3$.

The 2-dimensional linear groups behave rather badly over integral domains. First of all, their automorphisms are known only for some special domains. And secondly, exceptional automorphisms are quite common. Some form of the Euclidean algorithm seems to play a role in this branch of the theory. Very roughly speaking, the situation is this. Landin-Reiner [10], Borel [1] and Cohn [4] have number-theoretic results, Landin-Reiner [12] work over certain specialized Euclidean domains, Reiner [21] constructs exceptional automorphisms over the polynomial ring $k[X]$ and then finds all the automorphisms for this case, Cohn [3] works in those "GE_2-rings" which are also "k-rings with a degree function", and Dull [7] has the automorphisms over "GE_2-domains".

The linear theory is therefore essentially settled for $n \geqq 3$. So is the symplectic case. See Reiner [20] for **Z**, O'Meara [15] for the standard sympletic group over any integral domain when $n \geqq 4$, and Solazzi [25] for the symplectic congruence groups and their projective groups over integral domains when $n \geqq 6$. The symplectic theorems are essentially of expected type, although exceptional behavior seems possible in the 4-dimensional case in characteristic 2.

Borel [1] has found the automorphisms of certain groups of "integral points" of "connected almost simple k-split" algebraic groups satisfying certain conditions, with k an algebraic number field. As a special case of his theorem he obtains special cases of the linear and symplectic theorems stated above, namely, he derives the automorphisms of the linear and standard symplectic groups over the "Hasse domains" of an algebraic number field.

The automorphisms of the integral orthogonal groups and their congruence subgroups were determined by O'Meara [17], 1. over any local domain, and 2. over Hasse domains that are not totally real. This includes the ring of integers of an algebraic number field that is not totally real. Characteristic 2 is excluded. So are some small values of n. In all cases the automorphisms are essentially of expected type. A sketch of the proof, which depends on residual spaces, will be given shortly.

4. The method of involutions

Let us illustrate the method of involutions by applying it to $O_n(V)$. So V is an n-dimensional vector space over a field F of characteristic not 2, and B is a regular symmetric bilinear form. In order to simplify things we assume that V is anisotropic, i.e., that $B(x, x) = 0$ if and only if $x = 0$. However, the theorem remains true without this assumption.

THEOREM. *If* Λ *is an automorphism of* $O_n(V)$ *with* $n \geqq 3$, *then* Λ *has the expected form* $\Lambda = P_\chi \circ \Phi_g$.

PROOF. Each involution σ has the form $\sigma = -1_U \perp 1_W$ for some orthogonal splitting $V = U \perp W$ of V. If either U or W is a line we call σ an extremal involution. The first thing to do is to show that Λ preserves extremal involutions. Now Λ trivially preserves involutions, so we must find some group-theoretic way of separating the extremal involutions from the rest. If $n = 3$, the extremal involutions are the noncentral ones and we are through. So let $n \geqq 4$. Then a reasonably straightforward combinatorial argument on subspaces shows that, for any noncentral involution σ in $O_n(V)$,

$$\max_{\tau} \operatorname{card} C_*C_*(\sigma, \tau) = 8 \quad \text{if } \sigma \text{ extremal,}$$
$$= 16 \quad \text{if not.}$$

Here τ runs over all involutions in $O_n(V)$ that permute with σ, and C_* denotes the involutions in the centralizer in $O_n(V)$. But this is a group-theoretic description of extremal involutions, i.e., Λ induces a one-one correspondence of the set of extremal involutions onto itself. Now the extremal involutions essentially correspond to the lines of V via the splitting $\sigma = -1_U \perp 1_W$, and all lines of V are covered since V is assumed to be anisotropic. Therefore Λ induces a one-one correspondence $L \leftrightarrow L'$ of the set of lines of V onto itself. It can be shown that this correspondence satisfies the conditions of the so-called Fundamental Theorem of Projective Geometry which in turn asserts that the correspondence in question is induced by a semilinear isomorphism g of V onto V, i.e., $gL = L'$ for all lines L in V. Form Φ_g. We find that it is an automorphism of $O_n(V)$. Then $\Lambda \circ \Phi_g^{-1}$ is an automorphism whose resulting line correspondence turns out to be trivial. So $\Lambda \circ \Phi_g^{-1}$ cannot be far from the identity automorphism. In fact it is radial, i.e., $\Lambda \circ \Phi_g^{-1} = P_\chi$, i.e., $\Lambda = P_\chi \circ \Phi_g$.

5. The method of residual spaces

At this point of the talk we must start using, without further reference, several basic facts from algebraic number theory and the arithmetic theory of quadratic forms. Details can be found, for instance, in my book *Introduction to Quadratic Forms* [13].

Let F be any global field of characteristic not 2, Ω the set of nontrivial spots on F, S a Hasse set of spots, and

$$\mathfrak{o} = \bigcap_{\mathfrak{p} \in S} \mathfrak{o}(\mathfrak{p})$$

the corresponding Hasse domain. If, for example, F is an algebraic number field and S consists of all nonarchimedean spots, then S is a Hasse set of spots and $\mathfrak{o}$ is the ring of all algebraic integers in F. Next let V be an n-dimensional regular quadratic space over F with associated bilinear form B. As usual, $V_\mathfrak{p}$ denotes the localization of V at any given spot $\mathfrak{p}$ in Ω. We say that V is indefinite if $V_\mathfrak{p}$ is isotropic for some $\mathfrak{p}$ in $\Omega - S$. If $V_\mathfrak{p}$ is anisotropic at all $\mathfrak{p}$ in $\Omega - S$, we say that V is definite. Finally, let M be the free module $M = \mathfrak{o}x_1 + \cdots + \mathfrak{o}x_n$ where $x_1, \ldots, x_n$ is some base for V, and let $\mathfrak{a}$ be any integral ideal with $0 \subset \mathfrak{a} \subseteq \mathfrak{o}$. I am going to talk about the automorphisms of the group $G = \pm O_n(M; \mathfrak{a})$. The method of involutions cannot be used to study this group since, for instance, it can be shown that $\pm 1_V$ are the only involutions in G whenever $\mathfrak{a} \subset 2\mathfrak{o}$.

We have to exclude certain small values of n. Specifically, we assume throughout that

1. there is a complex spot in $\Omega - S$ and $n \geqq 7$ with $n \neq 8$, or
2. there is a discrete spot in $\Omega - S$ and $n \geqq 9$ with $n \neq 10$, or
3. ind $V_{\mathfrak{q}} \geqq 5$ at some $\mathfrak{q}$ in $\Omega - S$.

So, in particular, V will always be indefinite.

For any σ in $O_n(V)$ define the fixed space of σ as $P = \{x \in V \mid \sigma x = x\}$, the residual space of σ as the orthogonal complement $R = P^*$ of P in V, and the residual index of σ as res σ = dim R. We say that σ is regular if R is regular, degenerate if R is degenerate. Note that σ is a rotation if and only if res σ is an even integer. Call σ a plane rotation if res $\sigma = 2$, i.e., if R is a plane. We have

$$R = (\sigma - 1_V)V, \qquad P = \ker(\sigma - 1_V).$$

Whenever a σ in $O_n(V)$ is under discussion, the letter R will automatically refer to its residual space, the letter P to its fixed space.

We can now give the main idea behind the residual space approach to the automorphism problem. Try to show that an automorphism Λ of G preserves plane rotations, then try to set up a correspondence of planes onto planes, hence of lines to lines by intersecting planes, finally apply the Fundamental Theorem of Projective Geometry as in §4. In actual fact, we cannot prove that Λ preserves all plane rotations, just some of them. And Λ will not induce a correspondence of all planes to all planes, just for some of them. However, this correspondence of planes, restricted though it is, will be extensive enough to induce a correspondence of all lines onto all lines. The result will then follow from the Fundamental Theorem of Projective Geometry.

It is not too difficult to show that G enjoys the following properties.

PROPERTY $\mathcal{P}_1$. If Π is any indefinite anisotropic plane in V, there is a σ of infinite order in G such that $R = \Pi$.

PROPERTY $\mathcal{P}_2$. If Π is a degenerate plane in V, there are infinitely many σ's in G with $R = \Pi$.

PROPERTY $\mathcal{P}_3$. If U is a definite regular subspace of V, the subgroup of $O(U)$ that is induced by elements of G which leave U invariant is finite.

In particular, the indefinite anisotropic planes and the degenerate planes occur as residual spaces of elements of G. It will turn out that there are enough of these planes in V to allow us to perform all the necessary operations of the theory.

We now give a very sketchy outline of how the method of residual spaces can be applied to finding the automorphisms of the given group G. Let Λ be an automorphism of G.

The proof begins with a study of the derived chain D^kG, the essential point being that D^kG always contains plane rotations. Specifically, the following is proved.

5.1. *Suppose $k \geqq 0$. Let U be any indefinite regular ternary subspace of V. Then D^kG contains at least two plane rotations whose residual spaces are distinct planes in U.*

One of the consequences is the following irreducibility theorem.

5.2. *Let* $k \geqq 0$. *If* U *is any subspace of* V *with* $0 \subset U \subset V$, *there is a* σ *in* $D^k G$ *such that* $\sigma U \neq U$.

I have already mentioned that we cannot use Λ to establish a correspondence of all planes onto all planes. Accordingly we say that a plane Π in V behaves under Λ if there is at least one noninvolution σ in G with residual space Π such that $\Lambda\sigma$ is also a plane rotation. We have to show that enough planes behave under Λ. The following two complementary propositions are crucial and make the first move in that direction.

5.3. *Let* σ *be a regular plane rotation in* G *with* P *indefinite. For any*

$$\Sigma_0 \in D^2C(\sigma) - \operatorname{cen} D^2C(\sigma)$$

put

$$X = \langle \{\Sigma\Sigma_0\Sigma^{-1} \mid \Sigma \in C(\sigma)\} \rangle.$$

Then $D^2C(X) = 1_V$.

Here C denotes the centralizer in G.

5.4. *Let* σ *be a regular element of* G *such that* R *and* P *are indefinite with* $\dim R \geqq 3$ *and* $\dim P \geqq 3$. *Then there is at least one plane rotation* Σ_0 *whose residual space is contained in* P *such that*

$$\Sigma_0 \in D^2C(\sigma) - \operatorname{cen} D^2C(\sigma).$$

For any such Σ_0 *put*

$$X = \langle \{\Sigma\Sigma_0\Sigma^{-1} \mid \Sigma \in C(\sigma)\} \rangle.$$

Then $D^2C(X) \neq 1_V$.

In other words, these propositions put the first restrictions on what can happen to a plane rotation under Λ. Very roughly speaking they say that if σ is a certain type of plane rotation, then $\Lambda\sigma$ is either close to being a plane rotation (res $\Lambda\sigma \leqq 2$) or far from it (res $\Lambda\sigma \geqq n - 2$). Using these results and a long series of conjugacy, commutator and degeneracy arguments, it can be shown that

5.5. *At least one indefinite anisotropic plane behaves under* Λ.

So we have one good indefinite anisotropic plane R_1. We are going to move it around V until we find that all such planes are good. This is accomplished using the following. Let R_2 be any indefinite anisotropic plane in V. Proposition: if R_1 and R_2 span a regular ternary subspace of V, then R_2 also behaves under Λ. Proposition: if R_1 and R_2 are orthogonal, then R_2 also behaves under Λ. We conclude that

5.6. *All indefinite anisotropic planes in V behave under Λ,*

and then

5.7. *All degenerate planes in V behave under Λ.*

These arguments, incidentally, give one good reason why methods involving residual planes do not work when n is too small.

A little more good behavior is needed. Define $\mathcal{P}_\infty$ as the set of planes Π in V with the following property: there are infinitely many σ's in G with $R = \Pi$. Then

5.8. *All planes in $\mathcal{P}_\infty$ behave under Λ.*

We now have good behavior on enough planes in V. Use Λ to set up a one-one correspondence $R \leftrightarrow R'$ of $\mathcal{P}_\infty$ onto itself. Then show that every line L in V can be expressed $L = R_1 \cap R_2$ with R_1 and R_2 in $\mathcal{P}_\infty$. Define $L' = R_1' \cap R_2'$. This gives a well-defined one-one correspondence of the lines of V onto themselves. Show that the Fundamental Theorem of Projective Geometry applies and proceed more or less as in §4 to obtain Λ in the form $\Lambda = \mathrm{P}_\chi \circ \Phi_g$.

A somewhat similar result holds for any G that satisfies

$$\mathrm{O}_n^+(M; \mathfrak{a}) \subseteq G \subseteq \mathrm{O}_n(M).$$

A similar theory works for local domains of characteristic not 2.

References

1. A. Borel, *On the automorphisms of certain subgroups of semi-simple Lie groups*, Proc. Conference on Algebraic Geometry (Tata Institute, Bombay, 1968), pp. 43–73.

2. A. Borel and J. Tits, *On "abstract" homomorphisms of simple algebraic groups*, Proc. Conference on Algebraic Geometry (Tata Institute, Bombay, 1968) pp. 75–82.

3. P. M. Cohn, *On the structure of the GL_2 of a ring*, Inst. Hautes Études Sci. Publ. Math. No. 30 (1966), 5–53. MR **34** #7670.

4. ———, *Automorphisms of two-dimensional linear groups over Euclidean domains*, J. London Math. Soc. (2) **1** (1969), 279–292.

5. J. Dieudonné, *On the automorphisms of the classical groups*, Mem. Amer. Math. Soc. No. 2 (1951). MR **13,** 531.

6. ———, *La géométrie des groupes classiques*, 2nd rev. ed., Springer-Verlag, Berlin and New York, 1963. MR **28** #1239.

7. M. H. Dull, *On the automorphisms of the two-dimensional linear groups over integral domains*, Thesis, University of Notre Dame, 1969.

8. L. K. Hua and I. Reiner, *Automorphisms of the unimodular group*, Trans. Amer. Math. Soc. **71** (1951), 331–348. MR **13,** 328.

9. J. E. Humphreys, *On the automorphisms of infinite Chevalley groups*, Canad. J. Math. **21** (1969), 908–911.

10. J. Landin and I. Reiner, *Automorphisms of the Gaussian unimodular group*, Trans. Amer. Math. Soc. **87** (1958), 76–89. MR **21** #2012.

11. ———, *Automorphisms of the general linear group over a principal ideal domain*, Ann. of Math. (2) **65** (1957), 519–526. MR **19,** 388.

12. ———, *Automorphisms of the two-dimensional general linear group over a Euclidean ring*, Proc. Amer. Math. Soc. **9** (1958), 209–216. MR **21** #2013.

13. O. T. O'Meara, *Introduction to quadratic forms*, Die Grundlehren der math. Wissenschaften, Band 117, Academic Press, New York and Springer-Verlag, Berlin and New York, 1963. MR **27** #2485.

14. ———, *The automorphisms of the linear groups over any integral domain*, J. Reine Angew. Math. **223** (1966), 56–100. MR **33** #7427.

15. ———, *The automorphisms of the standard sympletic group over any integral domain*, J. Reine Angew. Math. **230** (1968), 104–138. MR **39** #4291.

16. ———, *The automorphisms of the orthogonal groups $\Omega_n(V)$ over fields*, Amer. J. Math. **90** (1968), 1260–1306. MR **38** #5940.

17. ———, *The automorphisms of the orthogonal groups and their congruence subgroups over arithmetic domains*, J. Reine Angew. Math. **238** (1969), 169–206.

18. ———, *Group-theoretic characterization of transvections using* CDC, Math. Z. **110** (1969), 385–394.

19. O. T. O'Meara and H. Zassenhaus, *The automorphisms of the linear congruence groups over Dedekind domains*, J. Number Theory **1** (1969), 211–221. MR **39** #4292.

20. I. Reiner, *Automorphisms of the symplectic modular group*, Trans. Amer. Math. Soc. **80** (1955), 35–50. MR **17**, 458.

21. ———, *A new type of automorphism of the general linear group over a ring*, Ann. of Math. (2) **66** (1957), 461–466. MR **20** #2380.

22. C. E. Rickart, *Isomorphisms of infinite-dimensional analogues of the classical groups*, Bull. Amer. Math. Soc. **57** (1951), 435–448. MR **13,** 532.

23. O. Schreier and B. L. van der Waerden, *Die Automorphismen der projektiven Gruppen*, Abh. Math. Sem. Univ. Hamburg **6** (1928), 303–322.

24. R. E. Solazzi, *The automorphisms of certain subgroups of $PGL_n(V)$*, Thesis, University of Notre Dame, 1969.

25. ———, *The automorphisms of the symplectic congruence groups* (preprint).

26. E. Spiegel, *On the automorphisms of the unitary group over a field of characteristic* 2, Amer. J. Math. **89** (1967), 43–50. MR **35** #263.

27. ———, *On the automorphisms of the projective unitary group over a field of characteristic* 2, Amer. J. Math. **89** (1967), 51–55. MR **35** #262.

28. R. Steinberg, *Automorphisms of finite linear groups*, Canad. J. Math. **12** (1960), 606–615. MR **22** #12165.

29. ———, *Lectures on Chevalley groups*, Yale University Lecture Notes, 1967.

30. C.-H. Wan, *On the automorphisms of linear groups over a non-commutative Euclidean ring of characteristic* $\neq 2$, Sci. Record **1** (1957), no. 1, 5–8. MR **20** #909.

31. ———, *On the automorphisms of linear groups over a non-commutative principal ideal domain of characteristic* $\neq 2$, Sci. Sinica **7** (1958), 885–933. MR **20** #5237b.

32. M. J. Wonenburger, *The automorphisms of the group of rotations and its projective group corresponding to quadratic forms of any index*, Canad. J. Math. **15** (1963), 302–303. MR **26** #3789.

33. ———, *The automorphisms of $U_n^+(k, f)$ and $PU_n^+(k, f)$*, Rev. Mat. Hisp.-Amer. (4) **24** (1964), 52–65. MR **28** #5115.

34. S.-J. Yan (Yen Shih-chien), *Linear groups over a ring*, Acta Math. Sinica **15** (1965), 455–468 Chinese Math.-Acta **7** (1965), 163–179. MR **36** #5237.

35. H. Zassenhaus, *Characterization of unipotent matrices*, J. Number Theory **1** (1969), 222–230. MR **39** #1568.

UNIVERSITY OF NOTRE DAME
NOTRE DAME, INDIANA

SKEW-SYMMETRIC FORMS FOR NUMBER FIELDS

KENKICHI IWASAWA

Let H be the 1-dimensional real cohomology group of a compact Riemann surface X. For x_1, x_2 in H, put

$$\langle x_1, x_2 \rangle = \int_X \omega_1 \wedge \omega_2,$$

where ω_1 and ω_2 denote real closed 1-forms on X representing the cohomology classes x_1 and x_2 respectively. Then $\langle x_1, x_2 \rangle$ defines a nondegenerate skew-symmetric bilinear form on the real vector space H, which plays an important role in the classical theory of algebraic functions. Following the analogy between algebraic number fields and algebraic function fields, we construct a skew-symmetric bilinear form for a number field which may be regarded as an analogue of the above $\langle x_1, x_2 \rangle$.

Let l be a fixed prime number and let Z_l and Q_l denote the ring of l-adic integers and the field of l-adic numbers respectively. We consider a number field K which is generated over a certain type of finite algebraic number field k by all l^n-th roots of unity for all $n \geqq 0$. Let L be the maximal unramified abelian l-extension over K, and F the field generated over K by all l^n-th roots, $n \geqq 0$, of all units in K. Since the Galois group G of the extension $L/L \cap F$ is an abelian profinite l-group, it may be considered as a Z_l-module. Hence, let V denote the tensor product of V and Q_l over Z_l. Then we can prove that V is a finite (even) dimensional vector space over Q_l and that it admits a nondegenerate skew-symmetric bilinear form as mentioned above.

V is obviously an l-adic representation space for the Galois group Γ of K/k. Examining the action of Γ on the bilinear form, we see that the characteristic polynomial of the linear transformation representing an element of Γ satisfies a functional equation of well-known type.

PRINCETON UNIVERSITY
PRINCETON, NEW JERSEY

K_2 OF GLOBAL FIELDS

B. J. BIRCH

I gave these lectures, on a topic which was at the time quite new to me, because I found the material interesting and hoped my audience would too. The exposition is derived from that of Bass and of Tate, in various lectures they gave in the spring of 1969, see for instance [2]; some of the material in §§5 and 6 is not available elsewhere.

1. I will always use the letter F for a field, with multiplicative group F^*, and $\mu(F)$ will be its subgroup of roots of unity.

DEFINITION 1. Let F be a field and G a multiplicative group. A G-valued *symbol* on F is a skew-symmetric bimultiplicative map $(\ ,\ )_x$: $F^* \times F^* \to G$ satisfying the identity

$$(a, 1 - a)_x = 1_G \quad \text{whenever } a \neq 0, 1;$$

the identity $(a, -a)_x = 1_G$ for all $a \in F^*$ is a consequence.

The G-valued symbols form a group $\mathrm{Symb}_G(F)$. We are mostly interested in the case $G = \mu$, when we denote the groups of symbols by Symb (G); it is actually more convenient to write μ additively as Q^+/Z, but in order to avoid confusion we won't do this.

To get K_2F, we stand on our heads.

DEFINITION 2. K_2F is the free group generated by objects $\{a, b\}$, one for each pair $(a, b) \in F^* \times F^*$, subject to the relations

$$\{a_1, a_2\} = \{a_1, b\}\{a_2, b\} \quad \text{for all } a_1, a_2, b \in F^*,$$

$$\{a, b\}\{b, a\} = 1 \quad \text{for all } a, b \in F^*, \text{ and } \{a, 1 - a\} = 1 \text{ wherever } a \neq 0, 1.$$

So the map $\{\ ,\ \}$: $F^* \times F^* \to K_2F$ is the *universal symbol* and $\mathrm{Symb}_G(F)$ is just the dual Hom (K_2F, G).

In case F has some topology, and G is discrete, denote by $\mathrm{Symb}'_G(F)$ the group of *continuous* symbols $F \to G$, and define K'_2 correspondingly: the elements of K'_2 are the $\{a, b\}'$, where $\{\ ,\ \}'$ is the "universal continuous symbol". There is an inclusion Symb $\supset$ Symb$'$, and a homomorphism $K_2 \to K'_2$.

2. Let us evaluate K_2F in some easy cases.

EXAMPLE 1. If F is finite, K_2F is trivial.

PROOF. Let u be a generator of F^*; every element of F^* is of form u^m, and $\{u^m, u^n\} = \{u, u\}^{mn}$ by bimultiplicativity. Also, $\{u, u\} = \{u, -u\}\{u, -1\} = \{u, -1\}$, and $\{u, -1\}^2 = 1$, so K_2F is annihilated by 2.
Choose $x, y \in F^*$ so that $ux^2 + uy^2 = 1$. Then

$$1 = \{ux^2, 1 - ux^2\} = \{ux^2, uy^2\} = \{u, u\}\{u, y\}^2\{x, u\}^2\{x, y\}^4 = \{u, u\}.$$

So K_2F is trivial.

EXAMPLE 2. K'_2C is trivial, since C^* is connected. The universal continuous symbol on R turns out to be

$$\begin{aligned}(a, b)_\infty &= +1 \quad \text{if either of } a, b > 0,\\ &= -1 \quad \text{if } a < 0 \text{ and } b < 0.\end{aligned}$$

EXAMPLE 3. Suppose F_v is a local field with discrete valuation v, integers $\mathfrak{o}$, units U, prime element π, and finite residue class field $\mathfrak{o}/\pi\mathfrak{o} = k$. There are natural maps $U \to k^* \to \mu(F)$ by $a \mapsto a^* \mapsto \tilde{a}$.
Define the tame Hilbert symbol $(\ ,\ )_v$: $F^* \times F^* \to \mu(F_v)$ by

$$(a\pi^i, b\pi^j)_v = (-1)^{ij}\tilde{a}^j/\tilde{b}^i \quad \text{whenever } a, b \in U;$$

this defines $(\ ,\ )_v$ for all pairs of $F^* \times F^*$, and it is easily verified that it is a symbol. Conversely, if $v(|\mu(F)|) = 0$, it is relatively easy to see that every continuous symbol is a power of $(\ ,\ )_v$; so $K'_2F_v \cong \mu(F_v)$, and dually Symb$'$ $F_v \cong \mu(F_v)$.
Generally, Calvin Moore [9] has proved that if F_v is a p-adic field, then the norm residue symbol is universal; so $K'_2F_v \cong \mu(F_v)$. (For an appropriate definition of the norm-residue symbol $(\ ,\ )_v$, see [4], particularly the Exercises; or, equally well, [11]. A typical example of a nontame symbol is the symbol $(\ ,\ )_2$ on Q_2 given by

$$(a, b)_2 = (-1)^{(a-1)(b-1)/4}, \qquad (a, 2)_2 = (-1)^{(a^2-1)/8} \quad \text{for } a, b \text{ odd.})$$

EXAMPLE 4. For each completion $Q_p (p = \infty, 2, 3, \ldots)$ of Q there is an embedding $Q \to Q_p$, giving a homomorphism $K_2Q \to K'_2Q_p$, and so a homomorphism

$$\lambda_Q\colon K_2Q \to \coprod_p K'_2Q_p \quad \text{by } \{a, b\} \to \{(a, b)_p\}.$$

(We get the direct sum $\coprod$ rather than the product $\prod$, since for any given a, b the norm residue symbol $(a, b)_p$ is trivial for almost all p.) We ask how far λ_Q differs from an isomorphism.

THEOREM 1 (TATE-GAUSS).

$$K_2Q \cong \{\pm 1\} \times k_3^* \times k_5^* \times k_7^* \times \cdots$$

by $\{a, b\} \to \{(a, b)_\infty, (a, b)_3, (a, b)_5, \ldots\}$. [*Note that* $(a, b)_2$ *has gone missing.*]

PROOF. Write $L(\leq p)$ for the subgroup of K_2F generated by $\{a, b\}$ with a, b products of ± 1 and primes $\leq p$; define $L(< p)$ similarly. We will show by induction on p that

$$L(\leq p) \cong \{\pm 1\} \times k_3^* \times \cdots \times k_p^* \text{ by } \{a, b\} \to \{(a, b)_\infty, (a, b)_3, \ldots, (a, b)_p\}.$$

Certainly the induction starts; the map of $L(< 2)$ given by $(\, , \,)_\infty$ is $1 - 1$, and $\{2, 2\} = \{2, -1\} = \{2, 1 - 2\} = 1$, so $L(\leq 2) = L(< 2)$. So it is enough to check the induction step, that $L(\leq p)/L(< p) \xrightarrow{\sim} k_p^*$; that is

If a symbol $(\, , \,)_x$ *is trivial on* $L(< p)$, *it acts as a power of the Hilbert symbol* $(\, , \,)_p$ *on* $L(\leq p)$.

For this, it is enough to verify that $(a, p)_x$ is multiplicative in the residue class of a modulo p, for $(a, p) = 1$; and for this it is enough to prove the following lemma:

LEMMA. *If* $0 \leq a, b, c < p$ *and* $ab \equiv c \ (p)$, *and* $(\, , \,)_x$ *is trivial on* $L(< p)$, *then* $(a, p)_x(b, p)_x = (c, p)_x$.

PROOF. Say $ab = dp + c$, then $0 \leq d < p$. If $d = 0$, O.K. Otherwise, if $d > 0$, then $1 = c/ab + dp/ab$, so $1 = \{c/ab, dp/ab\} = \{c/ab, p\} \bmod^\times L(< p)$, so $(c/ab, p)_x = 1$ as required.

After giving this nice proof of the structure of K_2Q, Tate realized that it was just the first proof of quadratic reciprocity, as expounded in Gauss' Disquisitiones [6]. To complete the proof of the reciprocity theorem, one notes that $(\, , \,)_2$ is a symbol which has not yet been accounted for, so it must be a product of the symbols $(\, , \,)_p$ for $p = \infty, 3, 5, 7, \ldots$. So there is a product formula

$$(*) \qquad (\, , \,)_2 = (\, , \,)_\infty^{s_\infty}(\, , \,)_3^{s_3}(\, , \,)_5^{s_5} \cdots.$$

We wish to show that s_∞ is odd, and that $s_p \equiv \frac{1}{2}(p - 1) \bmod (p - 1)$ for $p = 3, 5, 7, \ldots$. To check this, let p_0 be the first prime for which the formula goes wrong. Substituting $-1, -1$ into $(*)$, we see that s_∞ is odd. Substituting $-1, p$ into $(*)$ we see that $p_0 \not\equiv 3 \ (8)$; substituting $2, p$ into $(*)$ we see that $p_0 \not\equiv 5, 7 \ (8)$. To show that $p_0 \not\equiv 1 \ (8)$, we first prove that if p is a prime $\equiv 1 \ (8)$, then there is a prime q, $1 < q < p$, with $(p/q) = 1$ (this is not easy; see the Disquisitiones art. 129!); and then we substitute (p, q) into $(*)$.

We can sum all this up conveniently as

THEOREM 1B.

$$0 \to K_2Q \xrightarrow{\lambda_Q} \coprod K_2'Q_p \to \mu(Q) \to 0 \quad \textit{is exact.}$$

Here, $\mu(Q)$ is a pompous way of writing $\{\pm 1\}$; exactness at the right-hand end is the product formula of quadratic reciprocity, and exactness at the left-hand end is the theorem of Tate and Gauss.

EXAMPLE 5. Generally, if F is a global field (number field, or function field of a curve over a finite field) then for each completion we have a homomorphism $K_2F \to K_2'F_v$, so as above we have an exact sequence

$$0 \to ? \to K_2F \xrightarrow{\lambda_F} \coprod K_2'F_v \to ? \to 0,$$

where λ_F: $\{a, b\} \to \{(a, b)_v\}$.

The right-hand end of this sequence is fairly well understood. Calvin Moore not only identifies $K_2'F_v$ in terms of the norm residue symbol, but also identifies the cokernel as $\mu(F)$. So we have

THEOREM 2.

$$0 \to \ker \lambda_F \to K_2F \xrightarrow{\lambda_F} \coprod \mu(F_v) \to \mu(F) \to 0$$

is exact; exactness at the right-hand end is just the product formula for norm residue symbols.

3. It remains to identify $\ker \lambda_F$; I will talk about this in my second lecture. First, I must say something about why anyone ever thought of K_2F.

The idea goes back to Steinberg [14]. He defines a group $St_n(F)$ as the "covering group" of $SL_n(F)$; it is convenient to take natural embeddings $SL_n(F) \to SL_{n+1}(F)$, etc., and let $n \to \infty$; then the group we are interested in turns up in the exact sequence

$$0 \to K_2F \to StK \to SLF \to 0.$$

(Actually, letting $n \to \infty$ is unnecessary, since so long as $|F| > 5$, the situation is stable for $n \geq 3$.) Here, K_2F is in the centre of StF, and StF is maximal subject to connectedness and $\ker (St \to SL) \subseteq \text{Centre}\,(St)$.

Explicitly, St_nA may be defined for a general commutative domain A, as the free group on generators x_{ij}^a ($1 \leq i, j \leq n$, $a \in A$) subject to relations $x_{ij}^a x_{ij}^b = x_{ij}^{a+b}$, $[x_{ij}^a, x_{jk}^b] = x_{ik}^{ab}$ for $i \neq k$, $[x_{ij}^a, x_{kl}^b] = 1$ for $i \neq l$, $j \neq k$. There is a natural map $St_nA \to GL_nA$ by $x_{ij}^a \to I + ae_{ij}$, the image being the "elementary" subgroup $E_nA \subseteq GL_nA$; E_nA is in fact the commutator subgroup of GL_nA. We get an exact sequence

$$0 \to K_2A \to StA \to GL(A) \to GL(A)/E(A) \to 0;$$

this defines K_2A, and $GL(A)/E(A)$ could have been taken as a definition for K_1A; in case A is a field, $K_1F = F^*$. (Notation. $[x, y]$ denotes the commutator of x and y; and e_{ij} is the matrix with 1 in the i, j-place and 0 everywhere else.)

Alternatively: suppose GL is free on generators G, modulo relations R. Then

$$St \cong [G, G]/[G, R], \qquad K_2 \cong (R \cap [G, G])/[G, R].$$

Alternatively:

$$E(A) = H_1(GL(A), Z), \qquad K_2A = H_2(E(A), Z).$$

Steinberg's definition of St_n by generators and relations gives a mapping of $F^* \times F^*$ into his K_2 which is a symbol; but one must refer to Matsumoto's thesis [8] to establish that both definitions of K_2 are the same.

The proposal that the functor K_2 (which Calvin Moore had denoted by π_1) was the appropriate functor to act as the third term in the sequence $K_0, K_1, \ldots$ was made by Milnor (Princeton notes; cf. [16] and Swan [15]). He shows that it has two vital properties that commend it to cohomologists:

(1) There are long exact sequences.

(2) There are pairings $K_iA \times K_jA \to K_{i+j}A$ ($i + j \leq 2$) that make $\bigcup_{i=0}^{2} K_iA$ into the start of a commutative graded ring. (In the particular case $A = F$, $K_1F = F^*$ and we have the pairing $(a, b) \to \{a, b\}$.)

Several proposals have been made for stretching the sequence of K_i's further. Unhappily, practically none of them have been computed further than the verification that K_0A, K_1A, and K_2F are correct, and the properties (1) and (2) above. The best-known efforts seem to be

1. Nobile-Villamayor [10]. Take K_0A as the Grothendieck group on projective modules, as usual. Take $K_{i+1}A = K_iSA$, where SA is the "suspension ring of A", something like $SA = \{f \mid f \in A[x], f(0) = f(1)\}$.

2. Bass. Take $\mathcal{P}_n(A)$ as the category whose objects are $(P, \alpha_1, \ldots, \alpha_n)$ where P is a projective module and $\alpha_1, \ldots, \alpha_n$ are commuting automorphisms thereof. Then $K_n''A$ is the Grothendieck group formed from $\mathcal{P}_n(A)$, that is, a free abelian group generated by isomorphism classes of objects of $\mathcal{P}_n(A)$, subject to various relations; and then one makes $K_n''(A)$ into a graded commutative ring by natural explicit construction.

4. From now on, suppose that F is a global field, with completions F_v and ring of integers $\mathfrak{o}$. We have an exact sequence

$$0 \to \ker \lambda_F \to K_2F \to \coprod K_2'F_v \to \mu(F) \to 0,$$

and from now on we will be talking about $\ker \lambda_F$; more precisely, we will be talking about the conjecture that $\ker \lambda_F$ is finite and its order is essentially given by the numerator of the generalized Bernoulli number $B_2(F)$, when it makes sense. (Generalized Bernoulli numbers seem rather fashionable. In Fresnel's thesis [5], of which he described a more powerful version at this institute, they are used in the construction of p-adic L-functions; papers by Lang [7] and Barner [1] compute class invariants of real quadratic fields; and Siegel [13] defines generalized Bernoulli polynomials. From our point of view, it may be relevant to refer to Siegel [12], where he finds the volume of the fundamental domain of the Hilbert modular group—$B_2(F)$ occurs as a factor.)

Most results about $\ker \lambda_F$ are due to Bass and Tate. I will describe their results, more or less in chronological order.

THEOREM 3. *If F is a global field, then* $\ker \lambda_F$ *is finitely generated.*

To prove this, mimic the proof of Theorem 1, that is to say, the Gauss first proof. Order a set of generators of F^*, say $p_0, p_1, p_2, \ldots$, so that, for $n > n_0$, p_n has just one prime ideal factor that does not divide $p_0p_1 \ldots p_{n-1}$, and so that the sequence of norms is increasing. Let L_n be the subgroup of K_2F generated by $\{a, b\}$ with a, b products of $p_0, \ldots, p_n$; then, for each n, L_n is finitely generated.

To prove that ker λ_F is finitely generated, it is therefore enough to show that, for $n > n_1$, a symbol $(\ , \ )_x$ that acts trivially on L_{n-1} acts as a power of $(\ , \ )_n$ on L_n, where $(\ , \ )_n$ is the Hilbert symbol corresponding to the prime φ_n that divides p_n but not $p_0 \ldots p_{n-1}$. To prove this, Bass and Tate proved a nontrivial Minkowski-type theorem, as follows:

THEOREM. *Let U_{n-1} be the subgroup of F^* generated by $p_0, \ldots, p_{n-1}$; let ρ be the natural map $U_{n-1} \to (\mathfrak{o}/p_n)^*$. Then for* Norm $p_n > CD^{3/2}$, *with C a suitable constant, ρ is onto and* ker (ρ) *is generated by elements of the form* $1 + p_n u$ *with $u \in U_{n-1}$.*

THEOREM 4. *If F is a function field of a curve over a finite field k, then* ker λ_F *is finite.*

To prove this, one needs a proposition gleaned from the abstract theory (though it is a very down-to-earth sort of assertion) that Bass and Tate ascribe to each other.

Suppose G is a finite extension of F; then there is a norm map $\beta \mapsto N_{G/F}\beta$ of G to F. There is also a norm map N: $K_2G \to K_2F$ which fits together so that if $a \in F^*$ and $\beta \in G^*$ then $N\{a, \beta\}_G = \{a, N_{G/F}\beta\}_F$. Here, of course, we are distinguishing between elements of K_2G and K_2F by means of suffixes. From this we deduce easily

PROPOSITION 5. *Suppose $G = F(a^{1/m})$, and $b = N_{G/F}\beta$ with $\beta \in G$. Then $\{a, b\} \in (K_2F)^m$.*

For $\{a, b\}_F = \{a, N\beta\}_F = N\{a, \beta\}_G = (N\{a^{1/m}, \beta\}_G)^m \in (K_2F)^m$. (Unfortunately, I have not seen the nontrivial part of the proof of this proposition; its genesis is beyond reproach.)

PROOF OF THEOREM 4. Say k has characteristic p. If $F = k(x_1, \ldots, x_r)$, take G as $k(x_1^{1/p}, \ldots, x_r^{1/p})$ then $[G: F] = p$ and $N_{G/F}: \beta \to \beta^p$. If $a, b \in F$, then $a = \alpha^p$, $b = \beta^p$ with $\alpha, \beta \in G$; so $\{a, b\}_F = \{\alpha, \beta\}_G^{p^2} = (N\{\alpha, \beta\}_G)^p \in (K_2F)^p$. It follows that K_2F is p-divisible—so ker λ_F is finite and K_2F has no elements of order p.

There is a very pretty application to norm residues. Suppose that F is a number field with $m\sqrt{1} \in F$; then

$$\begin{array}{ccc}
\boxed{\{a, b\} \in (K_2F)^m} & \Rightarrow & \boxed{(a, b)_v \in (K_2'F_v)^m \text{ for all } v} \\
\Uparrow & & \Downarrow \\
\boxed{b \text{ is a norm from } F(m\sqrt{a})} & \Leftarrow & \boxed{b \text{ is a norm from } F_v(m\sqrt{a}) \text{ for all } v}
\end{array},$$

the implications being by turn trivial, local class field theory, Hasse's norm theorem, and Proposition 5. This merry-go-round raises various questions, which may be asked even without any hypotheses on F; for instance

(1) Does $(a, b) \in (K_2F)^m$ imply that a is a norm from $F(m\sqrt{b})$?

(2) Is b a norm from $F(m\sqrt{a})$ if and only if a is a norm from $F(m\sqrt{b})$?

(Tate and Serre have reminded me to refer to XIV of [11].)

5. Bass and Tate went on to compute K_2F in a number of cases; in the few cases they tried with F complex quadratic (for instance, $Q(\sqrt{-D})$ with $D = 1, 2, 6$) $\ker \lambda_F$ turned out to be trivial. When they took F as the function field of the pointless elliptic curve $y^2 + y = x^3 + x + 1$ over the field with 2 elements, it seemed however that $|\ker \lambda_F| = 5$. The other 4 elliptic curves over k_2 seemed to give $|\ker \lambda_F| = 7, 9, 11, 13$; helped by geometric insight, Tate suggested the hypothesis

$$|\ker \lambda_F| = (q - 1)(q^2 - 1)Z_F(-1)$$

when F is the function field of a curve of a finite field with q elements and $Z_F(s) = \prod((1 - \alpha_i q^{-s})/(1 - q^{-s})1 - q^{1-s})$ is the zeta-function of F.

It is natural to make the same conjecture for number fields,

$$|\ker \lambda_F| = (\text{trash})\, Z_F(-1),$$

at any rate if F is totally real so that $Z_F(-1)$ is useful. (Problem: What goes in its place if F is complex quadratic?) So A. O. L. Atkin and I tried to verify this. After a short while, Atkin had persuaded ATLAS to compute a number of cases, including $Q(w)$ for

$$w = \tfrac{1}{2}(1 + \sqrt{21}),\ \sqrt{7},\ \tfrac{1}{2}(1 + \sqrt{29}),\ \tfrac{1}{2}(1 + \sqrt{37}),\ \sqrt{11},\ \sqrt{19}$$

for which apparently $|\ker \lambda_F| = 2, 2, 3, 5, 7, 19$. All results so far obtained are consistent with the following:

CONJECTURE. Write Symb (F) for the group of symbols, and write Hilb (F) for the subgroup generated by tame Hilbert symbols (see Example 3). Suppose that F is either the function field of a curve over a finite field, or a totally real number field. Then

(A) $$|\text{Symb}\,(F)/\text{Hilb}\,(F)| = w'_F|Z_F(-1)|,$$

where

$$w'_F = |\mu(F)|^2 \prod_{[E:F]=2} \frac{|\mu(E)|}{|\mu(F)|}.$$

The left-hand side is essentially $|\ker \lambda_F|$, but we have made the conjecture appear neater by making the ramified norm-residue symbols contribute too; for instance, for $F = Q$, the left-hand side is 2, $w'_Q = 24$ and $Z_Q(-1) = -1/12$. If F is a function field over k_q, $w'_F = (q - 1)(q^2 - 1)$, consistent with Tate's original conjecture. We have not yet proved that $\frac{1}{2}w'_F Z_F(-1)$ is integral whenever F is totally real (computation suggests there should be Staudt-Clausen theorems). (Serre tells me that he can achieve this, combining Siegel's methods with the Gauss-Bonnet formula.)

In the calculations that led to the above conjecture, K_2F is treated as being defined by generators and relations: one looks at the subgroup L_n, for reasonable-sized n, and every time one has a relation $\prod p_i^{e_i} + \prod p_j^{f_j} = 1$, the identity $\{a, 1 - a\} = 1$ gives a relation between the $\{p_i, p_j\}$. Unfortunately, one never knows if there is lurking another independent relation, so such methods alone can only prove results of the type "left-hand side of (A) divides right-hand side of (A)". To complete the computation of ker λ_F, in a case where it is nontrivial, one needs a method of producing exotic symbols—that is to say, symbols that are not products of norm residue symbols.

An indirect method of testing the conjecture was suggested by Tate. He remarked that if p divides ker (λ_F), it is "almost" (i.e., apart from predictable cases) necessary that p divides the class number $h(F(p\sqrt{1}))$. This is consistent, since (by using the methods given in [3] in case $F = Q$) it is not hard to verify that p divides $w_F' Z_F(-1)$ (almost) implies p divides $h(F(p\sqrt{1}))$; and in fact the converse implication is (almost) true for $p = 2, 3$.

6. Finally, I describe a recipe for constructing exotic symbols; this recipe should enable us to calculate the 2 and 3-primary part of ker λ_F exactly, rather than just estimate it from above. We will try to show that an obvious necessary condition for an element of Symb (F) to have an mth root is also sufficient, at any rate if $m = 2^r$ or $m = 3^r$ and $3\sqrt{1} \in F$. It is easy to find the elements of Hilb (F) that satisfy this necessary and sufficient condition, so we obtain exotic symbols when we deserve them.

LEMMA. *If $m\sqrt{1} \in F$, and $(\ ,\ )_H \in (\text{Symb } F)^m$, then $(a, m\sqrt{1})_H = 1$ for all $a \in F^*$.*

This is trivial: say $(\ ,\ )_H = (\ ,\ )_x^m$, then $(a, m\sqrt{1})_H = (a, m\sqrt{1})_x^m = (a, 1)_x = 1$. We state the lemma because we want its converse.

HYPOTHESIS. *If $m\sqrt{1} \in F$ and $(\ ,\ )_H$ is a symbol on F such that $(a, m\sqrt{1})_H = 1$ for all $a \in F^*$, then there is a symbol $(\ ,\ )_x$ on F such that $(\ ,\ )_H = (\ ,\ )_x^m$.*

I have half a proof in the cases $m = 2$ and $m = 3$; I give it for $m = 2$; the case $m = 3$ is similar but nastier. Let S be the set of pairs (a, b) such that there exist x, y with $ax^2 + by^2 = 1$. For $(a, b) \in S$, *define*

$$(a, b)_x = (x, b)_H^{-1}(a, y)_H^{-1}(x, y)_H^{-2}.$$

Check by force that $(\ ,\ )_x$, so far as it is defined, depends only on a, b (and not on x, y), and it is bimultiplicative and skew symmetric; this is easy enough for anyone with a strong enough stomach. We have carefully arranged that $(ax^2, by^2)_x = (ax^2, 1 - ax^2)_x = 0$, and $(a, b)_H = (a, b)_x^2$.

One needs to extend the definition of $(\ ,\ )_x$ from S to $F^* \times F^*$ by using multiplicativity. Unhappily, here there is still a yawning gap, as I have not proved consistency of such an extension.

REFERENCES

1. K. Barner, *Über die Werte der Ringklassen-L-Funktionen reellquadratischer Zahlkörper an natürlichen Argumentstellen*, J. Number Theory **1** (1969), 28–64. MR **39** #139.

2. H. Bass, *K_2 and symbols. Algebraic K-theory and its geometric applications*, Lecture Notes in Math., no. 108, Springer-Verlag, Berlin and New York, 1969.

3. Z. I. Borevič and I. R. Šafarevič, *Number theory*, "Nauka", Moscow, 1964; English transl., Academic Press, New York, 1966. MR **30** #1080; MR **33** #4001.

4. J. W. S. Cassels and A. Fröhlich (editors), *Algebraic number theory*, Academic Press, New York, 1967.

5. J. Fresnel, *Nombres de Bernoulli et fonctions L p-adiques*, Ann. Inst. Fourier (Grenoble) **17** (1967), fasc. 2, 281–333. MR **37** #169.

6. C. F. Gauss, *Disquisitiones arithmeticae*, Fleischer, Leipzig, 1870; English transl., Yale Univ. Press, Conn., 1966. MR **33** #5545.

7. H. Lang, *Über eine Gattung elementar-arithmetischer Klasseninvarianten reell-quadratischer Zahlkörper*, J. Reine Angew. Math. **233** (1968), 123–175. MR **39** #168.

8. H. Matsumoto, *Sur les sous-groupes arithmétiques des groupes semi-simples déployés*, Ann. Sci. École Norm. Sup. (4) **2** (1969), 1–62. MR **39** #1566.

9. Calvin Moore, *Group extensions of p-adic and adelic linear groups*, Inst. Hautes Études Sci. Publ. Math. **39** (1969), 5–74.

10. A. Nobile and O. E. Villamayor, *Sur la K-théorie algébrique*, Ann. Sci. École Norm. Sup. (4) **1** (1968), 581–616. MR **39** #1526.

11. J.-P. Serre, *Corps locaux*, Publ. Inst. Math. Univ. Nancago, 8, Actualités Sci. Indust., no. 1296, Hermann, Paris, 1962. MR **27** #133.

12. C. L. Siegel, *The volume of the fundamental domain for some infinite groups*, Trans. Amer. Math. Soc. **39** (1936), 209–218.

13. ———, *Bernoullische Polynome und quadratische Zahlkörper*, Nachr. Akad. Wiss. Göttingen Math.-Phys. Kl. II **1968**, 7–38. MR **38** #2123.

14. R. Steinberg, *Générateurs, relations et revêtements de groupes algébriques*, Colloq. Théorie des Groupes Algébriques (Bruxelles, 1962) Librairie Universitaire, Louvain and Gauthier-Villars, Paris, 1962, pp. 113–127. MR **27** #3638.

15. R. G. Swan, *Algebraic K-theory*, Lecture Notes in Math. no. 76, Springer-Verlag, Berlin and New York, 1968.

16. J. Milnor, *Algebraic K-theory and quadratic forms*, Invent. Math. **9** (1970), 318–344.

MATHEMATICS INSTITUTE
OXFORD, ENGLAND

CLASS FORMATIONS

YUKIYOSI KAWADA[1]

This is an expository note concerning class formations defined by Artin and Tate. We introduce the notion of a class formation (§1–§6), and give several examples (§7–§13). Then we apply the cohomology theory of profinite groups developed by Tate to class formation theory (§14–§17).

1. Local class field theory

Let $\mathbf{Q}_p$ be the p-adic number field, Ω the algebraic closure of $\mathbf{Q}_p$. Let $\mathfrak{k} = \mathfrak{k}(\Omega/\mathbf{Q}_p)$ be the set of all subfields K of Ω containing $\mathbf{Q}_p$ such that $[K: \mathbf{Q}_p] < \infty$. Among the main theorems in local class field theory we have the following results.

(I) (Isomorphism theorem) Let $k, K \in \mathfrak{k}$ and let K/k be normal. Then

$$(1) \qquad G/G' \cong k^*/N_{K/k}K^* \qquad (G = G(K/k))$$

where $G(K/k)$ means the Galois group of K/k, $N_{K/k}$ the norm, k^* the multiplicative group of k and G' the commutator subgroup of G.

(II) (Nakayama map) We can choose a suitable 2-cocycle $f_{K/k}(\sigma, \tau)$ of G with values in K^* such that the isomorphism (1) is given by

$$(2) \qquad \sigma \bmod G' \mapsto \prod_{\tau \in G} f_{K/k}(\tau, \sigma)^{-1} \bmod N_{K/k}K^*.$$

We call $f_{K/k}$ a *canonical 2-cocycle* of K/k. We denote the inverse map of (2): $k^*/N_{K/k}K^* \to G/G'$ by

$$a \bmod N_{K/k}K^* \mapsto (a, K/k) \in G/G' \quad \text{for } a \in k^*.$$

---

[1] The author wishes to express his hearty thanks to Professor J. P. Serre and Professor A. Brumer for their kind advices.

Let $\Omega^a(k)$ be the maximal abelian extension of k (in Ω) and let $A(k) = G(\Omega^a(k)/k) = \lim_{\leftarrow} G(K/k)$ (where $[K: k] < \infty$ and K/k is abelian) be the compact Galois group.

(III) By choosing $f_{K/k}$ suitably for each abelian extension K/k we can define the *norm-residue symbol*

$$(a, k) = \lim_{\leftarrow} (a, K/k) \in A(k)$$

for $a \in k^*$. Then the map $a \mapsto (a, k)$ defines a continuous homomorphism $\Phi_k: k^* \to A(k)$ such that $\operatorname{Im} \Phi_k$ is dense in $A(k)$ and $\operatorname{Ker} \Phi_k = 1$.

(IV) (Existence theorem) For each open subgroup B of k^* there exists an open subgroup H of $A(k)$ such that $B = \Phi_k^{-1}(H)$. Let K be the subfield of Ω left fixed by H. Then $B = N_{K/k}K^*$ holds. B is called the subgroup associated to K. K/k is unramified if and only if B contains the unit group $U(k)$ of k.

We shall give a cohomological formulation of these results based on a class formation.

2. Class formation

Let k_0 be the ground field, Ω a fixed infinite separable normal algebraic extension of k_0 and $\mathfrak{k} = \mathfrak{k}(\Omega/k_0)$ the set of all subfields K of Ω containing k_0 such that $[K: k_0] < \infty$. Let $E(k)$ $(k \in \mathfrak{k})$ be a module (or a multiplicative abelian group) with the following properties:

$(\mathbf{CL}_1)$ For each pair $k, K \in \mathfrak{k}$ with $k \subset K$ an isomorphism $\varphi_{k,K}$ of $E(k)$ into $E(K)$ is defined such that for, $k \subset l \subset K$, $\varphi_{k,K} = \varphi_{l,K}, \circ \varphi_{k,l}$ holds.

$(\mathbf{CL}_2)$ If K/k is normal then $G = G(K/k)$ operates on $E(K)$ such that $\varphi_{k,K}(E(k)) = E(K)^G$ holds. Moreover, we assume that Galois theory holds for $\{\varphi_{l,K}E(l)\}$ $(k \subset l \subset K)$ in a natural way.

Hence if we put $E(\Omega) = \lim_{\to} E(K)$ $(K \in \mathfrak{k})$ and $G(\Omega/k) = \lim_{\leftarrow} G(K/k)$ (where $K \in \mathfrak{k}$, and K/k is normal), then $G(\Omega/k)$ operates on $E(\Omega)$ so that $\varphi_{k,\Omega} E(k) = E(\Omega)^{G(\Omega/k)}$ holds.

$(\mathbf{CL}_3)$ Let $k, K \in \mathfrak{k}$, K/k normal and $G = G(K/k)$. Then

$$H^1(G, E(K)) = 0, \qquad H^2(G, E(K)) \cong \mathbf{Z}/n\mathbf{Z}, \qquad n = [K: k].$$

We call then $\{E(k) \mid k \in \mathfrak{k}\}$ a *class formation* for $\mathfrak{k}$ and each module $E(k)$ the class module associated to k.

3. Canonical cohomology class

We consider two cases.

Case 1. $k \subset l \subset K$, K/k normal, $G = G(K/k)$ and $H = G(K/l)$.

Case 2. We assume further l/k is a Galois extension and $F = G(l/k)$.

In Case 1 $\operatorname{Res}_{G/H}: H^2(G, E(K)) \to H^2(H, E(K))$ is surjective and $\operatorname{Cor}_{H/G}: H^2(H, E(K)) \to H^2(G, E(K))$ is injective. In Case 2 $\operatorname{Inf}_{F/G}: H^2(F, E(l)) \to H^2(G, E(K))$ is injective. Then we can choose a generator $\xi_{K/k}$ of $H^2(G, E(K))$ for each normal extension K/k such that $\operatorname{Res}_{G/H} \xi_{K/k} = \xi_{K/l}$ in Case 1 and $\operatorname{Inf}_{F/G} \xi_{l/k} = [l: k]\xi_{K/k}$ in Case 2 hold. We call then $\xi_{K/k}$ the *canonical cohomology class* of K/k. By means of the canonical cohomology classes we can define the *invariant* of a cohomology class $\eta_{K/k} \in H^2(G, E(K))$ by $\text{inv}_{K/k}(\eta_{K/k}) = r/[K: k] \bmod \mathbf{Z}$ for $\eta_{K/k} = r\xi_{K/k}$. In Case 2 inv satisfies:

$$\text{inv}_{K/k}(\text{Inf}_{F/G}\eta_{l/k}) = \text{inv}_{l/k}(\eta_{l/k}) \bmod \mathbf{Z}.$$

Now we can define the cohomology group of the profinite group $G(\Omega/k)$ by

$$H^r(G(\Omega/k), E(\Omega)) = \varinjlim H^r(G(K/k), E(K)) \quad \text{for } K \in \mathfrak{k} \text{ and } r \geqq 0.$$

Then we have $H^1(G(\Omega/k), E(\Omega)) = 0$ and $H^2(G(\Omega/k), E(\Omega)) \cong B$, $B \subset \mathbf{Q}/\mathbf{Z}$ where the last isomorphism is given by $\eta \mapsto \text{inv}\ \eta$. If the degree $[\Omega: k_0]$ is divisible by p^∞ whenever p divides $[\Omega: k_0]$ then we have $H^r(G(\Omega/k), E(\Omega)) = 0$ for $r \geqq 3$. (See also §15.) Now we see that our definition of a class formation coincides with that given by Artin-Tate (Artin-Tate [1, Chapter 14]).

3A. Norm-residue map

By a theorem of Tate we know that for normal K/k with $G = G(K/k)$, $\hat{H}^r(G, E(K)) \cong \hat{H}^{r-2}(G, \mathbf{Z})$ holds for any $r \in \mathbf{Z}$, and the cup product with the canonical cohomology class $\xi_{K/k}$ gives this isomorphism: $\hat{H}^{r-2}(G, \mathbf{Z}) \ni \alpha \mapsto \alpha \cup \xi_{K/k} \in \hat{H}^r(G, E(K))$. In particular, for $r = 0$ this gives the isomorphism of $\hat{H}^{-2}(G, \mathbf{Z}) = G/G'$ onto $\hat{H}^0(G, E(K)) = E(K)^G/N_G(E(K)) = E(k)/N_G(E(K))$ by

$$\sigma \bmod G' \mapsto \prod_{\tau \in G} f_{K/k}(\tau, \sigma)^{-1} \bmod N_G E(K)$$

where $f_{K/k}$ is an arbitrary 2-cocycle in $\xi_{K/k}$. (Here $N_G a$ means the trace $\sum_{\sigma \in G} \sigma a$ as usual.) We denote the inverse map $E(k)/N_G E(K) \to G/G'$ by $a \bmod N_G E(K) \mapsto (a, K/k) \bmod G'$ for $a \in E(k)$.

Let $\Omega^a(k)$ be the maximal abelian extension of k in Ω, and let $A(k) = G(\Omega^a/k) = \varprojlim G(K/k)$ ($K \in \mathfrak{k}$, K/k abelian). Then we can define a symbol

$$(a, k) = \varprojlim (a, K/k) \quad \text{for } a \in E(k).$$

The map $a \mapsto (a, k)$ gives a homomorphism Φ_k of $E(k)$ into $A(k)$. We call Φ_k the *norm-residue map*. The image of Φ_k in $A(k)$ is dense, but the kernel of Φ_k is not 0 in general.

In order to describe the existence theorem in a class formation we need to assume some natural topology in each $E(k)$ with further axioms (cf. Artin-Tate [1, Chapter 14, §6], Serre [42, Chapter XI]). For a class formation see also Serre [42, Chapter XI], Lang [31, Chapter IX].

4. Class field theory

(E 0.1) The relation between local class field theory in §1 and class formation theory in §§2, 3 is obvious. Namely, if we take

$$E(k) = k^* \text{ (multiplicative group of } k)$$

for any $\mathfrak{p}$-adic number field k, then we can verify the axioms $\mathbf{CL}_1$–$\mathbf{CL}_3$. In this case we have

$$A(k)/\mathrm{Im}\,\Phi_k \cong \tilde{\mathbf{Z}}/\mathbf{Z}, \qquad \mathrm{Ker}\,\Phi_k = 1.$$

Here $\tilde{\mathbf{Z}} = \varprojlim_n \mathbf{Z}/n\mathbf{Z}$ (cf. Serre [42, Chapter XIII], Cassels-Fröhlich [4, Chapter VI]).

(E 0.2) For a formal power series field with a finite residue class field $k_0 = \mathbf{F}_p((X))$ and the maximal separable extension Ω of k_0 the situation is quite similar to (E 0.1).

(E 0.3) For global class field theory, i.e., for $k_0 = \mathbf{Q}$ and the algebraic closure Ω of $\mathbf{Q}$, let us take

$$E(k) = C_k = \text{the (multiplicative) idele class group of } k.$$

Then we can verify $\mathbf{CL}_1$–$\mathbf{CL}_3$ (cf. Artin-Tate [1, Chapters 5–8], Cassels-Fröhlich [4, Chapter VII]). In this case we have a natural topology in C_k by which C_k is a locally compact topological group. Let D_k be the connected component of 1 in C_k. Then C_k/D_k is a totally disconnected compact group. For the norm-residue map Φ_k: $C_k \to A(k)$ we have

$$\mathrm{Im}\,\Phi_k = A(k) \quad \text{and} \quad \mathrm{Ker}\,\Phi_k = D_k$$

so that $C_k/D_k \cong A(k)$ (algebraically and topologically) holds. The structure of D_k is given explicitly in Artin-Tate [1, Chapter 9]. Moreover, for normal K/k with $G = G(K/k)$ and $r \in \mathbf{Z}$, $\hat{H}^{2r+1}(G, D_K) = 1$ and $\hat{H}^{2r}(G, D_K) \cong$ (type $(2, 2, \ldots, 2)$ of order 2^μ) where μ is the number of archimedean primes of k which are ramified in K/k.

(E 0.4) For global class field theory where k_0 is a function field in one variable with a finite constant field $\mathbf{F}_q$ we can also take the idele class group C_k of k as $E(k)$. Then we can verify $\mathbf{CL}_1$–$\mathbf{CL}_3$ (Artin-Tate [1]). In this case we have

$$A(k)/\mathrm{Im}\,\Phi_k \cong \tilde{\mathbf{Z}}/\mathbf{Z} \quad \text{and} \quad \mathrm{Ker}\,\Phi_k = 1.$$

5. Topological class formation

Let k_0, Ω, $\mathfrak{k} = \mathfrak{k}(\Omega \mid k_0)$ and $\{E(k) \mid k \in \mathfrak{k}\}$ be as before and we assume that they satisfy the axioms $\mathbf{CL}_1$–$\mathbf{CL}_3$. We call $\{E(k) \mid k \in \mathfrak{k}\}$ a *topological class formation* if each $E(k)$ $(k \in \mathfrak{k})$ is a compact topological group, each $\varphi_{k,K}$ is continuous and each σ: $E(K) \to E(K)$ $(\sigma \in G(K/k)$, K/k normal) is continuous.

For a topological class formation the norm-residue map Φ_k: $E(k) \to A(k)$ $(k \in \mathfrak{k})$ is also continuous and $E(k)/\mathrm{Ker}\,\Phi_k \cong A(k)$ algebraically and topologically.

EXAMPLES. Let A be a discrete (or locally compact) abelian group. We denote by $\tilde{A}$ the projective limit of $\{A/A_\lambda\}$ where each A_λ is a subgroup (or an open subgroup) of A of finite index.

In case of local class field theory ((E 0.1) and (E 0.2)) we see that

$$E(k) = (k^*)^\sim$$

is a topological class formation so that Φ_k gives the isomorphism $E(k) \cong A(k)$. Similar situation holds in case (E 0.4), too.

In case of global class field theory (E 0.3) we see that

$$E(k) = C_k^0 \text{ (idele class group with volume 1)}$$

is a topological class formation.

Associated with a class formation we can define Weil groups. See Weil [50], Artin-Tate [1, Chapter 14], Lang [31, Chapter IX, 3] and also Kawada [21, III].

6. Construction of new class formations

Suppose that a class formation $\{E(k) \mid k \in \mathfrak{k}\}$ for $\mathfrak{k} = \mathfrak{k}(\Omega/k_0)$ is given. There are several ways to define a new class formation from $\{E(k)\}$.

(A) (New class formation for the same $\mathfrak{k}$) Suppose that a submodule $F(k)$ of $E(k)$ is given for each $k \in \mathfrak{k}$. In order that $\{E(k)^* = E(k)/F(k) \mid k \in \mathfrak{k}\}$ be a class formation for $\mathfrak{k}$ it is necessary and sufficient that

(i) $F(K) \cap \varphi_{k,K}E(k) = \varphi_{k,K}F(k)$ for $k \subset K$,

(ii) $F(K)^\sigma = F(K)$ for normal K/k and $\sigma \in G(K/k)$,

(iii) in case (ii) $\hat{H}^r(G, F(K)) = 0$ for $r \in \mathbf{Z}$.

EXAMPLE. Let us consider global class field theory. Assume that k is totally imaginary. Then for $E(k) = C_k$, $F(k) = D_k$ the above conditions (i), (ii), (iii) hold. Hence $\{E(k)^* = C_k/D_k\}$ satisfies $\mathbf{CL}_1$–$\mathbf{CL}_3$. In this case we have the isomorphism $\Phi_k^*\colon E(k)^* \cong A(k)$.

(B) (Restriction to a subfamily of $\mathfrak{k}$) Take $k_1 \in \mathfrak{k}$ and an infinite normal extension Ω_1 of k_1 in Ω_0. Then for the family $\mathfrak{k}_1 = \{k \mid k_1 \subset k \subset \Omega_1, k \in \mathfrak{k}\} = \mathfrak{k}_1(\Omega_1/k_1)$, $\{E(k) \mid k \in \mathfrak{k}_1\}$ is also a class formation.

EXAMPLE. Consider local class field theory: $E(k) = k^*$. Let k_1 be a $\mathfrak{p}$-adic number field and Ω_1 be the maximal unramified extension of k_1. Then $\{E(k) \mid k \in \mathfrak{k}_1\}$ is also a class formation. Take $F(k) = U(k)$ (the unit group of k). Then we can verify (i)–(iii) in (A). Hence $E(k)^* = k^*/U(k) \cong \mathbf{Z}$ $(k \in \Omega_1)$ is also a class formation for $\mathfrak{k}_1$.

(C) (Infinite algebraic extension of ground field k_0) Let k_1 be an infinite algebraic extension of k_0 contained in Ω. Let $\mathfrak{k}_1 = \mathfrak{k}_1(\Omega/k_1)$ be the family of all fields $k \subset \Omega$ such that $k_1 \subset k$ and $[k\colon k_1] < \infty$. If we assume that $\{E(k) \mid k \in \mathfrak{k}\}$ is a topological class formation, then we can construct a new class formation for $\mathfrak{k}_1$ by taking $E(k)^* = \lim_{\leftarrow} E(k_\lambda)$ where $k_\lambda \subset k$, $k_\lambda \in \mathfrak{k}$ and $k = \bigcup_\lambda k_\lambda$. This class formation $\{E(k)^* \mid k \in \mathfrak{k}_1\}$ is also topological. If $E(k) \cong A(k)$ for each $k \in \mathfrak{k}$ by the norm-residue map Φ_k for $\mathfrak{k}$, then we have also $E(k)^* \cong A(k)$ for each $k \in \mathfrak{k}_1$ by the new norm-residue map Φ_k^* (Kawada [21, IV]).

EXAMPLE. In local class field theory $E(k) = (k^*)^\sim$ gives a topological class formation. Hence over an infinite algebraic extension k_1 of $\mathbf{Q}_p$ we have a topological class formation by taking $E(k)^* = \lim_{\leftarrow} (k_\lambda^*)^\sim$ as above. In this case $E(k) \cong A(k)$ $(k \in \mathfrak{k})$ implies $E(k)^* \cong A(k)$ for $k \in \mathfrak{k}_1$. In global class field theory $E(k) = C_k^0$ gives a topological class formation and we can apply our method, too. If the ground field k_0 is totally imaginary we have another topological class formation by taking $E(k) = C_k/D_k$ for $k_0 \subset k$ such that $E(k) \cong A(k)$, and we have a simpler class formation theory over an infinite algebraic number field in this case.

7. Examples of a class formation (I)

(E1) Let k_0 be a prime field $\mathbf{F}_p$ of characteristic p, Ω the separable closure of $\mathbf{F}_p$. Take $E(k) = \mathbf{Z}$ for every $k \in \mathfrak{k} = \mathfrak{k}(\Omega/k_0)$ and let $\varphi_{k,K}E(k) \to E(K)$ be defined by $a \mapsto na$ ($k \subset K$, $[K: k] = n$). Moreover, let $G(K/k)$ operate trivially on $E(K)$ for normal K/k. Then $\{E(k)\}$ is a class formation and

$$A(k)/\mathrm{Im}\,\Phi_k \cong \tilde{\mathbf{Z}}/\mathbf{Z} \quad \text{and } \mathrm{Ker}\,\Phi_k = 0 \quad \text{for } k \in \mathfrak{k}.$$

The situation is similar in the following two cases:

(E 1.1) $k_0 = F((X))$, where F is an algebraically closed field of characteristic zero, and $\Omega = \bigcup_n F((X^{1/n}))$ is the algebraic closure of k_0.

(E 1.2) k_0 is a $\mathfrak{p}$-adic number field, Ω is the maximal unramified extension of k_0. (See §6 (B).) (cf. Serre [42, Chapter XIII, §§1, 2])

(E 2) (Kummer Extensions) Let k_0 be a field with the following properties:

(i) characteristic of k_0 is zero,

(ii) k_0 contains all the roots of unity,

(iii) for arbitrary finite normal extension K/k, (where $k_0 \subset k \subset K$ and $[K: k_0] < \infty$) $N_{K/k}K = k$ holds.

Let Ω be the algebraic closure of k_0. Then we have a topological class formation over $\mathfrak{k}$ by taking $E(k) = (k^* \otimes_{\mathbf{Z}} (\mathbf{Q}/\mathbf{Z}))^\wedge$, where $M^\wedge = \mathrm{Hom}\,(M, \mathbf{R}/\mathbf{Z})$ is the compact character group of M, $\varphi_{k,K}$: $E(k) \to E(K)$ is the conorm, i.e., $\varphi_{k,K}(\chi)(A) = \chi(N_{K/k}A)$ for $\chi \in E(k)$, $A \in K^* \otimes (\mathbf{Q}/\mathbf{Z})$ and $N_{K/k} = N \otimes 1$. Then we have $E(k) \cong A(k)$, i.e., $\mathrm{Im}\,\Phi_k = A(k)$ and $\mathrm{Ker}\,\Phi_k = 0$ (Kawada [19]). The situation is similar if the characteristic of k_0 is p ($\neq 0$) and Ω is the maximal l-extension of k_0 for $l \neq p$.

(E 2.1) Let k_0 be an infinite algebraic extension of $\mathbf{Q}_p$ which contains all the roots of unity. Then k_0 is a C_1-field (Lang [30]) and satisfies (iii). Hence we can apply (E 2). On the other hand we can also apply the class formation theory in §6 (C). These two topological class formations $\{E_1(k)\}$ and $\{E_2(k)\}$ are equivalent in the sense that there exists a homeomorphic isomorphism Ψ_k: $E_1(k) \to E_2(k)$ ($k \in \mathfrak{k}$) such that $\Psi_K \circ \overset{(1)}{\varphi}_{k,K} = \overset{(2)}{\varphi}_{k,K} \circ \Psi_k$ ($k \subset K$) and $\sigma \circ \Psi_K = \Psi_K \circ \sigma$ ($\sigma \in G(K/k)$ for normal K/k) hold. The isomorphism Ψ_k can be defined explicitly by means of Hilbert norm-residue symbol (Kawada [21, V]). Namely, let $k = \bigcup_n k_n$, $[k_n: \mathbf{Q}_p] < \infty$, where each k_n contains a primitive n-th root of unity. Then we have $E_1(k) = \lim_{\leftarrow} \bar{k}_n^*$ and $E_2(k) \cong (\bar{k}^* \otimes_{\mathbf{Z}} (\mathbf{Q}/\mathbf{Z}))^\wedge$. Let W be the group of all the roots of unity in k. Define a dual pairing $\bar{k}_n^*/\bar{k}_n^{*n} \times k_n^*/k_n^{*n} \to W$ by $(\alpha, \beta) \mapsto (\alpha, \beta/\mathfrak{p})_n$. This induces a dual pairing $E_1(k) \times E_2(k)^\wedge \to W$ by

$$(\lim \eta_n, \xi \otimes (r/s)) \mapsto \left(\frac{\eta_n, \xi^{rn/s}}{\mathfrak{p}}\right)_n$$

where $\eta = \lim \eta_n \in E_1(k)$ ($\eta_n \in \bar{k}_n^*$) and $\xi \otimes (r/s) \in E_2(k)^\wedge$ ($\xi \in k_n^*$). This gives the above isomorphism Φ_k: $E_1(k) \cong E_2(k)$.

(E 2.2) Let k_0 be an infinite algebraic extension of $\mathbf{Q}$ which contains all the roots of unity. Then we can apply (E 2) just as in case (E 2.1). The equivalence of two class formations (E 2) and (E 0.3) can be proved also by means of Hilbert norm-residue symbol (Kawada [21, V]).

Namely, let $k = \bigcup_n k_n$ ($[k_n\colon \mathbf{Q}] < \infty$), where each k_n contains a primitive n-th root of unity, and let W be the group of all the roots of unity in k. Since k_n $(n = 3, 4, \ldots)$ are totally imaginary we can take compact class modules $E_1(k) = \lim_{\leftarrow} C(k_n)/D(k_n)$ and $E_2(k) = (k \otimes_{\mathbf{Z}} (\mathbf{Q}/\mathbf{Z}))^\wedge$. Let J_n be the idele group of k_n. Then the locally compact group J_n/J_n^n is self-dual by

$$(\prod \alpha_{\mathfrak{p}} \bmod J_n^n, \prod \beta_{\mathfrak{p}} \bmod J_n^n)_n \mapsto \prod\nolimits_{\text{finite } \mathfrak{p}} \left(\frac{\alpha_{\mathfrak{p}}, \beta_{\mathfrak{p}}}{\mathfrak{p}}\right)_n \in W.$$

J_n/J_n^n contains a discrete subgroup $k_n^* J_n^n/J_n^n$ such that the factor group $J_n/k_n^* J_n^n$ is compact. Moreover, the discrete subgroup $k_n^* J_n^n/J_n^n$ of J_n/J_n^n is the annihilator of itself by this self-duality. Hence we have a dual pairing between the compact group $J_n/k_n^* J_n^n \cong C(k_n)/C(k_n)^n$ and the discrete group $k_n^* J_n^n/J_n^n \cong k_n^*/k_n^{*n}$. Now we have a dual pairing between $E_1(k)$ and $E_2(k)^\wedge$ by

$$(\eta, A) = (\eta_n, \xi^{rn/s})_n \in W,$$

where $\eta = \lim \eta_n$ ($\eta_n \in C(k_n)/D(k_n)$) and $A = \xi \otimes_{\mathbf{Z}} (r/s)$ ($\xi \in k_n^*$). This gives the above isomorphism $\Psi_k\colon E_1(k) \cong E_2(k)$.

8. Examples of a class formation (II) (function fields)

(E 3) Let k_0 be an algebraic function field in one variable with algebraically closed constant field F of characteristic zero and Ω the algebraic closure of k_0. Let $D(k)$ be the group of all fractional divisors $A = \prod_\nu P_\nu^{r_\nu}$ ($r_\nu \in \mathbf{Q}$, $r_\nu = 0$ a.a.ν) and $P(k)$ the group of all principal divisors. Let $\mathbf{D}(k) = D(k)/P(k)$ be the fractional divisor classes of k and put $E(k) = \mathbf{D}(k)^\wedge$. Then $\{E(k) \mid k \in \mathfrak{k}\}$ with the conorm map $\varphi_{k,K}\colon E(k) \to E(K)$ for $k \subset K$ is a class formation for $\mathfrak{k}$. The norm-residue map $\Phi_k\colon E(k) \to A(k)$ is surjective, but $\operatorname{Ker} \Phi_k = F(k) \neq 0$. (Kawada [21, VI].)

(E 3.1) Let $T\mathbf{D}(k)$ be the torsion subgroup of $\mathbf{D}(k)$. Since $\mathbf{D}(k)/T\mathbf{D}(k)$ is uniquely divisible $F(k) = (\mathbf{D}(k)/T\mathbf{D}(k))^\wedge$ is the component of 1 in $E(k) = \mathbf{D}(k)^\wedge$ such that $\hat{H}^r(G, F(K)) = 0$ for $r \in \mathbf{Z}$, and for normal K/k. Hence $E(k)^* = E(k)/F(k) = (T\mathbf{D}(k))^\wedge$ is a class formation for $\mathfrak{k}$ by §6 (A). Since k_0 satisfies (i)–(iii) of (E 2) we have another class formation $E(k)' = (k^* \otimes (\mathbf{Q}/\mathbf{Z}))^\wedge$ which is equivalent to $E(k)$ by $k^* \otimes (\mathbf{Q}/\mathbf{Z}) \cong T\mathbf{D}(k)$ (cf. M. Deuring, Math. Ann., **106** (1932)). See also (E 5.1).

(E 3.2) Let k_0 be as in (E 3) and Ω_ϕ be the maximal unramified extension of k_0. Let us denote $\mathfrak{k}_\phi = \mathfrak{k}(\Omega_\phi/k_0)$. Take $E(k)_\phi = (\mathbf{D}_\phi(k))^\wedge$ where $\mathbf{D}_\phi(k)$ means the divisor class group of k in the usual sense. Then we see that $\{E(k)_\phi\}$ is a class formation for $\mathfrak{k}_\phi$. In this case $\operatorname{Im} \Phi_k = A(k)$ but $\operatorname{Ker} \Phi_k \neq 0$ (Kawada-Tate [18]).

(E 3.3) Let k_0 be as in (E 3). Fix a finite set S ($\neq \phi$) of prime divisors of k_0. Let Ω_S be the maximal S-ramified extension of k_0, where k is called *S-ramified* (or unramified outside S) over k_0 if every prime divisor $P \notin S$ is unramified for k/k_0. Put $\mathfrak{k}_S = \mathfrak{k}(\Omega_S/k_0)$. Take $E(k)_S = (\mathbf{D}_S(k))^\wedge$ where $\mathbf{D}_S(k)$ means the S-fractional divisor class group of k. Then $\{E(k)_S\}$ is a class formation for $\mathfrak{k}_S$ (Kawada [21, VI]). In this case $\operatorname{Im} \Phi_k = A(k)$ and $\operatorname{Ker} \Phi_k$ is the connected component of $E(k)_S$, which is $\neq 0$. If we take $E(k)_S^* = (T\mathbf{D}_S(k))^\wedge$, then $\{E(k)_S^*\}$ is also a class formation for $\mathfrak{k}_S$ and satisfies $E(k)_S^* \cong A(k)$.

(E 3.4) Let us consider the case (E 3.3) for classical case, i.e., for the constant field **C**. For $k_0 \subset k$ let $S(k)$ denote the set of $s(k)$ prime divisors of k which are extensions of a prime divisor of k_0 contained in S. Let $R(k)$ be the Riemann surface of k and $R_S(k) = R(k) - S(k)$. Let $g(k)$ be the genus of $R(k)$. Let $E(k)$ be the one-dimensional integral homology group $H_1(R_S(k), \mathbf{Z})$ of $R_S(k)$ which is the direct sum of $2g(k) + s(k) - 1$ $(= [k: k_0](2g(k_0) + s(k_0) - 2) + 1)$ copies of $\mathbf{Z}$. Let $\varphi_{k,K}$: $E(k) \to E(K)$ be defined by $\gamma \mapsto V\gamma$ where $V\gamma$ is the covering path of $\gamma \in R_S(k)$ on the unramified covering surface $R_S(K)$ of $R_S(k)$. Then we can prove that $\{E(k)\}$ is a class formation for $\mathfrak{k}_S$. For the norm-residue map Φ_k we have Im Φ_k is dense in $A(k)$ and Ker $\Phi_k = 0$. (Kawada [21, VI]).

The equivalence of the class formations $E(k)$ and $E(k)_S$ defined in (E 3.3) is given by the pairing $H_1(R_S(k), Z) \times T\mathbf{D}_S(k) \to \mathbf{C}_1$ (where $\mathbf{C}_1$ means the set of all complex numbers with absolute value 1) defined by $(\eta, A)_S \mapsto \exp\left(\int_\eta d\log A\right)$ for 1-cycle η on $R_S(k)$ and $A \in T\mathbf{D}_S(k)$ (where $d\log A$ means the abelian differential of the third kind on $R_S(k)$ corresponding to a divisor A).

(E 3.5) In case (E 3.2) the situation is much complicated. Let $R(k)$ be the Riemann surface of $k \in \mathfrak{k}_\phi$ and $H_1(k)$ be the one-dimensional integral homology group of $R(k)$. Put $E(k) = H_1(k) + \mathbf{R}/\mathbf{Z}$. For $k \subset K$ we define $\varphi_{k,K}$ by $(\gamma, \lambda) \mapsto (V\gamma, \lambda)$ for $\gamma \in H_1(k)$ and $\lambda \in \mathbf{R}/\mathbf{Z}$ where $V\gamma$ means as in (E 3.4). If K/k is normal with $G = G(K/k)$, $\sigma \in G$ operates on $E(K)$ by σ: $(\gamma, \lambda) \mapsto (\sigma\gamma, \lambda + q(\sigma, \gamma))$. Here $q(\sigma, \gamma) = \int_\gamma dW(\sigma^{-1}P_0, P_0)/(2\pi i) \bmod \mathbf{Z}$ with the base point P_0 on $R(k)$ and $dW(P, Q)$ denotes the abelian differential of the third kind on $R(K)$ with poles of the first order at P, Q with the residues $+1$ and -1 respectively such that the integral $\int_\gamma dW(P, Q)$ is purely imaginary for any 1-cycle γ on $R(K)$. Then we can prove that $\{E(k)\}$ is a class formation for $\mathfrak{k}_\phi$ by showing that this class module $E(k)$ is the Pontrjagin dual of the locally compact Abelian group $\mathbf{D}(k)$. For the norm-residue map Φ_k, Im Φ_k is dense in $A(k)$ but Ker $\Phi_k = \mathbf{R}/\mathbf{Z}$, i.e. the connected component of 0 in $E(k)$ (Kawada-Tate [18], Kawada [19]).

(E 4) Class field theory over a function field in several variables with a finite constant field is developed by Lang [32] and Serre [40]. But the corresponding class formation theory is not yet developed until now (see also Serre [40, Chapter VI, n. 34]).

9. Examples of a class formation (III) (local fields)

Let k_0 be a local field, i.e., a complete field with respect to a discrete valuation, and $\bar{k}_0$ its residue field. Let Ω be the separable closure of k_0. We have local class field theory (E 0.1) and (E 0.2) in case $\bar{k}_0$ is a finite field $\mathbf{F}_q$ $(q = p^n)$ in both cases ch $(k) = 0$ and ch $(k) = p$. We have also generalized local class field theory by Moriya and Whaples [51]. It is proved that the local class field theory holds if and only if the residue class field $\bar{k}_0$ is quasi-finite. Here a field k is said to be quasi-finite if (i) k is perfect and (ii) the Galois group of the maximal separable extension of k is isomorphic to $\tilde{\mathbf{Z}}$. (Serre [42, Chapter XIII, §4, and Exercise 2 at page 203]).

(E 5) Let k_0 be a local field where the residue field $\bar{k}_0$ is algebraically closed. Lang [30] proved that k_0 is a C_1-field. Hence we can apply Kummer theory in (E 2) in case ch $(k_0) = 0$. Serre [41] gave another class formation theory in both cases (i) ch $(k_0) =$ ch $(\bar{k}_0) = 0$ or p and (ii) ch $(\bar{k}_0) = p$, ch $(k_0) = 0$. Namely, let U_k be the group of units of k with the structure of proalgebraic group over $\bar{k}$

and let $\pi_1(U_k)$ be the fundamental group of U_k. Put $E(k) = \pi_1(U_k)$. Then $\{E(k)\}$ is a class formation. Here Φ_K: $E(k) \cong A(k)$, so that $\operatorname{Im} \Phi_k = A(k)$ and $\operatorname{Ker} \Phi_k = 0$ hold.

(E 5.1) Let k_0 be an algebraic function field in one variable with algebraically closed constant field $\bar{k}_0$, where ch $(\bar{k}_0) = 0$ or p. Let Ω be the separable closure of k_0. Let $C_0(k)$ be the idele class group of degree 0 of k. Consider $C_0(k)$ as a pro-algebraic group and put $E(k) = \pi_1(C_0(k))$ for $k \in \Omega$. Then $\{E(k)\}$ is a class formation for $\mathfrak{k}$, and $E(k) \cong A(k)$ as in (E 5). (The case ch $(\bar{k}_0) = 0$ is considered in (E 3.1) already (Serre [41]).

10. Examples of a class formations (IV) (p-extensions in characteristic p)

(E 6) Let k_0 be a field of characteristic p. Let Ω be the maximal p-extension of k_0. For $k \in \mathfrak{k} = \mathfrak{k}(\Omega/k_0)$ let $V(k)$ be the additive group of all Witt vectors with components in k. Put $\wp x = x^p - x$ $(x \in V(k))$. Define $E(k) = W(k)^\wedge$ where $W(k) = (V(k)/\wp V(k)) \otimes_{\mathbf{Z}} (\mathbf{Q}_p/\mathbf{Z}_p))$ and $\varphi_{k,K}$ to be the cotrace. Then $\{E(k)\}$ is a topological class formation for Ω such that Φ_k: $E(k) \cong A(k)$ holds (Kawada [19]).

(E 6.1) Let k_0 be a formal power series field in one variable with a finite constant field and let Ω be the maximal p-extension of k_0. Then we have two kinds of class formations for $\mathfrak{k} = \mathfrak{k}(\Omega/k_0)$: $E_1(k)$ $(k \in \mathfrak{k})$ as in (E 0.2) and $E_2(k) = W(k)^\wedge$ $(k \in \mathfrak{k})$ as in (E 6). Define $\tilde{k} = \lim_{\leftarrow} (k^*/(k^*)^{p^n})$ which satisfies $\tilde{k}/k \cong \mathbf{Z}_p/\mathbf{Z}$. Then the compact multiplicative abelian group $\tilde{k}$ and the discrete additive group $W(k)$ are the dual groups in the sense of Pontrjagin and this proves the equivalence of two class formations. The duality is expressed by means of the residue-symbol defined by Witt (e.g., Kawada-Satake [20]).

(E 6.2) In a similar way we can prove the equivalence of two class formations (E 0.4) and (E 6) for the algebraic function fields in one variable with finite constant fields. This can be proved by the duality between the compact multiplicative abelian group $E_1(k) = \lim_{\leftarrow} J_k/k^* \cdot J_k^{p^n}$ (where J_k is the idele group of k) and the discrete additive group $W(k)$ (e.g., Kawada-Satake [20]).

(E 6.3) Let k_0 be an algebraic function field in one variable with algebraically closed constant field of characteristic p, Ω the maximal separable p-extension of k_0 and $\mathfrak{k} = \mathfrak{k}(\Omega/k_0)$. Let S be a finite set of prime divisors of k_0 and denote by $\mathfrak{k}_S$ the family of all S-ramified extensions of k_0 contained in Ω. Similarly we denote by $\mathfrak{k}_\phi$ the family of all unramified extensions of k_0 contained in Ω. Since $\mathfrak{k}_S$ and $\mathfrak{k}_\phi$ are subfamilies of $\mathfrak{k}$ we can apply the processes (A) and (B) in §6. Let $V_\phi(k)$ and $V_S(k)$ $(k \in \mathfrak{k})$ be the additive group of all unramified and S-ramified Witt-vectors x of k (i.e., $x \in \wp V(k_P)$ for every prime divisor P of k or for every $P \notin S(k)$ respectively). Then $E_\phi(k) = W_\phi(k)^\wedge$ $(W_\phi(k) = (V_\phi(k)/\wp V(k)) \otimes (\mathbf{Q}_p/\mathbf{Z}_p))$ and $E_S(k) = W_S(k)^\wedge$ $(W_S(k) = (V_S(k)/\wp V(k)) \otimes (\mathbf{Q}_p/\mathbf{Z}_p))$ make class formations for $\mathfrak{k}_\phi$ and $\mathfrak{k}_S$ respectively, such that the norm-residue map Φ_k is an isomorphism in each case (Kawada [21, VI]).

(E 6.4) For unramified p-extensions of an algebraic function field with algebraically closed constant field in (E 6.3) we can give another equivalent class formation over $\mathfrak{k}_\phi$ as follows. Let $D_0(k)$ be the divisor class group of degree zero of k. Let $D^{(n)}(k)$ be the subgroup of $D_0(k)$ consisting of all divisor classes A such that $A^{p^n} = 1$ and let $\bar{D}(k) = \lim_{\leftarrow} D^{(n)}$ which is isomorphic to a direct sum of ρ_k copies of $\mathbf{Z}_p$. Here ρ_k means the Hasse-invariant of k. It is proved by Cartier-Serre-Tamagawa that the compact group $\bar{D}(k)$ and the discrete additive group

$W_\Phi(k)$ are dual in the sense of Pontrjagin. This duality can be explicitly given by means of residue-symbol of Witt. Thus we can prove that $E(k) = \overline{D}(k)$ gives a topological class formation for $\mathfrak{k}_\Phi$ such that Φ_k: $E(k) = A(k)$ holds (Kawada [21, VI]).

11. Special class formation

Let k_0, Ω and $\mathfrak{k} = \mathfrak{k}(\Omega/k_0)$ be as before. Let $\{E(k) \mid k \in \mathfrak{k}\}$ be a topological class formation for $\mathfrak{k}$. We call $\{E(k)\}$ a *special class formation* if Φ_k: $E(k) \cong A(k)$ ($k \in \mathfrak{k}$) holds by the norm-residue map Φ_k.

EXAMPLES. In the above examples (E 0)–(E 6) we have special class formations in cases (E 0.1, 0.2, 0.4), (E 1, 1.1, 1.2), (E 2, 2.1, 2.2), (E 3.1, 3.4), (E 5, 5.1) and (E 6, 6.1, 6.2, 6.3, 6.4). But in cases (E 0.3) (global class field theory), (E 3.4) (unramified extensions of a function field in one variable with algebraically closed constant field of characteristic zero) and (E 4) (algebraic function fields in several variables) (cf. Serre [40, p. 159, Theorem 6]) it is impossible to define a special class formation for them.

Whether one can define a special class formation for $\mathfrak{k} = \mathfrak{k}(\Omega/k_0)$ or not depends only on the structure of the profinite Galois group $G = G(\Omega/k_0)$. Let H and U be the corresponding open subgroups of G to k and K ($k, K \in \mathfrak{k}$) respectively. Then we can identify $A(k)$ and $A(K)$ with H/H' and U/U' respectively, where $'$ means the commutator subgroup. If $k \subset K$, i.e., $H \supset U$ we have the transfer map: $V_{K/k}$: $A(k) \to A(K)$. Now, let $\{E(k) \mid k \in \mathfrak{k}\}$ be an arbitrary class formation for $\mathfrak{k}$. Then we have the following commutative diagram:

$$\begin{array}{ccc} E(k) & \xrightarrow{\varphi_{k,K}} & E(K) \\ \Psi_k \downarrow & & \downarrow \Psi_K \\ A(k) & \xrightarrow{V_{K/k}} & A(K) . \end{array}$$

Suppose that a profinite group G is given. To each open subgroup H of G let us associate $A(H) = H/H'$. Define $\varphi_{H,U}$: $A(H) \to A(U)$ for $H \supset U$ by $V_{H/U}$ (transfer map), and the operation σ: $A(H) \to A(H)$ by $\sigma \in H/U$ for normal subgroup U of H by the inner automorphism. We call G to be *malleable* (Brumer [2]) if $\{A(H), \varphi_{H,U}\}$ satisfy the axioms of a class formation, that is,

$\mathbf{M}_1$. $V_{H/U}$: $A(H) \to A(U)$ for $H \supset U$ is injective and if U is a normal subgroup of H its image is $A(U)^{H/U}$.

$\mathbf{M}_2$. If U is a normal subgroup of H then $H^1(H/U, A(U)) = 0$, and

$$H^2(H/U), A(U)) \cong Z/nZ$$

where $n = [H: U]$. The 2-cocycle $u_{H/U}$ of the group extension of $A(U)$ by H/U is a generator of $H^2(H/U, A(U))$.

THEOREM. *G is malleable if either $\mathbf{M}_1$ or $\mathbf{M}_2$ is satisfied*, i.e., *$\mathbf{M}_2$ implies $\mathbf{M}_1$ and conversely $\mathbf{M}_1$ implies $\mathbf{M}_2$* (Serre [40, p. 156, Theorem 5], Kawada [17]).

EXAMPLE 1. Let k be a field of characteristic zero which contains all the roots of unity. Then $A(k) \cong (k^* \otimes (\mathbf{Q}/\mathbf{Z}))^\wedge$ holds. Let Ω be the algebraic closure of k

and $G = G(\Omega/k)$. Then G is malleable, i.e., $E(k) = (k^* \otimes (\mathbf{Q}/\mathbf{Z}))^\wedge$ defines a special class formation for $\mathfrak{k} = \mathfrak{k}(\Omega/k)$ if and only if the condition (iii) in (E 2) holds (Kawada [17, II, Theorem 2]).

EXAMPLE 2. Let k be an algebraic function field in r variables ($r \geqq 1$) with algebraically closed constant field (algebraic closure of a finite field $\mathbf{F}_p$) and let Ω be separable closure of k. If we apply Theorem for extensions whose degree is prime to p, we see that the condition (iii) in (E 2) holds if and only if $r = 1$. Hence the Galois group $G = G(\Omega/k)$ is malleable if and only if $r = 1$ (Serre [40, Chapter VI, n. 34]).

EXAMPLE 3. Let k be an algebraic number field and let Ω be the maximal unramified p-extension of k. Then the condition $\mathbf{M}_1$ does not hold in this case, since the principal ideal theorem says that the transfer map $A(H) \to A(U)$ is trivial if U is the commutator subgroup of H. Hence the Galois group $G = G(\Omega/k)$ is not malleable.

If G is malleable then every closed subgroup H of G is also malleable. G is malleable if and only if its p-Sylow group G_p is malleable for any prime p.

For nonspecial class formations, e.g., (E 0.3) (global class field theory) and (E 3.4), we can define a compact topology in each $E(k)$ ($k \in \mathfrak{k}$) such that $E(k)/F(k) \cong A(k)$ holds with the connected component $F(k)$ of 0 of $E(k)$.

Now comes a problem whether we can define a class formation $\{E(k)\}$ suitably for any given k_0, Ω and $\mathfrak{k} = \mathfrak{k}(\Omega/k_0)$ or not. An answer is given by the following abstract class formation theory.

12. Abstract class formation

THEOREM. *Let k_0, Ω and $\mathfrak{k}$ be as before. Then we can associate a module $E(k)$ to each $k \in \mathfrak{k}$ suitably such that $\{E(k)\}$ is a class formation for $\mathfrak{k}$* (Grant-Whaples [8], Kawada [21]).

Assume for simplicity that $\mathfrak{k}$ is a countable set. Let $k_0 \subset k$ ($k \in \mathfrak{k}$) and take $k \subset L_n$ ($L_n \in \mathfrak{k}$) such that L_n/k is normal and $\Omega = \bigcup L_m$. Put $M_n(k) = (I(G_n) \otimes I(G_n))^{H_n}$ where $G_n = G(L_n/k_0)$, $H_n = G(L_n/k)$ and $I(G) = \sum_{\sigma \neq 1} \mathbf{Z}(\sigma - 1)$. Then we can define canonically a projective system $\{M_n(k)\}$ and its limit $E(k) = \lim_{\leftarrow} M_n(k)$. We can verify that $\{E(k)\}$ is a class formation. An equivalent construction is possible by taking inductive limit group of $M_n(k)$ instead of projective limit of $M_n(k)$.

This construction is not satisfactory enough in the sense that the relation to the known examples is not clear.

It is also an open problem to consider the relation between a result of Harrison [10] and class formation theory.

13. Free profinite group

An important example of a malleable group is given by the following theorem:

THEOREM (TATE). *Let G be a free profinite group or a free pro-p-group. Then G is malleable* (cf. e.g., Kawada [17, II, Theorem 6]).

A partial converse is given by

THEOREM. *Let G be a malleable pro-p-group which has a finite number of generators. In order that G is a free pro-p-group it is necessary and sufficient that $A(H) = H/H'$ has no p-torsion element for each open subgroup H of G* (Kawada [17, II, Theorem 7]). (See Appendix I.)

EXAMPLE 1. (E 0.1) Let k be a $\mathfrak{p}$-adic number field which does not contain any p-th primitive root of unity. Let Ω be the maximal p-extension of k and $G = G(\Omega/k)$. Then we know already that G is malleable and by assumption G has no p-torsion element. Hence G is a free pro-p-group (Šafarevič [36]).

EXAMPLE 2. (E 3.5) Let G be a pro-p-group with $2g$ generators and one fundamental relation $1 = [x_1, y_1] \ldots [x_g, y_g]$. Then G is not malleable, since $A(H)$ has no torsion element for any open subgroup H of G.

Among the examples of a special class formation (i.e., malleable profinite groups) we have the situation in Tate's theorem for (E 1, 1.1, 1.2), (E 2, 2.1, 2.2), (E 3.1, 3.5), (E 6, 6.1–6.4) and partially in (E 5):

THEOREM. *In the following cases the Galois group $G = G(\Omega/k)$ is a free pro-p-group*:

I. ch $(k) = p$.

(1) *k is an arbitrary field and Ω is the maximal p-extension of k. The number of free generators of G is* $[k\colon \wp k]$. (See e.g., Kawada [17, I, Theorem 1], Serre [44, Chapter II, §5].)

(2) *k is an algebraic function field in one variable with algebraically closed constant field and Ω is the maximal unramified or S-ramified extension* (Šafarevič [36], Kawada [20]).

II. ch $(k) = 0$ *(or different from p) and Ω is the maximal p-extension of k.*

(1) *k contains a primitive p-th root of unity and the condition* (iii) *in* (E 2) *holds* (Kawada [17, I, Theorem 1], Lang [31]). *The number of free generators of G is* $[k^*\colon (k^*)^p]$.

The following (2), (3), (4) *are special cases of* (1):

(2) (E 2). *k contains all the roots of unity and the condition* (iii) *in* (E 2) *holds.*

(3) *k is an algebraic function field in one variable with algebraically closed constant field* (Tsen).

(4) *k is an infinite algebraic extension of $\mathbf{Q}_p$ containing a p-th root of unity and* $[k\colon \mathbf{Q}_p] = p^\infty$.

(5) *k is a $\mathfrak{p}$-adic number field which does not contain any p-th root of unity* (Šafarevič [36]). *The number of free generators of G is* $[k\colon \mathbf{Q}_p] + 1$.

REMARK. (1) There is a remarkable result by Iwasawa [12] that $G = G(\Omega/k)$ is a (solvable) completion of a free group if k is the maximal abelian extension of an algebraic number field and Ω is the maximal solvable extension of k. (2) There is a classical result that $G = G(\Omega_S/k_0)$ in (E 3.4) is a free profinite group with $2g(k_0) + |S| - 1$ generators.

The next problem is to characterize malleable groups. This was done by Tate and Brumer by using cohomological dimension.

14. Cohomological dimension of a profinite group

Let G be a profinite group and let A be an abelian group on which G operates continuously. Then the cohomology group of G with coefficients in A is defined by $H^q(G, A) = \lim_{\rightarrow} H^q(G/U, A^U)$ ($q \geqq 0$) where U runs over all open subgroups of G. In particular, $H^0(G, A) = A^G$. Every profinite group G contains a closed pro-p-group G_p such that $[G: G_p]$ is prime to p. G_p is called a p-Sylow subgroup of G. Then Res: $H^q(G, A) \rightarrow H^q(G_p, A)$ is injective on the p-primary part. The *cohomological dimension* of G and the *strict cohomological dimension* of G are defined by

$\mathrm{cd}\,(G) \leqq n \Leftrightarrow H^r(G, A) = 0$ for all $r > n$ and for all torsion groups A,

$\mathrm{cd}_p\,(G) \leqq n \Leftrightarrow p$-primary part of $H^r(G, A) = 0$ for all $r > n$ and for all torsion group A,

$\mathrm{scd}\,(G) \leqq n \Leftrightarrow H^r(G, A) = 0$ for all $r > n$ and for all A,

$\mathrm{scd}_p\,(G) \leqq n \Leftrightarrow p$-primary part of $H^r(G, A) = 0$ for all $r > n$ and for all A.

We have the following properties:

(1) For a closed subgroup H of G, $\mathrm{cd}_p\,(H) \leqq \mathrm{cd}_p\,(G)$ and $\mathrm{scd}_p\,(H) \leqq \mathrm{scd}_p\,(G)$ hold. Equality holds if $[G: H]$ is prime to p, or if H is open and $\mathrm{cd}_p\,(G) < \infty$.

(2) For a closed normal subgroup N of G, $\mathrm{cd}_p\,(G) \leqq \mathrm{cd}_p\,(G/N) + \mathrm{cd}_p\,(N)$ holds.

(3) $\mathrm{cd}\,(G) = \sup_p (\mathrm{cd}_p\,(G))$, $\mathrm{scd}_p\,(G) = \sup_p (\mathrm{scd}_p\,(G))$, and

$$\mathrm{scd}_p\,(G) = \mathrm{scd}_p\,(G_p) = \mathrm{scd}\,(G_p), \qquad \mathrm{cd}_p\,(G) = \mathrm{cd}_p\,(G_p) = \mathrm{cd}\,(G_p).$$

(4) $\mathrm{cd}_p\,(G) \leqq \mathrm{scd}_p\,(G) \leqq \mathrm{cd}_p\,(G) + 1$ holds.

(5) Let $G = G_p$ be a pro-p-group. (i) $\mathrm{cd}_p\,(G_p) \leqq n$ if and only if

$$H^{n+1}(G_p, \mathbf{Z}/p\mathbf{Z}) = 0,$$

(ii) $\mathrm{scd}_p\,(G_p) \leqq n$ if and only if $\mathrm{cd}_p\,(G_p) \leqq n$ and for any open subgroup H of G_p, $H^{n+1}(H, \mathbf{Z}) = 0$.

(6) $\mathrm{cd}_p\,(G_p) = 0$ if and only if $G_p = 1$,

(7) $\mathrm{cd}_p\,(G_p) = \infty$ if G_p is a finite p-group,

(8) $\mathrm{cd}_p\,(G_p) = 1$ if and only if G_p is a free pro-p-group. Then $\mathrm{scd}_p\,(G_p) = 2$.

All the above results are proved by Tate. See Lang [31], Serre [44], Douady [8].

15. Profinite group G with scd $(G) = 2$

The following theorem was proved by Tate and Brumer [2]:

THEOREM. *Let G be a profinite group. In order that G is malleable it is necessary and sufficient that* $\mathrm{scd}\,(G) = 2$.

(Tate's proof is not published. Cf. Appendix II.)

Usually it is difficult to determine directly the scd of G.

EXAMPLES. We know already several examples of malleable groups.

(1) Let G be a free pro-p-group. Then G is both malleable (§13, Theorem) and $\mathrm{scd}\,(G) = \mathrm{scd}_p\,(G) = 2$ (§14, (8)).

In the following examples we put $G = G(\Omega/k_0)$.

(2) Let k_0 (ch $(k_0) = 0$ or p) be a local field with a finite residue field $\mathbf{F}_q$, Ω the maximal tamely ramified extension. Then G is malleable but G is not free.

(3) Let k_0 be as in (2) and Ω the separable closure of k_0. Then G is malleable by (E 0.1), cd $(G) = 2$ by §14, (1) and scd $(G) = 2$ (Serre [44, Chapter II]).

(4) Let k_0 be as in (2) and Ω the maximal separable p-extension of k_0. Then G is malleable. If either ch $(k_0) = p$, or ch $(k_0) = 0$ and k does not contain any primitive p-th root of unity then G is a free pro-p-group, but if ch $(k_0) = 0$ and k contains a primitive p-th root of unity then G has $[k_0\colon \mathbf{Q}_p] + 2$ generators and a unique fundamental relation (Kawada [17, I, Theorem 2]).

(5) Let k_0 be an algebraic number field. If k_0 is totally imaginary then G is malleable ((E 0.3) and §6, (A)) and hence scd $(G) = 2$. If k_0 is not totally real then $\mathrm{scd}_p\,(G) = 2$ for $p \neq 2$ (because the open subgroup H of G corresponding to $k_0(\zeta_p)$ has $\mathrm{scd}_p\,(H) = 2$ and $[G\colon H]$ is prime to p), and similarly $\mathrm{scd}_2\,(G) = \infty$.

(6) Let k_0 be an algebraic function field in r variables with a finite constant field and Ω the separable closure of k_0. If $r = 1$ then G is malleable. If $r > 1$ then G is not malleable (§11), i.e., scd $(G) \geqq 3$. On the other hand we have by a general theorem scd $(G) \leqq r + 1$ (see Serre [44, Chapter II, §4]). Actually it is supposed that scd $(G) = r + 1$ holds.

REMARK. Let k_0 be a C_i-field (Lang [30]), Ω the separable closure of k_0 and $G = G(\Omega/k_0)$. If k_0 is a C_1-field then cd $(G) \leqq 1$ (Serre [44, Chapter II, §4.5]). If k_0 is a C_r-field then whether cd $(G) \leqq r$ holds or not is an open problem (Serre).

(7) Let k_0 be an algebraic function field in one variable over $\mathbf{C}$, Ω the maximal unramified extension of k_0 and $G = G(\Omega/k_0)$. We know by (E 3.5) that G is not malleable.

In the above examples (2), (4), (7) we know the structure of G explicitly by giving its generators and (unique) fundamental relation. Thus we are led to the problem to determine whether such a group G is malleable or not, or to determine the cohomological dimension of G. We shall consider them in §16.

In particular, let G be a pro-p-group. Then $n = \dim_{\mathbf{Z}/p\mathbf{Z}} H^1(G, \mathbf{Z}/p\mathbf{Z})$ is the minimal number of generators of G and $r = \dim_{\mathbf{Z}/p\mathbf{Z}} H^2(G, \mathbf{Z}/p\mathbf{Z})$ is the number of fundamental relations. (See, e.g., Serre [44, Chapter I, §4].)

16. Profinite group with relations

Let Ω be an infinite separable normal algebraic extension of k_0 and $G = G(\Omega/k_0)$ be its profinite Galois group. The generators and relations of G for the examples (2), (4), (7) in §15 are as follows.

(2) Let k_0 be a local field and Ω the maximal tamely ramified extension of k_0. Then the profinite group G has two generators x, y with a unique relation $xyx^{-1}y^{-q} = 1$ where q is the number of elements of the residue field of k_0 (Iwasawa [13], Koch [25]).

(4) Let k_0 be a $\mathfrak{p}$-adic number field which contains a primitive p-th root of unity and Ω be the maximal p-extension of k_0. Then the pro-p-group G has $d + 2$ ($d = [k_0\colon Q_p]$) generators with a unique relation. If $p \neq 2$ the unique relation is

$$x_1^q[x_1, x_2]\ldots[x_{d+1}, x_{d+2}] = 1$$

where k_0 contains precisely a primitive q-th root of unity ($q = p^e$). If $p = 2$ and k_0 contains precisely a primitive 2^e-th ($e \geqq 2$) root of unity then the unique relation is the same as in the case $p \neq 2$ (with $q = 2^e$). If $p = 2$ and k_0 does not contain any primitive 4-th root of unity, then for odd d the unique relation is

$$x_1^2 x_2^4 [x_2, x_3] \ldots [x_{d+1}, x_{d+2}] = 1$$

and for even d the unique relation is of the form D_2^e ($s \neq \infty$ or $t \neq \infty$) given below (Demuškin [6], [7], Labute [28], Serre [43]).

(7) Let k_0 be an algebraic function field in one variable with the algebraically closed constant field of characteristic 0 and Ω the maximal unramified extension of k_0. Then the profinite group $G = G(\Omega/k_0)$ is the completion of a group with $2g$ generators $x_1, \ldots, x_g, y_1, \ldots, y_g$ and a unique relation $[x_1, y_1] \ldots [x_g, y_g] = 1$ (classical).

A pro-p-group G is called a *Demuškin group* if (1) $H^2(G, \mathbf{Z}/p\mathbf{Z}) \cong \mathbf{Z}/p\mathbf{Z}$, (2) $H^1(G, \mathbf{Z}/p\mathbf{Z})$ is of dimension n over $\mathbf{Z}/p\mathbf{Z}$, and (3) the cup-product $H^1(G) \times H^1(G) \to H^2(G) \cong \mathbf{Z}/p\mathbf{Z}$ is a nondegenerate bilinear form. We assume $n > 1$. G has n generators and a unique fundamental relation. The above examples (4), (7) are of this type and it can be proved that cd $(G) = 2$ (Serre [43]). Demuškin and Labute classified all these groups as follows.

Let q be such that $G/G' \cong (\mathbf{Z}_p)^{n-1} \times (\mathbf{Z}_p/q\mathbf{Z}_p)$.

(D_q) If $q \neq 2$ then the unique relation is

$$x_1^q [x_1, x_2] \ldots [x_{n-1}, x_n] = 1$$

where $q = p^f$ (> 2) or $q = p^\infty = 0$. G is determined by the invariants n and q.

(D_2^0) If $p = q = 2$ and n is odd then the unique relation is

$$x_1^2 x_2^k [x_2, x_3] \ldots [x_{2m}, x_{2m+1}] = 1$$

where $n = 2m + 1$ and $k = 2^s$ ($s \geqq 2$ or ∞). G is determined by the invariants $2m + 1$ and s.

(D_2^e) If $p = q = 2$ and n is even, then the unique relation is

$$x_1^{2+k} [x_1, x_2] x_3^h [x_3, x_4] \ldots [x_{2m-1}, x_{2m}] = 1$$

where $n = 2m$, $k = 2^s$, $h = 2^t$, s, t are $2, 3, \ldots, \infty$ and at least one of s, t is ∞. G is determined by the invariants $2m$, s and t.

THEOREM. *Let G be a Demuškin group. Then* $\mathrm{scd}_p\,(G) = 2$ *except the cases* D_q ($q = p^\infty = 0$) *or* D_2^e ($s = t = \infty$), D_2^0 ($s = \infty$). *In these exceptional cases* $\mathrm{scd}_p\,(G) = 3$.

This can be proved either by a direct computation in each case (Kawada [22]) or by using "module dualisant" of G (Serre [44, Chapter I, Proposition 31] and Labute [28]).

In general let G be a pro-p-group defined by a unique relation. In a paper [22] the author gave a method to determine whether G is malleable (i.e., scd $(G) \leqq 2$) or not and applied this method to several examples. Labute [29] gave a general criterion for cd $(G) \leqq 2$. (In case of a discrete group G with a single defining relation there is a well-known result by Lyndon [33] on $H^r(G, K)$ for arbitrary G-module K. E.g., $H^r(G, K) = 0$ holds for $r \geqq 3$ if G operates trivially on K and if the unique relation R of G cannot be expressed as $R = Q^q$ ($q > 1$).)

17. Extensions with given ramifications

Let k be an algebraic number field and S a finite set of places of k. An algebraic extension K of k is said to be *S-ramified* (or unramified outside S) if K/k is unramified for every place $P \notin S$. Let Ω be the maximal S-ramified algebraic extension of k and $\Omega^{(p)}$ the maximal S-ramified p-extension of k. The Galois group $G = G(\Omega/k)$ and $G^{(p)} = G(\Omega^{(p)}/k)$ are studied by many authors: Šafarevič [38], Tate [47], Poitou [34], [35], Brumer [3], Höchsmann [11], Koch [26], Uchida [49] and Takahashi [45], [46].

(I) Structure of $G^{(p)}$. Let $S = \emptyset$. This problem is the class field tower problem and is answered by Šafarevič (1964). There is an example that G is finite (O. Taussky, J. London Math. Soc., 12 (1937)). In case $S = \emptyset$ Šafarevič [38] obtained the minimal number of generators of $G^{(p)}$ and the number of relations among them. In some special cases he determined the relations explicitly. See also Koch [26], Brumer [3]. Brumer gave a necessary and sufficient condition that $G^{(p)}$ is a free pro-p-group. An example is the cyclotomic field $k = Q(\zeta_p)$, where p is a regular prime, and $S = \{\mathfrak{p}\}$ $(p \in \mathfrak{p})$.

(II) Cohomological dimension of $G^{(p)}$. Brumer [3] proved the following: Let S be a finite set of places of k containing all places above p, and assume that k is totally imaginary in case $p = 2$. Then $\mathrm{cd}_p\,(G^{(p)}) \leqq 2$.

It is conjectured that $\mathrm{scd}_p\,(G^{(p)}) = 2$ holds (Tate [47], Takahashi [45])[2]. This is proved in some cases where the explicit structure of $G^{(p)}$ is given (Koch [26], Brumer [3A]). But, as far as the author knows, this conjecture is not yet proved in general. It is known that $\mathrm{scd}_p\,(G^{(p)}) = 2$ is equivalent to the nonvanishing of the p-adic regulator of every K $(k \subset K \subset \Omega, [K\colon k] < \infty)$ (Brumer [3A], Takahashi [45]). The nonvanishing of the p-adic regulator of K is only proved for abelian extensions K of $\mathbf{Q}$ (Brumer [3B]).

In case $S = \emptyset$ we know already in §11 that $\mathrm{scd}_p\,(G^{(p)}) \geqq 3$ holds unless $G^{(p)} = \{1\}$. It is even conjectured that $\mathrm{scd}_p\,(G^{(p)}) = \infty$ holds.

For an algebraic function field k in one variable with algebraically closed constant field (not necessarily the complex number field) the following facts are known (Serre). Let Ω_ϕ be the maximal unramified extension of k. Then $\mathrm{scd}_p\,(G) = 3$ if $p \neq$ characteristic of k (assuming that the genus of $k \neq 0$). Let Ω_S be the maximal S-ramified extension of k $(S \neq \emptyset)$. Then $\mathrm{scd}_p\,(G) = 2$ if $p \neq$ the characteristic of k.

Appendix

(I) We have the following stronger result by Brumer:

Theorem. (a) *A profinite group G is p-malleable if and only if $H^2(H, \mathbf{Z}/p\mathbf{Z})^\wedge$ is naturally isomorphic to the p-torsion of H/H' for all open subgroups of G* (Brumer [2]).

(b) *Let G be a pro-p-group and malleable. Then G is a free pro-p-group if and only if G/G' is without torsion.*

---

[2]The proof of Lemma 2 in [45] is incomplete.

The proof follows from the following exact sequences. Consider

$$0 \to \mathbf{Z}/p\mathbf{Z} \to \mathbf{Q}_p/\mathbf{Z}_p \xrightarrow{p} \mathbf{Q}_p/\mathbf{Z}_p \to 0,$$

passing to cohomology gives

$$\begin{aligned} 0 \to H^1(H, \mathbf{Z}/p\mathbf{Z}) &\to H^1(H, \mathbf{Q}_p/\mathbf{Z}_p) \xrightarrow{p} H^1(H, \mathbf{Q}_p/\mathbf{Z}_p) \\ \to H^2(H, \mathbf{Z}/p\mathbf{Z}) &\to H^2(H, \mathbf{Q}_p/\mathbf{Z}_p) \xrightarrow{p} H^2(H, \mathbf{Q}_p/\mathbf{Z}_p). \end{aligned}$$

Take the dual. Then we get

$$0 \to H_2(H, \mathbf{Z}_p)/pH_2(H, \mathbf{Z}_p) \to H_2(H, \mathbf{Z}/p\mathbf{Z})^\wedge \to H/H' \xrightarrow{p} H/H'$$

(See Brumer [2]).

(II) We shall give here a proof of the theorem: "scd $(G) \leqq 2 \Leftrightarrow G$ is malleable" which the author learned from J. Tate in a conversation with him in 1962.

Let H, U be any open subgroups of G such that U is a normal subgroup of H. Then for any H-module A there is a spectral sequence of the following properties (see, e.g., G. Hochschild and J. P. Serre, Cohomology of group extensions, Trans. Amer. Math. Soc., **74** (1963), Chapter I, Proposition 7): the term $E_2^{p,q}$ is isomorphic to $H^p(H/U, H^q(U, A))$ and E_∞ is isomorphic to the graduated group associated with $H(H, A)$. Assume that scd $(G) \leqq 2$. This implies also scd $(U) \leqq 2$. Take $A = \mathbf{Q}/\mathbf{Z}$ on which H operates trivially. Then $H^j(U, \mathbf{Q}/\mathbf{Z}) \cong H^{j+1}(U, \mathbf{Z}) = 0$ for $j \geqq 2$. This means that $E_2^{p,q} = 0$ for $q \geqq 2$ and we have an exact sequence

$$\to H^{p+1}(H, \mathbf{Q}/\mathbf{Z}) \to E_2^{p,1} \xrightarrow{d^2} E_2^{p+2,0} \to H^{p+2}(H, \mathbf{Q}/\mathbf{Z}) \to$$

(see, e.g., R. Godement, Théorie des faisceaux, p. 85, Theorem 4.6.2). Hence from $H^{p+1}(H, \mathbf{Q}/\mathbf{Z}) = 0$ for $p \geqq 1$ follows that d^2: $E_2^{p,1} \cong E_2^{p+2,0}$, i.e.,

$$H^p(H/U, H^1(U, \mathbf{Q}/\mathbf{Z}) \cong H^{p+2}(H/U, H^0(U, \mathbf{Q}/\mathbf{Z})) \quad \text{for } p \geqq 1.$$

Since $H^0(U, \mathbf{Q}/\mathbf{Z}) \cong \mathbf{Q}/\mathbf{Z}$ and $H^1(U, \mathbf{Q}/\mathbf{Z}) = \text{Hom}\,(U, \mathbf{Q}/\mathbf{Z}) \cong (U/U')^\wedge$ (the character group of U/U') we have d^2: $H^p(H/U, (U/U')^\wedge) \cong H^{p+2}(H/U, \mathbf{Q}/\mathbf{Z})$. By the general duality formula: $H^p(H/U, \hat{A}) \cong \hat{H}^{-1-p}(H/U, A)$ we have d^2: $\hat{H}^{-p-3}(H/U, \mathbf{Z}) \cong \hat{H}^{-p-1}(H/U, U/U')$ for $p \geqq 1$. Moreover, d^2 is obtained by the cup-product with the 2-cocycle $u_{H/U}$ of the group extension of U/U' by H/U (Hochschild-Serre, Chapter III, Theorem 4). Hence we can apply the theorem of triples and we have $\hat{H}^n(H/U, \mathbf{Z}) \cong \hat{H}^{n+2}(H/U, U/U')$ for all $n \in \mathbf{Z}$ by means of the cup-product with $u_{H/U}$. This means that G is malleable.

Assume that G is malleable. Then in the above spectral sequence d_2: $E_2^{p,1} \to E_2^{p+2,0}$ is given by the cup-product with the canonical 2-cocycle in $H^2(H/U, U/U')$ which is an isomorphism $H^p(H/U, (U/U')^\wedge) \cong H^{p+2}(H/U, \mathbf{Q}/\mathbf{Z})$ for $p \geqq 1$ and is an epimorphism for $p = 0$. This implies $E_3^{p,0} = 0$ for $p \geqq 2$. Now by a general property (Hochschild-Serre, p. 126) that in the sequence $H^p(H/U, \mathbf{Q}/\mathbf{Z}) \cong E_2^{p,0} \to E_\infty^{p,0} \to H^p(H, \mathbf{Q}/\mathbf{Z})$ the composite map is the inflation map. Since $E_3^{p,0} \cong E_\infty^{p,0} = 0$ $(p \geqq 2)$ this inflation map is 0 (for every U). Hence $H^p(H, \mathbf{Q}/\mathbf{Z}) = \lim_{\to} H^p(H/U, \mathbf{Q}/\mathbf{Z}) = 0$ for $p \geqq 2$. This implies that scd $(G) \leqq 2$, Q.E.D.

REFERENCES

1. E. Artin and J. Tate, *Class field theory*, 1968, Benjamin, New York. MR **36** #6383.

2. A. Brumer, *Pseudocompact algebras, profinite groups and class formations*, J. Algebra **4** (1966), 442–470. MR **34** #2650.

3. ———, *Galois groups of extensions of algebraic number fields with given ramification*, Michigan Math. J. **13** (1966), 33–40. MR **33** #121.

3A. ———, *On the cohomological dimension of certain Galois groups* (to appear).

3B. ———, *On the units of algebraic number fields*, Mathematika **14** (1967), 121–124. (See also the references there.) MR **36** #3746.

4. J. W. S. Cassels and A. Fröhlich (Editors), *Algebraic number theory*, Academic Press, New York, 1967. MR **35** #6500.

5. C. Chevalley, *Class field theory*, Nagoya University, Nagoya, 1954. MR **16,** 678.

6. S. Demuškin, *The group of a maximal p-extension of a local field*, Izv. Akad. Nauk SSSR Ser. Mat. **25** (1961), 329–346 (Russian). MR **23** #A890.

7. ———, *Topological 2-groups with an even number of generators and a complete defining relation*, Izv. Akad. Nauk SSSR Ser. Mat. **29** (1965), 3–10; English transl., Amer. Math. Soc. Transl. (2) **66** (1968), 206–213. MR **31** #268.

8. A. Douady, *Cohomologie des groupes compacts totalement discontinus*, Séminaire Bourbaki 1959/60, exposés 189, Secrétariat mathématique, Paris, 1960. MR **23** #A2273.

9. K. Grant and G. Whaples, *Abstract class formations*, J. Fac. Sci. Univ. Tokyo Sect. I **11** (1965), 187–194. MR **32** #4117.

10. D. K. Harrison, *Abelian extensions of arbitrary fields*, Trans. Amer. Math. Soc. **106** (1963), 230–235. MR **26** #114.

11. K. Höchsmann, *Über die Gruppe der maximalen l-Erweiterung eines globalen Körpers*, J. Reine Angew. Math. **222** (1966), 142–147.

12. K. Iwasawa, *On solvable extensions of algebraic number fields*, Ann. of Math. (2) **58** (1953), 548–572. MR **15,** 509.

13. ———, *On Galois groups of local fields*, Trans. Amer. Math. Soc. **80** (1955), 448–469. MR **17,** 714.

14. ———, *A note on the group of units of an algebraic number field*, J. Math. Pures Appl. (9) **35** (1956), 189–192. MR **17,** 946.

15. ———, *An Γ-extensions of algebraic number fields*, Bull. Amer. Math. Soc. **65** (1959), 183–226. MR **23** #A1630.

16. Y. Kawada, *Algebraic number theory*, Kyōritsu Syuppan, 1957. (Japanese)

17. ———, *On the structure of the Galois group of some infinite extensions.* I, II, J. Fac. Sci. Univ. Tokyo Sect. I **7** (1954), 1–18, 87–106. MR **16,** 6.

18. Y. Kawada and J Tate, *On the Galois cohomology of unramified extensions of function fields in one variable*, Amer. J. Math. **77** (1955), 197–217. MR **16,** 799.

19. Y. Kawada, *Class formations.* I, Duke Math. J. **22** (1955), 165–177. MR **16,** 907.

20. Y. Kawada and I. Satake, *Class formations.* II, J. Fac. Sci. Univ. Tokyo Sect. I **7** (1956), 353–389. MR **19,** 380.

21. Y. Kawada, *Class formations.* III, J. Math. Soc. Japan **7** (1955), 453–490; IV, J. Math. Soc. Japan **9** (1957), 395–405; V, J. Math. Soc. Japan **12** (1906), 34–64; VI, J. Fac. Sci. Univ. Tokyo Sect. I **8** (1960), 229–262. MR **18,** 114; MR **20** #3117; MR **26** #4989; MR **26** #4990.

22. ———, *Cohomology of group extensions*, J. Fac. Sci. Univ. Tokyo Sect. I **9** (1963), 417–431. MR **27** #4848.

23. ———, *Abstract class formations*, Bol. Soc. Mat. São Paulo **15** (1960), 5–23. MR **31** #1248.

24. H. Koch, *Über Galoissche Gruppen von* $\mathfrak{p}$*-adischen Zahlkörpern*, Math. Nachr. **29** (1965), 77–111. MR **31** #2240.

25. ———, *Über die Galoissche Gruppe der algebraischen Abschliessung eines Potenzreihenkörpers mit endlichem Konstantenkörper*, Math. Nachr. **35** (1967), 323–327. MR **37** #5189.

26. ———, *l-Erweiterungen mit vorgegebenen Verzweigungsstellen*, J. Reine Angew. Math. **219** (1965), 30–61. MR **32** #4118.

27. ———, *Beweis einer Vermutung von Höchsmann aus der Theorie der l-Erweiterungen*, J. Reine Angew. Math. **225** (1967), 203–206. MR **35** #1569.

28. J. Labute, *Classification of Demushkin groups*, Canad. J. Math. **19** (1967), 106–132. MR **35** #1674.

29. ———, *Algèbres de Lie et pro-p-groupes définis par une seule relation*, Invent. Math. **4** (1967), 142–158. MR **36** #1581.

30. S. Lang, *On quasi algebraic closure*, Ann. of Math. (2) **55** (1952), 373–390. MR **13,** 726.

31. ———, *Rapport sur la cohomologie des groupes*, Benjamin, New York, 1967. MR **35** #2948.

32. ———, *Unramified class field theory over function fields in several variables*, Ann. of Math. (2) **64** (1956), 285–325. MR **18,** 672.

33. R. C. Lyndon, *Cohomology theory of groups with a single defining relation*, Ann. of Math. (2) **52** (1950), 650–665. MR **13,** 819.

34. G. Poitou, *Remarques sur l'homologie des groupes profinis*, Les Tendances Géom. en Algèbre et Théorie des Nombres, Centre National de la Recherche Scientifique, Paris, 1966, pp. 201–213. MR **36** #137.

35. ———, *Cohomologie galoisienne des modules finis*, Travaux et Recherches Mathématiques, no. 13, Dunod, Paris, 1967. MR **36** #2670.

36. I. Šafarevič, *On p-extensions*, Mat. Sb. **20 (62)** (1947), 351–363; English transl., Amer. Math. Soc. Transl. (2) **4** (1956), 59–72. MR **8,** 560.

37. ———, *Algebraic number fields*, Proc. Internat. Congress Math. (Stockholm, 1962) Inst. Mittag-Leffler, Djursholm, 1963, pp. 163–176; English transl., Amer. Math. Soc. Transl. (2) **31** (1963), 25–39. MR **28** #1; MR **34** #2569.

38. ———, *Extensions with given points of ramification*, Inst. Hautes Études Sci. Publ. Math. No. 18 (1963), 71–95 (295–316); English transl., Amer. Math. Soc. Transl. (2) **59** (1966), 128–149. MR **31** #1247.

39. J. P. Serre, *Classes des corps cyclotomiques*, Séminaire Bourbaki 1958/59, exposé 174, Secrétariat mathématique, Paris, 1959. MR **28** #1091.

40. ———, *Groupes algébriques et corps de classes*, Actualités Sci. Indust., no. 1264, Hermann, Paris, 1959. MR **21** #1973; MR **30** p. 1200.

41. ———, *Sur les corps locaux à corps résiduel algébriquement clos*, Bull. Soc. Math. France **89** (1961), 105–154. MR **26** #103.

42. ———, *Corps locaux*, Actualités Sci. Indust., no. 1296, Hermann, Paris, 1962. MR **27** #133.

43. ———, *Structure de certains pro-p-groupes*, Séminaire Bourbaki 1962/63, exposé 252, Secrétariat mathématique, Paris, 1964. MR **31** #4701.

44. ———, *Cohomologie galoisienne*, Lecture Notes in Math., no. 5, Springer-Verlag, Berlin and New York, 1965. MR **34** #1328.

45. T. Takahashi, *On extensions with given ramification*, Proc. Japan Acad. **44** (1968), 771–775.

46. ———, *Galois cohomology of finitely generated modules*, Tohoku Math. J. **21** (1969), 102–111.

47. J. Tate, *Duality theorems in Galois cohomology over number fields*, Proc. Internat. Congress Math. (Stockholm, 1962) Inst. Mittag-Leffler, Djursholm, 1963, pp. 288–295. MR **31** #168.

48. ———, *The cohomology groups of tori in finite Galois extensions of number fields*, Nagoya Math. J. **27** (1966), 709–719. MR **34** #7495.

49. K. Uchida, *On Tate's duality theorems in Galois cohomology*, Tohoku Math. J. **21** (1969), 92–101.

50. A. Weil, *Sur la théorie du corps de classes*, J. Math. Soc. Japan **3** (1951), 1–35. MR **13,** 439.

51. G. Whaples, *Generalized local class field theory.* I, Duke Math. J. **19** (1952), 505–517; II, Duke Math. J. **21** (1954), 247–255; III, Duke Math. J. **21** (1954), 583–586; V, Proc. Amer. Math. Soc. **8** (1957), 137–140. MR **14,** 140; MR **17,** 464; MR **19,** 834.

UNIVERSITY OF TOKYO
TOKYO, JAPAN

SOME CONJECTURES IN CLASS FIELD THEORY

J. A. SHALIKA

1. In these lectures, I want to furnish some evidence for certain conjectures concerning nonabelian extensions of number fields which generalize in a natural way the isomorphism theorem of abelian class field theory.

Let k be an algebraic number field, $\bar{k}$ an algebraic closure of k and let $\mathfrak{G}$ denote the Galois group of $\bar{k}$ over k with the usual topology. For each positive integer d and each ideal $\mathfrak{A}$ of k, let $\mathfrak{G}^*(d, \mathfrak{A})$ denote the set of equivalence classes of (continuous) irreducible unitary representations of $\mathfrak{G}$ of degree d and conductor $\mathfrak{A}$.

The only general theorem, which is essentially group-theoretic, concerning $\mathfrak{G}^*(d, \mathfrak{A})$ known is that this set is finite. The problem however is to obtain some "structural" results concerning $\mathfrak{G}^*(d, \mathfrak{A})$. We would like to describe this set "purely in terms of k" without reference to extension fields of k.

Put $\mathfrak{G}^*(d) = \bigcup_{\mathfrak{A}} \mathfrak{G}^*(d, \mathfrak{A})$ where the union is taken over all (integral) ideals of k. For $d = 1$, the problem is completely answered by abelian class field theory. In that case, $\mathfrak{G}^*(1) = \hat{\mathfrak{G}}_{ab}$ is simply the Pontrajagin dual of $\mathfrak{G}_{ab}$, where $\mathfrak{G}_{ab}$ denotes the Galois group of the maximal abelian extension of k. Moreover, if we denote by $k_A^\times$ the group of idèles of k and canonically imbed $k^\times$ (the multiplicative group of k) as a discrete subgroup of $k_A^\times$, then $\hat{\mathfrak{G}}_{ab}$ is isomorphic (canonically) to the torsion subgroup of the dual of

$$k_A^\times/k^\times\colon \hat{\mathfrak{G}} = \operatorname{Tor} \widehat{(k_A^\times/k^\times)}.$$

We begin by generalizing the right side of this equality.

For d a positive integer and a commutative ring R containing an identity, let $GL_d(R)$ denote the group of invertible $d \times d$ matrices with coefficients in R. Let k_A denote the ring of adeles of k and put, for fixed d,

$$G(A) = GL_d(k_A), \qquad G(k) = GL_d(k).$$

$G(A)$ has a natural structure as a locally compact unimodular group and $G(k)$

may be viewed in a natural manner as a discrete subgroup of $G(A)$. Also, denote by $\hat{G}(A)$ the set of equivalence classes of irreducible unitary representations of $G(A)$. We first define a subset of $\hat{G}(A)$ with certain invariance properties with respect to the discrete group $G(k)$.

For this purpose, denote by X the homogeneous space $G(k)\backslash G(A)$. Then, since $G(A)$ is unimodular and $G(k)$ is discrete, X carries an essentially unique invariant measure. We denote by $\mathfrak{H}$ the space of complex-valued square integrable functions on X with respect to this measure. $G(A)$ operates on $\mathfrak{H}$ as follows: for $x \in G(A)$, $f \in \mathfrak{H}$, $y \in X$, put $(T_x f)(y) = f(x^{-1}y)$. Then $T_x f \in \mathfrak{H}$, and the invariance of the above measure implies that for the map $x \mapsto T_x$ defines a unitary representation $(\Pi, \mathfrak{H})$ of $G(A)$ on the Hilbert space $\mathfrak{H}$. We then have a (direct integral) decomposition $(\Pi, \mathfrak{H}) = \int_{\xi \in \hat{Z}} (\Pi_\xi, \mathfrak{H}_\xi) d\xi$ over the characters ξ of the center Z of $G(A)$, (Z is the group of diagonal matrices in $G(A)$) where $\mathfrak{H}_\xi$ is the Hilbert space defined by the following two conditions:

(a) $f(zx) = \xi(z) f(x), \quad z \in Z, x \in X,$

(b) $$\int_{ZG(k)\backslash G(A)} |f(x)|^{|f^2(x)|^2} dx < \infty$$

where dx denotes the invariant measure on the homogeneous space $ZG(k)\backslash G(A)$.

The problem of "decomposing" Π_ξ into irreducible representations of $G(A)$ is the central problem of the general theory of automorphic functions. In this theory it is natural to introduce a certain stable subspace ${}^\circ\mathfrak{H}_\xi$ of $\mathfrak{H}_\xi$ ($\xi \in \hat{Z}$) called the space of *cusp forms*, whose definition we omit[1] and about which one has the following

THEOREM. [4] *For $\xi \in \hat{Z}$, let ${}^\circ\Pi_\xi$ denote the representation of $G(A)$ obtained by restricting Π_ξ to ${}^\circ\mathfrak{H}_\xi$. Then ${}^\circ\Pi_\xi$ decomposes into a discrete direct sum of irreducible unitary representations of $G(A)$ each occurring with finite multiplicity.*

For $\xi \in \hat{Z}$, denote by $\hat{X}_\xi$ the set of $\Pi \in \hat{G}(A)$ which occur in ${}^\circ\mathfrak{H}_\xi$ and put $\hat{X} = \bigcup_{\xi \in \hat{Z}} \hat{X}_\xi$. It is in terms of the representations Π of $G(A)$ in $\hat{X}$ that we wish to describe $\mathfrak{G}^*(d)$. That is we wish to define an injection $F_k : \mathfrak{G}^*(d) \hookrightarrow \hat{X}$ and to describe the image. For

$$d = 1, \quad X = \widehat{(k_A^\times / k^\times)}$$

and the map we are seeking is the one given by the abelian theory. At the moment, we are not in a position to define F_k for all d and $\mathfrak{A}$. We can do so for $d = 2$ and essentially all $\mathfrak{A}$ (assuming a conjecture of Artin concerning L-functions) and in general we can associate in a natural way to each irreducible unitary representation of $\mathfrak{G}$ with trivial conductor and degree d a unitary representation of $GL_d(k_A)$. A proof that this latter representation has the above invariance property with respect to $G(k)$ would then yield the "reciprocity law" for the maximal unramified extension of k.

---

[1]For the relation of these topics to the classical theory of automorphic function and the proofs of the theorem below see [4].

2. Definition of F_k for $d = 2$

The definition of F_k is local, i.e. for each place ν of k we define the local analogue $F_{k,\nu}$ of F_k and then take the appropriate product.

Let k_ν be the completion of k at ν and $\mathfrak{G}_\nu$ a fixed decomposition group at ν. Let $\mathfrak{G}^*(d, \nu)$ denote the set of equivalence classes of (continuous) unitary representations of $\mathfrak{G}_\nu$ of degree d. We have the obvious map

$$l_\nu : \mathfrak{G}^*(d) \rightarrow \mathfrak{G}^*(d, \nu)$$

given by restriction.

2.1. *Localization of $\hat{G}(A)$ and the Local Theory.* For ν finite, let O_ν denote the ring of integers of k_ν. Let G_ν and K_ν, for ν finite, denote respectively the locally compact groups $GL_d(k_\nu)$ and $GL_d(O_\nu)$. Then, for ν finite, K_ν is a maximal compact subgroup of G_ν. For ν finite, we say that an irreducible unitary representation $\Pi \in \hat{G}_\nu$ is of *class one* if there exists a nonzero vector v in the associated Hilbert space such that

$$\Pi(x)v = v \quad \text{for all } x \in K_\nu.$$

THEOREM. [4] *Suppose we are given, for each ν, a representation $\Pi_\nu \in \hat{G}_\nu$ such that for all places ν of k with the exception of a finite number Π_ν is of class one. Then*

(2.1.1) $\bigotimes_\nu \Pi_\nu$, *the topological tensor product (in an appropriate sense) exists;*

(2.1.2) *one obtains all irreducible unitary representations of $G(A)$ in this fashion;*

(2.1.3) *the components Π_ν are unique.*

It follows from the above discussion that we need only define the local maps $F_{k,\nu}: \mathfrak{G}^*(d, \nu) \hookrightarrow \hat{G}_\nu$ for then we may put, for $\pi \in \mathfrak{G}^*(d)$,

(2.14) $F_k(\pi) = \bigotimes_\nu F_{k,\nu}(l_\nu(\pi))$

provided almost all of the terms on the right are representations of class one. (Of course this way we only obtain a map of $\mathfrak{G}^*(d)$ into $\hat{G}(A)$ and not a priori into $\hat{X}$.) Before proceeding to the definition of $F_{k,\nu}$ for $d = 2$ we describe a condition on $F_{k,\nu}$ in general which to some extent determines it.

2.2. *Local L-functions.* For $\pi \in \mathfrak{G}^*(d, \nu)$, let $L_\nu(\pi, s)$ be the local L-function as defined by Langlands [7]. Up to a factor $W(\pi)$ of absolute value one called the *local root number* [3] this is essentially the same as that defined by Artin [1], [2].

Definition of the Group Theoretic L-Function. Let $M_d(k_\nu)$ denote the algebra of $d \times d$ matrices with coefficients in k_ν. Let $\mathcal{S}_\nu$ be the space of functions of Schwartz-Bruhat attached to the additive group of this algebra. For $f \in \mathcal{S}_\nu$, let $\hat{f}$ denote the Fourier transform of f. Then $\hat{f} \in \mathcal{S}_\nu$. For $\Pi \in \hat{G}_\nu$, $f \in \mathcal{S}_\nu$, $s \in \mathbf{C}$ with sufficiently large real part, put

$$\zeta(f, \Pi, s) = \int_{G_\nu} f(x)\Pi(x)|\det x|^s \, dx$$

where $|\ |$ denotes the normalized absolute value on k_ν, det denotes the determinant of $x \in G_\nu$ and dx is a fixed Haar measure on G_ν. We have the following conjecture and theorem due to Langlands and Jacquet.

CONJECTURE. (1) $s \mapsto \zeta(f, \Pi, s)$ has a continuation to a meromorphic function;

(2) there exists, for each $\Pi \in \hat{G}_\nu$, $s \in \mathbf{C}$, a complex valued merofunction, $\Gamma_\nu(\Pi, s)$, depending only on Π and s such that $\zeta(f, \Pi, s) = \Gamma_\nu(\Pi, s)\zeta(\hat{f}, \Pi^*, 1 - s)$ where Π^* denotes the representation contragredient to Π and the identity is understood in the sense of analytic continuation.

THEOREM. *The above conjecture is true for* $d = 2$.

With this in mind, it is natural to impose the following condition on $F_{k,\nu}$:

L.C.F.T. For $\pi \in \mathfrak{G}^*(d, \nu)$ $L_\nu(\pi, s) = \Gamma_\nu(F_{k,\nu}(\pi), s)$. We have then the following

THEOREM. ($d = 2$) *There exists a "natural" map* $F_{k,\nu}$ *from* $\mathfrak{G}^*(2, \nu)$ *into* $\hat{G}_\nu$ *satisfying the condition* L.C.F.T.

REMARK. There is a slight exception to this result for the places dividing 2 due to the temporary incompleteness of the theory of unitary representations of G_ν in this case.

In the following we will describe the local map $F_{k,\nu}$ for $d = 2$. To do this we first need to say something about $\mathfrak{G}^*(2, \nu)$.

2.3. *Description of* $\mathfrak{G}^*(2, \nu)$ *(odd conductor)*. We have the following elementary proposition concerning $\mathfrak{G}^*(2, \nu)$.

PROPOSITION. *Suppose* π *is an irreducible unitary representation of* $\mathfrak{G}_\nu$ *of degree* 2 *and odd conductor. Then* π *is a monomial representation, i.e. there exists a subgroup* H_ν *of* $\mathfrak{G}_\nu$ *necessarily of index* 2 *in* $\mathfrak{G}_\nu$ *together with a linear character* Ψ *of* H *such that* π *is the representation of* $\mathfrak{G}_\nu$ *induced by* Ψ: [2]

$$\pi = \operatorname*{ind}_{H_\nu \uparrow \mathfrak{G}_\nu} \Psi$$

(*we remark that* Ψ *is not unique in general*; *however the existence of* Ψ *is sufficient for what follows*).

Start with $\pi \in \mathfrak{G}^*(2, \nu)$ (of odd conductor) and choose Ψ and H_ν as in the proposition. Let L_ν denote the (quadratic) extension of k_ν determined by H_ν. Then Ψ is a linear character of the Galois group of L_ν and so by local class field theory we obtain a character ψ of $L_\nu^\times$, the multiplicative group of L_ν. Thus we expect to obtain irreducible unitary representations of $GL_2(k_\nu)$ parametrized by the duals of the multiplicative groups of quadratic extensions of k and indeed this is the case.

2.4. *A brief description of the irreducible unitary representations of* $GL_2(k_\nu)$. To begin with, consider the group $SL_2(k_\nu) = G'_\nu$.

The representations of G'_ν [4] may be naturally grouped into four series, as follows:

(1) the principal series,
(2) the discrete series of compact support,
(3) the special representation,
(4) the complementary series.

---

[2]There are examples to show that for ν even not all irreducible $\pi \in \mathfrak{G}^*(2, \nu)$ are of this form.

For the present purposes we need only consider the first two of these series and they may be described as follows [8]:

Let L_ν be a two dimensional commutative semisimple algebra over k_ν. Denote by N_ν and tr_ν respectively the relative norm and trace from L_ν to k_ν and by σ the nontrivial automorphism of L_ν over k_ν. Let ϕ be a nontrivial additive character of k. We first attach to each pair (ϕ, L_ν) a unitary representation, $\Pi(\phi, L_\nu)$, of G' acting on the Hilbert space $L_2(L_\nu)$:

DEFINITIONS. 1. for $f \in L_2(L_\nu)$, $u \in L_\nu$, put $\hat{f}(u) = \int_{L_\nu} f(v)\phi(\mathrm{tr}\,(uv^\sigma))dv$ then $\hat{f} \in L_2(L_\nu)$ and we normalize dv so that $f \mapsto \hat{f}$ is unitary.

2. $\kappa_\nu = \kappa(\phi, L_\nu)$ a certain constant of absolute value 1.

3. $f_b(u) = \phi(bN_\nu(u))$, $b \in k$, $u \in L_\nu$.

With this notation, we have

THEOREM 2.4.1. [8] *There exists a unique unitary representation*, $\Pi(\phi, L_\nu)$ *of* G' *on the Hilbert space* $L_2(L_\nu)$ *such that*

(a) $\Pi\left(\begin{pmatrix} 0 & 1 \\ -1 & 0 \end{pmatrix}\right) f = \kappa_\nu \hat{f}$,

(b) $\Pi\left(\begin{pmatrix} 1 & b \\ 0 & 1 \end{pmatrix}\right) f = f_b \cdot f, \forall b \in k.$

Now let N_ν^1 denote the subgroup of $L_\nu^\times$ consisting of elements of relative norm 1. Then, by a method essentially due to Kloosterman, we may define an action of N_ν^1 on $L_2(L_\nu)$ (in fact translation) which commutes with the action of G' defined by the above representation. Using this fact, we obtain, for each $\psi^1 \in \hat{N}_\nu^1$, what is essentially an irreducible representation of G'_ν. Passing to $GL_2(k_\nu)$ by a simple process we obtain, for each $\psi \in \hat{L}_\nu^\times$, finally an irreducible representation of $GL_2(k_\nu)$, independent of ϕ, which we denote by $\Pi(\psi, L_\nu)$.

2.5. *Definition of* $F_{k,\nu}$ *for* $d = 2$ *and odd conductor.* Start with $\pi \in \mathfrak{G}^*(2, \nu)$ and suppose π has odd conductor. There are two cases:

1. π is irreducible.

In this case, by §2.3, we may write $\pi = \mathrm{ind}_{H_\nu \uparrow \mathfrak{G}_\nu} \Psi$. Define L_ν and ψ in §2.3. We define $F_{k,\nu}(\pi) = \Pi(\psi, L_\nu)$. Then $F_{k,\nu}$ is well defined in the sense that if $\pi = \mathrm{ind}_{H'_\nu \uparrow \mathfrak{G}_\nu} \Psi'$ and L'_ν and Ψ' are the associated quadratic extension of k_ν and multiplicative character of L'_ν then the representations $\Pi(\psi, L_\nu)$ and $\Pi(\psi', L'_\nu)$ are unitary equivalent.

2. $\pi = \Psi_1 \oplus \Psi_2$ where Ψ_1 and Ψ_2 are linear characters of $\mathfrak{G}_\nu$. Let ψ_1, ψ_2 respectively be the corresponding characters of $k^\times$. Put $L_\nu = k_\nu \oplus k_\nu$ and define a character ψ of $L_\nu^\times$ by

$$\psi((x, y)) = \psi_1(x)\psi_2(y) \quad \text{for } x, y \in k^\times$$

Put $F_{k,\nu}(\pi) = \Pi(\psi, L_\nu)$. Thus we have defined the local $F_{k,\nu}$.[3] One can show, that for $\pi \in \mathfrak{G}^*(2)$, $F_{k,\nu}(l_\nu(\pi))$ is of class one for all but a finite number of places ν. This follows from the fact that only a finite number of places ν divide the conductor of π and standard facts from the representation theory of $GL_2(k_\nu)$. We may then define the global F_k by 2.1.4. We have thus constructed a map from

[3] $F_{k,\nu}$ is seen to satisfy L.C.F.T. by explicit calculation.

$\mathfrak{G}^*(2)$ into $\hat{G}_A$ (again with a reservation due to the even places). (The fact that F_k is injective follows trivially from the Tschebotareff density theorem.)

Let $\mathfrak{G}_w$ denote the Weil group [11] of k, and let χ denote a continuous linear character of $\mathfrak{G}_w$ (χ may be identified canonically with an idèle class character of k). For $\pi \in \mathfrak{G}^*(2)$, χ as above, one may consider π as a representation of degree 2 of $\mathfrak{G}_w$. Let $L(\pi\chi, s)$ denote the Artin-Hecke L-function [11] attached to $\pi\chi$. With the above definitions we then have

THEOREM [6] (JACQUET-LANGLANDS). *Fix $\pi \in \mathfrak{G}^*(2)$ (of odd conductor). Suppose that for all linear characters χ of $\mathfrak{G}_w$, the L-function $L(\Pi\chi, s)$ is entire (and satisfies mild growth conditions). Then $F_k(\pi) \in \hat{X}$, i.e. $F_k(\pi)$ occurs in the space of cusp forms; conversely, if $F_k(\pi) \in \hat{X}$, then $L(\pi\chi, s)$ is entire for all χ.*

REMARKS. 1. One also has the obvious analogue of the above theorem for function fields k of one variable. In that case it is known that the L-functions are entire. Thus if the characteristic of k is odd one obtains the reciprocity law for $\mathfrak{G}^*(2)$.

2. If π is monomial, it follows from the abelian theory that $L(\pi, s)$ is entire. In this way one obtains a construction of automorphic functions [8] which generalizes the classical procedure of Hecke and Maass.

2.6. *The Existence Theorem.* In this paragraph we state a conjecture which if proved would constitute a generalization of the existence theorem to $\mathfrak{G}^*(2)$.

We say that $\Pi_\nu \in \hat{G}_\nu$ of the form $\Pi(\psi_\nu, L_\nu)$ is of finite type if ψ_ν is of finite order. We say $\Pi = \bigotimes_\nu \Pi_\nu \in \hat{G}_A$ is of finite type if the orders of the corresponding ψ_ν are bounded independent of ν. Finally, we denote by $\hat{X}_f$ those Π in $\hat{X}$ of finite type. It is obvious that for $\pi \in \mathfrak{G}^*(2)$, $F_k(\pi)$ is of finite type.

CONJECTURE. F_k defines a bijective map from $\mathfrak{G}^*(2)$ to $\hat{X}_f$ (for ν odd, the obvious local analogue of this conjecture is true).

2.7. *Formulation as the Equality of L-functions.* We may formulate Theorem 2.5 and the preceding conjecture in terms of L-functions. For this we recall, in this special case the generalization of Hecke L-functions due to Godement [5] and Tamagawa [6]. Start with $\Pi \in \hat{X}_f$ and write $\Pi = \bigotimes_\nu \Pi_\nu$. As in 2.2 let $\Gamma_\nu(\Pi_\nu, s)$ be the (local) Zeta function attached to Π_ν. Then it is known [6] that for Π_ν of class one $\Gamma_\nu(\Pi_\nu, s)$ has the form

$$\Gamma_\nu(\Pi_\nu, s) = L_\nu(\Pi_\nu, s)/L_\nu(\Pi_\nu, 1 - s) \qquad 4$$

and $L_\nu(\Pi_\nu, s) = (1 - a_\nu q_\nu^s + b_\nu q_\nu^{2s})^{-1}$ where q_ν is the modulus of k_ν and a_ν and b_ν are certain constants.

If we put $\zeta(\Pi, s) = \prod_\nu L_\nu(\Pi_\nu, s)$ where the product is taken over those ν for which Π_ν is of class one, then it is known that $\zeta(\Pi, s)$ defines an entire function with a functional equation similar to that of the Hecke L-functions. The above Theorem 2.5 and conjecture are equivalent to the statement that the set of functions of the form $\zeta(\Pi, s)$ for $\Pi \in \hat{X}_f$ coincide with the Artin L-functions[5] $\zeta(\pi, s)$ for

---

[4]Here the Euler product is taken only over those places of k which do not divide the conductor of π.

[5]Here the Euler product is taken only over those places of k which do not divide the conductor of π.

$\pi \in \mathfrak{G}^*(2)$, the correspondence (for odd conductors) being given of course by $F_{k,\nu}$. The converse part of Theorem 2.5 is a simple consequence of §2.2 and the fact that the Zeta function of Godement-Tamagawa is entire.

3. The proof of Theorem 2.5 in a special case [8].

We assume that $\pi \in \mathfrak{G}^*(2)$ is of the form $\operatorname{ind}_{H \uparrow \mathfrak{G}} \chi$ where χ is a linear character of a subgroup H of $\mathfrak{G}$ and sketch the proof of the analogue of the first part of Theorem 2.5 for the group SL_2. Since degree $\pi = 2$, χ may be identified, as in the local case with a character of the idèle group $L_A^\times$ of a quadratic extension of k. Fix ϕ a nontrivial additive character of k_A whose restriction to k is trivial. As in the local case, one may associate to each pair (ϕ, L) a unitary representation, $\Pi(\phi, L)$, of $SL_2(k_A)$ on the Hilbert space $L_2(L_A)$ with the following properties:

(3.1) $\Pi\left(\begin{pmatrix} 0 & 1 \\ -1 & 0 \end{pmatrix}\right) f = \kappa \cdot \hat{f}$,

(3.2) $\Pi\left(\begin{pmatrix} 1 & b \\ 0 & 1 \end{pmatrix}\right) f = f_b \cdot f$

where $\kappa = \kappa(\phi, L)$, $\hat{f}$ and f_b are defined analogously as in the local case.

We also have

(3.3) $\kappa = \prod_\nu \kappa_\nu$

with κ_ν as defined in §2.4.

Let $\mathcal{S}(L_A)$ denote the Schwartz-Bruhat space attached to L_A. It follows easily from the definitions that $\mathcal{S}(L_A)$ is invariant under the operators of the representation $\Pi(\phi, L)$.

3.1. *Theta functions.* We wish to define a map commuting with the action of $SL_2(k_A)$ from $\mathcal{S}(L_A)$ into functions on $SL_2(k_A)$ invariant on the left by $SL_2(k)$. For this purpose we define a linear functional Θ on $\mathcal{S}(L_A)$ by

$$\Theta(f) = \sum_{\xi \in L} f(\xi) \qquad (f \in \mathcal{S}(L_A))$$

we have

PROPOSITION. Θ *is* $SL_2(k)$ *invariant,* i.e.

$$\Theta(\Pi(x)f) = \Theta(f) \qquad (f \in \mathcal{S}(L_A),\ x \in SL_2(k)).$$

(Here $\Pi(x)$ denotes the transformation of $\mathcal{S}(L_A)$ corresponding to x according to the representation $\Pi(\phi, L)$.)

Sketch of Proof. It follows readily from the definitions and the Poisson summation formula that

$$\Theta(\Pi(x)f) = c(x)\Theta(f) \qquad (x \in SL_2(k))$$

where c is a homomorphism of $SL_2(k)$ into the complex numbers whose restriction to the subgroup consisting of all elements of the form $\begin{pmatrix} 1 & x \\ 0 & 1 \end{pmatrix}$ $(x \in k)$ is trivial. Since however $SL_2(k)$ is generated by this subgroup and its conjugates we have $c(x) \equiv 1$. Q.E.D.

COROLLARY. $\kappa = \prod_{\nu} \kappa_{\nu} = 1.$

From the corollary one easily deduces the general quadratic reciprocity law.

Using Proposition 3.1, we obtain a map of $\mathcal{S}(L_A)$ into the space of cusp forms of $SL_2(k)$ as follows:

For $f \in \mathcal{S}(L_A)$, $x \in G'(A)$, put $(I(f))(x) = \Theta(\Pi(x)f)$. Then clearly $(I(f))(\gamma x) = I(f)(x)$, $\forall x \in G'(A)$, $\forall \gamma \in G'(k)$. Moreover one can show that $I(f)$ is indeed a cusp form. Thus I defines a map of $\mathcal{S}(L_A)$ into the space of cusp forms which clearly commutes with the action of $G'(A)$ on the respective spaces.

Let N_A^1 denote the subgroup of $L_A^\times$ consisting of elements of relative adele norm 1. As in the local case one can define an action of N_A^1 on $L_2(L_A)$ which commutes with the action of $G'(A)$ on that space. In this way one can "decompose" $\Pi(\phi, L)$ into essentially irreducible representations $\Pi(\phi, \psi, L)$ parametrized by the characters ψ of N_A^1 which are trivial on the subgroup $N^1 = N_A^1 \cap L$ of global elements of norm one. One obtains for each ψ the analogues Θ_ψ and I_ψ of Θ and I respectively with the relation

$$I_\psi(f)(x) = \Theta_\psi(\Pi(x)f) \qquad (x \in G'(A))$$

where f belongs to an appropriate dense subspace of the underlying Hilbert space of the representation $\Pi(\phi, \psi, L)$. One then shows that I_ψ extends to a bounded operator from that Hilbert space into the space of cusp forms. Taking ψ to be the restriction of χ to N_A^1, one obtains the desired conclusion.

Finally, I remark that a technique similar to that above may be used to prove Theorem 2.5. One defines an analogue of Θ_ψ for each $\pi \in \mathfrak{G}^*(2)$ (assuming Artin's conjecture) by "adelization" of the classical technique of Hecke relating Zeta-functions to Theta-series.

BIBLIOGRAPHY

1. E. Artin, *Über eine neue Art von L-Reihen*, Abh. Math. Sem. Univ. Hamburg **3** (1923), 89.

2. ———, *Zur Theorie der L-Reihen mit algemeinen Gruppen charakteren*, Abh. Math. Sem. Univ. Hamburg **8** (1931), 292–306.

3. B. Dwork, *On the Artin root number*, Amer. J. Math. **78** (1956), 444–472. MR **18,** 556.

4. I. M. Gelfand, M. I. Graev and I. Pjateckiĭ-Šapiro, *Generalized functions.* Vol. 6: *Theory of representations and automorphic functions*, "Nauka", Moscow, 1966; English transl., Saunders, Phila., Pa., 1969. MR **36** #3725; MR **38** #2093.

5. R. Godement, *Les fonctions ζ des algèbres simples.* I, II, Séminaire Bourbaki, 1958/59. Exposés 171, 176, Secrétariat mathématique, Paris, 1959. MR **28** #1091.

6. H. Jacquet and R. P. Langlands (to appear).

7. R. P. Langlands, Unpublished.

8. J. A. Shalika and S. Tanaka, *On an explicit construction of a certain class of automorphic forms*, Amer. J. Math. (to appear).

9. T. Tamagawa, *On the ζ-functions of a division algebra*, Ann. of Math. (2) **77** (1963), 387–405. MR **26** #2468.

10. A. Weil, *Über die Bestimmung Dirichletscher Reihen durch Funktionalgleichungen*, Math. Ann. **168** (1967), 149–156. MR **34** #7473.

11. ———, *Sur la théorie du corps de classes*, J. Math. Soc. Japan **3** (1951), 1–35. MR **13,** 439.

PRINCETON UNIVERSITY
PRINCETON, NEW JERSEY 08540

EXTENSIONS OF CYCLOTOMIC THEORY

THOMAS STORER

I. Introduction

With the exception of some nontrivial computational advances and an important application to the existence of combinatorial designs, the theory of cyclotomy in 1960 stood about where it had in Dickson's time 25 years earlier [1]; a collection of statements concerning the eth power class structure of $\mathbf{Z}p$, the integers modulo p for $p = ef + 1$ prime. Motivated by combinatorial considerations, Whiteman [10] in 1962 formulated a significant extension of this theory to $\mathbf{Z}_{pq} \cong_{(+)} \mathbf{Z}_p \oplus \mathbf{Z}_q$ (direct sum) for distinct odd primes p and q and, through a careful analysis of the domain-structure $\mathbf{Z}_{pq}$, succeeded in formulating eth power class structure results for $\mathbf{Z}_{pq}$ analogous to those for $\mathbf{Z}_p$ and $\mathbf{Z}_q$, when $e =$ g.c.d. $(p - 1, q - 1)$. The above results were generalized [5] to the field $GF(p^\alpha)$ and the domains $GD(p^\alpha q^\beta)$ in 1965, the latter approach being radically reformulated [7] in 1969 to a method which generalized to both $GD(N)$ and $\mathbf{Z}_N$ for $N \in \mathbf{N}_n$ (a natural number). The restriction $e =$ g.c.d. $(p - 1, q - 1)$ was dropped in [4], whence the eth power class structure theorems for domains was "completed" in that all results known for the original field $\mathbf{Z}_p$ are now known for the domains $GD(N)$. Finally, the Galois fields $GF(p^\alpha)$ are contained in a much wider class of algebraic structures, the commutative, inverse property, cyclic neofields (C.I.P. neofields) $\mathbf{N}_{p^\alpha}$, and in [9] it is shown that the eth power class structure of a distinguished subset of the C.I.P. neofields (i.e. those which are known) is determinable from the corresponding structure of $GF(p^\alpha)$. Thus a schematic "graph" of the extensions of cyclotomic theory (beginning with $\mathbf{Z}_p$ in 1950) is given by

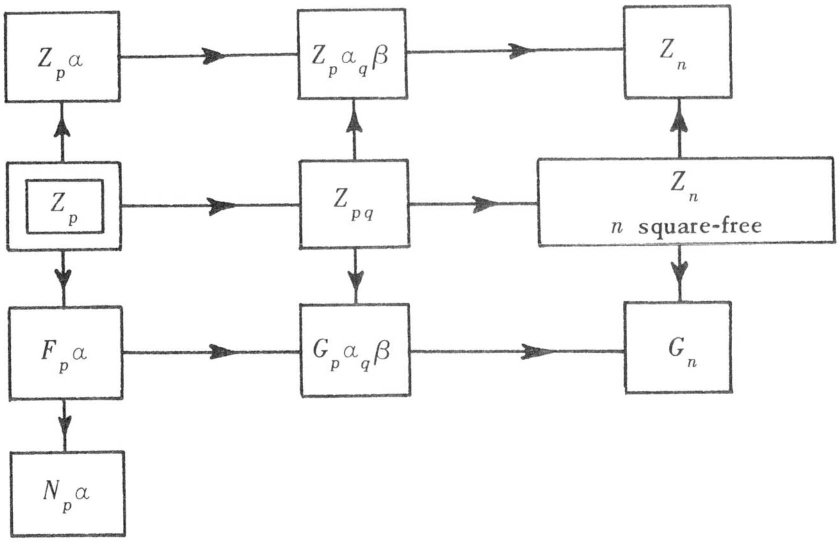

FIGURE 1

where $\mathbf{F}_{p^\alpha} = GF(p^\alpha)$ and $\mathbf{G}_{p^\alpha q^\beta} = GD(p^\alpha q^\beta)$ for convenience.

Here we shall discuss the nature of and difficulty involved in these various extensions, ignoring the combinatorial and geometrical motivations and applications which are involved. We begin with a recapitulation of the known theory (for $\mathbf{Z}_p$) in the early 1960's.

II. The integers modulo a prime

We begin with a generalizable notation for the parent structure (here $\mathbf{Z}_p$), introducing relevant parameters.

1. NOTATION. Let $q = ef + 1$ be an odd prime, and denote the (field of) integers modulo q by $\mathbf{F}_q$.

As we are principally interested in eth power class structure results, we next define the classes and introduce the arithmetic class-parameters.

2. CLASS STRUCTURE. Fix a generator (primitive root) x of $\mathbf{F}_q$ and define the *cyclotomic classes* relative to q, e, and x by

$$C_{q,i} = \{x^{es+i}: s = 0, 1, 2, \ldots, f - 1\} \qquad (i = 0, 1, 2, \ldots, e - 1).$$

For fixed i and j we define the *cyclotomic number* $(i, j)_q$ relative to q, e, and x to be the number of ordered pairs $(z_i, z_j) \in C_{q,i} \times C_{q,j}$ satisfying $z_i + 1 = z_j$ in $\mathbf{F}_q$, and the *cyclotomic matrix* $\Gamma_{q,e}$ to be the (e by e)-integral matrix whose ij-entry is the cyclotomic number $(i, j)_q$ for $0 \leq i, j \leq e - 1$.

Now cyclotomy, by its very nature, is intimately concerned with roots of unity, and the following definition is central enough to any generalization of the theory to warrant its special mention.

3. ROOT STRUCTURE. If for $N \in \mathbf{N}_n$ we introduce $\lambda_N = \exp\{2\pi i/N\}$, then by the *periods* $\eta_{q,k}$ relative to q, e, and x we shall mean

$$\eta_{q,k} = \sum_{b \in C_{q,k}} \lambda_q^b = \sum_{s=0}^{f-1} \lambda_q^{x^{es+k}}, \qquad (k = 0, 1, 2, \ldots, e-1).$$

As a remark, it is immaterial to the theory whether η_k is considered to be a complex number or an element of the integral group ring of $\mathcal{B} = \{\lambda_q^b : b \in \mathbf{F}_q\}$.

Finally, it is through relations between various arithmetic functions on $\mathbf{F}_q$ that one can actually evaluate the pertinent cyclotomic numbers for given q, e, and x, and thus determine the eth power class structure of the field. Thus we introduce the functions of Jacobsthal, Jacobi, and Lagrange as representative of these.

4. ARITHMETIC FUNCTIONS. For χ the quadratic character on F_q and $\alpha \in F_q$, let

$$\varphi_{q,e}(\alpha) = \sum_{y \in F_q} \chi(y)\chi(y^e + \alpha).$$

If e divides none of m, n, nor $m + n$, we define

$$F_{q,e}(\lambda_e^n) = \sum_{j=0}^{e-1} \lambda_e^{hj}\eta_{q,j}$$

and

$$R_{q,e}(m, n) = \sum_{k=0}^{e-1} \lambda_e^{nk} \sum_{h=0}^{e-1} \lambda_e^{-(m+n)h}(k, h)_q,$$

the last two being, again, relative to a fixed generator x.

Basic results needed for the determination of the cyclotomic constants are of two types: Parameter (elementary) relationships and functional relationships. For the former type we give in 5 the complete results for $\mathbf{F}_q$, while in 6 a representative sampling of the latter is displayed.

5. PARAMETER RELATIONS.

(i) $(i, j)_q = (i + me, j + ne)_q$ for all $m, n \in \mathbf{Z}$.

(ii) $(i, j)_q = (e - i, j - i)_q$.

(iii) $(i, j)_q = (j, i)_q$ if f is even,
$= (j + e/2, i + e/2)_q$ if f is odd.

(iv)

$$\sum_{i=0}^{e-1} (i, j)_q = f - \theta_{q,i} \quad \text{where } \theta_{q,i} = 1 \quad \text{if } -1 \in C_{q,i},$$
$$= 0 \quad \text{otherwise.}$$

6. FUNCTIONAL RELATIONS.

(i) $R_{q,e}(m, n) = F_{q,e}(\lambda_e^m)F_{q,e}(\lambda_e^n)/F_{q,e}(\lambda_e^{m+n})$.

(ii) $R_{q,e}(m, n)R_{q,e}(-m, -n) = q$.

(iii) If $l \in \mathbf{N}_n$ is defined by $x^l = 2 \in \mathbf{F}_q$, then

$$F_{q,e}(-1)F_{q,e}(\lambda_e^{2k}) = \lambda_e^{2lk}F_{q,e}(\lambda_e^k)F_{q,e}(-\lambda_e^k).$$

(iv) $\varphi_2(1) = R_4(1, 1) + R_4(-1, -1) = -2s$ where $q = s^2 + 4t^2$ with $s \equiv 1 \pmod 4$.

The relation in 6 (iii) is known as (one of) the "Jacobi identity", and 6 (iv) is intended to be typical of a class of known relations between the functions φ and R, and a quadratic decomposition of (a multiple of) q. Such relations as the latter are of primary interest to the current exposition only in §III below, where it is, for example, extremely useful in ascertaining which of the several quadratic decompositions of $q = p^\alpha$ is pertinent.

Finally, in terms of the above results (and various extensions and refinements of them) the entries of the cyclotomic matrices $\Gamma_{q,e}$ for $e = 2, 3, 4, 5, 6, 7, 8, 9, 10, 12, 14, 16, 18$ and 20 were (soon to be) determined. As a typical example (in keeping with the "typical" relation chosen for 6 (iv)) we present the cyclotomic numbers for $e = 4$ in the case f odd. Here the numbers in question are determined by the (4 by 4) cyclotomic matrix blank

A	B	C	D
E	E	D	B
A	E	A	E
E	D	B	E

FIGURE 2

together with the relations

$$\begin{aligned} 16A &= q - 7 + 2s, \\ 16B &= q + 1 + 2s - 8t, \\ 16C &= q + 1 - 6s, \\ 16D &= q + 1 + 2s + 8t, \\ 16E &= q - 3 - 2s \end{aligned}$$

where $q = s^2 + 4t^2$, with $s \equiv 1 \pmod 4$ and the sign of t depending upon the choice of generator x. Two things are of interest here; first, a prime q has at most one such binary quadratic decomposition in natural numbers and second, the congruence class of s modulo 4 is insured by the relation

$$s = 4(B + D) - (2f + 1) \equiv 1 \pmod 4.$$

With these few results, remarks, and examples, we assume that the state of the cyclotomic art in the early 1960's is now known and, disregarding historical accuracy in favor of suspense (and the establishment of a more general setting in which to fully appreciate the magnitude of the next significant contribution to the subject) we skip ahead several years to 1965 and the finite field [5].

III. The finite field

Here it is immediately noticed that, for p a prime, $\alpha \in \mathbf{N}_n$ and $q = p^\alpha$, the field $\mathbf{F}_q$ admits the notation 1, class structure definitions 2, and (consequently) parameter relations 5, of §II without change, the only critical factor being the existence of a generator x of $\mathbf{F}_q$ (properly speaking, of $\mathbf{F}_q - \{0\}$). Now, however, the symbol "λ_q^b" for $b \in \mathbf{F}_q$ of the root structure 3 no longer makes sense, and we must somehow obviate this deficiency if the theory is to be successfully extended to the present structure under consideration. In this regard we remark that what is obviously wanted is the introduction of a binary operation "multiplication" into the set of formal symbols $\mathcal{B} = \{\lambda_q^b : b \in \mathbf{F}_q\}$ in such a way that (i) $\mathcal{B}$ may be "identified" with the set $\{\lambda_q^n : n = 0, 1, 2, \ldots, q - 1\}$ and (ii) $\langle \mathcal{B}, \cdot \rangle$ becomes a group isomorphic to $\langle \mathbf{F}_q, + \rangle$. This is accomplished *via* the following three axioms:

AXIOM 1. *There is an identification "$\leftrightarrow$" such that* $\forall b \in \mathbf{F}_q : \exists n \in \{0, 1, 2, \ldots, q - 1\} : \lambda_q^b \leftrightarrow \lambda_q^n$ *and* $\lambda_q^0 \leftrightarrow 1$.

AXIOM 2. $\lambda_q^{b_1} = \lambda_q^{b_2}$ *if and only if* $b_1 = b_2 \in \mathbf{F}_q$.

AXIOM 3. $\lambda_q^{b_1}\lambda_q^{b_2} = \lambda_q^{b_1+b_2}$ *for all* $b_1, b_2 \in \mathbf{F}_q$.

Since $\mathbf{F}_q$ exists, there surely is such an identification, and it has been shown in [3; Part I, §4] that the results which follow are independent of the identification chosen.

We may now proceed formally as in §II to define the root structure 3 for the field $\mathbf{F}_q$ (noting that now the periods $\eta_{q,k}$ *must* be interpreted as elements of the integral group ring $\mathcal{B}\mathbf{Z}$ of $\mathcal{B}$) and to introduce the arithmetic functions 4 for $\mathbf{F}_q$. Now, in fact, the proofs of the functional relations 6 are valid as written under each identification satisfying the Axioms 1–3 above.

At this point one might be tempted to say that the passage from $\mathbf{F}_q$ for q an odd prime to $\mathbf{F}_q$ for q an odd prime power is, at least as far as the interests of cyclotomy are concerned, simply a matter of reinterpreting notation. That this is not quite true can be seen from the example for $e = 4$ and f odd of §II in the light of the present field structure. In this example, everything is absolutely true as stated (and identically proved) with the *proviso* that $q = p^\alpha = s^2 + 4t^2$ is *some* binary quadratic decomposition of q with $s \equiv 1 \pmod 4$, there being in general many of these. There is, however, precisely one such decomposition which is *proper*, and we now show that it is this representation to which we are necessarily led by the extended theory.

LEMMA. *Let* $q = p^\alpha$ *and* $p \equiv 1 \pmod 4$. *Then g.c.d.* $(s, p) = 1$.

PROOF. By 6 (iv) of §II we have

$$-2s = \varphi_2(1) = \sum_{y \in \mathbf{F}_q} \chi(y)\chi(y^2 + 1) = \sum_{y \in \mathbf{F}_q} (y^3 + y)^{(q-1)/2}$$

(Euler's Criterion)

$$= \sum_{y \in \mathbf{F}_q} \sum_{s=0}^{q-1/2} (q - 1) \Big/ \binom{2}{s} y^{3s} y^{(q-1)/2-s}$$

$$= \sum_{s=0}^{q-1/2} (q-1) \Big/ \binom{2}{s} \sum_{y \in \mathbf{F}_q} y^{2s+(q-1)/2} \equiv - \binom{(q-1)/2}{(q-1)/4} \pmod{q}$$
$$\equiv - \binom{(q-1)/2}{(q-1)/4} \pmod{p},$$

since

$$\sum_{y \in \mathbf{F}_q} y^{2s+(q-1)/2} = q - 1 \quad \text{if } s = (q-1)/4, \qquad = 0 \quad \text{otherwise.}$$

Now using the Lucas identity [2; pp. 417–420]: For $0 \leq a_i, b_i \leq p - 1$ with $i = 0, 1, 2, \ldots, \alpha - 1$ we have

$$\binom{a_0 + a_1 p + a_2 p^2 + \cdots + a_{\alpha-1} p^{\alpha-1}}{b_0 + b_1 p + b_2 p^2 + \cdots + b_{\alpha-1} p^{\alpha-1}} \equiv \binom{a_0}{b_0} \binom{a_1}{b_1} \binom{a_2}{b_2} \cdots \binom{a_{\alpha-1}}{b_{\alpha-1}} \pmod{p},$$

and noting that, for $i = 1, 2$,

$$\frac{q-1}{2^i} = \left(\frac{p-1}{2^i}\right)(1 + p + p^2 + \cdots + p^{\alpha-1})$$

we immediately have

$$-2s \equiv - \binom{(p-1)/2}{(p-1)/4}^{\alpha} \pmod{p}$$

whence, clearly, g.c.d. $(s, p) = 1$.

Again, this result should be taken as illustrative of a collection of results concerning the unique determination of the cyclotomic numbers for finite fields of proper prime-power order, and its proof as representative [8].

IV. The Galois domain

It was in 1962 that the first really important extension of the theory of cyclotomy appeared [10] in conjunction with a generalization and strengthening of a result concerning the existence of combinatorial designs. Here the parent algebraic structure is the Galois domain, and we give a formulation of the cyclotomic theory for domains analogous to that of §III for the finite field. The method given here is not that of the pioneering work [5], which was stated in terms of the particular domains $\mathbf{Z}_{pq}$, but is perhaps more easily generalizable to other structures; the details of the present approach may be found in [7], which appeared in 1969.

1. NOTATION. Let p_1 and p_2 be distinct odd primes, α_1 and α_2 nonzero natural numbers, and $q_1 = p_1^{\alpha_1} = ef_1 + 1$ and $q_2 = p_2^{\alpha_2} = ef_2 + 1$, where $e =$ g.c.d. $(q_1 - 1, q_2 - 1)$. For the Galois domain of order $q_1 q_2$ we shall write $\mathbf{G}_{q_1 q_2} = \mathbf{F}_{q_1} \oplus \mathbf{F}_{q_2}$ (direct sum) and we note that for $\alpha_1 = \alpha_2 = 1$ we have $\mathbf{G}_{q_1 q_2} \cong_{(+)} \mathbf{Z}_{q_1 q_2}$.

2. CLASS STRUCTURE. Let x_1 and x_2 be fixed generators of $\mathbf{F}_{q_1}$ and $\mathbf{F}_{q_2}$, respectively, and set $\mathbf{x} = (x_1, x_2)$ and $\mathbf{y} = (x_1, 1)$ in $\mathbf{G}_{q_1 q_2}$. The *cyclotomic classes* $C_{q_1 q_2, i}$ for $q_1 q_2$ and $\mathbf{x}$ (note, e is uniquely determined by q_1 and q_2 here) are now defined by

$$C_{q_1 q_2, i} = \{\mathbf{x}^s \mathbf{y}^i : s = 0, 1, 2, \ldots, d - 1\} \qquad (i = 0, 1, 2, \ldots, e - 1)$$

where $d = ef_1f_2$ and, in terms of these classes, the *cyclotomic numbers* $(i, j)_{q_1q_2}$ and *matrix* $\Gamma_{q_1q_2,e}$ for q_1q_2 and $\mathbf{x}$ are defined exactly as for $\mathbf{F}_q$.

We remark that, unlike the case for $\mathbf{F}_q$, replacement of the generator $\mathbf{x}$ of $\mathbf{G}_{q_1q_2}$ by a new generator $\mathbf{x}^*$ may no longer leave $C_{q_1q_2,0}$ fixed, and hence there may be several distinct cyclotomies definable on $\mathbf{G}_{q_1q_2}$ in the above manner; a simple computation shows that, in fact, this number is at most $\varphi(e)$—the Euler φ-function of e. In any case, the parameter relations for q_1q_2 for a fixed generator $\mathbf{x}$ are well-defined:

5. Parameter relations. (i) and (ii) are identical to those for $\mathbf{F}_q$ with q replaced by q_1q_2; (iii) and (iv) are replaced by

(iii)′ $(i, j)_{q_1q_2} = (j, i)_{q_1q_2}$ if ff' is odd,
$= (j + e/2, i + e/2)_{q_1q_2}$ if ff' is even.

(iv)′ $\sum_{j=0}^{e-1} (i, j)_{q_1q_2} = \frac{1}{e}[(q_1 - 2)(q_2 - 2) - 1] + \delta_i$

where $\delta_i = 1$ if $-1 \in C_{q_1q_2,i}$,
$= 0$ otherwise.

We now employ the same trick as for the field to make $\langle \mathcal{B}, \cdot \rangle$, where $\mathcal{B} = \{\lambda^{\mathbf{b}}_{q_1q_2} : \mathbf{b} \in \mathbf{G}_{q_1q_2}\}$, into a group isomorphic to $\langle \mathbf{G}_{q_1q_2}, + \rangle$. Here the axioms are these:

Axiom 1. *There is an identification "$\leftrightarrow$" such that* $\forall \mathbf{b} \in \mathbf{G}_{q_1q_2}$: $\exists n \in \{0, 1, 2, \ldots, q_1q_2 - 1\}$: $\lambda^{\mathbf{b}}_{q_1q_2} \leftrightarrow \lambda^n_{q_1q_2}$ *and* $\lambda^{\mathbf{0}}_{q_1q_2} \leftrightarrow 1$.

Axiom 2. (i) *If* $\mathbf{b} \in (\mathbf{F}_{q_1}, 0) \subset \mathbf{G}_{q_1q_2}$ *and* $\lambda^{\mathbf{b}}_{q_1q_2} \leftrightarrow \lambda^n_{q_1q_2}$, *then* $\lambda^n_{q_2} = 1$.
(ii) *If* $\mathbf{b} \in (0, \mathbf{F}_{q_2}) \subset \mathbf{G}_{q_1q_2}$ *and* $\lambda^{\mathbf{b}}_{q_1q_2} \leftrightarrow \lambda^n_{q_1q_2}$, *then* $\lambda^n_{q_1} = 1$.

Axiom 3. $\lambda^{\mathbf{b}_1}_{q_1q_2} = \lambda^{\mathbf{b}_2}_{q_1q_2}$ *if and only if* $\mathbf{b}_1 = \mathbf{b}_2 \in \mathbf{G}_{q_1q_2}$.

Axiom 4. $\lambda^{\mathbf{b}_1}_{q_1q_2}\lambda^{\mathbf{b}_2}_{q_1q_2} = \lambda^{\mathbf{b}_1+\mathbf{b}_2}_{q_1q_2}$ *for all* $\mathbf{b}_1, \mathbf{b}_2 \in \mathbf{G}_{q_1q_2}$.

Again, the results which follow are known to be independent of the identification chosen.

We may now proceed formally as before:

3. Root structure. The *periods* $\eta_{q_1q_2,k}$ relative to q_1q_2 and $\mathbf{x}$ are defined by

$$\eta_{q_1q_2,k} = \sum_{\mathbf{b} \in C_{q_1q_2,k}} \lambda^{\mathbf{b}}_{q_1q_2} = \sum_{s=0}^{d-1} \lambda^{\mathbf{x}^s\mathbf{y}^k}_{q_1q_2} \in \mathcal{B}\mathbf{Z}, \qquad (k = 0, 1, 2, \ldots, e - 1).$$

In discussing *arithmetic functions* on Galois domains we note that the Jacobsthal sum has not been defined in this setting; this of course may be done, but here it is our purpose to show that, in fact, the classical extensions of the arithmetic functions to domains is not critical (or even important) for the determination of the relevant class structure, and leads only to the interesting result that these functions split over the summand fields in an entirely straightforward, if surprising, manner.

4. Arithmetic functions. If e divides none of m, n, nor $m + n$, we define

$$F_{q_1q_2,e}(\lambda^n_e) = \sum_{j=0}^{e-1} \lambda^{nj}_e \eta_{q_1q_2,j}$$

and

$$R_{q_1q_2,e}(m,n) = \sum_{k=0}^{e-1} \lambda_e^{nk} \sum_{h=0}^{e-1} \lambda_e^{-(m+n)h}(k,h)_{q_1q_2},$$

relative to the fixed generator $\mathbf{x}$.

Here the methods of [10] will produce the functional relations 6 (i) and 6 (ii) of §II for domains by replacement of q by q_1q_2 in these identities but is apparently not strong enough to produce any analogue of the Jacobi identity 6 (iii) without the introduction of powerful new ideas.

To that end we mention a result [7] of an essentially new type, relating the eth power class structure of $\mathbf{G}_{q_1q_2}$ corresponding to a fixed generator $\mathbf{x} = (x_1, x_2)$ to the eth power class structures on $\mathbf{F}_{q_1}$ and $\mathbf{F}_{q_2}$ with respect to their generators x_1 and x_2, respectively. This result is

THEOREM 1. *Given $\mathbf{F}_{q_1}$ and $\mathbf{F}_{q_2}$ with fixed generators x_1 and x_2, respectively, $e = \text{g.c.d.}\,(q_1 - 1, q_2 - 1)$, and the corresponding cyclotomic numbers $(i,j)_{q_1}$ and $(i',j')_{q_2}$; the cyclotomic numbers for $\mathbf{G}_{q_1q_2}$ with fixed generator $\mathbf{x} = (x_1, x_2)$ are given by*

$$(i,j)_{q_1q_2} = \sum_{m,n=0}^{e-1} (i+m, j+n)_{q_1}(m,n)_{q_2}.$$

Thus the cyclotomies and class structures for the Galois domain $\mathbf{G}_{q_1q_2}$ are explicitly known as soon as the corresponding problem has been solved for the summand fields.

By way of example, we present the cyclotomic numbers for $e = 4$ in the case ff' even. Here the (4 by 4) matrix blank is identical to that in §II (since the parameter relations 5 (ii) and (iii) are identical) and, setting $\mathring{M} = \frac{1}{4}[(q_1 - 2)(q_2 - 2) - 1]$, the constant replacing f in 5 (iv)$'$ for domains, we find that

$$\begin{aligned} 8A &= 2\mathring{M} + 3 - s, \\ 8B &= 2\mathring{M} - 1 - s - 4t, \\ 8C &= 2\mathring{M} - 1 + 3s, \\ 8D &= 2\mathring{M} - 1 - s + 4t, \\ 8E &= 2\mathring{M} + 1 + s \end{aligned}$$

where $q_1q_2 = s^2 + 4t^2$ is determined by the representations of $q_1 = s_1^2 + 4t_1^2$ and $q_2 = s_2^2 + 4t_2^2$ giving rise to the cyclotomic numbers on the summand fields $\mathbf{F}_{q_1}$ and $\mathbf{F}_{q_2}$. Note that if these representations of q_1 and q_2 are proper, then the two proper representations of q_1q_2 are

$$q_1q_2 = (s_1s_2 \pm 4t_1t_2)^2 + 4(s_1t_2 \mp s_2t_1)^2,$$

in which case the $\varphi(4) = 2$ cyclotomies on $\mathbf{G}_{q_1q_2}$ are distinct. Theorem 1 may, in fact, be used to easily show that in this case the two cyclotomic numbers $(0,0)_{q_1q_2}$ for these cyclotomies are unequal.

Immediate corollaries of Theorem 1 include the complete analogues of the parameter relations (5) and the functional relations 6 (i)–(iii); note that 6 (iv) is no longer relevant. In particular, the analogue of the Jacobi identity 6 (iii) for domains is

COROLLARY 1. *If* l_1 *and* l_2 *are defined by* $x_1^{l_1} = 2 \in \mathbf{F}_{q_1}$ *and* $x_2^{l_2} = 2 \in \mathbf{F}_{q_2}$, *then*

$$F_{q_1q_2,e}(-1)F_{q_1q_2,e}(\lambda_e^{2k}) = \lambda_e^{2(l_1-l_2)k}F_{q_1q_2,e}(\lambda_e^k)F_{q_1q_2,e}(-\lambda_e^k).$$

The proof of Corollary 1 follows from the Jacobi identity on each summand field and the fact that $F_{q_1q_2,e}$ splits over the summand fields as follows:

COROLLARY 2. *If* $m_1, m_2 \in \mathbf{Z}$ *satisfy* $m_2p_1 + m_1p_2 = 1$, *and* $\beta_1, \beta_2 \in \mathbf{N}_n$ *are defined by* $m_i \in C_{q_1q_2,\beta_i}$ *for* $i = 1, 2$, *then*

$$F_{q_1q_2,e}(\lambda_e^n) = \lambda_e^{(\beta_2-\beta_1)n}F_{q_1,e}(\lambda_e^n)F_{q_2,e}(\lambda_e^{-n}).$$

Finally, the functional relation 6 (i) for $\mathbf{F}_{q_1}$ and $\mathbf{F}_{q_2}$, together with Corollary 2 above, implies that the $R_{q_1q_2,e}$ split over the summand fields in an entirely similar manner:

COROLLARY 3. $R_{q_1q_2,e}(m, n) = R_{q_1,e}(m, n)R_{q_2,e}(-m, -n)$.

As concluding remarks to this section on domains, it should be mentioned that the above theory has been extended in two directions. The problems which have been solved are:

(i) Let $\mathbf{x}$ now be any unit in $\mathbf{G}_{q_1q_2}$, define $C_{q_1q_2,0} = \langle \mathbf{x} \rangle$ the multiplicative subgroup generated by $\mathbf{x}$, and let $C_{q_1q_2,i}$ be the cosets of $C_{q_1q_2,0}$ in the group of units of $\mathbf{G}_{q_1q_2}$, in some order. For all i and j, determine the cyclotomic numbers (i.e. the class structure) of this system [4].

(ii) Given distinct odd primes $p_1, p_2, \ldots, p_n$ and natural numbers $\alpha_1, \alpha_2, \ldots, \alpha_n$, let $q_i = p_i^{\alpha_i}$ for each i, and choose a fixed generator of each $\mathbf{F}_{q_i}$. Now let $\mathbf{x} = (x_1, x_2, \ldots, x_n) \in \mathbf{G}_{q_1q_2\cdots q_n}$ and define $C_{q_1q_2\cdots q_n,0} = \langle \mathbf{x} \rangle$. Then, if $C_{q_1q_2\cdots q_n,i}$ are the multiplicative cosets of $C_{q_1q_2\cdots q_n,0}$ in $\mathbf{G}_{q_1q_2\cdots q_n}$ in some order, determine the cyclotomic numbers for this system [7].

The methods of solution of problems (i) and (ii) above can be used together to develop a complete cyclotomic theory for $\mathbf{G}_n$ and any $n \in \mathbf{N}_n$. We remark that, for n square-free $\mathbf{G}_n \cong_{(+)} \mathbf{Z}_n$, and this fact may be used to extend the complete cyclotomic theory for $\mathbf{Z}_n$, n square-free, to $\mathbf{Z}_n$ for any $n \in \mathbf{N}_n$ [6].

V. Neofields

A *neofield* $\mathbf{N}_v$ of order v is a triple $\langle \mathbf{N}_v, +, \cdot \rangle$, where $\langle \mathbf{N}_v - \{0\}, \cdot \rangle$ is a group, $\langle \mathbf{N}_v, + \rangle$ is a loop, and both distributive laws hold. We shall restrict our attention in this section to neofields $\mathbf{N}_v$ for which the multiplicative group is cyclic and the additive loop is commutative and enjoys the *inverse property* (I.P.); that is

$$(a + b) + (-b) = a \quad \text{for all } a, b \in \mathbf{N}_v;$$

such neofields are usually termed C.I.P. neofields. Note that inverses are unique, so that writing $(-b)$ in the above definition of I.P. makes sense. While there is as yet no complete characterization of C.I.P. neofields, it is known [9] that there exist *proper* (i.e. not the field) C.I.P. neofields $\mathbf{N}_v$ of every prime-power order $V = p^\alpha \geq 11$; there are none for $v < 11$. In order to discuss the class structure of the known prime-power C.I.P. neofields, we outline their construction.

Let $\mathbf{F}_{p^\alpha} = \langle S, +, \cdot \rangle$ with $S = \{0, 1, a, a^2, \ldots, a^{p^\alpha - 2}\}$ be a presentation of the field of order $p^\alpha \geq 11$, whose addition table is specified by the function $T(x) = 1 + x$ for $x \in S$ (alternatively, we may specify $\mathbf{F}_{p^\alpha}$ by $\langle S, T\rangle$, since for all presentations of $\mathbf{F}_{p^\alpha}$ the group $\langle S - \{0\}, \cdot\rangle$ is to be cyclic). We define the one-to-one unary functions T' and T_0 on S in terms of T by

$$T'(x) = -1 + x, \qquad T_0(x) = T(x) \quad \text{if } x = 0, -1 \in S,$$
$$= x/T(x) \quad \text{otherwise.}$$

It is easily shown that, for all $x \in S$, we have $(T'T_0)^3(x) = x$, and we write the *orbit* of x as

$$\theta(x) = \{x, (T'T_0)(x), (T'T_0)^2(x)\}$$

provided the cardinality of this latter set is not equal to 1. The following theorem provides the construction of C.I.P. neofields of every prime-power order $p^\alpha \geq 11$:

THEOREM 2. *If $\mathbf{F}_v = \langle S, T\rangle$ is a presentation of the field of order $v \geq 11$, then there is a function T_* on S satisfying*

(i) *$T_* \not\equiv T$ and $T_* \not\equiv T_0$ on S,*

(ii) *For each $x \in S$, either $T_*(x) = T(x)$ or $T_*(x) = T_0(x)$,*

(iii) *If T_* agrees with T (or T_0) at a point $x \in S$, then T_* agrees with T (or T_0) everywhere on $\{\theta(x) \cup \theta(x^{-1})\}$. Further, if T_* is any function on S satisfying conditions (i)–(iii) above, then $\langle S, T_*\rangle$ is a presentation of a C.I.P. neofield $\mathbf{N}_v$ of order v.*

At this point several remarks are in order: (1) It is known that some (in fact, "most") of the C.I.P. neofields arising from Theorem 2 are proper; it is conjectured that all are. (2) It is not known if every C.I.P. neofield of prime-power order is realizable from Theorem 2. (3) There are (cyclic) C.I.P. neofields of non-prime-power order (e.g. $v = 14$) for which no reasonable construction is known.

We are now in a position to discuss the class structure of those C.I.P. neofields arising from Theorem 2.

1. NOTATION. Let $q = p^\alpha = ef + 1$ be an odd prime power, and denote the field of order q or a neofield of order q constructed from Theorem 2 by $\mathbf{N}_q$.

2. CLASS STRUCTURE. Fix a generator x of $\mathbf{N}_q$ and define the *cyclotomic classes* relative to q, e, and x by

$$C_{q,i} = \{x^{es+i}: s = 0, 1, 2, \ldots, f - 1\} \qquad (i = 0, 1, 2, \ldots, e - 1).$$

The *cyclotomic numbers* and *matrix* for q, e, and x are now defined for this system exactly as in the class structure (2) of §II, and the parameter relations (5) of that section are valid for this system as written (since none of the proofs involved entails the associativity of addition). Thus the cyclotomic matrix blanks for (all) C.I.P. neofields are identical to those for the fields.

The root structure (3), arithmetic functions (4), and functional relations (6) are now superfluous to the determination of the class structure of the neofields $\mathbf{N}_v$ and, in fact, this very class structure can in this case be used to introduce and study these concepts; in fact one can obtain a good "measure" of the degree of nonassociativity of a neofield in this way, and hence the theory of cyclotomy for these structures may provide a desirable classification scheme.

We now use the parameter relations 5 (ii) and (iii) to generate all cyclotomic numbers $\Theta(i, j)_q$ *associated with* a given cyclotomic number $(i, j)_q$ for a given neofield $\mathbf{N}_q$ and e, x, and note that $|\Theta(i, j)_q| = 1, 2, 3$, or 6. The following theorem in effect says that the cyclotomic problem for any $\mathbf{N}_q$ is solved if and only if the problem is solved for the corresponding field $\mathbf{F}_q$:

THEOREM 3. *Let $\mathbf{N}_q = \langle S, T_* \rangle$ be a fixed C.I.P. neofield constructed from the presentation $\langle S, T \rangle$ of the field $\mathbf{F}_q$ of order $q = p^\alpha = ef + 1$ by Theorem 2. Let $X = \{x_1, x_2, \ldots, x_n\}$ be an ordered system of (distinct) representatives for the orbit pairs $\theta(x) \cup \theta(x^{-1})$ from S on which $T_* = T_0$, and let $|X_{i,j}|$ be the number of elements $x_k \notin \theta(1)$ from X for which $x_k \in C_i$ and $T(x_k) \in C_j$. Finally, let*

$$\mathfrak{A} = \{A_{i,j}: (i, j)_q \text{ is a representative of } \Theta(i, j)_q\}$$

and

$$\mathfrak{A}' = \{A'_{i,j}: (i, j)_q \text{ is a representative of } \Theta(i, j)_q\}$$

be the equivalent cyclotomic numbers for $\mathbf{F}_q$ and $\mathbf{N}_q$, respectively, with respect to e and a fixed generator x. Then we have

(1) $A'_{i,j} = A_{i,j}$ *if*

a. $|\Theta(i, j)_q| < 3$,

b. $|\Theta(i, j)_q| = 3$ *and* $(e/2, 0)_q \in \Theta(i, j)_q$ *if qf is even,* $(e/2, e/2)_q \in \Theta(i, j)_q$ *if qf is odd.*

c. $|\Theta(i, j)_q| = 6$ *and* $(2i, i)_q \in \Theta(i, j)_q$ *for some* i.

(2) *Otherwise*

$$A'_{i,j} = A_{i,j} + \frac{6}{|\Theta(i, j)_q|} \sum_{(l,m) \in \Theta(i,j)_q} \{|X_{l,m}| - |X_{l,l-m}| + \delta_{l,m}\}$$

where

$$\begin{aligned} \delta_{i,j} &= 1 \quad \text{if } \theta(1) \cap X \neq \emptyset,\ i = 0,\ 2 \in C_j, \text{ and } j \neq 0, e/2 \\ &= -1 \quad \text{if } \theta(1) \cap X \neq \emptyset,\ i = 0,\ 2 \in C_{e-j}, \text{ and } j \neq 0, e/2 \\ &= 0 \quad \text{otherwise.} \end{aligned}$$

As our final example of the case $e = 4$ and (now) vf odd, we discuss this case in the setting of Theorem 3 and the C.I.P. neofield. Again the (4 by 4) matrix is identical to that in §II (since the parameter relations 5 (ii) and (iii) are identical); the remaining inequivalent cyclotomic numbers are given by

$$\begin{aligned} 16A &= q - 7 + 2s, \\ 16B &= q + 1 + 2s - 8t + 32\epsilon, \\ 16C &= q + 1 - 6s, \\ 16D &= q + 1 + 2s + 8t - 32\epsilon, \\ 16E &= q - 3 - 2s \end{aligned}$$

where

$$\epsilon = \sum_{(l,m)\in\Theta(0,3)_q} \{|X_{l,m}| - |X_{l,l-m}|\}$$

and $q = p^\alpha = s^2 + 4t^2$ with $s \equiv 1 \pmod 4$ is that representation of q determined by the cyclotomic structure on $\mathbf{F}_q$; the sign of t again depends upon the choice of generator x.

REFERENCES

1. L. E. Dickson, *Cyclotomy, higher congruences and Waring's problem*, Amer. J. Math. **57** (1935), 391–424.

2. E. Lucas, *Théorie des nombres.* Tome I: *Le calcul des nombres entiers, le calcul des nombres rationnels, la divisibilité arithmétique*, Librairie Scientifique et Technique Albert Blanchard, Paris, 1961. MR **23** #A828.

3. T. Storer, *Cyclotomy and difference sets*, Lectures in Advanced Mathematics, no. 2, Markham, Chicago, Ill., 1967. MR **36** #128.

4. ———, *A complete cyclotomic theory for* $\mathbf{Z}_{p_0p_1}$, Duke Math J. (to appear).

5. ———, *A family of difference sets*, Dissertation, University of Southern California, 1965.

6. ———, *A note on the arithmetic structure of the integers modulo N*, J. Combinatorial Theory **5** (1968), 207–209. MR **37** #5184.

7. ———, *On the arithmetic structure of Galois domains*, Acta Arith. **15** (1969), 139–159.

8. ———, *On the unique determination of the cyclotomic numbers for Galois fields and Galois domains*, J. Combinatorial Theory **2** (1967), 296–300. MR **35** #1572.

9. T. Storer and E. C. Johnsen, *Combinatorial structures in loops.* II: *Difference sets*, (Submitted for publication).

10. A. L. Whiteman, *A family of difference sets*, Illinois J. Math. **6** (1962), 107–121. MR **25** #3003.

UNIVERSITY OF MICHIGAN
ANN ARBOR, MICHIGAN 48104

REDUCIBILITY OF LACUNARY POLYNOMIALS

A. SCHINZEL

Reducibility in this lecture means reducibility over the rational field Q and all the polynomials and rational functions considered are supposed to have integral coefficients. For a given polynomial $F(x_1, \ldots, x_k)$, $|F|$ is the maximum degree of F with respect to x_i and $\|F\|$ is the sum of squares of the coefficients of F.

The first result concerning reducibility of lacunary polynomials is due to Vahlen and gives a necessary and sufficient condition for the reducibility of $ax^n + b$. This result was generalized by Capelli to reducibility in arbitrary number field (for an exposition see [6]) which in virtue of Galois theory permits to decide about reducibility of $F(x^n)$, where F is a fixed polynomial and n is a variable. It does not permit however to describe how $F(x^n)$ factorizes. In a paper of mine already published [4] I have proved the following theorem

THEOREM 1. *For any irreducible polynomial F, which does not divide $x^\delta - x$ ($\delta > 0$) and any integer $n > 0$ there exists an integer ν such that*

(i) $0 < \nu < c_1(F)$,

(ii) $n = \nu u$,

(iii) $F(x) \overset{\text{can}}{=} \text{const}\, F_1(x)^{e_1} F_2(x)^{e_2} \cdots F_s(x)^{e_s}$ *implies*

$$F(x^n) \overset{\text{can}}{=} \text{const}\, F_1(x^u)^{e_1} F_2(x^u)^{e_2} \cdots F_s(x^u)^{e_s}$$

(the symbol $\overset{\text{can}}{=}$ means that on the right hand side occurs a product of irreducible nonconstant polynomials relatively prime in pairs). $c_1(F)$ is a computable constant.

Note a certain formal analogy with a result of Gourin [7]. Since the proof is accessible I shall only mention that it is based on the properties of a function $e(\alpha, \Omega)$, which to a given non-zero element α of a field Ω ascribes the maximal exponent e such that $\alpha = \zeta\beta^e$, where $\beta \in \Omega$ and ζ is a root of unity, or 0 if the maximum does not exist. In order to estimate $c_1(F)$ one needs a bound for $e(\alpha, Q(\alpha))$ where α is a zero of F. Such a bound can be deduced from a quantitative version of the old theorem of Kronecker: if an algebraic integer has all its conjugates on the unit circle then it is a root of unity. In a quantitative form we assume that an integer $\alpha \neq 0$ is not a root of unity and give a lower bound for the maximum absolute value of the conjugates. In the general case the best such bound due to P. E. Blanksby and H. L. Montgomery is $1 + 1/(30|F|^2 (\log 6|F|))$ but if α^{-1} is not conjugate to α Cassels [2] has given the bound $1 + 1(10|F|)$, which seems to be near the truth. If we use this bound we get easily

$$e(\alpha, Q(\alpha)) \ll |F| \log \|F\|.$$

Of course this is not the only possibility of estimating $e(\alpha, Q(\alpha))$; a much better bound would follow from an estimate for the absolute value of the product of the zeros of F lying outside the unit circle. Unfortunately the relevant problem of Lehmer remains unsolved since 1933. So much about the proof of Theorem 1, now consider its generalizations.

It is clear that the theorem cannot be extended to cyclotomic polynomials F since for such F and suitable n the number of irreducible factors of $F(x^n)$ can be arbitrarily large. Therefore if we wish to avoid the assumption of irreducibility it is natural to introduce $KF(x)$—the "kernel" of $F(x)$ free of cyclotomic factors and powers of x. Then Theorem 1 remains valid with the operation K applied to the left hand side of (iii). In order to generalize the theorem to several variables I need the following notation.

If $\phi(x_1, \ldots, x_k) = f(x_1, \ldots, x_k)\prod_{i=1}^k x_i^{\alpha_i}$ where f is a polynomial, $(f(x_1, \ldots, x_k), x_1, \ldots, x_k) = 1$ and the α_i are integers, then $J\phi(x_1, \ldots, x_k) = f(x_1, \ldots, x_k)$. Let $J\phi(x_1, \ldots, x_k) \stackrel{\text{can}}{=} \text{const} \prod_{\sigma=1}^s f(x_1, \ldots, x_k)^{e_\sigma}$. I set

$$K\phi(x_1, \ldots, x_k) = \text{const} \prod\nolimits_1 f_\sigma(x_1, \ldots, x_k)^{e_\sigma},$$
$$L\phi(x_1, \ldots, x_k) = \text{const} \prod\nolimits_2 f_\sigma(x_1, \ldots, x_k)^{e_\sigma},$$

where $\prod_1$ is extended over these f_σ which do not divide $J(x_1^{\delta_1}, \ldots, x_k^{\delta_k} - 1)$ for any $[\delta_1, \ldots, \delta_k] \neq \overline{0}$, $\prod_2$ is extended over all f_σ such that $Jf_\sigma(x_1^{-1}, \ldots, x_k^{-1}) \neq \pm f_\sigma(x_1, \ldots, x_k)$. The leading coefficients of $K\phi$ and $L\phi$ are assumed equal to that of $J\phi$. In particular for $k = 1$, $L\phi(x)$ equals $J\phi(x)$ deprived of all its monic irreducible reciprocal factors. We have $KJ = JK = K$, $LJ = JL = L$, $LK = KL = L$; the last formula requires a proof. The maximal absolute value of the elements of a vector $\bar{v}$ or of a matrix $\mathbf{M}$ is denoted by $h(\bar{v})$ or $h(\mathbf{M})$, respectively. The elements are supposed integral throughout. $\mathbf{M}^T$ and $\mathbf{M}^A$ is the matrix transposed and adjoint to $\mathbf{M}$, respectively. Now the proper generalization of Theorem 1 would be the following

CONJECTURE. For any polynomial $F(x_1, \ldots, x_k)$ and any vector

$$\bar{n} = [n_1, \ldots, n_k] \neq \overline{0}$$

such that $F(x^{n_1}, \ldots, x^{n_k}) \neq 0$ there exist a matrix $\mathbf{N} = [\nu_{ij}]_{i \le r; j \le k}$ of rank r and a vector $\bar{v} = [v_1, \ldots, v_r]$ such that

(i) $h(\mathbf{N}) \le c_2(k, r, |F|, \|F\|)$,

(ii) $\bar{n} = \bar{v}\mathbf{N}$,

(iii) $KF(\prod_{i=1}^{r} y_i^{\nu_{i1}}, \ldots, \prod_{i=1}^{r} y_i^{\nu_{ik}}) \overset{\text{can}}{=} \text{const} \prod_{\sigma=1}^{s} F_\sigma(y_1, \ldots, y_r)^{e_\sigma}$
implies

$$KF(x^{n_1}, \ldots, x^{n_k}) \overset{\text{can}}{=} \text{const} \prod_{\sigma=1}^{s} KF_\sigma(x^{v_1}, \ldots, x^{v_r})^{e_\sigma}.$$

Unfortunately, this I can prove only for $k = 2$ under the assumption

$$KF(x_1, x_2) = LF(x_1, x_2)$$

and postpone it to the end of my lecture. In full generality I can only prove

THEOREM 2. = *Conjecture with K replaced by L.*

Before I proceed to the proof I indicate two consequences which permit one perhaps to better understand the meaning of the theorem.

COROLLARY 1. *For any polynomial $f(x) \neq 0$ the number of its irreducible nonreciprocal factors except x counted with their multiplicities does not exceed $c_3(\|f\|)$.*

PROOF. Let $Jf(x) = a_0 + \sum_{j=1}^{k} a_j x^{n_j}$, where $a_j \neq 0$, n_j distinct > 0. Set in Theorem 2

$$F(x_1, \ldots, x_k) = a_0 + \sum_{j=1}^{k} a_j x_j.$$

We have $k \leq \|F\| - 1 = \|f\| - 1$. By Theorem 2 the number l of irreducible factors of $Lf(x)$ equals the number of irreducible factors of

$$LF\left(\prod_{i=1}^{r} y_i^{\nu_{i1}}, \prod_{i=1}^{r} y_i^{\nu_{i2}}, \ldots, \prod_{i=1}^{r} y_i^{\nu_{ik}}\right),$$

hence $l \leq 2kh(N) \leq 2\|f\| \max_{r \leq k < \|f\|} c_2(k, r, 1, \|f\|) = c_3(\|f\|)$.

COROLLARY 2. *If $k \geq 2$, $a_j \neq 0$ $(0 \leq j \leq k)$ then either $L(a_0 + \sum_{j=1}^{k} a_j x^{n_j})$ is irreducible or there is a vector $\bar{\gamma}$ such that*

$$0 < h(\bar{\gamma}) < c_4\left(\sum_{j=0}^{k} a_j^2\right)$$

and $\bar{\gamma}\bar{n} = 0$.

PROOF. Set again in Theorem 2:

$$F(x_1, \ldots, x_k) = a_0 + \sum_{j=1}^{k} a_j x_j.$$

It is an easy consequence of Capelli's general theorem that if the rank of $\mathbf{N}$ is k, $JF(\prod_{i=1}^{k} y_i^{\nu_{i1}}, \ldots, \prod_{i=1}^{k} y_i^{\nu_{ik}})$ is irreducible, on the other hand if the rank of $\mathbf{N}$ is less than k then the existence of $\bar{\gamma}$ follows from (ii).

The proof of Theorem 2 is based on 5 main lemmata.

LEMMA 1 (STRAUS [4]). *Let $x_1, \ldots, x_n$ be real numbers and let m be the dimension of the vector space spanned by $\{x_i - x_j, (i, j = 1, \ldots, n)\}$ over Q. Let m' be the dimension of the rational vector space spanned only by those $x_i - x_j$ for which $x_i - x_j \neq x_k - x_l$ where $(i, j) \neq (k, l)$. Then $m' = m$.*

LEMMA 2. *Let k_i $(0 \leq i \leq l)$ be an increasing sequence of integers. Let $k_{j_p} - k_{i_p}$ $(1 \leq p \leq p_0)$ be all the numbers which appear only once in the double sequence $k_j - k_i$ $(0 \leq i \leq j \leq l)$. Suppose that for each p*

$$k_{j_p} - k_{i_p} = \sum_{q=1}^{k} c_{pq} n_q,$$

where c_{pq} are integers, $|c_{pq}| \leq c$. Then either there exist matrices $\mathbf{K} = [\kappa_{qi}]_{q \leq k; i \leq l}$ and $\mathbf{\Lambda} = [\lambda_{qt}]_{q \leq k; t \leq k}$ and a vector $\bar{u}$ such that

(1) $[k_1 - k_0, \ldots, k_l - k_0] = \bar{u}\mathbf{K}, \bar{n} = [n_1, \ldots, n_k] = \bar{u}\mathbf{\Lambda}$,

(2) $h(\mathbf{K}) \leq c_5(k, l, c)$,

(3) $|\mathbf{\Lambda}| \neq 0, h(\mathbf{\Lambda}) \leq 2^l$

or there exists a vector $\bar{\gamma}$ such that

$$\bar{\gamma}\bar{n} = 0 \quad \text{and} \quad 0 < h(\bar{\gamma}) \leq c_6(k, l, c).$$

PROOF. By the assumption for each pair $\langle i, j \rangle$, where $0 \leq i \leq j \leq l$ and $\langle i, j \rangle \neq \langle i_p, j_p \rangle$ $(1 \leq p \leq p_0)$ there exists a pair $\langle g_{ij}, h_{ij} \rangle \neq \langle i, j \rangle$ such that $k_j - k_i = k_{h_{ij}} - k_{g_{ij}}$. Let us consider the system of linear homogeneous equations

$$x_0 = 0,$$

$$x_j - x_i - x_{h_{ij}} + x_{g_{ij}} = 0, \qquad \langle i, j \rangle \neq \langle i_1, j_1 \rangle, \ldots, \langle i_{p_0}, j_{p_0} \rangle,$$

$$x_{j_p} - x_{i_p} - \sum_{q=1}^{k} c_{pq} y_q = 0 \qquad (1 \leq p \leq p_0),$$

satisfied by $x_i = k_i - k_0$ $(0 \leq i \leq l)$, $y_q = n_q$ $(1 \leq q \leq k)$.

Let $\mathbf{A}$ be the matrix of the system obtained from the above by cancelling the first equation and substituting $x_0 = 0$ in the others, $\mathbf{B}$ be the matrix of the coefficients of the x's, $-\mathbf{\Gamma}$ the matrix of the coefficients of the y's so that $\mathbf{A} = \mathbf{B} \mid -\mathbf{\Gamma}$ in the sense of juxtaposition. We assert that the system has at most k linearly independent solutions. Indeed, if we had $k + 1$ such solutions $\bar{a}_1, \ldots, \bar{a}_{k+1}$ then taking $\xi_1, \ldots, \xi_{k+1}$ as real numbers rationally independent, we should find a set of reals $\sum_{m=1}^{k+1} a_{mi}\xi_m$ $(0 \leq i \leq l)$, where all the differences would span over the rationals a space of dimension $k + 1$, while the differences, occurring only once,

$$\sum_{m=1}^{k=1} (a_{m,j_p} - a_{m,i_p})\xi_m = \sum_{m=1}^{k+1} \xi_m \sum_{q=1}^{k} c_{pq} a_{m,1+q} = \sum_{q=1}^{k} c_{pq} \left(\sum_{n=1}^{k+1} a_{m,1+q}\xi_m \right)$$

would span a space of dimension at most k contrary to Lemma 1.

It follows that the rank of A is $l + \rho$, where $0 \leq \rho < k$. If the rank of $\mathbf{B}$ is l, let $\mathbf{\Delta}$ be a nonsingular submatrix of $\mathbf{B}$ of degree l. Solving the system by means of Cramer formulae we find a system of k linearly independent integral solutions which can be written horizontally in the form $\mathbf{K}' \mid \mathbf{\Lambda}'$, where elements of $\mathbf{K}'$ are determinants obtained from $\mathbf{\Delta}$ by replacing one column by a column of $\mathbf{\Gamma}$ and $\mathbf{\Lambda}' = D\mathbf{I}_k$, $D = |\mathbf{\Delta}|$, $\mathbf{I}_k$ is the identity matrix.

By Hadamard's inequality

$$|D| \leq 2^l, \qquad h(\mathbf{K}') \leq c_7(l, c).$$

From $\mathbf{K}' \mid \mathbf{\Lambda}'$ we obtain by a standard method (see [1, Appendix A]) a fundamental system of integral solutions $\mathbf{K} \mid \mathbf{\Lambda}$ satisfying (2), (3). Since the system is fundamental there exists an integral vector $\bar{u}$ satisfying (1).

If the rank of $\mathbf{B}$ is less than l we find a system of $k - \rho$ linearly independent integral solutions in the form $\mathbf{K}' \mid \mathbf{\Lambda}'$ where $h(\mathbf{\Lambda}') \leq c_8(k, l, c)$ and the rank of $\mathbf{\Lambda}' < k$. By a well-known lemma [1, Lemma 3, Chapter VI] there exists an integral vector $\bar{\gamma} \neq \bar{0}$ such that $\mathbf{\Lambda}'\bar{\gamma}^T = \bar{0}$ and $h(\bar{\gamma}) \leq c_6(k, l, c)$. Since $\bar{n} = \bar{u}'\mathbf{\Lambda}'$ we get $\bar{\gamma}\bar{n} = \bar{n}\bar{\gamma}^T = \bar{u}'\mathbf{\Lambda}'\bar{\gamma}^T = 0$ ($\bar{u}'$ not necessarily integral).

LEMMA 3 ($L3_k$). *Let $P(x_1, \ldots, x_k) \neq 0$, $Q(x_1, \ldots, x_k) \neq 0$ be polynomials and $(P, Q) = G$. For any vector $\bar{n} = [n_1, \ldots, n_k]$ we have either*

$$(LP(x^{n_1}, \ldots, x^{n_k}), LQ(x^{n_1}, \ldots, x^{n_k})) = \text{const } LG(x^{n_1}, \ldots, x^{n_k})$$

or there exists a vector $\bar{\beta}$ such that

(4) $\bar{\beta}\bar{n} = 0$,

(5) $0 < h(\bar{\beta}) < c_9(P, Q)$.

LEMMA 4 ($L4_k$). *For any polynomial $F(x_1, \ldots, x_k) \neq 0$, any vector $\bar{n} = [n_1, \ldots, n_k]$ and any irreducible factor $f(x)$ of $LF(x^{n_1}, \ldots, x^{n_k})$ either there exist a matrix $\mathbf{\Lambda} = [\lambda_{qt}]$ of degree k, a vector $\bar{u} = [u_1, \ldots, u_k]$ and a polynomial $T(z_1, \ldots, z_k)$ such that*

$$|\mathbf{\Lambda}| \neq 0, \qquad h(\mathbf{\Lambda}) \leq c_{10}(\|F\|) = 2^{\|F\|},$$

$$\bar{n} = \bar{u}\mathbf{\Lambda},$$

$$T(z_1, \ldots, z_k) \mid F\left(\prod_{q=1}^{k} z_q^{\lambda_{q1}}, \ldots, \prod_{q=1}^{k} z_q^{\lambda_{qk}}\right),$$

$$f(x) = \text{const } LT(x^{u_1}, \ldots, x^{u_k})$$

or there exists a vector $\bar{\gamma}$ such that

$$\bar{\gamma}\bar{n} = 0, \qquad 0 < h(\gamma) < c_{11}(F). \tag{6}$$

Since $L3_1$ is obvious, the proof is performed in two steps.

IMPLICATION $L3_k \rightarrow L4_k$. (Proof based on an idea of Ljunggren [3]). Let $F(x_1, \ldots, x_k) = \sum_{i=0}^{I} a_i x_1^{\alpha_{i1}} \ldots x_k^{\alpha_{ik}}$, where a_i are integers $\neq 0$ and the vectors $\bar{\alpha}_i$ are all different. Let further

$$F(x^{n_1}, \ldots, x^{n_k}) = f(x)g(x),$$

where f and g have integral coefficients (if necessary we may change $f(x)$ by a constant factor without impairing the assertion of the lemma). We set

$$f(x^{-1})g(x) = \sum_{i=0}^{l} c_i x^{k_i} \qquad (c_i \text{ integers} \neq 0,\ k_0 < k_1 < \cdots < k_l)$$

and consider two expressions for $F(x^{n_1}, \ldots, x^{n_k})F(x^{-n_1}, \ldots, x^{-n_k})$:

$$F(x^{n_1}, \ldots, x^{n_k})F(x^{-n_1}, \ldots, x^{-n_k}) = \sum_{i=0}^{I} a_i^2 + \sum_{0 \leq i,j \leq I} a_i a_j x^{\bar{n}\bar{\alpha}_j - \bar{n}\bar{\alpha}_i},$$

$$(f(x^{-1})g(x))(f(x)g(x^{-1})) = \sum_{i=0}^{l} c_i^2 + \sum_{0 \leq i,j \leq l;\, i \neq j} c_i c_j x^{k_j - k_i}.$$

If for any pair $\langle i, j \rangle$

(7) $$i \neq j \quad \text{and} \quad \bar{n}\bar{\alpha}_j - \bar{n}\bar{\alpha}_i = 0$$

we have (6) with $h(\bar{\gamma}) \leq |F|$.

If no pair $\langle i, j \rangle$ satisfies (7) it follows that $F(x^{n_1}, \ldots, x^{n_k}) \neq 0$,

$$\sum_{i=0}^{l} c_i^2 = \sum_{i=0}^{I} a_i^2 = \|F\|, \qquad l < \|F\|,$$

each number $k_j - k_i$ which appears only once in the double sequence $k_j - k_i$ $(0 \leq i \leq j \leq l)$ has a value $\sum_{q=1}^{k} n_q d_q$ with $|d_q| \leq |F|$. Applying Lemma 2 with $c = |F|$ we find matrices $\mathbf{K} = [\kappa_{qt}]$, $\mathbf{\Lambda} = [\lambda_{qt}]$ and a vector $\bar{u}$ satisfying (2), (3) and

$$k_i - k_0 = \sum_{q=1}^{k} \kappa_{qi} u_q, \qquad h(\mathbf{K}) < c_5(k, \|F\|, |F|)$$

or a vector $\bar{\gamma}$ satisfying (6) with $h(\bar{\gamma}) < c_6(k, \|F\|, |F|)$. Set

$$P(z_1, \ldots, z_k) = \sum_{i=0}^{I} a_i \prod_{q=1}^{k} z_q^{\sum_{t=1}^{k} \lambda_{qt}\alpha_{it}},$$

$$Q(z_1, \ldots, z_k) = J \sum_{i=0}^{l} c_i \prod_{q=1}^{k} z_q^{\kappa_{qi}}.$$

We get from $L3_k$ that either

$$(LP(x^{u_1}, \ldots, x^{u_k}), LQ(x^{u_1}, \ldots, x^{u_k})) = \text{const}\, LG(x^{u_1}, \ldots, x^{u_k})$$

or $\bar{\beta}\bar{u} = 0$ with $\bar{\beta}$ satisfying (5). In the former case

$$Lg(x) = \text{const}\,(LF(x^{n_1}, \ldots, x^{n_k}), Lf(x^{-1})g(x)) = \text{const}\,(LP(x^{u_1}, \ldots, x^{u_k})), LQ(x^{u_1}, \ldots, x^{u_k}) = \text{const}\, LG(x^{u_1}, \ldots, x^{u_k}),$$

$$f(x) = \frac{LF(x^{n_1}, \ldots, x^{n_k})}{Lg(x)} = \frac{LP(x^{u_1}, \ldots, x^{u_k})}{\text{const}\, Lg(x^{u_1}, \ldots, x^{u_k})} = \text{const}\, LT(x^{u_1}, \ldots, x^{u_k}),$$

where $T = PG^{-1}$.

In the latter case we have $\bar{\gamma}\bar{n} = 0$ with $\gamma = \bar{\beta}\mathbf{\Lambda}^A$ and $h(\bar{\gamma}) \leq kh(\bar{\beta})h(\mathbf{\Lambda}^A) < kc_9(P, Q)(k - 1)^{(k-1)/2}h(\mathbf{\Lambda}) < c_{11}(F)$.

IMPLICATION $L4_k \to L3_{k+1}$. Let $P = GT$, $Q = GU$, let R_j be the resultant of T, U with respect to x_j. If

$$(LP(x^{n_1}, \dots, x^{n_{k+1}}), LQ(x^{n_1}, \dots, x^{n_{k+1}})) \neq \text{const}\, LG(x^{n_1}, \dots, x^{n_{k+1}})$$

then there exists an irreducible polynomial $f(x)$ such that

$$f(x) \mid (LT(x^{n_1}, \dots, x^{n_{k+1}}), LU(x^{n_1}, \dots, x^{n_{k+1}})).$$

Clearly for each $j \leq k + 1$,

$$f(x) \mid R_j(x^{n_1}, \dots, x^{n_{k+1}})$$

where x^{n_j} does not occur among the arguments of R_j. By $L4_k$ either there exist a matrix $\mathbf{\Lambda}_j$, a vector $\bar{u}_j$ and a polynomial T_j such that

$$|\mathbf{\Lambda}_j| \neq 0, \qquad h(\mathbf{\Lambda}_j) \leq 2^{\|R_j\|},$$

$$[n_1, \dots, n_{j-1}, n_{j+1}, \dots, n_{k+1}] = \mathbf{\Lambda}_j\bar{u}_j, \tag{8}$$

$$T_j \mid R_j\left(\prod_{q=1}^{k} z_q^{\lambda_{q1}}, \dots, \prod_{q=1}^{k} z_q^{\lambda_{qk}}\right), \qquad f(x) = \text{const}\, T_j(x^{u_{j1}}, \dots, x^{u_{jk}}) \tag{9}$$

or

$$\bar{\gamma}_j[n_1, \dots, n_{j-1}, n_{j+1}, \dots, n_{k+1}] = 0 \quad \text{with } 0 < h(\bar{\gamma}_j) < c_{11}(R_j).$$

In the latter case we have $\bar{\beta}\bar{n} = 0$, where

$$0 < h(\bar{\beta}) \leq \max c_{11}(R_j).$$

In the former case we set $\bar{u}_{k+1} = \bar{v} = [v_1, \dots, v_k]$, find

$$f(x) = \text{const}\, LT_{k+1}(x^{v_1}, \dots, x^{v_k}),$$
$$Jf(x^{-1}) = \text{const}\, LT_{k+1}(x^{-v_1}, \dots, x^{-v_k})$$

and

$$\frac{Jf(x^{-1})}{f(x)} = \frac{LT_{k+1}(x^{-v_1}, \dots, x^{-v_k})}{LT_{k+1}(x^{v_1}, \dots, x^{v_k})} = \frac{JT_{k+1}(x^{-v_1}, \dots, x^{-v_k})}{JT_{k+1}(x^{v_1}, \dots, x^{v_k})}. \tag{10}$$

Let

$$T_{k+1}(z_1, \dots, z_k) = \sum_{i=0}^{I} a_i z_1^{\alpha_{i1}} z_2^{\alpha_{i2}} \dots z_k^{\alpha_{ik}},$$

where $a_i \neq 0$ $(0 \leq i \leq I)$ and the vectors $\bar{\alpha}_i$ are all different.

Let $\bar{\alpha}_i\bar{u}$ takes its minimum for $i = m$, maximum for $i = M$. Since

$$Jf(x^{-1}) = \text{const}\, f(x)$$

we get from (10)

$$d(x) = a_m JT_{k+1}(x^{-v_1}, \dots, x^{-v_k}) - a_M JT_{k+1}(x^{v_1}, \dots, x^{v_k}) \neq 0.$$

The lowest term in $d(x)$ is of the form $ax^{\gamma v}$, where $\bar{\gamma} = \bar{\alpha}_i - \bar{\alpha}_m$ or $\bar{\alpha}_M - \bar{\alpha}_i$, so that

$$a \neq 0; \qquad \bar{\gamma}\bar{v} > 0 \tag{11}$$

and by (9)

$$h(\bar{\gamma}) \leq k|R_{k+1}|h(\mathbf{A}_{k+1}). \tag{12}$$

It follows that

$$\frac{Jf(x^{-1})}{f(x)} = \frac{a_M}{a_m} + \frac{a}{a_m^2} x^{\bar{\gamma}\bar{v}} \bmod x^{\bar{\gamma}\bar{v}+1}. \tag{13}$$

By (8)

$$|\mathbf{A}_{k+1}|\bar{\gamma}\bar{v} = (\bar{\gamma}\mathbf{A}_{k+1}^A)[n_1, \ldots, n_k]$$

and since $\bar{\gamma}' = \bar{\gamma}\mathbf{A}_{k+1}^A \neq \bar{0}$, we have for some $j \leq k$, $\gamma_j' \neq 0$. Applying (8) and (9) we find as above

$$\frac{Jf(x^{-1})}{f(x)} = \frac{b_N}{b_n} + \frac{b}{b_n^2} x^{\bar{\delta}\bar{v}_j} \bmod x^{\bar{\delta}_j\bar{v}_j+1} \tag{14}$$

with

$$b \neq 0, \qquad \bar{\delta}\bar{v}_j > 0, \tag{15}$$

$$h(\bar{\delta}) \leq k|R_{j+1}|h(\mathbf{A}_{j+1}). \tag{16}$$

It follows from (10), (11), (13), (14), (15) that $\bar{\gamma}\bar{v} = \bar{\delta}\bar{v}_j$, and setting $\bar{\delta}' = \bar{\delta}\mathbf{A}_j^A$ we obtain

$$\sum_{i=1}^{j-1} (|\mathbf{A}_j|\gamma_i' - |\mathbf{A}_{k+1}|\,\delta_i')n_i + |\mathbf{A}_j|\gamma_j' n_j + \sum_{i=j+1}^{k} (|\mathbf{A}_j|\gamma_i' - |\mathbf{A}_{k+1}|\,\delta_{i-1}')n_i - |\mathbf{A}_{k+1}|\,\delta_k' n_{k+1} = 0$$

which is the desired equality (4). Inequality (5) follows from (12) and (16).

LEMMA 5. *For any polynomial $F(x_1, \ldots, x_k)$ and any vector $\bar{n} = [n_1, \ldots, n_k]$ such that $F(x^{n_1}, \ldots, x^{n_k}) \neq 0$ there exists a matrix $\mathbf{M} = [\mu_{ij}]$ of degree k and a vector $\bar{v} = [v_1, \ldots, v_k]$ such that*

$$|\mathbf{M}| \neq 0, \qquad h(\mathbf{M}) < c_{12}(k, \|F\|), \tag{17}$$

$$\bar{n} = \bar{v}\mathbf{M} \tag{18}$$

and either

$$LF\left(\prod_{i=1}^{k} y_i^{\mu_{i1}}, \ldots, \prod_{i=1}^{k} y_i^{\mu_{ik}}\right) \stackrel{\text{can}}{=} \text{const} \prod_{\sigma=1}^{s} F_\sigma(y_1, \ldots, y_k)^{e_\sigma}$$

implies

$$LF(x^{n_1}, \ldots, x^{n_k}) \overset{\text{can}}{=} \text{const} \prod_{\sigma=1}^{s} LF_\sigma(x^{v_1}, \ldots, x^{v_k})^{e_\sigma} \tag{19}$$

or there exists a vector $\bar{\gamma}$ *such that*

$$\bar{\gamma}\bar{n} = 0, \qquad 0 < h(\bar{\gamma}) < c_{13}(F). \tag{20}$$

OUTLINE OF THE PROOF. Let S be the set of all integral matrices $\mathbf{\Lambda} = [\lambda_{qt}]$ of degree k satisfying

$$|\mathbf{\Lambda}| \neq 0, \qquad h(\mathbf{\Lambda}) \leq 2^{\|F\|}, \tag{21}$$

$$\bar{n} = \bar{u}\mathbf{\Lambda}. \tag{22}$$

Integral vectors $\bar{m}$ such that for all $\mathbf{\Lambda} \in S$ and a suitable integral vector $\bar{v}_{\mathbf{\Lambda}}$, $\bar{m} = \bar{v}_{\mathbf{\Lambda}}\mathbf{\Lambda}$ form a module $\mathfrak{M}$, say.

By (21) for any $\mathbf{\Lambda} \in S$, $|\mathbf{\Lambda}|$ divides $c_{12}(k, \|F\|) = \mu$. Clearly vectors

$$[\mu, 0, \ldots, 0], \ldots, [0, \ldots, 0, \mu]$$

belong to $\mathfrak{M}$. It follows by a standard method (see [1, Appendix A]) that $\mathfrak{M}$ has a basis $\bar{\mu}_1, \ldots, \bar{\mu}_k$ such that

$$\mathbf{M} = \begin{bmatrix} \bar{\mu}_1 \\ \bar{\mu}_2 \\ \vdots \\ \bar{\mu}_k \end{bmatrix},$$

satisfies (17). Since $\bar{n} \in \mathfrak{M}$, $\mathbf{M}$ satisfies also (18). In order to prove the alternative (19) or (20) we set

$$F\left(\prod_{i=1}^{k} y_i^{\mu_{i1}}, \ldots, \prod_{i=1}^{k} y_i^{\mu_{ik}}\right) \overset{\text{can}}{=} \text{const} \prod_{\sigma=1}^{s} F_\sigma(y_1, \ldots, y_k)^{e_\sigma}. \tag{23}$$

If $(LF_\rho(x, \ldots, x^{v_k}), LF_\tau(x^{v_1}, \ldots, x^{v_k})) = 1$ then by Lemma 3

$$\bar{\beta}\bar{v} = 0 \quad \text{with } h(\bar{\beta}) < c_9(F_\rho, F_\tau) < c_{14}(F, \mathbf{M})$$

and

$$\bar{\gamma}\bar{n} = 0 \quad \text{with } \bar{\gamma} = \bar{\beta}\mathbf{M}^A,\ h(\bar{\gamma}) < c_{13}(F).$$

Assume, therefore, that for all distinct $\rho, \tau \leq s_1$

$$(LF_\rho(x^{v_1}, \ldots, x^{v_k}), LF_\tau(x^{v_1}, \ldots, x^{v_k})) = 1 \tag{24}$$

and let $f(x)$ be any irreducible factor of $LF(x^{n_1}, \ldots, x^{n_k})$. By Lemma 4 either (20) holds or there exist a matrix $\mathbf{\Lambda} = [\lambda_{qt}]$ of degree k, a vector $\bar{u} = [u_1, \ldots, u_k]$ satisfying (21), (22) and a polynomial T such that

$$T(z_1, \ldots, z_k) \mid F\left(\prod_{q=1}^{k} z_q^{\lambda_{q1}}, \ldots, \prod_{q=1}^{k} z_q^{\lambda_{qk}}\right), \tag{25}$$

$$f(x) = \text{const}\, LT(x^{u_1}, \ldots, x^{u_k}). \tag{26}$$

Since $\mathbf{\Lambda} \in S$ and by the choice of $\mathbf{M}$: $\bar{\mu}_1, \ldots, \bar{\mu}_n \in \mathfrak{M}$ we have for some vectors $\bar{\vartheta}_1, \ldots, \bar{\vartheta}_n$: $\bar{\mu}_i = \bar{\vartheta}_i \mathbf{\Lambda}$, thus

$$\mathbf{M} = \boldsymbol{\theta}\mathbf{\Lambda}, \tag{27}$$

$$\boldsymbol{\theta} = \begin{bmatrix} \bar{\vartheta}_1 \\ \vdots \\ \bar{\vartheta}_n \end{bmatrix},$$

$$\bar{u} = \bar{v}\boldsymbol{\theta}, \tag{28}$$

$$W(y_1, \ldots, y_k) = JT\left(\prod_{i=1}^{k} y_i^{\vartheta_{i1}}, \ldots, \prod_{i=1}^{k} y_i^{\vartheta_{ik}}\right).$$

We have by (25) and (27)

$$W(y_1, \ldots, y_k) \mid F\left(\prod_{i=1}^{k} y_i^{\mu_{i1}}, \ldots, \prod_{i=1}^{k} y_i^{\mu_{ik}}\right),$$

by (26) and (28)

$$f(x) = \text{const}\, LW(x^{v_1}, \ldots, x^{v_k}).$$

Since $f(x)$ is irreducible, the last two formulae imply in view of (23)

$$f(x) = \text{const}\, LF_\rho(x^{v_1}, \ldots, x^{v_k}) \quad \text{for some } \rho \le s_1 \tag{29}$$

and since $Jf(x^{-1}) \ne \pm f(x)$ we have $\rho \le s$. By (24)

$$\left(f(x), \prod_{\sigma=s+1}^{s_1} LF(x^{v_1}, \ldots, x^{v_k})^{e_\sigma}\right) = 1$$

and because of the arbitrariness of $f(x)$

$$\left(LF(x^{n_1}, \ldots, x_n^{n_k}), \prod_{\sigma=s+1}^{s_1} LF_\sigma(x^{v_1}, \ldots, x^{v_k})^{e_\sigma}\right) = 1.$$

Since

$$LF(x^{n_1}, \ldots, x^{n_k}) = \text{const} \prod_{\sigma=1}^{s_1} LF_\sigma(x^{v_1}, \ldots, x^{v_k})^{e_\sigma},$$

it follows that

$$LF(x^{n_1}, \ldots, x^{n_k}) = \text{const} \prod_{\sigma=1}^{s} LF(x^{v_1}, \ldots, x^{v_k})^{e_\sigma}.$$

Moreover none of the $LF(x^{v_1}, \ldots, x^{v_k})$ $(\sigma \le s)$ is reducible since taking as $f(x)$ any of its irreducible factors we would obtain from (29) a contradiction with (24).

It remains to prove that none of $LF(x^{v_1}, \ldots, x^{v_k})$ $(\sigma \le s)$ is constant unless (20) holds and this is done in a straightforward but rather lengthy manner.

Proof of Theorem 2. The theorem is true for $k = 1$ by Lemma 5. Assume it is true for polynomials in $k-1$ variables and consider $F(x_1, \dots, x_k)$. By Lemma 5 either there exist a matrix $\mathbf{M}$ and a vector $\bar{v}$ with the properties (17)–(19) or there exists a vector $\bar{\gamma}$ satisfying (20). In the former case the theorem holds with $r = k$, in the latter case $\bar{n}$ belongs to the module $\mathfrak{N}$ of integral vectors perpendicular to $\bar{\gamma}$. It is easy to see that $\mathfrak{N}$ has a basis which written in the form of a matrix $\mathbf{\Delta} = [\delta_{tj}]_{t<k;j\le k}$ satisfies

$$h(\mathbf{\Delta}) \le (k-1)\, h(\bar{\gamma}), \tag{30}$$

$$\text{rank of } \mathbf{\Delta} = k-1, \tag{31}$$

$$\bar{n} = \bar{m}\mathbf{\Delta}, \quad \bar{m} \text{ integral} \ne \bar{0}, \tag{32}$$

$$F'(z_1, \dots, z_{k-1}) = JF\left(\prod_{t=1}^{k-1} z_t^{\delta_{t1}}, \prod_{t=1}^{k-1} z_t^{\delta_{t2}}, \dots, \prod_{t=1}^{k-1} z_t^{\delta_{tk}}\right). \tag{33}$$

It follows easily

$$|F'| \le 2(k-1)|F|h(\mathbf{\Delta}), \qquad \|F'\| \le \|F\|^2. \tag{34}$$

By the inductive assumption there exist a matrix $\mathbf{N}' = [\nu'_{it}]_{i\le r;t<k}$ and a vector $\bar{v} = [v_1, \dots, v_r]$ such that

$$h(\mathbf{N}') \le c_2(k-1, r, |F'|, \|F'\|), \tag{35}$$

$$\text{rank of } \mathbf{N}' = r, \tag{36}$$

$$\bar{m} = \bar{v}\mathbf{N}', \tag{37}$$

$$LF'\left(\prod_{i=1}^{r} y_i^{\nu'_{i1}}, \dots, \prod_{i=1}^{r} y_i^{\nu'_{ik-1}}\right) \overset{\text{can}}{=} \text{const} \prod_{\sigma=1}^{s} F_\sigma(y_1, \dots, y_r)^{e_\sigma}$$

implies

$$LF'(x^{m_1}, \dots, x^{m_{k-1}}) \overset{\text{can}}{=} \text{const} \prod_{\sigma=1}^{s} F_\sigma(y_1, \dots, y_r)^{e_\sigma}.$$

Set

$$\mathbf{N} = \mathbf{N}'\mathbf{\Delta}. \tag{38}$$

It follows from (31) and (36) that $\mathbf{N}$ is of rank r. By (30), (34), (35) and (38)

$$h(\mathbf{N}) \le (k-1)^2 h(\bar{\gamma}) h(\mathbf{N}') \le c_2(k, r, |F|, \|F\|)$$

which gives (i).

By (32) and (37): (ii) $\bar{n} = \bar{v}\mathbf{N}$. By (33) and (38)

$$LF'\left(\prod_{i=1}^{r} y_i^{\nu'_{i1}}, \dots, \prod_{i=1}^{r} y_i^{\nu'_{ik-1}}\right) = LF\left(\prod_{i=1}^{r} y_i^{\nu_{i1}}, \dots, \prod_{i=1}^{r} y_i^{\nu_{ik}}\right)$$

and by (32) and (33)

$$JF'(x^{m_1}, \ldots, x^{m_{k-1}}) = JF(x^{n_1}, \ldots, x^{n_k}),$$

which proves (iii).

By a laborious calculation one can obtain the following estimation for

$$c_2(k, r, |F|, \|F\|)$$

$$\begin{aligned} c_2(k, r, |F|, \|F\|) &\leq \exp 9k2^{\|F\|-5} \quad \text{if } r = k, \\ &\leq \exp (5.2^{\|F\|^2-4} + 2\|F\| \log |F|^\star) \quad \text{if } r + k = 3, \\ &\leq \exp_{(k-r)(k+r-3)}(8k|F|^{\star\|F\|-1} \log \|F\|) \quad \text{otherwise,} \end{aligned}$$

where $|F|^\star = \max(2, (|F|^2) + 2)^{1/2}$.

Now I turn to the special case of the conjecture formulated at the beginning and related to the operation K. It is the following

THEOREM 3. *For any polynomial* $F(x_1, x_2)$ *such that* $KF(x_1, x_2) = LF(x_1, x_2)$ *and any vector* $\bar{n} = [n_1, n_2] \neq \bar{0}$ *such that* $F(x^{n_1}, x^{n_2}) \neq 0$ *there exist a matrix* $\mathrm{N} = [\nu_{ij}]_{i\leq r; j\leq 2}$ *of rank* r *and a vector* $\bar{v} = [v_1, v_2]$ *such that*

(i) $h(\mathrm{N}) < c_{15}(F)$,

(ii) $\bar{n} = \bar{v}\mathrm{N}$,

(iii) $KF(\prod_{i=1}^r y_i^{\nu_{i1}}, \prod_{i=1}^r y_i^{\nu_{i2}}) \overset{\text{can}}{=} \text{const} \prod_{\sigma=1}^s F_\sigma(y_1, y_2)^{e_\sigma}$ *implies*

$$KF(x^{n_1}, x^{n_2}) \overset{\text{can}}{=} \text{const} \prod_{\sigma=1}^{s} KF(x^{v_1}, x^{v_r})^{e_\sigma}.$$

This theorem has the following consequences.

COROLLARY 3. *If* a, b, c, n, m *are integers,* $n > m > 0$, $abc \neq 0$ *then either* $K(ax^n + bx^m + c)$ *is irreducible or*

$$n/(n, m) < c_{16}(a^2 + b^2 + c^2).$$

COROLLARY 4. *For any integer* $a \neq 0$ *and any polynomial* $f(x)$ *satisfying* $f(0) \neq 0$, $f(x) \not\equiv -a$ *there exists infinitely many* n *such that* $K(ax^n + f(x))$ *is irreducible.*

Both corollaries are deduced in [4] from a special case of Theorem 3. Corollary 4 is related to the following

THEOREM 4. *For any integers* $a \neq 0$, $b \neq 0$ *and any polynomial* $f(x)$ *satisfying* $f(0) \neq 0$, $f(1) \neq -a - b$ *there exist infinitely many pairs* n, m *such that* $ax^n + bx^m + f(x)$ *is irreducible.*

Proof of this theorem is too long to be given here. Instead I outline a proof of Theorem 3.

LEMMA 6. *If* $KF(x_1, x_2) = LF(x_1, x_2)$ *and* $[n_1, n_2] \neq \bar{0}$ *then either*

$$KF(x^{n_1}, x^{n_2}) = LF(x^{n_1}, x^{n_2})$$

or

$$\frac{\max \{|n_1|, |n_2|\}}{(n_1, n_2)} < c_{17}(F).$$

PROOF. Suppose that $KF(x^{n_1}, x^{n_2}) \neq LF(x^{n_1}, x^{n_2})$ and let $f(x)$ be an irreducible factor of $KF(x^{n_1}, x^{n_2})/(LF(x^{n_1}, x^{n_2}))$. There exists an irreducibe factor $G(x_1, x_2)$ of $KF(x_1, x_2)$ such that $f(x) \mid G(x^{n_1}, x^{n_2})$. Since $KF(x_1, x_2) = LF(x_1, x_2)$ we have $H(x_1, x_2) = JG(x_1^{-1}, x_2^{-1}) \neq \pm G(x_1, x_2)$, on the other hand $f(x) \mid H(x^{n_1}, x^{n_2})$. Let R_i be the resultant of G and H with respect to x_{3-i}. Clearly $f(x) \mid R_i(x^{n_i})$. Applying Theorem 1 we find that

$$|f|/n_i = \mu_i/\nu_i, \quad \text{where } \nu_i \leq c_1(R_i) \text{ and } \mu_i \leq |R_i|\nu_i.$$

Hence

$$\frac{n_1}{n_2} = \frac{\mu_2\nu_1}{\mu_1\nu_2} \quad \text{and} \quad \frac{\max\{|n_1|, |n_2|\}}{(n_1, n_2)} \leq c_1(R_1)c_1(R_2) \max\{|R_1|, |R_2|\} < c_{17}(F).$$

PROOF OF THEOREM 3. If

$$\frac{\max\{|n_1|, |n_2|\}}{(n_1, n_2)} > \max\{c_{13}(F), c_{17}(F)\}$$

we set $\mathbf{N} = \mathbf{M}$, where $\mathbf{M}$ is the matrix of Lemma 5. Clearly $\mathbf{N}$ is of rank 2,

$$h(\mathbf{N}) \leq c_{12}(2, |F|),$$
$$\bar{n} = \bar{v}\mathbf{N}.$$

Moreover

$$KF(y_1^{\nu_{11}}y_2^{\nu_{21}}, y_1^{\nu_{12}}y_2^{\nu_{22}}) \stackrel{\text{can}}{=} \text{const} \prod_{\sigma=1}^{s} F_\sigma(y_1, y_2)^{e_\sigma} \tag{39}$$

implies in virtue of the identity $LK = L$

$$LF(y_1^{\nu_{11}}y_2^{\nu_{21}}, y_2^{\nu_{12}}y_2^{\nu_{22}}) \stackrel{\text{can}}{=} \text{const} \prod_{\sigma=1}^{s_0} F_\sigma(y_1, y_2)^{e_\sigma},$$

where $JF_\sigma(y_1^{-1}, y_2^{-1}) \neq \pm F_\sigma(y_1, y_2)$ for $\sigma \leq s_0$ exclusively; in virtue of Lemma 5

$$LF(x^{n_1}, x^{n_2}) = \text{const} \prod_{\sigma=1}^{s_0} LF_\sigma(x^{v_1}, x^{v_2})^{e_\sigma}.$$

By Lemma 6, $KF(x^{n_1}, x^{n_2}) = LF(x^{n_1}, x^{n_2})$ thus $KF_\sigma(x^{v_1}, x^{v_2}) = LF_\sigma(x^{v_1}, x^{v_2})$ $(\sigma \leq s_0)$ and we get

$$KF(x^{n_1}, x^{n_2}) = \text{const} \prod_{\sigma=1}^{s_0} KF_\sigma(x^{v_1}, x^{v_2})^{e_\sigma}.$$

It remains to prove $s_0 = s$. Supposing contrarywise that

$$F_s(y_1, y_2) = \pm JF_s(y_1^{-1}, y_2^{-1})$$

we obtain

$$D(z_1, z_2) = JF_s(z_1^{\nu_{22}} z_2^{-\nu_{21}}, z_1^{-\nu_{12}} z_2^{\nu_{11}}) = \pm JF_s(z_1^{-\nu_{22}} z_2^{\nu_{21}}, z_1^{\nu_{12}} z_2^{-\nu_{11}}).$$

On the other hand, by (39), $F_s(y_1, y_2)$ divides $f(y_1^{\mu_{11}} y_2^{\mu_{21}}, y_1^{\mu_{12}} y_2^{\mu_{22}})$ where $G(x_1, x_2)$ is a certain irreducible factor of $KF(x_1, x_2)$. Since $KF(x_1, x_2) = LF(x_1, x_2)$ we have $JG(x_1^{-1}, x_2^{-1}) \neq \pm G(x_1, x_2)$ and

$$(JG(z_1^{|\mathbf{N}|} z_2^{|\mathbf{N}|}), JG(z_1^{-|\mathbf{N}|}, z_2^{-|\mathbf{N}|})) = 1.$$

On substituting $y_1 = z_1^{\nu_{22}} z_2^{-\nu_{21}}$, $y_2 = z_1^{-\nu_{12}} z_2^{\nu_{11}}$ we infer that $D(z_1, z_2)$ divides $JG(z_1^{|\mathbf{N}|}, z_2^{|\mathbf{N}|})$ and $JG(z_1^{-|\mathbf{N}|}, z_2^{-|\mathbf{N}|})$, thus $D(z_1, z_2) = \text{const}$ and since the substitution is invertible ($|\mathbf{N}| \neq 0$) $F_s(y_1, y_2) = \text{const}$, a contradiction.

Suppose now that

$$\frac{\max\{|n_1|, |n_2|\}}{(n_1, n_2)} < \max\{c_{13}(F), c_{17}(F)\}.$$

We set

$$F'(x) = JF(x^{n_1/(n_1,n_2)}, x^{n_2/(n_1,n_2)})$$

and apply Theorem 1 in its general form to polynomial F' and the number (n_1, n_2). It follows that for a certain integer ν, $0 < \nu < c_1(F')$, $(n_1, n_2) = \nu v$ and

$$KF'(x^\nu) \overset{\text{can}}{=} \text{const} \prod_{\sigma=1}^{s} F_\sigma(x)^{e_\sigma}$$

implies

$$KF'(x^{(n_1,n_2)}) \overset{\text{can}}{=} \text{const} \prod_{\sigma=1}^{s} F_\sigma(x^v)^{e_\sigma}.$$

The matrix

$$\mathbf{N} = \left[\frac{n_1}{(n_1, n_2)}, \frac{n_2}{(n_1, n_2)}\right]\nu$$

has all the properties asserted in the theorem. Full details will appear in Acta Arithmetica XVI.

Note added in proof. Since the lecture was delivered, progress has been made with the problem of Lehmer concerning the product $p(F)$ of the zeros of a monic irreducible polynomial F, lying outside the unit circle. Blanksby and Montgomery have proved that if F is not cyclotomic then $p(F) > 1 + 1/(52|F| \log 6|F|)$ (to appear in Acta Arithmetica XVIII) and C. J. Smyth has proved that if F is not reciprocal then $p(F) > \sqrt{5}/2$ (written communication). The last result leads to a simple direct proof of Corollary 1.

REFERENCES

1. J. W. Cassels, *An introduction to Diophantine approximation*, Cambridge Tracts in Math. and Math. Phys., no. 45, Cambridge Univ. Press, New York, 1957. MR **19**, 396.

2. ———, *On a problem of Schinzel and Zassenhaus*, J. Math. Sci. **1** (1966), 1–8. MR **35** #159.

3. W. Ljunggren, *On the irreducibility of certain trinomials and quadrinomials*, Math. Scand. **8** (1960), 65–70. MR **23** #A1627.

4. A. Schinzel, *On the reducibility of polynomials and in particular of trinomials*, Acta Arith. **11** (1965), 1–34. MR **31** #4783.

5. E. G. Straus, *Linear dependence in finite sets of numbers*, Acta Arith. **11** (1965), 198–200.

6. N. G. Tschebotarew, *Grundzuge der Galoisschen Theorie*, Noordhoff, Groningen, 1950. MR **12**, 666.

7. E. Gourin, *On irreducible polynomials in several variables which become reducible when the variables are replaced by powers of themselves*, Trans. Amer. Math. Soc. **32** (1931), 485–501.

INSTITUTE OF MATHEMATICS, PAN
WARSAW, POLAND

QUADRATIC FORMS OVER FIELDS

A. PFISTER

1. Theorems of Cassels

1. Let K be any commutative field of characteristic not 2 and let $x_1, \ldots, x_n$ be indeterminates over K. A *quadratic form* φ of dimension n over K is a homogeneous polynomial of degree 2

$$\varphi(x) = \sum_{i,k=1}^{n} a_{ik}x_i x_k, \qquad a_{ik} \in K.$$

Without loss of generality we will always assume that $a_{ik} = a_{ki}$. We also use the matrix notation:

$$x = \begin{pmatrix} x_1 \\ \vdots \\ x_n \end{pmatrix}, \qquad x' = (x_1, \ldots, x_n), \qquad A = (a_{ik}) = A', \qquad \varphi(x) = x'Ax.$$

φ induces a symmetric *bilinear form*

$$\langle x, y \rangle = \langle x, y \rangle_\varphi = x'Ay.$$

Two n-dimensional quadratic forms $\varphi(x) = x'Ax$, $\psi(x) = x'Bx$ over K are called *equivalent* (over K) if there is a nonsingular linear transformation T of the vector-space K^n (i.e. an $n \times n$ matrix with entries in K) such that $\psi(x) = \varphi(Tx)$, or equivalently $B = T'AT$. We write: $\psi \cong \varphi$, $B \cong A$. It is well known (and easy to prove) that every quadratic form is equivalent to a *diagonal form* $\varphi(x) = \sum_{i=1}^{n} a_i x_i^2$. The latter is denoted by $\varphi = (a_1, \ldots, a_n)$. From now on all quadratic forms φ will be considered to be diagonal and to be *nonsingular*, i.e. all $a_i \neq 0$.

We say that φ *represents* an element $b \in K$ over K if there is a vector u with components in K and $\varphi(u) = b$. For $b = 0$ this is always possible with $u = 0$.

φ is called *isotropic* (over K) if $\varphi(x) = 0$ has a solution $x = u \neq 0$. Otherwise φ is called *anisotropic*.

The following results are well known:

PROPOSITION 1. *φ isotropic $\Rightarrow \varphi \cong (1, -1, a_3, \ldots, a_n)$ for suitable $a_3, \ldots, a_n \in K^*$.*

PROPOSITION 2. *$\varphi \cong (a_1, \ldots, a_n)$ represents $b \in K^* \Leftrightarrow \varphi \oplus (-b) \cong (a_1, \ldots, a_n, -b)$ is isotropic.*

COROLLARY. *φ isotropic $\Rightarrow \varphi$ is universal, i.e. represents every $b \in K^*$.*

In 1963 J. W. S. Cassels found interesting results concerning sum of squares in rational function fields. They can easily be generalized to yield the theorems presented here.

THEOREM 1. *Let x be an indeterminate over K and let $p(x) \in K[x]$ be a polynomial in x. If the quadratic form $\varphi = (a_1, \ldots, a_n)$ over K represents $p(x)$ in the field $K(x)$ then it represents $p(x)$ in the ring $K[x]$.*

PROOF. By assumption

$$a_1(f_1/f_0)^2 + \cdots + a_n(f_n/f_0)^2 = p \tag{1}$$

with $f_i = f_i(x) \in K[x]$ $(i = 0, \ldots, n)$ and $f_0 \neq 0$.

We exclude two trivial cases:

(i) If $p = 0$ multiply (1) by f_0^2.

(ii) If φ is isotropic we may assume that $a_1 = 1$, $a_2 = -1$ and have

$$1\left(\frac{p+1}{2}\right)^2 - 1\left(\frac{p-1}{2}\right)^2 + a_3 0^2 + \cdots + a_n 0^2 = p.$$

In the remaining cases we may assume that (1) is a representation of p in which the degree $\gamma(f_0)$ of f_0 is minimal. We assume $\gamma(f_0) > 0$ and derive a contradiction. Dividing f_i by f_0 we get polynomials g_i $(i = 1, \ldots, n)$ such that

$$\gamma(f_i - f_0 g_i) < \gamma(f_0) \qquad (i = 1, \ldots, n). \tag{2}$$

Putting $g_0 = g_0(x) = 1$ and $\gamma(0) = -\infty$, (2) also holds for $i = 0$. For abbreviation we introduce the quadratic form $\psi = (-p, a_1, \ldots, a_n)$ in $K(x)$, i.e.

$$\psi(u) = -pu_0^2 + a_1u_1^2 + \cdots + a_nu_n^2,$$

and the corresponding bilinear form

$$\langle u, v\rangle = -pu_0v_0 + a_1u_1v_1 + \cdots + a_nu_nv_n.$$

Putting

$$f = \begin{pmatrix} f_0 \\ \vdots \\ f_n \end{pmatrix}, \qquad g = \begin{pmatrix} g_0 \\ \vdots \\ g_n \end{pmatrix},$$

(1) becomes $\langle f, f\rangle = \psi(f) = 0$.

Now f and g are linearly independent over $K(x)$, for otherwise $f = f_0 \cdot g$ and $\psi(g) = 0$, contradicting the minimality of $\gamma(f_0)$. Therefore $h = \langle g, g\rangle f - 2\langle f, g\rangle g \neq 0$, but

$$\psi(h) = \langle g, g\rangle^2 \psi(f) - 4\langle f, g\rangle^2 \langle g, g\rangle + 4\langle f, g\rangle^2 \psi(g) = 0$$

by (1). Since φ is anisotropic in K and hence in $K(x)$, and since $h \neq 0$ we have $h_0 \neq 0$. From the construction of h we note that the components h_i of h are polynomials in x. Now

$$h_0 = \langle g, g\rangle f_0 - 2\langle f, g\rangle = \frac{1}{f_0}\psi(f_0 \cdot g - f) = \frac{1}{f_0}\sum_1^n a_i(f_i - f_0 g_i)^2.$$

By (2),

$$\gamma(h_0) = -\gamma(f_0) + \gamma\left(\sum_1^n a_i(f_i - f_0 g_i)^2\right) < -\gamma(f_0) + 2\gamma(f_0) = \gamma(f_0).$$

This gives the desired contradiction and hence the result.

2. Remark. The corresponding result is no longer true in the case of two or more variables. The existence of counter-examples goes back to Hilbert (1888) but an easy counter-example was only recently found by T. S. Motzkin (1967). Other counter-examples are due to R. M. Robinson (1969).

Counter-example by Motzkin: Let

$$p(x, y) = 1 + x^2(x^2 - 3)y^2 + x^2y^4 \in R[x, y].$$

Then

$$\begin{aligned} p(x, y) &= \frac{(1 - x^2y^2)^2 + x^2(1 - y^2)^2 + x^2(1 - x^2)^2y^2}{1 + x^2} \\ &= \frac{(1 + x^2 - 2x^2y^2)^2 + (x(1 - x^2)y)^2 + (x(1 - x^2)y^2)^2 + (x^2(1 - x^2)y^2)^2}{(1 + x^2)^2}. \end{aligned}$$

This shows that $p(x, y)$ is represented by the form $\varphi = (1, 1, 1, 1)$ over the ring $R(x)[y]$ and hence over the field $R(x, y)$.

However, $p(x, y)$ cannot be represented as a sum of (any number of) squares of polynomials $f_i(x, y) \in R[x, y]$. In fact, if

$$p(x, y) = \sum_1^n f_i(x, y)^2$$

then the f_i are of degree ≤ 3 and are bounded on the lines $x = 0$ and $y = 0$ since $p(0, y) = p(x, 0) = 1$. This shows that $f_i(x, y) = a_i + xyl_i(x, y)$ where a_i is a real constant and l_i is a linear polynomial. But then the coefficient of the term x^2y^2 in $\sum_1^n f_i^2$ is $\sum_1^n l_i(0, 0)^2 \geq 0$ whereas it should be -3.

3. The first application of Theorem 1 is a lemma which we shall need later.

LEMMA. *Let $\varphi = (a_1, \ldots, a_m)$ be a quadratic form over K and let $x_1, \ldots, x_n$ be indeterminates over K. Suppose that φ represents $p(x_1, \ldots, x_n) \in K[x_1, \ldots, x_n]$ over $K(x_1, \ldots, x_n)$ and let $c_1, \ldots, c_n$ be arbitrary elements in K. Then φ represents $p(c_1, \ldots, c_n)$ over K.*

PROOF. Using Theorem 1 this is a trivial induction on n.

THEOREM 2. *Let $d, a_1, \ldots, a_m \in K^*$ and assume that $\varphi = (a_1, \ldots, a_m)$ represents the polynomial $d + a_1x^2$ over $K(x)$. Then either φ is isotropic over K or $\varphi' = (a_2, \ldots, a_m)$ represents d over K.*

PROOF. Suppose φ is anisotropic. By Theorem 1

$$a_1f_1^2 + \cdots + a_mf_m^2 = d + a_1x^2 \tag{1}$$

where $f_i(x) \in K[x]$. Comparing terms of highest degree on both sides of (1) we see that all f_i must be linear in x.

Let

$$f_i(x) = b_i + c_ix \qquad (i = 1, \ldots, m).$$

Now at least one of the equations $b_1 + c_1x = \pm x$ is solvable with, say $x = c$, $c \in K$. Then (1) implies

$$\sum_2^m a_i(b_i + c_ic)^2 = d,$$

i.e. φ' represents d over K.

THEOREM 3. *Let $\varphi = (a_1, \ldots, a_m)$, $\psi = (b_1, \ldots, b_n)$ be quadratic forms over K, φ anisotropic. Suppose that $x_1, \ldots, x_n$ are indeterminates over K and that φ represents*

$$\psi(x) = \sum_1^n b_ix_i^2 \quad \text{over} \quad K(x) = K(x_1, \ldots, x_n).$$

Then φ contains ψ, i.e. $\varphi \cong (b_1, \ldots, b_n, \ldots)$. In particular $n \leq m$.

PROOF. By induction on m. We may start with $m = 0$ where the result is trivial. $m > 0$: From the Lemma φ represents b_1 over K. Hence

$$\varphi \cong (b_1, \ldots) = (b_1) \oplus \varphi'$$

and this represents $b_1x_1^2 + (b_2x_2^2 + \cdots + b_nx_n^2)$ over $K(x_2, \ldots, x_n)(x_1)$. Applying Theorem 2 to $K(x_2, \ldots, x_n)$ with

$$d = b_2x_2^2 + \cdots + b_nx_n^2$$

we see, on noting that φ is anisotropic over $K(x_2, \ldots, x_n)$, that φ' represents d over $K(x_2, \ldots, x_n)$. By induction hypothesis φ' contains $\psi' = (b_2, \ldots, b_n)$, so φ contains ψ.

4. A field K is called (formally) *real* if, whenever $a_1, \ldots, a_n \in K$ $(n \geq 1)$ and $\sum_1^n a_i^2 = 0$, then $a_1 = \cdots = a_n = 0$. If K is real then $K(x_1, \ldots, x_n)$ is real too. Taking

$$\varphi = \underbrace{(1, \ldots, 1)}_{(n-1) \text{ times}}, \qquad \psi = \underbrace{(1, \ldots, 1)}_{n \text{ times}}$$

in Theorem 3 we have the following

COROLLARY. *Let K be real, let $x_1, \ldots, x_n$ be indeterminates over K. Then $x_1^2 + \cdots + x_n^2$ is not a sum of $n - 1$ squares in the field $K(x_1, \ldots, x_n)$. Similarly: $1 + x_1^2 + \cdots + x_n^2$ is not a sum of n squares in $K(x_1, \ldots, x_n)$.*

2. Multiplicative quadratic forms

1. We introduce some more notations for quadratic forms. Let $\varphi \cong (a_1, \ldots, a_m)$, $\psi \cong (b_1, \ldots, b_n)$, $a \in K^*$. We define the following operations on the set of equivalence classes of quadratic forms:
addition $\varphi \oplus \psi \cong (a_1, \ldots, a_m, b_1, \ldots, b_n)$,
multiplication $\varphi \otimes \psi \cong (\ldots, a_ib_j, \ldots)_{i=1,\ldots,m;\, j=1,\ldots,n}$,
scalar multiplication $a\varphi \cong (aa_1, \ldots, aa_m)$,
iterated addition $r \times \varphi \cong \underbrace{\varphi \oplus \cdots \oplus \varphi}_{r \text{ times}}$ for an integer $r \geq 1$.

Let now φ be a quadratic form of dimension n and let x, y be indeterminate vectors with components $x_1, \ldots, x_n, y_1, \ldots, y_n$. Let $K(x, y)$ denote the rational function field $K(x_1, \ldots, x_n, y_1, \ldots, y_n)$.

DEFINITION. (a) φ is called *multiplicative* (over K) if there exists a vector z with components $z_1, \ldots, z_n \in K(x, y)$ such that

$$\varphi(x)\varphi(y) = \varphi(z). \tag{1}$$

(b) φ is called *strictly multiplicative* (over K) if in addition z can be chosen to depend linearly on y, i.e. if there is a matrix T_x with entries from $K(x)$ (automatically nonsingular) such that

$$\varphi(x)\varphi(y) = \varphi(T_xy). \tag{2}$$

2. This concept is a generalization of the well-known *composition forms* for which (1) holds with a vector z depending bilinearly on x and y. It is known that composition forms exist only in dimensions 1, 2, 4, 8 whatever the field K is (char $K \neq 2$). In contrast to this result we will now prove that strictly multiplicative quadratic forms exist for all dimensions 2^k $(k = 0, 1, \ldots)$.

THEOREM 1. *The form $\varphi = (1, a_1) \otimes \cdots \otimes (1, a_k)$ is strictly multiplicative for arbitrary $k \geq 0$ and $a_1, \ldots, a_k \in K^*$.*

We need an easy

LEMMA. *Suppose that the two-dimensional form (a, b) represents an element $c \in K^*$. Then $(a, b) \cong (c, abc) \cong c(1, ab)$.*

The proof of Theorem 1 is now by induction on k. $k = 0 \Rightarrow \varphi = (1)$: trivial $k > 0$. We have to show: φ strictly multiplicative, $a \in K^* \Rightarrow \varphi \otimes (1, a)$ strictly multiplicative. By (2) we have the following equivalences over the field $K(x, y)$

$$\varphi \oplus a\varphi \cong \varphi(x)\varphi \oplus a\varphi(y)\varphi \cong (\varphi(x), a\varphi(y)) \otimes \varphi$$

(by the lemma, applied to $c = \varphi(x) + a\varphi(y) \in K(x, y)$)

$$\cong (\varphi(x) + a\varphi(y))(1, a\varphi(x)\varphi(y)) \otimes \varphi \cong (\varphi(x) + a\varphi(y))(\varphi \oplus a\varphi(x)\varphi(y)\varphi)$$

$$\cong (\varphi(x) + a\varphi(y))(\varphi \oplus a\varphi).$$

From this we see immediately that $\varphi \otimes (1, a) \cong \varphi \oplus a\varphi$ is strictly multiplicative.

REMARK. This short proof is due to Witt.

COROLLARY 1. *For $n = 2^k$ the form $\varphi(x) = \sum_1^n x_i^2$ is strictly multiplicative, i.e. $\varphi \cong \varphi(x)\varphi$ over $K(x)$. In particular φ represents the element $\varphi(x)\varphi(y)$ over $K(x, y)$.*

COROLLARY 2. *If $n = 2^k$ then $G_n = G_n(K) = \{a \in K^* \mid a$ is a sum of n squares in $K\}$ is a subgroup of K^*.*

PROOF. Let $a = \varphi(u)$, $b = \varphi(v)$ where u, v have components in K. Since φ represents $\varphi(x)\varphi(y)$ over $K(x, y)$ we deduce from the Lemma of §1 that φ represents $\varphi(u)\varphi(v)$ over K, so that $ab \in G_n$. Furthermore $a^{-1} = a^{-2}a = \varphi(a^{-1}u) \in G_n$.

3. In order to get a complete classification of multiplicative and strictly multiplicative quadratic forms, we have to combine Theorem 1 with Theorem 3 of §1 and with

WITT'S CANCELLATION THEOREM. *Let φ, φ_1, φ_2 be (nonsingular) quadratic forms. Then $\varphi \oplus \varphi_1 \cong \varphi \oplus \varphi_2 \Rightarrow \varphi_1 \cong \varphi_2$.*

Short proof: We may assume that $\dim \varphi = 1$, $\varphi = (a)$. Let $\dim \varphi_1 = \dim \varphi_2 = n$. By hypothesis there exist an element $t \in K$, vectors $u, v \in K^n$ and an $n \times n$-matrix T with entries in K such that

$$ax^2 + \varphi_2(y) = a(tx + u'y)^2 + \varphi_1(vx + Ty).$$

At least one of the two equations $\pm x = tx + u'y$ has a solution $x = w'y$ with $w \in K^n$. Then

$$\varphi_2(y) = \varphi_1((vw' + T)y).$$

The matrix $vw' + T$ must be nonsingular, hence $\varphi_2 \cong \varphi_1$.

THEOREM 2. (a) *Every anisotropic multiplicative form φ over K is of type $\varphi \cong (1, a_1) \otimes \cdots \otimes (1, a_k)$, and hence strictly multiplicative.*

(b) *An isotropic form φ is always multiplicative; φ is strictly multiplicative if and only if φ is totally isotropic, i.e.*

$$\varphi \cong i \times (1, -1), \qquad i \geq 1.$$

PROOF. (a) φ multiplicative $\Rightarrow \varphi$ represents $\varphi(x)\varphi(y)$ over $K(x)(y) \Rightarrow \varphi$ contains $\varphi(x)\varphi$ over $K(x)$ by Theorem 3 of §1 since φ is anisotropic $\Rightarrow \varphi \cong \varphi(x)\varphi$ over $K(x)$, i.e. φ is strictly multiplicative. Also φ represents 1 over K. Let $k \geq 0$ be the maximal integer such that φ contains a form $\psi \cong (1, a_1) \otimes \cdots \otimes (1, a_k)$. Assume that $\varphi \cong \psi \oplus \varphi'$, $\varphi' \cong (b, \ldots)$ and let z be an indeterminate vector of length 2^k. Then

$$\psi \oplus \varphi' \cong \varphi \cong \psi(z)\varphi \cong \psi(z)\psi \oplus \psi(z)\varphi' \cong \psi \oplus \psi(z)\varphi' \quad \text{over } K(z),$$

since φ and ψ are strictly multiplicative. From Witt's Theorem we have $\varphi' \cong \psi(z)\varphi'$; in particular φ' represents $b\psi(z)$ over $K(z)$. Applying Cassels' Theorem again we see that φ' contains $b\psi$ whence φ contains $\psi \oplus b\psi \cong \psi \otimes (1, b)$: Contradiction. So we must have $\varphi' = \emptyset$, $\varphi \cong \psi$.

(b) φ isotropic $\Rightarrow \varphi$ universal $\Rightarrow \varphi$ represents $\varphi(x)\varphi(y)$ over $K(x, y) \Rightarrow \varphi$ multiplicative. $\varphi \cong i \times (1, -1) \Rightarrow \varphi \cong \varphi(x)\varphi$ over $K(x)$ since $(1, -1) \cong f(x)(1, -1)$ for any nonzero element $f(x) \in K(x)$.

Conversely, assume $\varphi \cong i \times (1, -1) \oplus \varphi'$ with $i \geq 1$, φ' anisotropic, and φ strictly multiplicative. Then by applying Witt's Theorem to

$$i \times (1, -1) \oplus \varphi' \cong \varphi \cong \varphi(x)\varphi \cong i \times \varphi(x)(1, -1) \oplus \varphi(x)\varphi' \quad \text{over } K(x)$$

we conclude $\varphi' \cong \varphi(x)\varphi'$ over $K(x)$. If now $\dim \varphi' > 0$, $\varphi' = (b, \ldots)$ then φ' contains $b\varphi$ which is impossible since $\dim \varphi' < \dim \varphi$.

4. *Applications to nonreal fields.* Let K be a nonreal field, char $K \neq 2$. Then there is a minimal number $s = s(K)$, called Stufe, such that

$$-1 = e_1^2 + \cdots + e_s^2, \qquad e_i \in K^*.$$

THEOREM 3. *s is a power of 2.*

PROOF. Assume $n = 2^m$, $n \leq s < 2n$. Then $1 + e_1^2 + \cdots + e_s^2 = 0$, $1 + e_1^2 + \cdots + e_{n-1}^2 \neq 0$. Hence

$$(1 + e_1^2 + \cdots + e_{n-1}^2)^2 + (1 + e_1^2 + \cdots + e_{n-1}^2)(e_n^2 + \cdots + e_s^2) = 0.$$

By Corollary 2 the second summand is a sum of n squares in K. Dividing by the first summand we find that -1 is already a sum of n squares which gives the result.

THEOREM 4 (without proof). *Let R be a real field, let $x_1, \ldots, x_n$ be indeterminates and $2^m \leq n < 2^{m+1}$. Let K be the quadratic extension of $R(x_1, \ldots, x_n)$ determined by the equation $y^2 + x_1^2 + \cdots + x_n^2 = 0$. Then $s(K) = 2^m$.*

5. *Some results on the structure of the Witt group.* By Witt's Theorem any quadratic form φ over K can be written as $\varphi \cong i \times (1, -1) \oplus \varphi_0$ where $i \geq 0$ and φ_0 is anisotropic (or $\varphi_0 = \emptyset$). The index i is uniquely determined by φ, φ_0 is determined up to equivalence and is called the anisotropic kernel of φ. Two forms φ, ψ are called similar if their kernels φ_0, ψ_0 are equivalent. Since $\varphi \oplus (-\varphi) \cong \dim \varphi \times (1, -1)$ the similarity classes $\bar{\varphi}$ of quadratic forms φ form a commutative group W^+ with zero-element $\overline{(1, -1)} = \bar{\emptyset}$. Together with $\otimes$ W^+ is a ring W.

Using the theory of multiplicative forms one can deduce the following results:

(1) If an element $\bar{\varphi} \in W^+$ has finite order then its order is a power of 2.

(2) The elements of finite order in W^+ coincide with the nilpotent elements in W.

(3) Every zero-divisor in W has even dimension.

(4) If $s(K) = s < \infty$ then W is a 2-group of exponent $2s$.

(5) If K is real then for any real algebraic closure $K_\alpha \supset K$ we have a natural map $i_\alpha\colon W \to W_\alpha = W(K_\alpha) \cong Z$. Let $i = \prod_\alpha i_\alpha\colon W \to \prod_\alpha W_\alpha$. Then $\ker i$ is exactly the torsion subgroup of W^+.

(6) $\bar{\varphi} \in W$ is a unit if and only if

$$\dim \varphi \text{ odd} \quad \text{for } K \text{ nonreal},$$
$$i_\alpha(\bar{\varphi}) = \pm 1 \quad \text{for all } \alpha \quad \text{for } K \text{ real}.$$

3. Positive definite rational functions

1. Let $f(x) = g(x)/h(x) \in R(x) = R(x_1, \ldots, x_n)$ be a rational function in n variables with real coefficients. f is called *positive definite*, if $f(a) \geq 0$ for all $a = (a_1, \ldots, a_n) \in R^n$ with $h(a) \neq 0$. Hilbert's problem no. 17 is the following: Is every positive definite function f a sum of squares in $R(x)$?

This is trivial for $n = 1$ and was proved for $n = 2$ by Hilbert in 1893. He showed that in this case every positive definite function is a sum of 4 squares. The general n-variable case was settled by E. Artin in 1926. However, his proof is purely abstract and does not give an estimate for the number of squares needed. In this quantitative direction the first progress was recently made by J. Ax who showed that positive definite functions in 3 variables are sums of 8 squares. At the same time he simplified the proof for 2 variables and he gave a precise conjecture for the n-variable case, namely that every positive definite function should be a sum of 2^n squares.

One of the main ideas of Ax's proof is to make use of a well-known theorem of Tsen (1936) which has been rediscovered by Lang (1952):

THEOREM 1. *Let C be an algebraically closed field, let K be a field of transcendence degree n over C. Let f be a form of degree d in more than d^n variables with coefficients in K. Then f has a nontrivial zero in K.*

Before I state and prove the main theorem of this section, I will give a lemma on multiplicative quadratic forms.

LEMMA. *Suppose $\varphi \cong (1, a_1) \otimes \cdots \otimes (1, a_n)$ over K and write φ in the form $\varphi \cong (1) \oplus \varphi'$. Let b_1 be any nonzero element of K which is represented by φ'. Then there exist elements $b_2, \ldots, b_n \in K^*$ such that $\varphi \cong (1, b_1) \otimes \cdots \otimes (1, b_n)$.*

PROOF. By induction on n.

The case $n = 1$ is trivial. Let now $\psi \cong (1, a_1) \otimes \cdots \otimes (1, a_n) \cong (1) + \psi'$, $\varphi \cong \psi \otimes (1, a)$ and suppose that the lemma is true for n. Since $\varphi' \cong \psi' \oplus a\psi$ the element b_1 is of the form $b_1' + ab$ where b_1' is represented by ψ', b by ψ. If $b = 0$ then the result follows from the induction hypothesis

$$\psi \cong (1, b_1') \otimes (1, b_2) \otimes \cdots \otimes (1, b_n) \quad \text{for suitable } b_2, \ldots, b_n.$$

If $b \neq 0$ then $\psi \cong b\psi$ so we may assume $b = 1$ and $b_1' \neq 0$. Hence

$$\varphi \cong \psi \otimes (1, a) \cong (1, b_1') \otimes (1, a) \otimes (1, b_2) \otimes \cdots \otimes (1, b_n).$$

But

$$(1, b_1') \otimes (1, a) \cong (1, b_1'a) \oplus (b_1', a) \cong (1, b_1'a) \oplus b_1(1, b_1'a) \cong (1, b_1) \otimes (1, b_1'a)$$

since $b_1 = b_1' + a$ is represented by (b_1', a). Putting $b_1'a = b_{n+1}$ we have the result.

2. **Theorem 2.** *Let R be a real closed field, let K be an extension field of transcendence degree n over R. Let $\varphi \cong (1, a_1) \otimes \cdots \otimes (1, a_n)$ be a strictly multiplicative quadratic form of dimension 2^n over K and let $b \neq 0$ be a totally positive element of K, i.e. an element which can be represented as a finite sum of squares (over K). Then φ represents b over K.*

Proof. (a) We may suppose that φ is anisotropic since otherwise φ is universal which immediately gives the result. Also the case $b = b_1^2$ is trivial. We will first treat the case $b = b_1^2 + b_2^2$, $b_1 b_2 \neq 0$. By Tsen's theorem we know that φ is universal over the field $K(i) = K(\sqrt{-1})$. If $K = K(i)$ then the result follows. If not, then $\beta = b_1 + ib_2$ generates $K(i)$ over K and φ represents β over $K(i)$. This shows that there are vectors u, v with components in K such that

$$\varphi(u + \beta v) = \beta, \qquad v \neq 0.$$

Hence $\varphi(u) + 2\beta\langle u, v\rangle_\varphi + \beta^2\varphi(v) = \beta$. Comparing with $\beta^2 - 2b_1\beta + b = 0$ we find

$$\varphi(u) - b\varphi(v) = 0 \qquad (\text{and } 2\langle u, v\rangle_\varphi + 2b_1\varphi(v) = 1).$$

Since φ is multiplicative this gives the result.

(b) We will now suppose that Theorem 2 holds for all forms φ of the given type and for all elements $b \in K^*$ which are sums of k squares ($k \geq 2$) and will proceed by induction on k. Up to a square factor a sum of $k + 1$ squares looks like $c = 1 + b$ where b is a sum of k squares. Putting $\varphi = (1) \oplus \varphi'$ the induction hypothesis gives $b = b_1^2 + b_2$ where b_2 is represented by φ', and without restriction $b_2 \neq 0$. We want to show that φ represents c.

Consider the form $\psi = \varphi \otimes (1, -c) = (1) \oplus \varphi' \oplus (-c\varphi) = (1) \oplus \psi'$. ψ' represents $b_2 - c = (b - b_1^2) - (1 + b) = -1 - b_1^2$. By the lemma we have therefore

$$\psi \cong (1, -1 - b_1^2) \otimes \chi \text{ with a form } \chi \cong (1, c_1) \otimes \cdots \otimes (1, c_n).$$

Applying the induction hypothesis to χ we have χ represents $1 + b_1^2$. Hence $\psi \cong \varphi \oplus (-c\varphi)$ is isotropic. Therefore $\varphi(u) - c\varphi(v) = 0$ with nonzero vectors u, v over K. Since φ is anisotropic and multiplicative it follows that φ represents c over K.

This concludes the proof of Theorem 2.

Using Artin's Theorem and putting $a_1 = \cdots = a_n = 1$ we get the

COROLLARY. *Every positive definite function in $R(x_1, \ldots, x_n)$ is a sum of 2^n squares.*

3. There are several problems arising in connection with Theorem 2 and the Corollary.

PROBLEM A. Is there a similar result if R is replaced by Q, i.e. is any positive definite function in $Q(x_1, \ldots, x_n)$ a sum of $c(n)$ squares where $c(n)$ depends only on n? It is known that $c(0) = 4$ (Theorem of Lagrange) and that $c(1) \leq 8$ (Theorem of Landau), but it is not yet known whether $c(2)$ is finite. In order to prove the finiteness of $c(n)$ along the lines given here, one would have to have a weaker form of Tsen's Theorem for quadratic forms over $Q(i)(x_1, \ldots, x_n)$. For instance one can show the following: $c(n) \leq 2^{n+2}$ under the assumption that r quadratic forms in more than $4r$ variables over $Q(i)$ always have a nontrivial common zero ($r = 1, 2, \ldots$). Unfortunately it is rather unlikely that this assumption is true.

PROBLEM B. Is the bound 2^n in the Corollary best possible?

This is obviously true for $n = 0$ and $n = 1$ (since $1 + x^2$ is not a square in $R(x)$), but still open for $n \geq 2$. The best known lower bound, given by the Corollary at the end of §1, is $n + 1$.

We will investigate this problem somewhat further on the special example of §1. So let

$$f(x, y) = 1 + x^2(x^2 - 3)y^2 + x^2y^4 \in R[x, y].^2$$

If f is a sum of 3 squares in $R(x, y)$ so it is already a sum of 3 squares in $R(x)[y]$ (by Theorem 1 of §1), i.e. we have

$$f(x, y) = \sum_{i=1}^{3} (a_i + b_iy + c_iy^2)^2$$

where $a_i, b_i, c_i \in R(x)$. This gives the following system of quadratic equations

$$\sum_1^3 a_i^2 = 1, \qquad \sum a_ib_i = 0, \qquad \sum (b_i^2 + 2a_ic_i) = x^2(x^2 - 3),$$
$$\sum b_ic_i = 0, \qquad \sum c_i^2 = x^2.$$

Given any solution of this system one can apply an orthogonal transformation in 3-dimensional space over $R(x)$ to it and one gets another solution. Therefore we may assume that $a_1 = 1, a_2 = a_3 = 0$. The other equations then reduce to

$$b_1 = 0, \qquad b_2^2 + b_3^2 = x^2(x^2 - 3) - 2c_1,$$
$$b_2c_2 + b_3c_3 = 0, \qquad c_2^2 + c_3^2 = x^2 - c_1^2.$$

c_2, c_3 cannot both be zero since then $c_1 = \pm x$ and $x^4 - 3x^2 \mp 2x$ positive definite, which is not the case (put $x = \pm 1$). So we may assume that $c_2 \neq 0$ and get

$$b_2^2 + b_3^2 = (b_3c_3/c_2) + b_3^2 = (b_3/c_2)^2(c_2^2 + c_3^2).$$

---

[2]It has since been shown that $f(x, y)$ is not a sum of 3 squares in $R(x, y)$. See a forthcoming paper by Cassels, Ellison and the author.

Hence

$$(b_3/c_2)^2(x^2 - c_1^2) = x^2(x^2 - 3) - 2c_1.$$

Finally, putting $c_1 = x\xi$ and $(b_3x/c_2)(1 - \xi^2) = \eta$, we have

$$\eta^2 = (1 - \xi^2)(x^4 - 3x^2 - 2x\xi)$$

with $\xi, \eta \in R(x)$. This is the equation of an elliptic curve Γ over the field $R(x)$. There are clearly 3 "rational" points on this curve, namely the points with $\eta = 0$. But these do not lead to a representation of f as a sum of 3 squares since $\eta = 0$ implies $c_1 = \pm x$ or $2c_1 = x^2(x^2 - 3)$. The first two cases have already been excluded, the last case would imply

$$4(c_2^2 + c_3^2) = 4x^2 - 4c_1^2 = 4x^2 - x^4(x^2 - 3)^2 \geq 0$$

for all x. If one could show that Γ has no other "rational" points then f could not be a sum of 3 squares.

REFERENCES

0. A. Pfister, *Zur Darstellung definiter Funktionen als Summe von Quadraten*, Invent. Math. **4** (1967) 229–237, and the literature mentioned there. MR **36** #5095.

1. T. S. Motzkin, *The arithmetic-geometric inequality*, Proc. Sympos. on Inequalities (Wright-Patterson Air Force Base, Ohio, 1965) Academic Press, New York, 1967, pp. 205–224. MR **36** #6569.

2. A. Pfister, *Quadratische Formen in beliebigen Körpern*, Invent. Math. **1** (1966), 116–132. MR **34** #169.

3. R. M. Robinson, *Some definite polynomials which are not sums of squares of real polynomials* (to appear).

4. C. Tsen, *Quasi-algebraisch-abgeschlossene Funktionenkörper*, J. Chinese Math. Soc. **1** (1936), 81–92.

GEORG-AUGUST-UNIVERSITAT ZU GOTTINGEN
GOTTINGEN, FEDERAL REPUBLIC OF GERMANY

A METAMATHEMATICAL APPROACH TO SOME PROBLEMS IN NUMBER THEORY

JAMES AX

There are at least two underlying reasons for what some may regard as the surprising connections between logic and number theory.

First of all, there is a common set of objects studied in each discipline, namely polynomials $f(X_1, \ldots, X_n)$ in several variables with coefficients in the rational integers $\mathbf{Z}$. An important problem in number theory—Hilbert's 10th—is that of deciding when $f = 0$ has a solution in rational integers, i.e. giving a decision procedure for statements of the form

$$(*) \qquad \exists X_1 \ldots \exists X_n, f(X_1, \ldots, X_n) = 0$$

interpreted in $\mathbf{Z}$.

As Julia Robinson has explained in her lectures, the deepest progress on this question, which has been in a negative direction, was obtained through the metamathematical considerations of first Gödel and then Davis, Putnam and Robinson. These considerations involved generalizing the problems to those of the form

$$(**) \qquad Q_1 X_1 \ldots Q_n X_n, f(X_1, \ldots, X_n) = 0$$

where each Q_i is either $\exists$ or $\forall$.

Such statements comprise the most general elementary statements (about integral domains) and for them Gödel proved the nonexistence of a decision procedure. Davis later showed essentially the same result holds when Q_i is $\exists$ for $i \neq 2$. Further approximations to Hilbert's 10th include the nonexistence of a decision procedure for (*) when X_1 is restricted to the powers of 2 or to the primes.[1]

A second connection between logic and number theory has already been alluded to: elementary statements such as (**) have an interpretation in any commutative

[1]This problem has recently been resolved along these lines by Yu. V. Mattiasevich.

ring. Moreover it is sometimes possible to deduce from the truth of (**) for certain (rings) A, the truth of (**) for certain other (rings) B. Such *transfer principles* have an obvious utility and this suggests the approach mentioned in the title of these lectures.

An important tool for reducing these metamathematical principles to algebraic isomorphisms is that of ultraproducts. If F^i, $i \in I$, is a collection of fields indexed by a set I, then the *ultraproducts* of the F^i are merely the residue class fields of the commutative ring $F = \prod_{i \in I} F^i$. In order to get a better understanding of these ultraproducts we define for each $f = (f^i) \in F$, $z(f) = \{i \mid f^i = 0\}$, and for each ideal A of F we set $z(A) = \{z(f) \mid f \in A\}$.

DEFINITION. A *filter* on I is a nonempty set of subsets of I such that

(i) $\phi \notin D$;

(ii) $s, t \in D \rightarrow s \cap t \in D$;

(iii) $s \in D, s \subseteq t \subseteq I \rightarrow t \in D$.

EXAMPLE. Let D be the filter of cofinite subsets of I, i.e. assuming I infinite D consists of all subsets s of I such that $I - s$ is finite.

It is straightforward to verify that the correspondence $A \rightarrow z(A)$ identifies the ideals of F with the filters on I is inclusion preserving and so makes correspond maximal ideals of A with maximal filters which are called *ultrafilters*. As an example, for each $i_0 \in I$, $\{s \mid i_0 \in s \subseteq I\}$ is an ultrafilter on I, called a *principal ultrafilter* since it corresponds to a principal maximal ideal. An ultrafilter D is nonprincipal iff D contains the filter of cofinite subsets of I. If the F^i are merely sets or groups we can still define $\prod F^i/D$ to be $\prod F^i$ modulo the identification of $(f^i)_i$ and $(g^i)_i$ when $\{i \mid f^i = g^i\} \in D$.

From now on we assume I countably infinite. Then we have the following

CARDINALITY PROPERTY. If for every positive integer n, $\{i \in I \mid \#F_i \leq n\}$ is finite then for every nonprinciple ultrafilter D on I, $\#(F/D) \geq 2^{\aleph_0}$. Here F/D is really F modulo the maximal ideal corresponding to D.

As an application of the ultraproduct construction, let us give a simple proof of Hilbert's Nullstellensatz taken in the following form.

THEOREM 1. *Let k be an algebraically closed field and $f_1, \ldots, f_m \in k[X_1, \ldots X_n]$. If*

$$(\dagger) \qquad f_1 = \cdots = f_m = 0$$

has a solution over some extension l of k, then ($\dagger$) *has a solution over k.*

PROOF. We can assume k, l countable and l algebraically closed. Let D be a nonprincipal ultrafilter on I, which exists by Zorn's lemma. Set

$$L = l^I/D \supseteq l \supseteq k \quad \text{and} \quad K = k^I/D \supseteq k.$$

Since l and k are algebraically closed, so are L and K. By the cardinality property of ultraproducts $\#L = \#K = 2^{\aleph_0}$. It follows from Steinitz's isomorphism theorem that L and K are isomorphic over k. Now ($\dagger$), having a solution in l, has one in L and hence in K. Then for some $i \in I$, the coordinates of representatives of this solution comprise a solution of ($\dagger$) in k, thereby completing the proof.

In the course of the proof we used that certain properties of k and l held for K

and L. Although these are easy to verify directly, the underlying property of ultraproducts we are using is the

PERSISTENCE PROPERTY. An elementary statement E holds for $\prod_{i\in I} F_i/D \leftrightarrow \{i \in I \mid E \text{ holds } F_i\} \in D$.

As a second illustration of this method, this time we prove a result which had only been known in special cases.

THEOREM 2. *If V is an algebraic variety and $V \xrightarrow{\varphi} V$ an injective morphism, then φ is surjective.*

PROOF. For simplicity, we consider only the most interesting case: when V is a subvariety of affine n-space over the complex numbers $\mathbf{C}$. Then V is given by $f_1 = \cdots = f_m = 0$ where $f_1, \ldots, f_m \in \mathbf{C}[X_1, \ldots, X_n]$ and φ is given by $g_1, \ldots, g_n \in \mathbf{C}[X_1, \ldots, X_n]$.

Let N be a positive integer such that

$$m, n, \deg f_\mu, \deg g_\nu \leq N.$$

Then there exists an elementary statement E_N which holds in a field $C \leftrightarrow$ for every choice of $m, n \leq N$ and of polynomials of degree $\leq N\, f_1, \ldots, f_m, g_1, \ldots, g_n \in C[X_1, \ldots, X_n]$ such that $g_1, \ldots, g_n$ induces an injective mapping φ from $V = \{x \in C^n \mid f_1(x) = \cdots = f_m(x) = 0\}$ to V, φ is surjective. We want to prove E_N holds for $\mathbf{C}$.

Now E_N holds for C when C is finite. From this it follows that E_N holds for $\tilde{\mathbf{F}}_p$, the algebraic closure of the field with p-elements $\mathbf{F}_p$. By the persistence property, E_N holds for $C = \prod_{p\in P} \tilde{\mathbf{F}}_p/D$ where D is any nonprincipal ultraproduct on the set P of primes. Now, as before, C is algebraically closed and $\#C = 2^{\aleph_0}$. Moreover since the elementary statement $\exists X p_0 X - 1 = 0$ hold for $\tilde{\mathbf{F}}_p$ for $p \neq p_0$, it holds for C, i.e. char $C = 0$. By the above mentioned isomorphism theorem $C \approx \mathbf{C}$ and this proves the result.

Our third illustration is of more interest to number theorists. In joint work with Kochen we established a partial result on Artin's Conjecture:

Every form of degree d in $n > d^2$ variables has a nontrivial zero over the p-adic numbers $\mathbf{Q}_p$, i.e. $\mathbf{Q}_p$ is $C_2(d)$.

This statement is now known to be false in general. However we established the following result.

THEOREM 3. *There exists $p_0(d)$ such that for all $p \geq p_0(d)$ every form of degree d in $n > d^2$ variables has a nontrivial zero over $\mathbf{Q}_p$.*

The idea of the proof is to use the fact, due independently to Carlitz and Lang, that the analogous statement is true in $S_p = \mathbf{F}_p((t))$, the field of formal power series over $\mathbf{F}_p$. Set

$$n = d^2 + 1, \qquad m = \binom{d+n-1}{n-1},$$

and let A_d be the elementary statement

$$\forall X_1 \ldots \forall X_m \exists Y_1 \ldots \exists Y_n X_1 Y_1^d + X_2 Y_1^{d-1} Y_2 + \cdots + X_m Y_n^d = 0$$

and

$$(Y_1 \neq 0 \text{ or } Y_2 \neq 0 \text{ or } \dots \text{ or } Y_n \neq 0).$$

Then we want to prove that for $p \geq p_0(d)$ A_d is true in $\mathbf{Q}_p$. We know A_d is true in S_p for all p. Thus it suffices to prove the following transfer principle.

For every elementary statement E, $\{p \mid E \text{ is true in } \mathbf{Q}_p\}$ and $\{p \mid E \text{ is true in } S_p\}$ differ by a finite set.

Let $s \subseteq P$ be an infinite set for which, say, E is true in S_p and false in $\mathbf{Q}_p$ for $p \in s$. By Zorn's lemma, there exists a nonprincipal ultrafilter D on P for which $s \in D$. By the persistence property s is true in $S = \prod_{p \in P} S_p/D$ and false in $Q = \prod_{p \in P} \mathbf{Q}_p/D$.

Thus the principle, and hence our result on Artin's Conjecture is reduced to proving the following result.

THEOREM 4. $Q \approx S$.

Now $\mathbf{Q}_p$ and S_p have natural valuations which induce valuations on Q and S; the isomorphism of Theorem 4 is a value isomorphism. Although this is not needed for Theorem 3, it is essential to carry along the valuations in the proof of Theorem 4 which is of an inductive nature.

To prove Theorem 4, we first abstract certain common valuation-theoretic properties in the proposition below. We then state an isomorphism theorem to the effect that any two valued fields sharing these properties are isomorphic. The definitions of the valuation-theoretic terms occurring in the statements of the proposition and the isomorphism theorem can be found in the appendix to this lecture. This appendix actually constitutes a short self-contained course in valuation theory leading up to the proof of the isomorphism theorem.

PROPOSITION. *Let T denote either Q or S. Then*

(a) T *is* $\prod_{p \in P} \mathbf{Z}/D$*-valued;*

(b) $\bar{T} \approx \prod_{p \in P} \mathbf{F}_p/D$;

(c) $\operatorname{char} \bar{T} = 0$;

(d) $\# T = 2^{\aleph_0}$;

(e) *T has a cross-section;*

(f) *T is Henselian;*

(g) *T is $\aleph_1$-pseudo-complete.*

For the definition of the terms involved in: (a) cf. Definition 2 of §2; (e) cf. Definition 2 of §7; (f) cf. Definition 1 of §4; (g) cf. Definition 1 of §7 and Definitions 4, 5 of §6.

PROOF. For brevity we consider only the case $T = Q$.

(a) If $x^* \in T - 0$ there exists $x^p \in Q_p - 0$ such that $x = (x^p)_p \in \prod_{p \in P} \mathbf{Q}_p$ represents x^* modulo (the maximal ideal corresponding to) D. Then $(\operatorname{ord} x^p)_p \in \prod_{p \in P} Z$ represents an element $\operatorname{ord}^* x^*$ in $\prod_{p \in P} Z/D = G$. It is easy to see that $\operatorname{ord}^* x^*$ is independent of the choice of x, that the group G has the structure of an ordered abelian group with $x^* < y^*$ if $\{p \mid x^p < y^p\} \in D$, and ord* is a G-valuation of T.

(b) With the notation of (a) let ord* $x^* \geq 0$. Then we can assume ord $x^p \geq 0$ for all p so that $(\bar{x}^p)_p \in \prod_{p \in P} \mathbf{F}_p$. The element $\bar{x}^*$ of $\prod_{p \in P} \mathbf{F}_p/D$ represented by $(\bar{x}^p)_p$ depends only on x^* and the map $x^* \longrightarrow \bar{x}^*$ induces the desired isomorphism.

(c) follows from (b) as in the proof of Theorem 2.

(d) follows from the cardinality property.

(e) Define φ: $G \rightarrow T - 0$ as follows. Let $g^* \in G$. Then there exists $g^p \in \mathbf{Z}$ such that g^* is represented by $(g^p)_p \in \prod_{p \in P} \mathbf{Z}$. Set

$$\varphi(g^*) = (p^{g^p})_p^*, \text{ the element of } T \text{ represented by } (p^{g^p})_p \in \prod_{p \in P} \mathbf{Q}_p.$$

(f) follows from the persistence property (and Corollary 2 to Proposition 11 of §4). It is also easily seen directly by the usual method of picking representatives.

(g) Let $(x_i^*)_{i<w}$ be a pseudo-Cauchy sequence of elements of T. The idea is that if $x_i^p \in \mathbf{Q}_p$ is such that

(α) $(x_i^p)_p \in \prod_p \mathbf{Q}_p$ represents x_i^*,

(β) ord $(x_j^p - x_i^p) =$ ord $(x_{i+1}^p - x_i^p)$ for $i < j < w$

then $(x_p^p)_p \in \prod_p Q_p$ represents a pseudo-limit $x^* \in T$ of $(x_i^*)_{i<w}$. Indeed we want to show ord* $(x^* - x_i^*) =$ ord* $(x_{i+1}^* - x_i^*)$, i.e. $s = \{p \mid$ ord $(x_p^p - x_i^p) =$ ord $(x_{i+1}^p - x_i^p)\} \in D$. But ord $(x_p^p - x_i^p) =$ ord $(x_{i+1}^p - x_i^p)$ if $i < p$ so that s is cofinite and hence a member of D, since D is nonprinciple. Thus it only remains to find x_i^p satisfying α and β. This is accomplished by induction on i. For $i = 0, 1$ we let $(x_i^p)_i$ be an arbitrary representative of x_i^*. Now assume $1 \leq j < w$, and that we have chosen $x_i^p \in \mathbf{Q}_p$ so that $(x_i^p)_p$ represents x_i^* for $i \leq j$ and ord $(x_{i+1}^p - x_i^p) =$ ord $(x_j^p - x_i^p)$ if $i < j$. Let $(y^p)_p$ be an arbitrary representative of x_{j+1}^*. Let $s = \{p \mid$ ord $(y^p -$ ord $x_j^p) >$ ord $(x_j^p - x_{j-1}^p)\}$. Since $(x_i^*)_i$ is a pseudo-Cauchy sequence, ord $(x_{j+1}^* - x_j^*) >$ ord $(x_j^* - x_{j+1}^*)$ which means $s \in D$. Define $x_{j+1}^p = y^p$ if $p \in S$; $x_{j+1}^p = x_j^p$ otherwise. This definition of the x_i^p satisfies α and β, thereby completing the proof.

ISOMORPHISM THEOREM. *Let V and V' be Henselian, $\aleph_1$-pseudo-complete valued fields of cardinality $\aleph_1$ with cross-sections which have isomorphic value groups and isomorphic residue class fields of characteristic zero. Then V is value isomorphic to V.*

Thus we have reduced a certain diophantine statement—Theorem 3—to proving an isomorphism theorem, which is carried out in the appendix (cf. Theorem B of §7). We note that the above Isomorphism Theorem together with the proposition yields Theorem 4 only if we assume the continuum hypothesis $2^{\aleph_0} = \aleph_1$. However for the application of Theorem 3, this hypothesis is easily removed.

A fourth illustration of the procedure of reducing diophantine type statements to algebraic isomorphisms of ultraproducts occurs in the theory of finite fields. If F is a field let Abs (F) denote the subfield of elements of F algebraic over its prime field. Then the following isomorphism theorem holds.

THEOREM 5. *Let F and F' be nonprincipal ultraproducts of finite fields. Then $F \approx F'$ if and only if* Abs $(F) \approx$ Abs (F').

This is proved in "The elementary theory of finite fields," Annals of Mathematics, vol. 88, 1968, pp. 239–271.

THEOREM A. *Let d be a positive integer, and let p be a prime. For each m, let Q_{p^m} be the unramified extension of the p-adic numbers of degree m. Then the set of m such that Q_{p^m} is $C_2(d)$ differs by a finite set from some $a(n,W)$.*

It is hoped that these examples persuade the reader of the value of a metamathematical approach in certain situations.

APPENDIX

The Isomorphism Theorem for Valuation Theory

CONTENTS

This appendix provides an account of the isomorphism theorem which (a) avoids the unnecessary complications of the original paper[2] and (b) proceeds in a relatively leisurely and self-contained manner to the main result. The reasons for (a) are evident; (b) seemed worthwhile in view of the necessity of dealing with nonclassical valuations (nonArchimedean value groups). The expert in valuation theory can begin with §6.

1. Valuation subrings of a field

DEFINITION 1. A local domain is an integral domain R with a unique maximal ideal M. R/M is called the residue class field of R, denoted by $\overline{R}$.

REMARK. R is local $\Leftrightarrow$ the set of nonunits of R forms an ideal.

DEFINITION 2. If F is an integral domain and $\mathbf{p}$ a prime ideal of R, then $R_{\mathbf{p}}$ is the subring of the quotient field of R consisting of all a/b with $a \in R$ and $b \in R - \mathbf{p}$.

LEMMA 1. *$R_{\mathbf{p}}$ is a local domain with maximal ideal $\mathbf{p}R_{\mathbf{p}}$. The inclusion $R \to R_{\mathbf{p}}$ induces an inclusion $R/\mathbf{p} \to R_{\mathbf{p}}/\mathbf{p}R_{\mathbf{p}}$. This inclusion is an isomorphism $\Leftrightarrow$ $\mathbf{p}$ is maximal.*

PROOF. To prove the first part we observe that $a/b \in R_{\mathbf{p}}$ is a unit $\Leftrightarrow a \notin \mathbf{p}$. Hence the set of nonunits of $R_{\mathbf{p}}$ is $\mathbf{p}R_{\mathbf{p}}$ which is easily seen to be an ideal. It is

[2] J. Ax and S. Kocher, *Diophantine problems over local fields I*, Amer. J. Math., 87 (1965), 605–630.

clear that we have an inclusion $R/\mathbf{p} \to R_{\mathbf{p}}/\mathbf{p}R_{\mathbf{p}}$ such that $R_{\mathbf{p}}/\mathbf{p}R_{\mathbf{p}}$ is the quotient field of $R/\mathbf{p}$. Hence the inclusion is a surjection $\Leftrightarrow$ $R/\mathbf{p}$ is a field $\Leftrightarrow$ $\mathbf{p}$ is a maximal ideal.

DEFINITION 3. If R is a subring of a field F and $x \in F$, then x is said to be integral over $R \Leftrightarrow \exists$ monic $f \in R[X]$ such that $f(x) = 0$.

LEMMA 2. *Let R be a local subdomain of a field F, with maximal ideal M. If $x \in F$ is integral over R, then $1 \notin MR[x]$.*

PROOF. Suppose false i.e. $1 = \sum_{i=0}^{k} m_i x^i$ where $m_i \in M$. Since $1 - m_0$ is a unit in R, we may assume $m_0 = 0$. Choose $f(X) = \sum_{j=0}^{n} r_j X^j$ a monic polynomial in $R[X]$ of minimal degree such that $f(x) = 0$. Since the Lemma is trivial if $x \in R$, we may assume $n > 1$. Observe also that we may take $k < n$. Then we have

$$x^n = r_{n-1}x^{n-1} + \cdots + r_0 \sum_{i=1}^{k} m_i x^i$$

and after dividing by x we obtain an integral equation for x of smaller degree, a contradiction.

DEFINITION 4. A subring T of a field F is said to be a valuation subring if for all $x \in F$, $x \notin T \Rightarrow x^{-1} \in T$.

EXAMPLE 1. If F is a field, then F is a valuation subring of F, called the trivial valuation subring of F.

EXAMPLE 2. If $F = Q$, the rationals, then is a valuation subring of $Q \Leftrightarrow \mathbf{Q} = Z_p$ for some prime ideal p of Z, the integers.

EXAMPLE 3. If k is a field and $k(t)$ a simple transcendental extension of k, then $\mathbf{Q}$ is a valuation subring of $k(t)$ containing $k \Leftrightarrow \mathbf{Q} = K[t]$ for some prime ideal p of $k[t]$ or $\mathbf{Q} = k[1/t]_{(1/t)}$.

LEMMA 3. *A subring of a field F is a valuation subring of $F \Leftrightarrow$ the ideals of T are linearly ordered $\Rightarrow T$ is a local domain.*

PROOF. Now suppose I_1 and I_2 are two ideals of a valuation subring T, and suppose there exists $a \in I_1 - I_2$. Let $b \in I_2$. Then either $a/b \in T$ or $b/a \in T$. But $a/b \in T$ implies $a = (a/b) \cdot b \in I_2$, which is false. Hence we have $b/a \in T$, from which it follows that $b = (b/a) \cdot a \in I_1$. Therefore $I_2 \subseteq I_1$. Conversely, suppose the ideals of T are linearly ordered and let a/b be in the quotient field of T, where $a, b \in T$. Then either $aT \subseteq bT$ or $bT \subseteq aT$. In the first case we have $a/b \in T$ and in the second $b/a \in T$. Finally, if the ideals of T are linearly ordered, T can have only one maximal ideal so T is local.

PROPOSITION 1. *Let M be a maximal ideal of a subring R of a field F. Then $\exists$ a valuation subring $\mathcal{O}$ of F such that $R \subset \mathcal{O}$, $M \subset \mathfrak{M}$ = maximal ideal of $\mathcal{O}$ and (with respect to the natural inclusion, $R/M \to \mathcal{O}/\mathfrak{M}$) $\mathcal{O}/\mathfrak{M}$ is algebraic over R/M.*

PROOF. Apply Zorn's Lemma to the set $\mathfrak{A} = \{(S, N)$ S is a subring of F, N is a maximal ideal of S, $R \subseteq S$, $M \subseteq N$, S/N is algebraic over $R/M\}$ ordered by

$(S_1, N_1) < (S_2, N_2) \Leftrightarrow S_1 \subseteq S_2$ and $N_1 \subseteq N_2$, to obtain a pair $(\mathcal{O}, \mathfrak{M})$ which is maximal. By Lemma 1, $\mathcal{O}$ is local. We want to show $\mathcal{O}$ is a valuation subring of F. So suppose $x \in F - \mathcal{O}$. By the maximality of $(\mathcal{O}, \mathfrak{M})$ we must have $\mathfrak{M}\mathcal{O}[x] + x\mathcal{O}[x] = \mathcal{O}[x]$ (for otherwise $\mathcal{O}[x]/(\mathfrak{M}\mathcal{O}[x] + x\mathcal{O}[x] \approx \mathcal{O}/\mathfrak{M})$ i.e.

$$1 = \sum_{i=0}^{s} m_i x^i + x \sum_{j=0}^{t} y_j x^j, \qquad m_i \in \mathfrak{M},\ y_j \in \mathcal{O}.$$

We may assume $m_0 = 0$, so we have

$$1 = \sum_{i=1}^{r} z_i x^i, \qquad z_i \in \mathcal{O},$$

or

$$x^{-r} = \sum_{i=1}^{r} z_i (x^{-1})^{r-i}.$$

Therefore x^{-1} is integral over $\mathcal{O}$, so by Lemma 2, $1 \notin \mathfrak{M}\mathcal{O}[x^{-1}]$. Choosing $\mathfrak{N}$ to be a maximal ideal in $\mathcal{O}[x^{-1}]$ containing $\mathfrak{M}\mathcal{O}[x^{-1}]$, we have that $(\mathcal{O}[x^{-1}], \mathfrak{N})$ is in $\mathcal{A}$. Hence $\mathcal{O}[x^{-1}] = \mathcal{O}$ and $x^{-1} \in \mathcal{O}$ and we are done.

DEFINITION 5. If R is a ring, R^* = units of R.

COROLLARY 1 (TO PROPOSITION 1). *If R is a subring of a field F then $T = \{x \in F \mid x \text{ integral}/R\} = \bigcap_{\mathcal{O} \in S} \mathcal{O}$, where S is the set of all valuation subrings of F such that $R \subset \mathcal{O}$ and $M = R \cap \mathfrak{M}(\mathcal{O})$ is a maximal ideal of R and $\mathcal{O}/\mathfrak{M}(\mathcal{O})$ is algebraic over R/M.*

PROOF. Let $x \in T$ and $\mathcal{O} \in S$. Then x is integral over $\mathcal{O}$. If $x \notin \mathcal{O}$, then $x^{-1} \in \mathfrak{M}(\mathcal{O})$. But then $x^n = a_{n-1}x^{n-1} + \cdots + a_0$, $a_i \in \mathcal{O}$, implies $1 = a_{n-1}x^{-1} + \cdots + a_0 x^{-n} \in \mathfrak{M}(\mathcal{O})$, a contradiction. Hence

$$T \subseteq \bigcap_{\mathcal{O} \in S} \mathcal{O}.$$

Let $x \in F - T$. Then $1/x \notin (R[1/x])^*$ or else we would have $(1/x)\sum_{i=0}^{n} r_i (1/x)^i = 1$, $r_i \in R$, which implies $x^{n+1} = \sum_{i=0}^{n} r_i x^{n-i}$, but x is not integral over R. Therefore $\exists$ a maximal ideal N of $R[1/x]$ such that $1/x \in N$.

Apply Proposition 1 to $R[1/x]$ and N to get a valuation ring such that $R \subset R[1/x] \subset R$, $R \cap N \subset N \subset \mathfrak{M}(\mathcal{O})$, and $\mathcal{O}/\mathfrak{M}(\mathcal{O})$ is algebraic over $R[1/x]/N = R/R \cap N$. Then $1/x \in N \Rightarrow 1/x \in \mathfrak{M}(\mathcal{O}) \Rightarrow x \notin \mathcal{O}$, so $x \notin \bigcap_{\mathcal{O} \in S} \mathcal{O}$.

COROLLARY 2-DEFINITION 6. *$\{x \in F \mid x \text{ integral}/R\}$ is a subring of F called the integral closure of R in F.*

COROLLARY 3. (CHEVALLEY'S PLACE EXTENSION THEOREM). *Every homomorphism of a subring R of a field F into an algebraically closed field C can be extended to a valuation subring of F.*

PROOF. Let $\varphi: R \to C$ be such a homomorphism. Since we can always extend φ to $R_{\ker \varphi}$, we may assume that R is a local ring and that $\ker \varphi = M$, the maximal ideal of R. Thus we have

$$\varphi = R \xrightarrow{\pi} R/M \xrightarrow{i} C$$

where π is the canonical projection and i is an embedding of R/M in C.

By Proposition 1, we can choose a valuation subring $\mathcal{O}$ of F such that $R \subseteq \mathcal{O}$, $M \subseteq \mathfrak{M}(\mathcal{O})$, and $\mathcal{O}/\mathfrak{M}(\mathcal{O})$ is algebraic over R/M. Hence we have a commutative diagram.

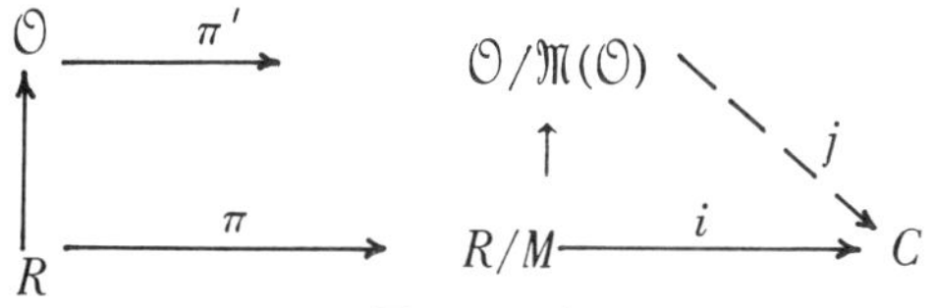

FIGURE 1

The map j exists since $\mathcal{O}/\mathfrak{M}(\mathcal{O})$ is algebraic over R/M and C is algebraically closed. Then $j_0\pi'$ is the extension of φ we want.

2. Valuations

DEFINITION 1. An ordered abelian group is an abelian group with a linear ordering such that $x, y > 0 \Rightarrow x + y > 0$.

EXAMPLES.

1. $\{0\}$, **Z**, **Q**, and **R** = reals with their usual ordering.
2. A subgroup of an ordered group inherits an ordered group structure.
3. Let I be well ordered and A_i for $i \in I$ ordered abelian groups. Then $\prod A_i$ is ordered by:

$$a < b \Leftrightarrow a \neq b \text{ and let } a_{l_0} < b_{l_0} \text{ where } l_0 = \text{smallest } i \text{ such that } a_i \neq b_i.$$

4. An abelian group A can be ordered $\Leftrightarrow A$ is torsion free. Note that if A is torsion free, $A \subset A \otimes \mathbf{Q} \subset \mathbf{Q}^I$ for some I.

Let F be a field and A an ordered abelian group.

DEFINITION 2. An *A-valuation of F* is a surjection $\text{ord}: F \to A \cup \{\infty\}$ such that,

(i) $\text{ord} f = \infty \Leftrightarrow f = 0$,

(ii) $\text{ord} f_1 f_2 = \text{ord} f_1 + \text{ord} f_2$,

(iii) $\text{ord} (f_1 + f_2) \geq \min (\text{ord} f_1, \text{ord} f_2)$

where it is understood that $\forall a \in A \cup \{\infty\}$, $\infty + a = \infty$ and $\infty \geq a$.

REMARK. If $\text{ord} f_1 \neq \text{ord} f_2$, $\text{ord} (f_1 + f_2) = \min (\text{ord} f_1, \text{ord} f_2)$.

PROOF. Suppose $\text{ord} (f_1 + f_2) >_{\neq} \min (\text{ord} f_1, \text{ord} f_2)$, and say $\text{ord} f_1 >_{\neq} \text{ord} f_2$. Then

$$\text{ord} (f_2) = \text{ord} ((f_1 + f_2) - f_1) \geq \min \{\text{ord} (f_1 + f_2), \text{ord} (f_1)\} >_{\neq} \text{ord} (f_2),$$

a contradiction.

PROPOSITION 2. *If* ord *is an A-valuation of F then* $\mathcal{O} = \{x \in F \mid \text{ord}\, x \geq 0\}$ *is a valuation subring of F. Conversely, if* $\mathcal{O}$ *is a valuation subring of F then* $A = F^*/\mathcal{O}^*$ *is ordered by*

$$f_1\mathcal{O}^* > f_2\mathcal{O}^* \Leftrightarrow f_1f_2^{-1} \in \mathcal{O} - \mathcal{O}^*$$

and the canonical epimorphism $F^* \to A$ *combined with* $0 \to \infty$ *defines an* A*-valuation*, ord, *of* F *such that* $\mathcal{O} = \{x \in F \mid \text{ord}\, x \geq 0\}$.

PROOF. It is immediate from the definition of ord that $\mathcal{O}$ is a subring of F. Let $x \in F - \mathcal{O}$, i.e. ord $x < 0$. Then ord $x^{-1} = -\text{ord}\, x > 0$, so $x^{-1} \in \mathcal{O}$ and $\mathcal{O}$ is a valuation subring of F. The maximal ideal of $\mathcal{O} = \{x \in F \mid \text{ord}\, x > 0\}$.

Conversely, if $\mathcal{O}$ is a valuation subring of F, define A and $>$ as in the statement of the Proposition. Write the operation in A additively, i.e. $f_1\mathcal{O}^* + f_2\mathcal{O}^* = f_1f_2\mathcal{O}^*$. Since $f \in \mathcal{O}$ or $f^{-1} \in \mathcal{O}$ we have $f\mathcal{O}^* \geq 0$ or $-f\mathcal{O}^* \geq 0$, so $>$ does define a linear ordering on A.

Then (i) and (ii) of the definition of a valuation are trivially satisfied and (iii) is satisfied since

$$\text{ord} f_1 \geq \text{ord} f_2 \Rightarrow f_1f_2^{-1} \in \mathcal{O} \Rightarrow (f_1 + f_2)f_2^{-1} = f_1f_2^{-1} + 1 \in \mathcal{O}$$
$$\Rightarrow \text{ord}\,(f_1 + f_2) \geq \text{ord}\,(f_2).$$

DEFINITION 3. If ord_i is an A_i-valuation of F_i for $i = 1, 2$ then a value-monomorphism (resp.: value-isomorphism) from F_1 to F_2 is a field monomorphism (resp.: isomorphism) $\varphi: F_1 \to F_2$ such that $\text{ord}_1 l > \text{ord}_1 f \leftrightarrow \text{ord}_2 \varphi(l) > \text{ord}_2 \varphi(f)$. If $F_1 \subset F_2$ then we say F_2 is a (valued field) extension of $F_1 \leftrightarrow$ the inclusion $F_1 \to F_2$ is a value-monomorphism.

DEFINITION 4. Let $F_1 \subset F_2$ be fields with valuations ord_1 and ord_2 respectively and valuation rings $\mathcal{O}_1 = \mathcal{O}\,(\text{ord}_1)$ and $\mathcal{O}_2 = \mathcal{O}\,(\text{ord}_2)$.

Then ord_2 extends $\text{ord}_1 \leftrightarrow$ there is an order monomorphism $\varphi: \text{ord}_1(F_1^*) \to \text{ord}_2(F_2^*)$ such that $\text{ord}_2 \mid F_1 = \varphi_0\, \text{ord}_1$. We say $\mathcal{O}_2$ is an extension of $\mathcal{O}_1 \leftrightarrow \mathcal{O}_2 \cap F_1 = \mathcal{O}_1$.

PROPOSITION 3. *Let* $F_1 \subset F_2$ *be valued fields with valuations* ord_1 *and* ord_2 *and valuation rings* $\mathcal{O}_1$ *and* $\mathcal{O}_2$ *respectively. Then the following are equivalent*:

(i) ord_2 *extends* ord_1

(ii) F_2 *is an extension of* F_1

(iii) $\mathcal{O}_2 \supset \mathcal{O}_1$ *and* $\mathfrak{M}(\mathcal{O}_2) \supset \mathfrak{M}(\mathcal{O}_1)$

(iv) $\mathcal{O}_2$ *extends* $\mathcal{O}_1$.

PROOF. (i) $\Leftrightarrow$ (ii) and (ii) $\Rightarrow$ (iii) are immediate from the definitions. If (iii) holds, then clearly $\mathcal{O}_2 \cap F_1 \supseteq \mathcal{O}_1$. Conversely, $x \in F_1 - \mathcal{O}_1 \Rightarrow x^{-1} \in \mathfrak{M}(\mathcal{O}_1) \Rightarrow x^{-1} \in \mathfrak{M}(\mathcal{O}_2) \Rightarrow x \notin \mathcal{O}_2 \Rightarrow x \notin \mathcal{O}_2 \cap F_1$, so $\mathcal{O}_2 \cap F_1 \subseteq \mathcal{O}_1$. Finally, if (iv) holds we may define $\varphi: F_1^*/\mathcal{O}_1^* \to F_2^*/\mathcal{O}_2^*$ which is easily seen to yield a value-monomorphism.

The following proposition is an immediate consequence of Propositions 1 and 3.

PROPOSITION 4. *If* $F_1 \subset F_2$ *every valuation of* F_1 *extends to* F_2.

PROPOSITION 5. *Let* $\mathcal{O}$ *be a valuation subring of* F *and* R *a subring such that* $\mathcal{O} \subset R \subset F$. *Then:*

(a) R *is a valuation subring of* F;

(b) $\mathfrak{M}(R)$ *is a prime ideal* $\mathbf{p}$ *of* $\mathcal{O}$;

(c) $R = \mathcal{O}_{\mathbf{p}}$.

PROOF. (a) $x \in F - R \Rightarrow x \in F - \mathcal{O} \Rightarrow x^{-1} \in \mathcal{O} \Rightarrow x^{-1} \in R$.

(b) It suffices to show $\mathfrak{M}(R) \subseteq \mathcal{O}$. But $x \in \mathfrak{M}(R) \Rightarrow x^{-1} \notin R \Rightarrow x^{-1} \notin \mathcal{O} \Rightarrow x \in \mathcal{O}$.

(c) Since $x \in \mathcal{O} - \mathbf{p} \Rightarrow x \in R^*$, we certainly have $\mathcal{O}_{\mathbf{p}} \subseteq R$. Conversely, let $x \in R$. If $x \in \mathcal{O}$, then $x \in \mathcal{O}_p$. Otherwise, $x \notin \mathcal{O} \Rightarrow x \notin \mathfrak{M}(R) \Rightarrow 1/x \in \mathcal{O} - \mathfrak{M}(R) \Rightarrow x = 1/(1/x) \in \mathcal{O}_{\mathbf{p}}$.

PROPOSITION 6. *Let* V_i, $i \in I$, *be a finite set of valuation subrings of* F *with maximal ideals* $\mathfrak{M}_i = \mathfrak{M}(V_i)$. *Set* $D = \bigcap_{i \in I} V_i$, $\mathbf{p}_i = D \cap \mathfrak{M}_i$. *Then* $V_i = D_{\mathbf{p}_i}$.

PROOF (NAGATA). Clearly $D_{\mathbf{p}_i} \subseteq V_i$. Let $a \in V_{i_1}$. Set $I_1 = \{i \in I \mid a \in V_i\}$, and for $i \in I_1$ let α_i denote the residue class of a in $\overline{V}_i = V_i/\mathfrak{M}_i$. Since I_1 is finite we may choose a prime number r such that for all $i \in I_1$, $r >$ char $\overline{V}_i$ and α_i is not a primitive rth root of unity.

Set $b = 1 + a + a^2 + \cdots + a^{r-1}$. Then for $i \in I_1$, $\bar{b} =$ the residue class of $b \bmod \mathfrak{M}_i = 1 + \alpha_i + \cdots + \alpha^{r-1}$. If $\alpha_i = 1$, then $\bar{b} = r \neq 0$ by choice of r. If $\alpha_i \neq 1$, then $\bar{b} = (1 - \alpha_i^r)/(1 - \alpha_i) \neq 0$ since α_i is not a primitive rth root of unity. Hence b is a unit in V_i, i.e. $1/(1 + a + \cdots + a^{r-1}) \in V_i^*$, for all $i \in I_1$.

For $i \in I - I_1$, and $j \in Z_{\geq 0}$ we have

$$\frac{a^j}{1 + a + \cdots + a^{r-1}} = \left(\frac{1}{a}\right)^{r-1-j} \frac{1}{(1/a)^{r-1} + (1/a)^{r-2} + \cdots + 1}.$$

Since $a \notin V_i$, $1/a \in \mathfrak{M}_i$. Thus $1 + (1/a) + \cdots + (1/a)^{r-1} \in V_i^*$ and

$$\left(\frac{1}{a}\right)^{r-1-j} \frac{1}{1 + (1/a) + \cdots + (1/a)^{r-1}} \in V_i \quad \text{for } j \leq r - 1.$$

In particular for $j = 0, 1$ we have

$$\frac{1}{1 + a + \cdots + a^{r-1}}, \qquad \frac{a}{1 + a + \cdots + a^{r-1}} \in V_i \quad \text{for } i \in I - I_1.$$

Thus we have

$$\frac{1}{1 + a + \cdots + a^{r-1}}, \qquad \frac{a}{1 + a + \cdots + a^{r-1}} \in D$$

and

$$\frac{1}{1 + a + \cdots + a^{r-1}} \notin \mathbf{p}_{i_1}.$$

Hence

$$a = \frac{a}{(1 + a + \cdots + a^{r-1})} \Big/ \frac{1}{(1 + a + \cdots + a^{r-1})} \in D_{\mathbf{p}_{i_1}}.$$

COROLLARY. *If also $V_i \not\subset V_j$ for $i \neq j$ then:*
(i) *the $\mathbf{p}_i$ are the maximal ideals of D;*
(ii) *If $a_i \in V_i$ for $i \in I$ then $\exists a \in D$ such that*

$$a - a_i \in \mathfrak{M}_i, \qquad \forall i \in I.$$

PROOF. (i) Since by the proposition $\mathbf{p}_i \not\subset \mathbf{p}_j$ for $i \neq j$, it suffices to show that if $\mathfrak{A}$ is an ideal of D, then $\mathfrak{A} \subseteq \mathbf{p}_j$ for some $j \in I$. Suppose false for some $\mathfrak{A}$. Choose $a_j \in \mathfrak{A} - \mathbf{p}_j$ for every $j \in I$. Also choose $b_{ij} \in \mathbf{p}_i - \mathbf{p}_j$ for $i \neq j$. (The b_{ij} exist by hypothesis.)

Set $c_j = \prod_{i \neq j} b_{ij}$. Then $c_j \in \bigcap_{i \neq j} \mathbf{p}_i - \mathbf{p}_j$.
Set $d = \sum_{j \in I} a_j c_j$. Then $d \notin \mathbf{p}_i \ \forall i \in I$.
Thus $d \in \bigcap_{i \in I} V_i^* = D^*$. But $d \in \mathfrak{A}$, an ideal in D, a contradiction.
(ii) Since the $\mathbf{p}_i$ are pairwise comaximal, the canonical map $D \to \bigoplus_{i \in I} D/\mathbf{p}_i$ is a surjection. Since the $\mathbf{p}_i$ are maximal, the map $D/\mathbf{p}_i \to D\mathbf{p}_i/\mathbf{p}_i D\mathbf{p}_i = V_i/\mathfrak{M}_i$ is an isomorphism (Lemma 1, §1); and the conclusion follows.

PROPOSITION 7. *Let F/E be an extension of valued fields so that we have natural inclusions $\bar{E} \to \bar{F}$, $\operatorname{ord} E^* \to \operatorname{ord} F^*$. Let $\omega_1, \ldots, \omega_f \in \mathcal{O}(F)$ be such that $\bar{\omega}_1, \ldots, \bar{\omega}_f$ are $\bar{E}$-linearly independent and let $\Pi_1, \ldots, \Pi_l \in F^*$ be such that $\operatorname{ord} \Pi_1, \ldots, \operatorname{ord} \Pi_l$ represent distinct cosets of $\operatorname{ord} F^*/\operatorname{ord} E^*$. Then $\forall a_{\varphi,\epsilon} \in \mathbf{F}$, $\operatorname{ord} (\sum_{\varphi,\epsilon} a_{\varphi\epsilon}\omega_\varphi \Pi_\epsilon) = \min_{\varphi,\epsilon} \operatorname{ord} a_{\varphi,\epsilon}\Pi_\epsilon$. In particular, the $\omega_\varphi \Pi_\epsilon$ are linearly independent over E.*

PROOF. If $a_{\varphi\epsilon} = 0$ for all φ, ϵ the conclusion is obvious. Otherwise, we may assume that $a_{\varphi,\epsilon} \neq 0$ for *any* pair (φ, ϵ) (since min ord $a_{\varphi\epsilon}\pi_\epsilon$ is attained for (φ, ϵ) with $a_{\varphi\epsilon} = 0 \Leftrightarrow a_{\varphi\epsilon} = 0$ for all (φ, ϵ)).
Fix ϵ. Choose $\varphi(\epsilon)$ such that $\operatorname{ord} a_{\varphi(\epsilon)\epsilon} = \min_\varphi \operatorname{ord} a_{\varphi\epsilon}$. Then $\operatorname{ord} \sum_\varphi a_{\varphi\epsilon}\omega_\varphi = \operatorname{ord} a_{\varphi(\epsilon)\epsilon} + \operatorname{ord} \sum_\varphi a_{\varphi(\epsilon)\epsilon}^{-1} a_{\varphi\epsilon}\omega_\varphi$. Now the coefficients of the $\bar{\omega}_\varphi$ in $\sum_\varphi a_{\varphi(\epsilon)\epsilon}^{-1} a_{\varphi\epsilon}\omega_\varphi$ are elements of $\bar{E}$ and are not all zero (the coefficient of $\bar{\omega}_{\varphi(\epsilon)}$ is 1), so this element $\neq 0$ in $\bar{F}$. Hence $\operatorname{ord} \sum_\varphi a_{\varphi(\epsilon)\epsilon}^{-1}\omega_\varphi = 0$ so $\operatorname{ord} \sum_\varphi a_{\varphi\epsilon}\omega_\varphi = \operatorname{ord} a_{\varphi(\epsilon)\epsilon}$. Moreover, since $\operatorname{ord} \sum_\varphi a_{\varphi\epsilon}\omega_{\varphi\epsilon} \operatorname{ord} E^*$,

$$\operatorname{ord} \prod_\epsilon \sum a_{\varphi\epsilon}\omega_\varphi \neq \operatorname{ord} \prod_{\epsilon_1} \sum a_{\varphi\epsilon_1}\omega_\varphi \quad \text{for } \epsilon \neq \epsilon_1$$

by definition of the $\prod_\epsilon$. Therefore

$$\begin{aligned} \operatorname{ord} \sum_\epsilon \prod_\epsilon \sum_\varphi a_{\varphi\epsilon}\omega_\varphi &= \min_\epsilon \left\{\operatorname{ord} \prod_\epsilon \sum_\varphi a_{\varphi\epsilon}\omega_\varphi\right\} \\ &= \min_\epsilon \left\{\operatorname{ord} \prod_\epsilon a_{\varphi(\epsilon)\epsilon}\right\} \\ &= \min_{\varphi,\epsilon} \operatorname{ord} \prod_\epsilon a_{\varphi\epsilon}. \end{aligned}$$

DEFINITION 5. $e = e(F/E) = (\operatorname{ord} F^* \colon \operatorname{ord} E^*)$
$f = f(F/E) = [\bar{F} \colon \bar{E}]$
$n = n(F/E) = [F \colon E]$

COROLLARY. $ef \leq n$.

3. Topology defined by a valuation

DEFINITION 1. Let ord be an A-valuation of F, $a \neq \{0\}$. Then the *topology of F defined by ord* has as neighborhood base of 0, the sets $\{f \in F \mid \operatorname{ord} f > a\}$ for $a \in A$.

DEFINITION 2. If λ is a limit ordinal, a Cauchy sequence (of length λ) is a function b from the ordinals $< \lambda$ to F such that $\forall \alpha \in A \exists \rho < \lambda \forall \sigma, \tau$ with $\rho < \sigma < \tau < \lambda$ we have ord $(b(\tau) - b(\sigma)) > \alpha$.

DEFINITION 3. F is *complete* $\Leftrightarrow$ every Cauchy sequence has a limit.

REMARK. Since every limit ordinal has a cofinal subset of order type of cardinal, F is complete $\Leftrightarrow$ every Cauchy sequence of length a cardinal has a limit.

LEMMA. *Let F be an A-valued field and let $\aleph$ be the smallest cardinal such that A has a well-ordered cofinal subset of cardinality $\aleph$. Then F is complete $\Leftrightarrow$ every Cauchy sequence of length $\aleph$ has a limit.*

PROOF. First, it is easy to see that a sequence of length $< \aleph$ is Cauchy $\Leftrightarrow$ it is eventually constant. Secondly, if $\{x_\tau\}$, $\tau < \mathfrak{A}$ is a Cauchy sequence of length $\mathfrak{A} > \aleph$, identify $\aleph$ with the well-ordered cofinal subset of A and define $\{\tau(\sigma)\}_{\sigma<\aleph}$ recursively such that

(i) $\tau(\tau) > \tau(\rho)$ for all $\rho < \sigma$; and

(ii) $\tau \geq \tau(\sigma) \Rightarrow$ ord $(x_\tau - x_{\tau(\sigma)}) > \sigma$. Then $\{x_{\tau(\sigma)}\}_{\sigma<\aleph}$ is a Cauchy sequence and $\lim x_{\tau(\sigma)} = x \Rightarrow \lim x_\tau = x$.

PROPOSITION 8. *If F is a valued field $\exists!$ (up to F value-isomorphism) valued field $\hat{F}$ extending F such that $\hat{F}$ is complete and F is dense in $\hat{F}$.*

PROOF. We first prove the existence of F. If ord $(F^*) = A$, let $\aleph$ be the cardinality of a smallest well-ordered cofinal subset of A.

Define $\mathfrak{F} = \{\{x_\sigma\}_{\sigma<\aleph} \mid$ Cauchy sequence in F of length $\aleph\}$. $\mathfrak{F}$ is a ring under the operations $\{x_\sigma\} + \{y_\sigma\} = \{x_\tau + y_\tau\}$, $\{x_\sigma\} \cdot \{y_\sigma\} = \{x_\sigma y_\sigma\}$.

Define $\mathfrak{N} = \{\{x_\sigma\}_{\sigma<\aleph} \in \mathfrak{F} \mid \forall \alpha \in A, \exists \tau < \aleph, \forall \sigma \geq \tau, \operatorname{ord}(x_\sigma) > \alpha\}$. $\mathfrak{N}$ is an ideal in $\mathfrak{F}$. We assert that if $\{x_\sigma\}_{\sigma<\aleph} \in \mathfrak{F} - \mathfrak{N}$ then eventual ord x_σ exists. Indeed there exists $\beta \in A$ such that ord $x_\sigma < \beta$ for all $\sigma < \aleph$. Choose $\tau < \aleph$ such that $\sigma \geq \tau \Rightarrow$ ord $(x_\sigma - x_\tau) > \beta$. Then for $\sigma > \tau$, ord $x_\sigma =$ ord $((x_\sigma - x_\tau) + x_\tau) =$ ord x_τ. We also assert that $\mathfrak{F}/\mathfrak{N}$ is a field. For it $\{x_\sigma\}_{\sigma<\aleph} \in \mathfrak{F} - \mathfrak{N}$, then $\exists \tau < \aleph$ such that $\sigma \geq \tau \Rightarrow x_\sigma \neq 0$; define $\{y_\sigma\}_{\sigma<\aleph}$ by

$$\begin{aligned} y_\sigma &= 0 && \sigma < \tau, \\ &= x_\sigma^{-1} && \sigma \geq \tau. \end{aligned}$$

Then $\{y_\sigma\}_{\sigma<\aleph}$ is Cauchy, since ord $(y_\sigma - y_\rho) =$ ord $(x_\rho - x_\sigma)/x_\sigma x_\rho =$ ord $(x_\rho - x_\sigma) =$ (constant) for σ, ρ sufficiently large. Thus $\{y_\sigma\}\{x_\sigma\} = 1 \bmod \mathfrak{N}$.

Denote $\mathfrak{F}/\mathfrak{N}$ by $\hat{F}$ and define a valuation, ord, on $\hat{F}$ by ord $(\{x_\sigma\}_{\sigma<\aleph} + \mathfrak{N}) =$ eventual ord x_σ for $\{x_\sigma\} \in \mathfrak{F} - \mathfrak{N}$, and ord $(\mathfrak{N}) = \infty$. It is easy to check that ord is well defined and a valuation.

Define a value-monomorphism $\varphi\colon F \to \hat{F}$ by $\varphi(x) = \{x_\sigma\}_{\sigma<\aleph} + \mathfrak{N}$, where $x_\sigma = x$, $\forall\sigma < \aleph$. Then $\varphi(F)$ is dense in $\hat{F}$ since if $\{x_\sigma\}_{\sigma<\aleph} + \mathfrak{N}$ is an element of $\hat{F}$, $\lim_\sigma \varphi(x_\sigma) = \{x_\sigma\}_{\sigma<\aleph} + \mathfrak{N}$.

If $\{x_\sigma\}_{\sigma<\aleph}$ is a Cauchy sequence in F it is easy to show that $\{\varphi(x_\sigma)\}_{\sigma<\aleph}$ has a limit in $\hat{F}$, namely $\{x_\sigma\}_{\sigma<\aleph} + \mathfrak{N}$.

Finally, since $\varphi(F)$ is dense in $\hat{F}$ and every Cauchy sequence in $\varphi(F)$ converges in $\hat{F}$, it follows that every Cauchy sequence in $\hat{F}$ converges.

DEFINITION 4. $\hat{F}$ is called the *completion* of F.

DEFINITION 5. $\mathbf{Q}_p = \hat{\mathbf{Q}}$, with respect to the p-adic valuation of $\mathbf{Q}$ (= valuation with valuation subring $\mathbf{Z}_{(p)}$).

S_p = completion of $\mathbf{F}_p(t)$ with respect to the valuation subring $\mathbf{F}_p[t]_{(t)}$ ($\mathbf{F}_p$ = field with p elements).

LEMMA 1. *If $A \subset B$ are ordered abelian groups and if $(B\colon A) = n < \infty$ then A is cofinal in B, i.e. $\forall b \in B$, $\exists a \in A$ with $a > b$.*

PROOF. If $b > 0$, then $nb > b$ and $nb \in A$. If $b < 0$, then $-nb > b$ and $-nb \in A$.

LEMMA 2. *Let F/E be a finite extension of valued fields with E complete. If $\omega_i, \ldots, \omega_n \in F$ are E-linearly independent then $f^{(\rho)} = \sum_{\nu=1}^{n} a_\nu^{(\rho)}\omega_\nu$, $\rho < \lambda$ is Cauchy in $F \Leftrightarrow a_\nu^{(\rho)}$, $\rho < \lambda$ is Cauchy in E for $\nu = 1, \ldots, n$.*

PROOF. ($\Leftarrow$) is clear.

($\Rightarrow$) we prove by induction on n.

For $n = 1$, it is clear, since $a_i^{(\rho)}\omega_1$ is Cauchy $\Leftrightarrow a_1^{(\rho)}\omega_1/\omega_1$ is Cauchy. (Note $\omega_1 \neq 0$ by linear independence.)

Assume we have proved the lemma for $n - 1$ but that it is false for n, i.e. $\exists\alpha \in \operatorname{ord} E^*$ such that for all $\rho < \lambda$, $\exists\sigma_\rho$ such that $\rho < \sigma_\rho < \lambda$ and

$$\operatorname{ord}\,(a_n{}^{(\sigma_\rho)} - a^{(\rho)}) \leq \alpha.$$

Now $f^{(\sigma_\rho)} - f^{(\rho)}$ is a null sequence and hence Cauchy. Therefore

$$\frac{f^{(\sigma_\rho)} - f^{(\rho)}}{a_n^{(\sigma_\rho)} - a_n^{(\rho)}} = \sum_{\nu=1}^{n-1} \frac{a_\nu^{(\sigma_\rho)} - a_\nu^{(\rho)}}{a_n^{(\sigma_\rho)} - a_n^{(\rho)}}\,\omega_\nu + \omega_n$$

is null and Cauchy (since $\operatorname{ord}\,(a_n^{(\sigma_\rho)} - a_n^{(\rho)}$ is bounded).

Thus we get that

$$g^{(\rho)} = \sum_{\nu=1}^{n-1} \frac{a_\nu^{(\sigma_\rho)} - a_\nu^{(\rho)}}{a_n^{(\sigma_\rho)} - a_n^{(\rho)}}\,\omega_\nu$$

is Cauchy. Hence b the inductive hypothesis and the completeness of E, $g^{(\rho)} \to \sum_{\nu=1}^{n-1} a_\nu\omega_\nu$ for some $a_\nu \in E$.

Therefore $g^{(\rho)} + \omega_n \to \sum_{\nu=1}^{n-1} a_\nu\omega_\nu + \omega_n$.

But we know $g^{(\rho)} + \omega_n \to 0$. This contradicts the linear independence of the ω_i.

DEFINITION 6. An ordered abelian group A is *archimedean* $\Leftrightarrow$ for any positive $a, b \in A$, $\exists n \in \mathbf{Z}_{>0}$ such that $n > b$.

PROPOSITION 9. *Let E,* ord *be complete with* ord E^* *archimedean. If $F \mid E$ is a finite extension of fields, there exists a unique extension of* ord *to F.*

PROOF. By Lemma 2, every extension of ord to F determines the same topology on F. So we will be done if we can show that any extension is completely determined by the topology it defines on F.

Let ord be any extension to F of the valuation on E. It follows from Lemma 1, that since ord E^* is archimedean so is ord F^*. But ord F^* archimedean implies $-n$ ord $x \to \infty$ if ord $x < 0$ i.e. $x^{-n} \to 0$. Conversely, it is true in general that ord $x \geq 0$ implies x^{-n} does not approach 0. Hence $\mathcal{O}_{\text{ord}} = F - \{x \in F \mid x^{-n} \to 0\}$ and the proof is complete.

4. Algebraic extensions of valued fields

Let $\mathcal{O}$ be a valuation subring of E, F/E a normal extension, and V a valuation subring of F extending $\mathcal{O}$. Then it is easy to see that if $\sigma \in \mathcal{G}(F/E)$, σV is also an extension of $\mathcal{O}$. We now prove that every extension of $\mathcal{O}$ can be gotten in this way.

PROPOSITION 10. *Let $\mathcal{O}$ be a valuation subring of E and let F/E be a finite normal (not necessarily separable) extension. If V and W are extensions of $\mathcal{O}$ to F then $\exists \sigma \in \mathcal{G}(F \mid E)$ such that $\sigma V = W$. Also $V \not\subset W$ unless $V = W$.*

PROOF. We prove the second statement first. Suppose $V \subset W$. Then $\mathfrak{M}(W) \subset V$ by Proposition 5 and we have

$$\bar{E} = \mathcal{O}/\mathfrak{M}(\mathcal{O}) \hookrightarrow V/\mathfrak{M}(W) \hookrightarrow W/\mathfrak{M}(W) = \bar{F}.$$

But $[\bar{E}: \bar{F}] \leq [E: F] < \infty$, so $V/\mathfrak{M}(W)$ is a field. Hence $\mathfrak{M}(W)$ is maximal in V, $\mathfrak{M}(W) = \mathfrak{M}(V)$, and $W = V$.

Thus there are no nontrivial inclusions among the σV, τW for $\sigma, \tau \in \mathcal{G}(F/E) = G$. Now let $H = \{\sigma \in G \mid \sigma V = V\}$ and $K = \{\sigma \in G \mid \sigma W = W\}$ and choose $\{\sigma_i \mid i = 1, \ldots, n\}$ and $\{\tau_j \mid j = 1, \ldots, m\}$ to be complete sets of representatives of distinct cosets of G/H and G/K respectively. If $\sigma_i V \not\subset \tau_j W$ for any i, j, then we may apply the Corollary to Proposition 6 to choose $a \in \bigcap_{i=1}^{n} \sigma_i V \bigcap_{j=1}^{m} \tau_j W$ such that

$$a - 1 \in \sigma_i \mathfrak{M}(V) = \mathfrak{M}(\sigma_i V) \qquad i = 1, \ldots, n$$

and

$$a \in \tau_j \mathfrak{M}(W) = \mathfrak{M}(\tau_j W) \qquad j = 1, \ldots, m.$$

But then $\sigma(a) \in \mathfrak{M}(V) + 1$, $\forall \sigma \in \mathcal{G}(F/E)$ and $\sigma(a) \in \mathfrak{M}(W)$ $\forall \sigma \in \mathcal{G}(F/E)$. So $N_{F/E}(a) - 1 \in \mathfrak{M}(V) \cap E = \mathfrak{M}(\mathcal{O})$ and $N_{F/E}(a) \in \mathfrak{M}(W) \cap E = \mathfrak{M}(\mathcal{O})$, which is impossible since $\mathfrak{M}(\mathcal{O})$ is an ideal. Therefore $\sigma_i V \subset \tau_j W$ for some i, j thus $\sigma_i V = \tau_j W$ or $\tau_j^{-1} \sigma_i V = W$.

COROLLARY. *Let F/E be normal and $\mathcal{O}_1$, $\mathcal{O}_2$ valuation subrings of F such that $\mathcal{O}_1 \cap E = \mathcal{O}_2 \cap E$. Then $\exists \sigma \in \mathcal{G}(F \mid E)$ such that $\sigma \mathcal{O}_1 = \mathcal{O}_2$.*

PROOF. Let $\mathfrak{N}$ be the set of all finite, normal subextensions of F/E. For $N \in \mathfrak{N}$, let $S(N) = \{\sigma \in \mathcal{G}(N/E) \mid \sigma(\mathcal{O}_1 \cap N) = \mathcal{O}_2 \cap N\}$. $S(N) \neq \emptyset$ be Proposition 10. $\mathfrak{N}$ is a directed set under the ordering given by inclusion. If $N, N' \in \mathfrak{N}$ and $N \subset N'$ define $\varphi_N^{N'}: S(N') \to S(N)$ by $\varphi_N^{N'}(\sigma) = \sigma \mid N$. To prove the corollary it suffices to show $\emptyset \neq \lim_{\leftarrow} S(N) = \{g \in \Pi S(N) \mid \varphi_N^{N'}(g(N')) = g(N) \forall N \leq N'\}$. (For if $g \in \lim_{\leftarrow} S(N)$, $\sigma \in \mathcal{G}(F/E)$ defined by $\sigma \mid N = g(N)$ for all $N \in \mathfrak{N}$ has the desired property.)

For any $M \in \mathfrak{N}$ define

$$A_M = \left\{g \in \Gamma \quad S(N) \middle| \varphi_N^{N'}(g(N')) = g(N)\ \forall N \leq N' \leq M\right\}.$$

Then $\lim_{\leftarrow} S(N) = \bigcap_{M \in \mathfrak{N}} A_M$. Give $S(N)$ the discrete topology. Since the $S(N)$ are compact, so is $\prod_{N \in \mathfrak{N}} S(N)$. But the A_M are closed in $\prod S(N)$ and have the finite intersection property. Therefore $\emptyset \neq \bigcap_{M \in \mathfrak{N}} A_M = \lim_{\leftarrow} S(N)$.

DEFINITION 1. A valued field E is Henselian $\Leftrightarrow f, g, h \in \mathcal{O}[X]$, monic and such that $\bar{f} = \bar{g}\bar{h}$ and $\bar{1} \in \langle \bar{g}, \bar{h} \rangle \bar{E}[X]$ implies $\exists$ monic $g_1, h_1 \in \mathcal{O}[X]$ such that $f = g_1 h_1$, $\bar{g} = \bar{g}_1$, $\bar{h} = \bar{h}_1$.

DEFINITION 2. A valued field E, ord has the *uniqueness property* $\Leftrightarrow$ there exists one and only one extension of ord to every finite algebraic extension of E.

REMARK. If E has the uniqueness property there exists one and only one extension of ord to any algebraic extension of E.

EXAMPLES. By Proposition 9, if E is complete and ord E^* is Archimedean, E has the uniqueness property. This applies to $\mathbf{Q}_p$ and S_p (by their Definition 5 of §3).

PROPOSITION 11. *E is Henselian $\Leftrightarrow$ E has the uniqueness property.*

PROOF. ($\Rightarrow$) Assume E is Henselian, F/E is finite and ord extends in two inequivalent ways to F. Since we may extend the two inequivalent valuations to a normal closure of F/E we may assume F/E is normal. Also since every valuation on F_s extends uniquely to F (because $G(F/F_s) = \{1\}$) we may assume F/E is separable.

Let ord denote one extension of the valuation on E to F. We are assuming there exists $\sigma \in G(F/E)$ and $\alpha \in F$ such that ord $\sigma(\alpha) \neq$ ord α. By replacing α by $\alpha^{e(F/E)}$ we may assume ord $\tau(\alpha) \in$ ord E^* for all $\tau \in G(F/E)$. (We still have ord $\sigma(\alpha) \neq$ ord α since an ordered group is torsion-free.)

By choosing $\beta \in E^*$ such that ord $\beta = \min\{\text{ord}\, \tau(\alpha) \mid \tau \in G(F/E)\}$ and replacing α by $\alpha\beta^{-1}$, we may assume $\min\{\text{ord}\, \tau(\alpha) \mid \tau \in G(F/E)\} = 0$.

Let $S = \{a' \in F \mid \alpha'$ is E-conjugate to α and ord $\alpha' = 0\}$ and let $T = \{\alpha'' \in F \mid \alpha''$ is E-conjugate to α and ord $\alpha'' > 0\}$. Let $j > 0$ be the number of elements in S. Then if $f(x)$ is the minimal polynomial of α over E, we have f is irreducible,

$$f(X) = \sum_{i=0}^{n} a_{n-i} X^i = \prod_{\alpha' \in S} (X - \alpha') \prod_{\alpha'' \in T} (X - \alpha'')$$

and

$$\text{ord}\, a_{n-1} = 0 \quad \text{if } i = n - j$$
$$> 0 \quad \text{if } i < n - j \text{ and ord } a_{n-i} \geq 0 \text{ for all } i$$

(since a_{n-i} = the $(n - i)$th symmetric polynomial in the elements of $S \cup T$).

Thus $\bar{f}(X) = \sum_{i=n-j}^{n} \bar{a}_{n-i} X^i$

or

$$(*) \qquad \bar{f}(X) = X^{n-j} \sum_{i=n-j}^{n} \bar{a}_{n-i} X^{i-(n-j)}.$$

Moreover $0 < j < n$ (since $S \neq \emptyset$, $T \neq \emptyset$) and $\bar{a}_{n-j} \neq 0$. Hence (*) is a factorization of $\bar{f}$ into monic relatively prime factors; since E is Henselian and f is irreducible this is a contradiction. ($\Leftarrow$) Suppose we are given f, g, $h \in O[X]$ monic such that $\bar{f} = \bar{g}\bar{h}$ with $(\bar{g}, \bar{h}) = \bar{1}$.

Let F be a splitting field of f over E and let $\mathcal{O}'$ be the unique extension of $\mathcal{O}$ to F. Write $f(X) = \prod_{i=1}^{n} (X - \alpha_i)$ in $\mathcal{O}'[X]$. (The α_i are in $\mathcal{O}'$ since $\mathcal{O}'$ is the integral closure of $\mathcal{O}$ in F.) Index the α_i such that $\bar{g}(X) = \prod_{i=1}^{m} (X - \bar{\alpha}_i)$, $h(X) = \prod_{i=m+1}^{n} (X - \bar{\alpha}_i)$. Set $g_1(X) = \prod_{i=1}^{m} (X - \alpha_i)$, $h_1(X) = \prod_{i=m+1}^{n} (X - \alpha_i)$. Then we have $\bar{g}_1 = \bar{g}$, $\bar{h}_1 = \bar{h}$ and $f = g_1 h_1$. It remains to show that $g_1, h_1 \in \mathcal{O}[X]$.

Let $\sigma \in G(F/E)$. Then by the uniqueness property for E, $\sigma\mathcal{O}' = \mathcal{O}'$, $\sigma\mathfrak{M}(\mathcal{O}') = \mathfrak{M}(\mathcal{O}')$, so that σ induces an $\bar{E}$-automorphism of $\bar{F}$, denoted by $\bar{\sigma}$.

Now σ induces a permutation of the α_i, so that if $i \leq m$, $\sigma\alpha_i = \alpha_j$ for some $j \leq n$. Suppose $j > m$; then we would have $0 = \bar{\sigma}\bar{g}(\bar{\alpha}_i) = \bar{g}(\bar{\sigma}\bar{\alpha}_i) = \bar{g}(\bar{\alpha}_j)$ which contradicts the fact that $\bar{h}(\bar{\alpha}_j) = 0$ and $(\bar{g}, \bar{h}) = 1$. Hence $\sigma\alpha_i = \alpha_j$ for $j \leq m$ and therefore $\sigma g_1 = g_1$, for all $\sigma \in G(F/E)$. Thus the coefficients of g are in the fixed field of $G(F/E)$ i.e. they are purely inseparable over E. But then if $\rho =$ char E, there exists a positive integer τ such that $f^{\rho^\tau} = g_1^\rho h_1^\rho$ in $E[X]$. It follows immediately, by unique factorization in $E[X]$ and the fact that $g_1^{\rho^\tau}$ and $h_1^{\rho^\tau}$ are relatively prime, that $g_1^{\rho^\tau}$ and $h_1^{\rho^\tau}$ are p^τ-th powers in $E[X]$. Thus g_1, h_1 are in $E[X]$ and hence in $\mathcal{O}[X]$.

COROLLARY 1. *If E is Henselian and F/E is algebraic then F is Henselian (with respect to the unique extension to F to the valuation on E).*

COROLLARY 2. $\mathbf{Q}_p$ *and* $\mathbf{S}_p$ *are Henselian.*

PROPOSITION 12. *If* (F, ord) *is Henselian and E is relatively separable algebraically closed in F (i.e. E has no proper separable algebraic extensions contained in F), then* (E, ord $|_E$) *is Henselian.*

PROOF. Let $\mathcal{O}$ be the valuation ring of ord. Suppose f, g, $h \in (\mathcal{O} \cap E)[X]$, monic, such that $\bar{f} = \bar{g}\bar{h}$ in $\bar{E}[X]$ with $(g, h) = 1$. Then $f = g_1 h_1$ in $F[X]$ since F is Henselian. Since the roots of f are algebraic over E, so are the coefficients g_1 and h_1. Thus, since the coefficients are in F, they are purely inseparable over E. Hence, by the same argument as in the proof of Proposition 11, it follows that $g_1, h_1 \in E[X]$.

DEFINITION 3. Let F/E be an extension of valued fields. Then F is a *Henselization of* $E \Leftrightarrow F$ is Henselian and for all extensions F'/E such that F' is Henselian there exists a unique E-value monomorphism: $F \to F'$.

PROPOSITION 13. *Every valued field E has a Henselization F which is unique up to E-value-isomorphism and which is separable algebraic over E.*

PROOF. The uniqueness of F is clear from the definition. To prove the existence of F, let $\tilde{E}^s$ be a separable algebraic closure of E and choose an extension, $\mathcal{O}$, of the valuation on E to $\tilde{E}^s$. Define $H = \{\sigma \in G(\tilde{E}^s/E) \mid \sigma(\mathcal{O}) = \mathcal{O}\}$ and let F be the fixed field of H, with valuation subring $= \mathcal{O} \cap F$.

First we show F is Henselian. By the same considerations as in the first part of the proof of Proposition 11 it suffices to show that the valuation on F extends uniquely to any finite Galois extension N/F. Clearly also we may take $N \subseteq \tilde{E}^s$; then $\mathcal{O} \cap N$ is one extension of $\mathcal{O} \cap F$, and it will suffice to show that $\sigma(\mathcal{O} \cap N) = \mathcal{O} \cap N$ for all $\sigma \in \mathcal{G}(N/F)$. But consider the map: $H \to \mathcal{G}(N/F)$ given by restriction of the elements of H to N. The image of this map has fixed field F and therefore it equals $\mathcal{G}(N/F)$, since N/F is finite. Hence if $\sigma \in \mathcal{G}(N/F)$, $\exists \tau \in H$ such that $\tau \mid N = \sigma$; so $\sigma(\mathcal{O} \cap N) = \tau(\mathcal{O} \cap N) = \tau(\mathcal{O}) \cap N = \mathcal{O} \cap N$.

Now let F'/E be an Henselian extension of valued fields. We wish to show there exists a unique E-value-monomorphism: $F \to F'$. By Proposition 12 we may assume F'/E is separable algebraic. Thus we may also assume $F' \subseteq \tilde{E}^s$. Let $\mathcal{O}'$ be the unique extension of the valuation on F' to $\tilde{E}^s$. Then $\mathcal{O}'$ and $\mathcal{O}$ are both extensions of the valuation on E, so $\exists \lambda \in \mathcal{G}(\tilde{E}^s/E)$ such that $\lambda\mathcal{O} = \mathcal{O}'$. We will show that λ is the map we want. Since F' is Henselian, for all $\sigma \in \mathcal{G}(\tilde{E}^s/E)$ we have $\sigma\mathcal{O}' = \mathcal{O}'$ i.e. $\sigma\lambda\mathcal{O} = \lambda\mathcal{O}$ i.e. $\lambda^{-1}\sigma\lambda\mathcal{O} = \mathcal{O}$ i.e. $\lambda^{-1}\sigma\lambda \in H$. So we have $\mathcal{G}(\tilde{E}^s/F') \subseteq \lambda H \lambda^{-1}$. Hence the fixed field of $\lambda H \lambda^{-1} \subseteq$ fixed field of $\mathcal{G}(\tilde{E}^s/F')$ or $\lambda F \subseteq F'$. Thus $(\lambda \mid F)$ is an E-monomorphism: $F \to F'$ and since $\lambda(\mathcal{O} \cap F) = \lambda(\mathcal{O}) \cap \lambda F = \mathcal{O}' \cap \lambda F$; $(\lambda \mid F)$ is a value-monomorphism.

Finally, we want to show $(\lambda \mid F)$ is unique. Suppose ρ: $F \to F'$ is an E-value-monomorphism. Extend ρ to $\tilde{E}^s$. Then $\lambda(\mathcal{O} \cap F) = \mathcal{O}' \cap \lambda F$, $\rho(\mathcal{O} \cap F) = \mathcal{O}' \cap \rho F$ which implies $\lambda^{-1}(\mathcal{O}') \cap F = \mathcal{O} \cap F = \rho^{-1}(\mathcal{O}') \cap F$. Thus, since F is Henselian, $\lambda^{-1}(\mathcal{O}') = \mathcal{O} = \rho^{-1}(\mathcal{O}')$, from which it follows that $\lambda\mathcal{O} = \rho\mathcal{O}$ or $\rho^{-1}\lambda\mathcal{O} = \mathcal{O}$. Consequently $\rho^{-1}\lambda \in H$ and $(\rho \mid F) = (\lambda \mid F)$ since F is the fixed field of H.

DEFINITION 4. Let F/E be an extension of valued fields. Then F/E is *immediate* $\Leftrightarrow e(F/E) = 1 = f(F/E)$.

PROPOSITION 14. *Let* (F, ord) *be a finite Galois extension of* E. *Set* $H = \{g \in G = \mathcal{G}(F/E) \mid \mathrm{ord} \circ g = \mathrm{ord}\}$, *and let* $S =$ *fixed field of* H. *Then*

(a) ord *is the unique extension of* ord $\mid S$ *to* F. *Valuing* S, E *by* ord $\mid S$, ord $\mid E$ *respectively, we have*:

(b) $f(S/E) = 1$,

(c) $e(S/E) = 1$

so that S/E *is immediate.*

PROOF. (a). Any extension of ord $\mid S$ to F is of the form ord $\circ\, g$ for some $g \in \mathcal{G}(F/S) = H$.

(b) Write G as the union of disjoint cosets of H: $G = \bigcap_{\beta=1}^{b} g_\beta H$ and let $g_1 = 1_F$. Observe that $g_1 \mid S, \ldots, g_b \mid S$ are all the distinct E-monomorphisms of S into F. Also, if $\mathcal{O}$ is the valuation subring of F associated with ord, $\mathcal{O} \cap S = g_\beta^{-1}(\mathcal{O}) \cap S \Leftrightarrow \beta = 1$. Hence we may apply the Corollary to Proposition 6 to choose $s \in S$ such that ord $(s - 1) > 0$ and ord $g_\beta(s) > 0$ for $\beta > 1$.

Now let $\alpha \in \mathcal{O} \cap S$. We want to find $a \in E$ such that ord $(a - \alpha) > 0$. By the Corollary to Proposition 6 we may assume ord $g_\beta(\alpha) \geq 0$ for all β (since $\mathcal{O} \cap S/\mathfrak{M}(\mathcal{O}) \cap S \cong D/\mathfrak{M}(\mathcal{O}) \cap S \cap D$, where $D = \bigcap_{\beta=1}^{b} g_\beta^{-1}(\mathcal{O}) \cap S$). Thus replacing α by $s\alpha$ we may assume ord $g_\beta(\alpha) > 0$, for $\beta > 1$.

Let $a = T_{S/E}(\alpha)$. Then

$$\operatorname{ord}(a-\alpha) = \operatorname{ord}\left(\sum_{\beta=1}^{b} g_\beta(\alpha) - \alpha\right) = \operatorname{ord}\left(\sum_{\beta=2}^{b} g_\beta(\alpha)\right) \geq \min_{\beta \geq \alpha} \operatorname{ord} g_\beta(\alpha) > 0.$$

(c) Let $\alpha \in S^*$. We want to find $a \in E^*$ such that ord α = ord a. Let s be as in (b). Since ord $s = 0$, ord $g_\beta(s) > 0$ for $\beta > 1$, we can choose $m \in \mathbf{Z}_{>0}$ such that ord $s^m\alpha \neq$ ord $g_\beta(s^m\alpha)$ for $\beta > 1$. Hence, replacing α by $s^m\alpha$ we may assume ord $\alpha \neq$ ord $g_\beta(\alpha)$ for $\beta > 1$.

Let $f(X) = \sum_{i=0}^{n} r_{n-i}X^i = \prod_\beta (X - g_\beta(\alpha)) \in E[X]$. Now $\pm r_R$ = the kth symmetric polynomial in the $g_\beta(\alpha)$. Hence, by definition of v, ord r_v = ord $\prod_{\beta \in V} g_\beta(\alpha) \neq \infty$ and ord r_{v+1} = ord $\alpha \prod_{\beta \in V} g_\beta(\alpha)$.

Thus ord (r_{v+1}/r_v) = ord (α). So $a = r_{v+1}/r_v$ is the element we want.

REMARK. $e(F/E)$ and $f(F/E)$ are *multiplicative* i.e.

$$e(E_2/E) = e(E_2/E_1)e(E_1/E), \qquad f(E_2/E) = f(E_2/E_1)f(E_1/E)$$

if $E \subset E_1 \subset E_2$ is a tower of valued fields. This is clear since the degree of a field extension and the index of a subgroup in a group are multiplicative.

COROLLARY 1. *If* $\operatorname{ord}_1, \ldots, \operatorname{ord}_g$ *are all the extensions of* ord $\mid E$ *to* F, *then* $e = e(F/E, \operatorname{ord}_i), f = f(F/E, \operatorname{ord}_i)$ *are independent of* i, *and* $efg \leq n(F/E)$.

PROOF. Given i, j by Proposition 10, $\exists \sigma \in \mathcal{G}(F/E)$ such that $\operatorname{ord}_i \sigma = \operatorname{ord}_j$. Hence $\operatorname{ord}_j F^* = \operatorname{ord}_i \sigma F^* = \operatorname{ord}_j F^*$. Moreover, σ induces an automorphism

$$\bar{\sigma}\colon \mathcal{O}_{\operatorname{ord}_i}/\mathfrak{M}(\mathcal{O}_{\operatorname{ord}_i}) \to \mathcal{O}_{\operatorname{ord}_j}/\mathfrak{M}(\mathcal{O}_{\operatorname{ord}_j})$$

so the first statement is proved.

Since e, f are multiplicative,

$$\begin{aligned} efg &= e(F/S, \operatorname{ord})e(S/E, \operatorname{ord} \mid S)f(F/S, \operatorname{ord})e(S/E, \operatorname{ord} \mid S)g \\ &= e(F/S, \operatorname{ord})f(F/S, \operatorname{ord})g \quad \text{by Proposition 14,} \\ &\leq n(F/S, \operatorname{ord})g \quad \text{by Proposition 7,} \\ &= n(F/E, \operatorname{ord}) \quad \text{(since the distinct extensions of } \mathcal{O} \cap E \text{ to } F \\ &\qquad \text{are exactly the } g_\beta(\mathcal{O})\text{; so } g = b = n(S/E)). \end{aligned}$$

COROLLARY 2. *If* F' *is the Henselization of* E, *then* F'/E *is immediate.*

PROOF. Extend ord on F' to $\tilde{E}^s$. It suffices to show that $E \subset E' \subset F'$, E'/E finite, implies E'/E immediate. Let F be the Galois closure of E'. (E'/E is separable since F'/E is.) Let $H = \{g \in \mathcal{G}(F/E)\colon \operatorname{ord} \mid F \circ g = \operatorname{ord} \mid F\}$ and let S = the fixed field of H. If we can show $E' \subset S$ we will be done. To do this it suffices to show that $g \in H \Rightarrow \exists \sigma \in \mathcal{G}(\tilde{E}^s/E)$ such that ord $\circ\, \sigma$ = ord and $\sigma \mid F = g$. (Then σ fixes E', since $E' \subset F'$ = fixed field of $\{\sigma \in \mathcal{G}(\tilde{E}^s/E)\colon \operatorname{ord} \circ \sigma = \operatorname{ord}\}$.)

Let τ be any extension of g to $\tilde{E}^s$. Then $(\operatorname{ord} \circ \tau) \mid F = \operatorname{ord} \mid F$. Thus $\exists \rho \in \mathcal{G}(\tilde{E}^s/F)$ such that ord $\circ\, \tau$ = ord $\circ\, \rho$, or ord $\circ\, \tau\rho^{-1}$ = ord. So $\sigma = \tau\rho^{-1}$ works since $\tau\rho^{-1} \mid F = \tau \mid F = g$.

5. Henselian fields

REMARK. Let F, ord be Henselian. Let ord again denote an extension of ord to $\tilde{F}$. Then we know any other extension is of the form ord $\circ\, \sigma$ for some $\sigma \in (\tilde{F}, F)$ and as F is Henselian, ord and ord $\circ\, \sigma$ are equivalent. We claim they are equal. Indeed there exists an order-preserving automorphism φ of $B = \operatorname{ord} \tilde{F}^*$ such that φ restricted to $A = \operatorname{ord} F^*$ is the identity and such that $\operatorname{ord} \circ \sigma(\alpha) = \varphi \circ \operatorname{ord} \alpha$ $\forall \alpha \in \tilde{F}$. It suffices to show that φ is the identity.

Let $b \in B$. Then $\exists n \in \mathbf{Z}_{>0}$ such that $nb \in A$. Thus $n\varphi(b) = \varphi(nb) = nb$ and so $\varphi(b) = b$ as asserted.

In particular, if F'/F is finite and $\alpha \in F'$ then

$$\operatorname{ord} N_{F'/F}(\alpha) = \operatorname{ord}\Big(\prod_\sigma \sigma\alpha\Big) = p^i \sum_\sigma \operatorname{ord} \sigma\alpha = \Big(p^i \sum_\sigma \operatorname{ord} \alpha\Big) = n(F'/F) \operatorname{ord} \alpha$$

where σ runs over the distinct F-monomorphisms of F' into $\tilde{F}$ and p^i is the inseparability degree of F'/F.

DEFINITION. If F/E is a finite extension let $d(F/E) = n(F/E)/e(F/E)f(F/E)$. d is multiplicative since n, e, and f are multiplicative.

PROPOSITION 15. *Let* (E, ord) *be Henselian, and* $\operatorname{char} \bar{E} = p$. *(We allow* $p = 0$.*) Let* $F'/F/E$ *be finite extensions. Then if* $p \neq 0$, $d(F'/F) = p^m$, *for some* $m \in \mathbf{Z}_{\geq 0}$*; if* $p = 0$, $d(F'/F) = 1$.

PROOF. (ARTIN). Since $ef \leq n$, $d \in Q_{\geq 1}$. We prove the proposition in a series of steps:

(1) If q is a prime and $q \mid e(F'/F)$, then $q \mid n(F'/F)$. Let $\alpha \in F'^*$. Then $n(F'/F) \operatorname{ord} \alpha = \operatorname{ord} N_{F'/F}(\alpha) \in \operatorname{ord} F^*$ (because F is Henselian). Therefore $n(F'/F)$ annihilates $\operatorname{ord} F'^*/\operatorname{ord} F^*$. So if $q \mid e(F'/F)$, there exists an element of order q in $\operatorname{ord} F'^*/\operatorname{ord} F^*$, so $q \mid n(F'/F)$.

(2) If q is a prime and $q \mid f(F'/F)$, then $q \mid n(F'/F)$. First we claim $\exists \beta \in \bar{F}'$ such that $q \mid [\bar{F}(\beta) \colon \bar{F}]$. Indeed, let $\bar{F}_s'$ be the separable part of $\bar{F}'/\bar{F}$; then $\bar{F}_s'$ is simple and if $q \neq p$, $q \mid [\bar{F}_s' \colon \bar{F}]$. If $q = p \neq 0$, and $\bar{F}_s' \neq \bar{F}'$, let $\beta \in \bar{F}' - \bar{F}_s'$; $\beta^{p^r} \in \bar{F}_s'$ for some $r > 0$ and it follows that

$$p \mid [\bar{F}(\beta) \colon \bar{F}] = [\bar{F}(\beta) \colon \bar{F}(\beta^{p^r})][\bar{F}(\beta^{p^r}) \colon \bar{F}].$$

Choose $b \in \mathcal{O}(F')$ such that $\bar{b} = \beta$. Let $H(X)$ be the monic irreducible polynomial for b over F. $H \in \mathcal{O}(F)[X]$ since the coefficients of H are integral over $\mathcal{O}(F)$, which is integrally closed. Since F is Henselian and H is irreducible, $\bar{H}$ must be primary i.e. $\bar{H} = \bar{J}^a$ for some monic irreducible $\bar{J} \in \bar{F}[X]$ and some $a \in \mathbf{Z}_{>0}$. Then $0 = \bar{H}(b) = \bar{H}(\beta) = \bar{J}^a(\beta) \Rightarrow \bar{J}$ is the monic irreducible polynomial for $\beta/\bar{F}$. Thus, since $q \mid [\bar{F}(\beta) \colon \bar{F}] = \deg \bar{J}$, and $\deg \bar{J} \mid \deg H = [F(b) \colon F]$, and $[F(b) \colon F] \mid n(F'/F)$, we are done.

(3) If $n(F'/F)$ is a prime $q \neq p$, then $d(F' \mid F) = 1$. For if $d(F'/F) > 1$, then we must have $e(F'/F) = f(F'/F) = 1$. Let $\alpha \in F' - F$ and let $g(X)$ be the minimal polynomial for α over F. Say $g(X) = X^q + a_{q-1}X^{q-1} + \cdots + a_0$. Replacing α by $\alpha - (a_{q-1}/q)$ we may assume $a_{q-1} = 0$. Since $e(F'/F) = 1$,

replacing α by α/b, where $b \in F$ and ord b = ord α, we may assume ord $\alpha = 0$, which implies ord $a_0 = 0$ (since F Henselian $\Rightarrow$ all the conjugates of α have ord value 0). Since $f(F'/F) = 1$, $\exists c \in F$ such that $\bar{\alpha} = \bar{c} \neq 0$. Thus $0 = \bar{g}(\bar{\alpha}) = \bar{g}(\bar{c})$. Since F is Henselian, and g is irreducible, we have $\bar{g}(X) = (X - \bar{c})^q$. Thus $0 = \bar{a}_{q-1} = \bar{q}\bar{c}$, contradicting $q \neq \text{char } \bar{E}$ and $\bar{c} \neq 0$.

(4) It suffices to prove the proposition for separable extensions. Indeed, if char $E \neq 0$, then char $\bar{E} = p$ implies char $E = p$. Thus $n(F'/F'_s) = p^r$ for some $r \geq 0$, and—by (1) and (2)—$ef(F'/F'_s) = p^t$, for some $t \leq r$. Thus, since d is multiplicative, to prove $d(F'/F)$ is a power of p it remains only to show $d(F'_s/F)$ is a power of p.

(5) F'/F separable $\Rightarrow d(F'/F) = p^t$ if char $\bar{E} = p \neq 0$,
$= 1$ if char $\bar{E} = p = 0$.

Let q be a prime $\neq p$ such that $n(F'/F) = q^i q'$ $(i \geq 1, q \nmid q')$. It suffices to show $ef(F'/F) = q^i q''$ $(q \nmid q'')$. Let N = Galois closure of F'/T. Let S be a q-Sylow subgroup of $\mathcal{G}(N/F)$ such that $S \cap \mathcal{G}(N/F')$ is a q-Sylow subgroup of $\mathcal{G}(N/F')$. Let E_0 = fixed field of S, and E_i the fixed field of $S \cap \mathcal{G}(N/F')$. Then we have the following situation:

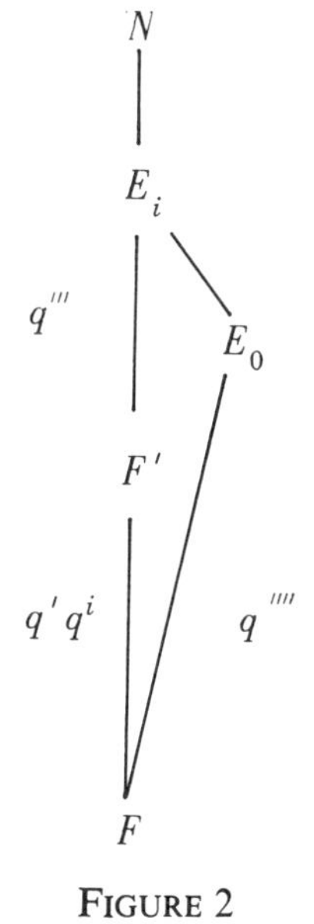

FIGURE 2

The indicated degrees q''' and q'''' are prime to q by definition of E_i and E_0. $[E_i: E_0]$ is a power of q. Hence $[E_i: E_0] = q^i$. So we can find extensions $E_0 \subset E_1 \subset \cdots \subset E_i$ each of degree q; thus by (3), $ef(E_j/E_{j-1}) = q$ for $1 \leq j \leq i$. Hence $n(E_i/E_0) = q^i = ef(E_i/E_0)$ by multiplicativity.

Now we have $ef(E_i/E_0)ef(E_0/F) = ef(E_i/F')ef(F'/F)$. As $n(E_0/F) = q''''$ and $n(E_i/F') = q'''$ are prime to q, it follows from (1) and (2) that $ef(E_0/F)$ and $ef(E_i/F')$ are prime to q. Thus the power of q in $ef(F'/F)$ is the same as that in $ef(E_i \mid E_0)$, namely q^i.

COROLLARY. *Let F/E be an extension of valued fields and suppose* char $E = 0$. *Then F is a Henselization of $E \Leftrightarrow F$ is Henselian and F/E is immediate algebraic $\Leftrightarrow F$ is a maximal algebraic immediate extension of E.*

PROOF. The first $\Rightarrow$ has been proved earlier. Now let F be Henselian and F/E immediate algebraic. Let F' be a Henselization of E such that $E \subset F' \subset F$. Then for any finite extension K of F' with $K \subset F$, $n(K/F') = ef(K/F') = 1$ so $K = F'$. Hence $F = F'$ so F is a Henselization of E. Furthermore if F'' is an immediate algebraic extension of F, then $n(F''/F) = ef(F''/F) = 1, \Rightarrow F'' = F$, so F is a maximal algebraic immediate extension of E.

Conversely, let F be a maximal algebraic immediate extension of E. Let F'/F be a Henselization of F. But then F'/F is immediate and algebraic, so $F' = F$ and F is Henselian.

PROPOSITION 16. *Let* (E, ord) *be Henselian and* char $\bar{E} = 0$. *Let A be any subfield of E maximal with respect to the property that* ord $(A^*) = 0$. *Then* $\bar{E} = \bar{A}$ $(\cong A)$.

PROOF. Since char $\bar{E} = 0$, char $E = 0$, so $\mathbf{Q} \subseteq E$. $\mathbf{Q}$ must have the trivial valuation since $\mathbf{Q}$ with any nontrivial valuation has residue class field of nonzero characteristic. Thus by Zorn's lemma there exists a maximal subfield, A, with the property that ord $(A) = 0$.

Suppose the proposition false and let $\beta \in \bar{E} - \bar{A}$.

Case 1. Suppose β is transcendental/$\bar{A}$. Then let $b \in \mathcal{O}(E)$ be such that $\bar{b} = \beta$. Then for any $f \in A[X]$, -0 ord $f(b) = 0$. (Otherwise ord $f(b) > 0 \Rightarrow \bar{f}(\bar{b}) = 0 \Rightarrow \beta$ algebraic/A.) But then ord $(A(b)^*) = 0$, contrary to the maximality of A.

Case 2. Suppose β is algebraic/$\bar{A}$. Let $J \in A[X]$, be monic and such that $\bar{J}$ is the monic irreducible polynomial for β/A. Then $\bar{J} = (X - \beta)h(X)$ with $h(X)$ monic in $\bar{E}[X]$. Since char $\bar{E} = 0$, $((X - \beta), h(X)) = 1$. So since E is Henselian $\exists b \in \mathcal{O}(E)$ such that $\bar{b} = \beta$ and $J(b) = 0$. Therefore b is algebraic/A. But the trivial valuation has only the trivial extension to algebraic extensions, so ord $(A(b)^*) = 0$, which is again contrary to the maximality of A.

6. Algebraically complete fields

DEFINITION 1. F is algebraically complete $\Leftrightarrow n(F'/F) = e(F'/F)f(F'/F)$ for all finite extensions F'/F.

REMARK. Proposition 15 says F Henselian and char $F = 0$ implies F algebraically complete. On the other hand, if F is algebraically complete, then since the Henselization of F is an immediate algebraic extension, F is Henselian.

DEFINITION 2. Let $E \subset K$, $L \subset M$ be fields. Then K is said to be *linearly disjoint* from L over E if in KL every set of elements of K that is linearly independent over E is also linearly independent over L.

$E \subset K$ is said to be *regular* if K is linearly disjoint from E over $\tilde{E}$, where E is the algebraic closure of $\tilde{E}$.

REMARKS. It is easily shown that K is linearly disjoint from L over $E \Leftrightarrow L$ is linearly disjoint from K over E.

$E \subset K$ regular $\Rightarrow E$ is relatively algebraically closed in K; and E is relatively algebraically closed in K, char $E = 0 \Rightarrow E$ is regular in K.

DEFINITION 3. Let B be an abelian group and A a subgroup. A is said to be *pure* in B if for any $a \in A$ and any $n \in \mathbf{Z}_{\geq 0}$ if $a \in nB$, then $a \in nA$.

REMARK. If B is a torsion-free group, then A is pure in $B \Leftrightarrow B/A$ is torsion free.

PROPOSITION 17. *Let* (F, ord) *be algebraically complete, let* t *be a transcendental element over* F, *and suppose* ord *extends to* $F(t)$ *such that* $\bar{F}$ *is regular in* $\bar{F}(t)$ *and* ord F^* *is pure in* ord $F(t)^*$. *Then* ord *extends from* $F(t)$ *to* $\tilde{F}(t)$ *uniquely, and for any* $\alpha \in \tilde{F}$ *there exists* $a \in F$ *such that* $\text{ord}\,(t - a) \geq \text{ord}\,(t - \alpha)$.

PROOF. Extend ord in any way to $F(t)$, and denote the extension by ord. Let $\alpha \in \tilde{F}$. Choose $1 = \gamma_1, \ldots, \gamma_e$ in $F(\alpha)^*$ such that $\text{ord}\,\gamma_1, \ldots, \text{ord}\,\gamma_e$ represent all the distinct cosets of $\text{ord}\,F(\alpha)^*/\text{ord}\,F^*$; and choose $1 = \omega_1, \ldots, \omega_f$ in $F(\alpha)$ such that $\bar{\omega}_1, \ldots, \bar{\omega}_f$ are a basis for $\bar{F}(\alpha)/\bar{F}$. Since F is algebraically complete, $n(F(\alpha)/F) = ef = e(F(\alpha)/F)f(F(\alpha)/F)$, so that α has a unique representation of the form

$$\alpha = \sum_{\epsilon,\varphi} a_{\epsilon\varphi}\gamma_\epsilon\omega_\varphi \quad \text{with } a_{\epsilon\varphi} \in F.$$

Since $\bar{F}$ is regular in $\bar{F}(t)$, $\bar{\omega}_1, \ldots, \bar{\omega}_f$ are linearly independent over $\bar{F}(t)$. Hence $f(F(t, \alpha)/F(t)) \geq f(F(\alpha)/F)$. Since $\text{ord}\,F^*$ is pure in $\text{ord}\,F(t)^*$ we have that $\text{ord}\,F(t)^* \cap \text{ord}\,F(\alpha)^* = \text{ord}\,F^*$ (because $e(F(\alpha)/F)\,\text{ord}\,F(\alpha)^* \subseteq \text{ord}\,F^*$). Thus we have (inside $\text{ord}\,F(t, \alpha)^*$), $(\text{ord}\,F(t)^* + \text{ord}\,F(\alpha)^*)/\text{ord}\,F(t)^* \cong \text{ord}\,F(\alpha)^*/(\text{ord}\,F(t)^* \cap \text{ord}\,F(\alpha)^*) = \text{ord}\,F(\alpha)^*/\text{ord}\,F^*$. Thus

$$e(F(t, \alpha)/F(t)) \geq e(F(\alpha)/F).$$

But then we have

$$n(F(\alpha)/F) = n(F(t, \alpha)/F(t)) \geq ef(F(t, \alpha)/F(t)) \geq ef(F(\alpha)/F) = n(F(\alpha)/F),$$

so the inequalities become equalities. In particular

$$n(F(t, \alpha)/F(t)) = ef(F(t, \alpha)/F(t))$$

for every $\alpha \in \tilde{F}$ implies the extension of ord on $F(t)$ to $\tilde{F}(t)$ is unique.

Moreover, we have

$$\text{ord}\,(\alpha - t) = \text{ord}\left(\sum_{\epsilon\varphi} a_{\epsilon\varphi}\gamma_\epsilon\omega_\varphi - t\right) = \text{ord}\left(\sum b_{\epsilon\varphi}\gamma_\epsilon\omega_\varphi\right),$$

where $b_{11} = a_{11} - t$ and $b_{\epsilon\varphi} = a_{\epsilon\varphi}$ for $(\epsilon, \varphi) \neq (1, 1)$, and since the $\gamma_\epsilon\omega_\varphi$ are linearly independent over $F(t)$, we have by Proposition 7 that

$$\text{ord}\,(\alpha - t) = \min_{\varphi,\epsilon} \text{ord}\,b_{\epsilon\varphi}\gamma_\epsilon\omega_\varphi \leq \text{ord}\,b_{11}\gamma_1\omega_1 = \text{ord}\,(a_{11} - t).$$

So take $a = a_{11}$.

PROPOSITION 18. *Suppose further that* $\text{ord}\,F(t)^* \supsetneq \text{ord}\,F^*$. *Then there exists* $b \in F$ *such that for* $s = t - b$ *we have* $\text{ord}\,s \notin \text{ord}\,F^*$ *and* $\text{ord} \sum_{i=1}^n a_i s^i = \min_i (\text{ord}\,a_i + i\,\text{ord}\,s)$, *where* $a_i \in F$.

PROOF. By hypothesis there exists $H(X) \in F[X]$ such that $\operatorname{ord} H(t) \notin \operatorname{ord} F^*$. We may take H to be monic irreducible. Then in $\operatorname{ord} \tilde{F}(t)^*$, $\operatorname{ord} H(t) = \operatorname{ord} \prod_{i=1}^{h} (t - \alpha_i)$ for some $\alpha_i \in \tilde{F}$. Thus $\operatorname{ord} H(t) = h \operatorname{ord} (t - \alpha_1)$ by the uniqueness of the extension of ord to $\tilde{F}(t)$. Hence $n \operatorname{ord} (t - \alpha_1) \notin \operatorname{ord} F^*$ for any $n \in \mathbf{Z}_{>0}$, since $\operatorname{ord} F^*$ is pure in $\operatorname{ord} F(t)^*$.

By Proposition 17, $\exists b \in F$ such that $\operatorname{ord} (t - b) \geq \operatorname{ord} (t - \alpha_1)$. If $\operatorname{ord} (t - b) > \operatorname{ord} (t - \alpha_1)$, then $\operatorname{ord} (\alpha_1 - b) = \operatorname{ord} ((t - b) - (t - \alpha_1)) = \operatorname{ord} (t - \alpha_1)$ which is a contradiction, since $e(F(\alpha_1)/F) \operatorname{ord} (\alpha_1 - b) \in \operatorname{ord} F^*$. Therefore $\operatorname{ord} (t - b) = \operatorname{ord} (t - \alpha_1) \notin \operatorname{ord} F^*$. So we have proved the first statement of the proposition.

For the second, observe that $\operatorname{ord} a_i + i \operatorname{ord} s \neq \operatorname{ord} a_j + j \operatorname{ord} s$ for $i \neq j$, for otherwise $(i - j) \operatorname{ord} s = \operatorname{ord} a_j - \operatorname{ord} a_i \in \operatorname{ord} F^*$.

DEFINITION 4. Let F be a valued field and λ a limit ordinal, then a *λ-pseudo-Cauchy sequence* in F is a set $\{f_\rho \mid \rho < \lambda\}$ such that $\operatorname{ord} (f_\tau - f_\rho) > \operatorname{ord} (f_\rho - f_\sigma)$ whenever $\sigma < \rho < \tau < \lambda$.

REMARK. If $\{f_\rho\}_{\rho<\lambda}$ is pseudo-Cauchy, then $\operatorname{ord} (f_\rho - f_\sigma) = \operatorname{ord} (f_{\sigma+1} - f_\sigma)$ for any $\rho > \sigma$ (since $\operatorname{ord} (f_\rho - f_\sigma) = \operatorname{ord} ((f_\rho - f_{\sigma+1}) + (f_{\sigma+1} - f_\sigma))$).

DEFINITION 5. If $\{f_\rho\}_{\rho<\lambda}$ is a pseudo-Cauchy sequence, $f \in F$ is said to be a *pseudo-limit* of $\{f_\rho\}_{\rho<\lambda} \Leftrightarrow \operatorname{ord} (f - f_\rho) = \operatorname{ord} (f_{\rho+1} - f_\rho)$ for every $\rho < \lambda$.

PROPOSITION 19. *Let $F(t)/F$ be an immediate extension. Then there exists a pseudo-Cauchy sequence $\{f_\rho\}_{\rho<\lambda}$ in F with t as a pseudo-limit but with no pseudo-limits in F.*

If in addition, F is algebraically complete, then for any such sequence $\{f_\rho\}_{\rho<\lambda}$ and any $H(X) \in F[X]$, $\operatorname{ord} H(f_\rho)$ *is eventually constant and we have* $\operatorname{ord} H(t) =$ *eventual* $\operatorname{ord} H(f_\rho)$.

PROOF. Let $R = \{\operatorname{ord} (t - f) \mid f \in F\}$. We assert that R does not have a largest element. Indeed, let $\operatorname{ord} (t - f) \in R$. Since $F(t)/F$ is immediate, $\exists g \in F$ such that $\operatorname{ord} (t - f) = \operatorname{ord} g$ and $\exists h \in F$ such that $[(t - f)g^{-1}]^- = \bar{h}$. Then we have $\operatorname{ord} ((t - f)g^{-1} - h) > 0$ or $\operatorname{ord} (t - (f + gh)) > \operatorname{ord} g + \operatorname{ord} (t - f)$. Hence there exists a well-ordered subset of R, $\{\operatorname{ord} (t - f_\rho) \mid \rho < \lambda\}$ of order type a limit ordinal λ such that $\operatorname{ord} (t - f_\rho) > \operatorname{ord} (t - f_\sigma)$ for $\lambda > \rho$ and such that for any $f \in F$, $\exists \rho_f < \lambda$ such that $\operatorname{ord} (t - f) < \operatorname{ord} (t - f_\rho)$ whenever $\lambda > \rho > \rho_f$. Then we have $\operatorname{ord} (f_\sigma - f_\rho) = \operatorname{ord} ((t - f_\rho) - (t - f_\rho)) = \operatorname{ord} (t - f_\rho)$ for all $\rho < \sigma < \lambda$. Thus $\{f_\rho\}_{\rho<\lambda}$ is pseudo-Cauchy and t is a pseudo-limit. If $f \in F$ is a pseudo-limit of $\{f_\rho\}_{\rho<\lambda}$ then $\operatorname{ord} (f - f_\rho) = \operatorname{ord} (f_{\rho+1} - f_\rho) = \operatorname{ord} (t - f_\rho)$, so $\operatorname{ord} (t - f) = \operatorname{ord} ((t - f_\rho) + (f_\rho - f)) \geq \operatorname{ord} (t - f_\rho)$ for all $\rho < \lambda$, a contradiction.

Now assume F is algebraically complete and let $H(X) \in F[X]$. Since ord on $F(t)$ extends uniquely to $\tilde{F}(t)$ it is clear that it suffices to prove the last statement of the proposition for $H(X)$ of the form $(Z - \alpha)$ where $\alpha \in \tilde{F}$. By Proposition 17, $\exists a \in F$ such that $\operatorname{ord} (t - a) \geq \operatorname{ord} (t - \alpha)$. Also we know $\exists \rho_0 < \lambda$ such that $\operatorname{ord} (t - f) > \operatorname{ord} (t - a) \geq \operatorname{ord} (t - \alpha)$ whenever $\rho_0 < \rho < \lambda$.

Hence $\operatorname{ord} H(t) = \operatorname{ord} (t - \alpha) = \operatorname{ord} ((t - f_\rho) + (f_\rho - \alpha)) = \operatorname{ord} (f_\rho - \alpha)$ for all $\rho > \rho_0$.

7. Proof of the isomorphism theorem

DEFINITION 1. Let $\aleph$ be an infinite cardinal. F, a valued field, is said to be $\aleph$-*pseudo-complete* $\Leftrightarrow$ every pseudo-Cauchy sequence of length $\lambda \leq \aleph$ has a pseudo-limit in F.

DEFINITION 2. A cross-section of a valued field (F, ord) is a homomorphism π: ord $F^* \to F^*$ such that ord $\circ\, \pi = 1_{\text{ord}\, F^*}$.

Our aim is to prove the following theorem.

THEOREM A. *Let V and V' be Henselian G-valued fields with valuations* ord *and* ord$'$ *respectively and cross-sections π and π'. Assume $\bar{V} \cong \bar{V}' = C$ and that* char $C = 0$. *Suppose further that* $\dim V/C = \dim V'/C = \aleph$ *where $\aleph$ is an infinite cardinal $> \aleph_0$, (= the smallest infinite cardinal) and that V and V' are $\aleph$-pseudo-complete. Then V is value-isomorphic to V'.*

REMARK. Since V is Henselian and char $C = 0$ we may assume $C \subseteq V$. We choose a particular embedding of C in V and then we assume $\aleph = \dim V/C$ (= transcendence degree of V over C).

DEFINITION 3. Let (V, ord) be a valued field with cross-section π. $a \in V$ is called a 1-unit if ord $a = 0$ and $\bar{a} = \bar{1}$. The product of 1-units is a 1-unit.

An element $a \in V$ is called normalized (with respect to π) if $a\pi(-\text{ord}\, a)$ is a 1-unit. Note that the product of normalized elements is normalized. Also if $C \subseteq F \subseteq V$ then for all $\gamma \in \text{ord}\, F^* \exists f \in F^*$ such that f is normalized and ord $f = \gamma$.

LEMMA 1. *Let V be a Henselian valued field with* char $V = 0$. *If $a \in V$ is a 1-unit, then for any $n \in \mathbf{Z}_{>0}$ there is a unique $b \in V$ such that $b^n = a$ and b is a 1-unit.*

PROOF. Consider $X^n - a$. In $\bar{V}[x]$ we have

$$X^n - \bar{a} = X^n - \bar{1} = (X - \bar{1})(X^{n-1} + X^{n-2} + \cdots + X + \bar{1}) = (X - \bar{1})u(X).$$

Since char $\bar{V} = 0$, $u(\bar{1}) = n \neq 0$. Thus $X - \bar{1}$ and $u(X)$ are relatively prime, so since V is Henselian there exists a unique $b \in V$ such that $b^n = a$ and $\bar{b} = \bar{1}$.

PROPOSITION 20. *Let V be a Henselian G-valued field with $\bar{V} = \bar{C}$,* ord $(C^*) = 0$, *and* char $C = 0$. *Suppose we have $C \subseteq F \subseteq V$. Then F is relatively algebraically closed in $V \Leftrightarrow$* ord F^* *is pure in G and F is Henselian.*

PROOF. ($\Rightarrow$) By Proposition 12, F is Henselian. Let $\gamma \in G$, $n \in \mathbf{Z}_{>0}$ and assume $n\gamma \in \text{ord}\, F^*$. Let $g \in V$ be such that ord $g = \gamma$ and let $f \in F$ be such that ord $f = n\gamma$. Then ord $fg^{-n} = 0$ so $\exists c \in C \subseteq F$ such that $fg^{-n}c$ is a 1-unit. By Lemma 1 $\exists w \in V$ such that $w = 1$ and $w^n = fg^{-n}c$. Thus we have $(gw)^n = fc \in F$, and since F is relatively algebraically closed in V, $gw \in F$. Therefore $\gamma = \text{ord}\, g = \text{ord}\, gw \in \text{ord}\, F^*$. ($\Leftarrow$) Let $\tilde{F}^r$ be the relative algebraic closure of F in V. Then $C \subseteq \bar{\tilde{F}}^r \subseteq \bar{V} = C$, and ord $\tilde{F}^*/\text{ord}\, F^*$ is torsion so ord $\tilde{F}^{r*} = \text{ord}\, F^*$. Thus $\tilde{F}^r$ is an immediate algebraic extension of F so since char $C = 0$ and F is Henselian, $\tilde{F}^r = F$.

LEMMA 2. *Let F be valued by* ord *and let $C \subseteq F$ be such that* $\operatorname{ord}(C^*) = 0$. *Then* $\dim(F/C) \geq \operatorname{rank}_{\mathbf{Z}} \operatorname{ord} F^*$.

PROOF. Let $\gamma_1, \ldots, \gamma_n \in \operatorname{ord} F^*$ be linearly independent over $\mathbf{Z}$ and let $f_\nu \in F^*$ be such that $\operatorname{ord} f_\nu = \gamma_\nu$. It suffices to show that the f_ν are algebraically independent over C. Suppose $H \in C[X_1, \ldots, X_2]$, say $H = \sum_j a_{j_1 \ldots j_n} X_1^{j_1} \ldots X_n^{j_n}$. Then $\operatorname{ord} H(f_1, \ldots, f_n) = \min_j \sum_{\nu=1}^n j_\nu \gamma_\nu$ $(\neq \infty)$ because the $\sum_{\nu=1}^n j_\nu \gamma_\nu$ are distinct. This completes the proof.

We will now prove two propositions which will constitute the induction step in the proof of the theorem.

PROPOSITION 21. *Let V, V' and C be as in Theorem* A. *Suppose we have F such that $C \subseteq F \subseteq V$ and $C \subseteq F \subseteq V'$, F is relatively algebraically closed in V and V', and* $\operatorname{ord} | F = \operatorname{ord}' | F$. *Suppose further that* $\dim(F/C) < \aleph$ *and that* $\exists t \in V - F$ *such that* $\operatorname{ord} F(t)^* = \operatorname{ord} F^*$. *Then there is an* ord-*F-monomorphism* $\rho \colon \tilde{F}(t)^r \to V'$, *where $\tilde{F}(t)^r$ is the relative algebraic closure of $F(t)$ in V. Where we use* ord-*monomorphism to mean that ρ is a monomorphism with* $\operatorname{ord}' \circ \rho = \operatorname{ord}$.

PROOF. By Proposition 19 there exists a pseudo-Cauchy sequence $\{f_\sigma\}_{\sigma<\tau}$ with t as a pseudo-limit and no pseudo-limits in F. Clearly $\tau \leq \#(\operatorname{ord} F^*)$. We claim $\#(\operatorname{ord} F^*) < \aleph$. Indeed by Lemma 2, $\# \operatorname{ord} F^* \leq \aleph_0 \operatorname{rank}_{\mathbf{Z}} \operatorname{ord} F^* \leq \aleph_0 \dim(F/C) < \aleph_0 \aleph = \aleph$.

Since V' is $\aleph$-pseudo-complete $\{f_\sigma\}_{\sigma<\tau}$ has a pseudo-limit $t' \in V'$. Define an F-isomorphism $\lambda \colon F(t) \to F(t')$ by $\lambda(t) = t'$. Then for any $H \in F[X]$, $\operatorname{ord} H(t) =$ eventual $\operatorname{ord} H(f_\sigma) =$ eventual $\operatorname{ord}' H(f_\sigma) = \operatorname{ord}' H(t)$. So λ is an ord-F-isomorphism. Now $\tilde{F}(t)^r/F(t)$ is immediate algebraic (since $\operatorname{ord} F(t)^* = \operatorname{ord} F^*$ is pure in G by Proposition 20), and hence $\tilde{F}(t)^r$ is a Henselization of $F(t)$ (by the corollary to Proposition 15).

Similarly, the relative algebraic closure of $F(t')$ in V' (denoted $\tilde{F}(t')^r$) is a Henselization of $F(t')$. Therefore λ extends to an ord-F-isomorphism $\rho \colon \tilde{F}(t)^r \to \tilde{F}(t')^r \subset V'$.

PROPOSITION 22. *Let V, V', C be as in Theorem* A, *and suppose we have F such that $C \subseteq F \subseteq V$ and $C \subseteq F \subseteq V'$, F is relatively algebraically closed in V and V' and* $\operatorname{ord} | F = \operatorname{ord}' | F$. *Suppose $t \in V - F$ such that* $\operatorname{ord} t = \tau \notin \operatorname{ord} F^* = A$ *and t is normalized. Assume that for any $\alpha \in A$ $\exists f \in F$ such that* $\operatorname{ord}(f\pi(-\alpha) - 1) > 0$, $\operatorname{ord}'(f\pi'(-\alpha) - 1) > 0$. *Then $\exists$ an* ord-*F-monomorphism* $\rho \colon \tilde{F}(t)^r \to V'$ *such that for any* $\beta \in \operatorname{ord} \tilde{F}(t)^{r*}$ $\exists b \in \tilde{F}(t)^r$ *such that* $\operatorname{ord}(b\pi(-\beta) - 1) > 0$ *and* $\operatorname{ord}'(\rho(b)\pi'(-\beta) - 1) > 0$.

PROOF. We do not have, as in Proposition 21, that $\operatorname{ord} F(t)^* (= A + \mathbf{Z}\tau)$ is pure in G. Our aim therefore will be to construct an algebraic extension Λ of $F(t)$ such that $\operatorname{ord} \Lambda^*$ is pure in G. If we can define an ord-F-morphism: $\Lambda \to V'$ then we can extend it to $F(t)$ by the same argument as in Proposition 21. To construct Λ, let $S = \{n \in \mathbf{Z}_{>0} \mid \exists \gamma \in G - A,\ \alpha \in A \text{ with } \tau = n\gamma + \alpha\}$. For $n \in S$, pick $\gamma_n \in G - A$, $\alpha_n \in A$ such that $\tau = n\gamma_n + \alpha_n$. Choose $a_n \in F$ such that $\operatorname{ord}(a_n \pi(-\alpha_n) - 1) > 0$ and $\operatorname{ord}'(a_n \pi'(-\alpha_n) - 1) > 0$. Choose $t' \in V'$

such that $\operatorname{ord}'(t'\pi'(-\tau) - 1) > 0$ (for example, take $t' = \pi'(\tau)$). We must have $b_n \in \Lambda$ such that $\operatorname{ord} b_n = \gamma_n$ and b_n is algebraic over $F(t)$. The natural choice for b_n would be $(t_n\pi(-\alpha_n))^{1/n}$, but we do not know that $\pi(\alpha_n)$ is in F. However we do have $a_n \in F$ which is just as good for our purposes. So we proceed in a series of steps as follows:

(1) $\exists/b_n \in V$, $b_n' \in V'$ such that $t/a_n = b_n^n$, $t'/a_n = b_n'^n$ and $\operatorname{ord}(b_n\pi(-\gamma_n) - 1) > 0$, $\operatorname{ord}'(b_n'\pi'(-\gamma_n) - 1) > 0$.

PROOF. $t/a_n\pi(\gamma_n)^n$ is a 1-unit since

$$\begin{aligned}\operatorname{ord}(t/a_n\pi(\gamma_n)^n - 1) &= \operatorname{ord}((t\pi(\tau)\pi(\alpha_n))/(\pi(\tau)a_n\pi(\alpha_n)\pi(\gamma_n)^n) - 1)\\ &= \operatorname{ord}((t/\pi(\tau))(\pi(\alpha_n)/a_n) - 1) > 0\end{aligned}$$

(since $t/\pi(\tau)$ and $\pi(\alpha_n)/a_n$ are 1-units). Hence $\exists \mid c_n \in V$ such that $c_n^n = t/a_n\pi(\gamma_n)^n$ and $\operatorname{ord}(c_n - 1) > 0$. Then set $b_n = \pi(\gamma_n)c_n$. Then $b_n^n = t/a_n$, $\operatorname{ord}(b_n\pi(-\gamma_n) - 1) = \operatorname{ord}(c_n - 1) > 0$, and the uniqueness of c_n implies the uniqueness of b_n. We construct b_n' similarly.

(2) If $m, n \in S$ are such that $n = mk$, then

$$b_m = b_n^k r_{m,n}, \qquad b_m' = b_n'^k r_{m,n}, \quad \text{with } r_{m,n} \in F.$$

PROOF. Set $r = r_{m,n} = b_m/b_n^k \in V^*$. Then $r^m = (b_m/b_n^k)^m = b_m^m/b_n^n = (t/a_m)/(t/a^n) = a_n/a_m \in F$. Since F is relatively algebraically closed in V, $r \in F$. Moreover, since b_m and b_n are normalized so is r. Thus r is the unique normalized mth root of a_n/a_m in F. (We have uniqueness since $r/\pi(k\gamma_n - \gamma_m)$ is a normalized mth root of the 1-unit $a_n/(a_m\pi(n\gamma_n - m\gamma_m))$, which is unique by Lemma 1.) Since $b_m'/b_n'^k$ is also a normalized mth root of a_n/a_m, $b_m'/b_n'^k = r$.

(3) If m divides n, then $F(b_m) \subseteq F(b_n)$. This is immediate from (2).

(4) For $n \in S$, define an F-isomorphism $\varphi_n\colon F(b_n) \to F(b_n')$ by $\varphi_n(b_n) = b_n'$. Then φ_n is an ord-isomorphism, and if $m \mid n$, then $\varphi_n \mid F(b_m) = \varphi_m$.

PROOF. Since $\operatorname{ord} b_n = \operatorname{ord}' b_n' = \gamma_n \notin A$, Proposition 18 implies that φ_n is an ord-isomorphism. Furthermore

$$\varphi_n(b_m) = \varphi_n(b_n^k r_{m,n}) = \varphi_n(b_n)^k r_{m,n} = b_n'^k r_{m,n} = b_m' = \varphi_m(b_m).$$

(5) Set $\Lambda = F(\{b_n \mid n \in S\})$, $\Lambda' = F(\{b_n' \mid n \in S\})$. Define $\varphi\colon \Lambda \to \Lambda'$ by $\varphi \mid F(b_n) = \varphi_n$. Then φ is an ord-F-isomorphism and $\operatorname{ord}\Lambda^* = B$; the divisible closure of $B = \operatorname{ord} F(t)^*$ in G. Moreover, for all $\beta \in B'$ $\exists l \in \Lambda$ such that $\operatorname{ord}(l\pi(-\beta) - 1) > 0$ and $\operatorname{ord}'(\varphi(l)\pi'(-\beta) - 1) > 0$.

PROOF. To show φ is well defined it suffices to show that $m, n \in S \Rightarrow [m, n] \in S$. Here $[m, n]$ is the least common multiple of m and n. Suppose therefore that $\tau = m\gamma_m + \alpha_m$ and $\tau = n\gamma_n + \alpha_n$. Let $d = (m, n)$, the greatest common divisor of m and n. Then $m = dr$, $n = ds$, $(r, s) = 1$ and $[m, n] = drs$. Since $(r, s) = 1$, $\exists u, v \in \mathbf{Z}$ such that $\mu r + \nu s = 1$. Then

$$\begin{aligned}\tau = (\mu r + \nu s)\tau = \mu r\tau + \nu st &= \mu\, drs\gamma_n + \nu\, drs\gamma_m + \mu r\alpha_n + \nu s\alpha_m\\ &= [m, n](\mu\gamma_n + \nu\gamma_m) + (\mu r\alpha_n + \nu s\alpha_m).\end{aligned}$$

Now let $\beta \in B$, the divisible closure of B. Then $\exists n \in \mathbf{Z}_{>0}$ such that $n\beta \in B$, so $\exists m \in \mathbf{Z}$ and $\alpha \in A$ such that $n\beta = m\tau + \alpha$. If $m = 0$, then $\beta \in A$ since A is pure in G. So we can assume $m > 0$. Let $d = (m, n)$. Then $\alpha \in dG$ and since A is pure in G, $\alpha = d\alpha'$ for some $\alpha' \in A$. Thus $\beta(n/d) = \tau(m/d) + \alpha'$, so we may assume $(m, n) = 1$. Then we have $\mu n\beta = \mu m\tau + \mu\alpha = (1 - \nu n)\tau + \mu\alpha = \tau - \nu n\tau + \mu\alpha$, or $\tau = n(\mu\beta + \nu\tau) - \mu\alpha$, so $n \in S$. Therefore $n\gamma_n + \alpha_n = n(\mu\beta + \nu\tau) - \mu\alpha$, whence $\mu\alpha - \alpha_n \in nG \cap A$. Thus $\exists\alpha'' \in A$ such that $\mu\alpha - \alpha_n = n\alpha''$, so $\gamma_n = \mu\beta - \nu\tau - \alpha''$. Therefore, we have $\mu\beta \in \operatorname{ord} \Lambda^*$. Similarly $\nu\beta \in \operatorname{ord} \Lambda'$, so $\beta \in \operatorname{ord} \Lambda^*$.

To prove the last statement of (5) it suffices to consider β of the form $m\gamma_s + \alpha$ with $m \in \mathbf{Z}$, $s \in S$ and $\alpha \in A$. Let $a \in F$ be normalized and such that $\operatorname{ord} a = \alpha$. Then $\operatorname{ord}(b_s^m a\pi(-\beta) - 1) > 0$ and $\operatorname{ord}'(b_s'^m a\pi'(-\beta) - 1) = \operatorname{ord}'(a(b_s^m - a) \times \pi'(-\beta) - 1) > 0$ since the product of normalized elements is normalized.

(6) φ extends to an ord-F-monomorphism ρ: $\tilde{F}(t)^r \to V'$, such that $\forall\beta \in \operatorname{ord} \tilde{F}(t)^{r*} \exists b \in \tilde{F}(t)^r$ such that $\operatorname{ord}(b\pi(-\beta) - 1) > 0$ and $\operatorname{ord}'(b\pi'(-\beta) - 1) > 0$.

Proof. $\tilde{F}(t)^r$ (resp. $\tilde{F}(t')^r$) is an immediate algebraic Henselian extension of Λ (resp. Λ'), and therefore a Henselization of Λ (resp. Λ') since char $C = 0$. Hence φ extends. Furthermore since $\operatorname{ord} \tilde{F}(t)^{r*} = B'$, we can choose $b \in \Lambda$ as in (5).

This completes the proof of Proposition 22.

Definition 4. Let V and V' be G-valued fields with cross-sections π and π' respectively. Let $F \subseteq V$, $F' \subseteq V'$. Then an ord-isomorphism φ: $F \to F'$ is a norm-isomorphism $\Leftrightarrow$ for every $\alpha \in \operatorname{ord} F^* \exists f \in F$ such that $\operatorname{ord}(f\pi(-\alpha) - 1) > 0$ and $\operatorname{ord}'(\varphi(f)\pi'(-\alpha) - 1) > 0$, i.e. f normalized implies $\varphi(f)$ normalized.

We are now ready to begin the

Proof of Theorem A. Let $B = \{t_\lambda \mid 1 \leq \lambda < \aleph\}$, $B' = \{t'_\lambda \mid 1 \leq \lambda < \aleph\}$ be transcendence bases for V and V' respectively, well ordered by the ordinals less than $\aleph$.

We will define inductively for every $\lambda < \aleph$ relatively algebraically closed normalized subfields $F_\lambda \subset V$, $F'_\lambda \subset V'$ and norm-isomorphisms φ_λ: $F_\lambda \to F'_\lambda$ satisfying the following conditions:

(i) $F_0 = F'_0 = C$; $\varphi_0 =$ identity on C.

(ii) If $\rho < \lambda$, then $F_\rho \subseteq F_\lambda$ and $\varphi_\rho = \varphi_\lambda \mid F_\rho$.

(iii) (a) If λ is even (i.e. λ is a positive even integer or a limit ordinal plus a positive even integer), then $F_\lambda = F_{\lambda-1}(t_\nu)$, where $\nu = \nu(\lambda) =$ smallest ordinal $\nu < \aleph$ such that $t_\nu \notin F_{\lambda-1}$.

(iii) (b) If λ is odd, then $F'_\lambda = F'_{\lambda-1}(t'_{\nu'})$ where $\nu' = \nu'(\lambda) =$ smallest ordinal $\nu' < \aleph$ such that $t'_{\nu'} \notin F'_{\lambda-1}$.

(iii) (c) If λ is a limit ordinal, then $F_\lambda = \bigcup_{\rho>\lambda} F_\rho$; φ_λ is such that $\varphi_\lambda \mid F_\rho = \varphi_\rho$.

Suppose now that $0 < \sigma < \aleph$, and we have defined F_λ, F'_λ, φ_λ for $\lambda < \sigma$ satisfying the above conditions. We see inductively that $\dim F_\lambda/C = \dim F'_\lambda/C = \#\lambda$.

Case 1: σ is a limit ordinal. Let $F_\sigma = \bigcup_{\lambda<\sigma} F_\lambda$; define φ_σ such that $\varphi_\sigma \mid F_\lambda = \varphi_\lambda$. Since the F_λ are relatively algebraically closed and normalized and the φ_λ are norm-isomorphisms, F_σ and φ_σ have the same properties.

Case 2: σ is even. Choose $\nu = \nu(\sigma)$ to be the smallest ordinal $\nu < \aleph$ such that $t_\nu \notin F_{\sigma-1}$. Let $F_\sigma = \tilde{F}_{\sigma-1}(t_\nu)^r$. Assume first that $F_\sigma/F_{\sigma-1}$ is immediate, i.e.

ord $F_\sigma^* = \text{ord } F_{\sigma-1}^*$. By Proposition 21, $\varphi_{\sigma-1}$ extends to an ord-monomorphism $\varphi_\sigma: F_\sigma \to V'$. Since $F_\sigma/F_{\sigma-1}$ is immediate and $\varphi_{\sigma-1}$ is a norm-isomorphism, φ_σ is also a norm-isomorphism. The definition of φ_σ in the proof of Proposition 21 shows that $F_\sigma' = \varphi_\sigma(F_\sigma)$ is relatively algebraically closed in V'. Thus F_σ, F_σ' and φ_σ satisfy our conditions.

Now suppose $F_\sigma/F_{\sigma-1}$ is not immediate. Then ord $F_\sigma^*/\text{ord } F_{\sigma-1}(t_\nu)^*$ is torsion, and ord $F_{\sigma-1}^*$ is pure in G (Proposition 20), so ord $F_{\sigma-1}^* \underset{\neq}{\subset} \text{ord } F_{\sigma-1}(t_\nu)^*$. Then by Proposition 18, there exists $b \in F_{\sigma-1}$ such that ord $(t_\nu - b) \notin \text{ord } F_{\sigma-1}^*$. Hence we can choose $c \in C^* \subseteq F_{\sigma-1}^*$ such that ord $(c(t_\nu - b)) \notin \text{ord } F_{\sigma-1}^*$ and $c(t_\nu - b)$ is normalized. (Choose $\bar{c}^{-1} = [(t_\nu - b)\pi(-\text{ord }(t_\nu - b)]^-$.) Thus we are in a position to apply Proposition 22 and get an extension of $\varphi_{\sigma-1}$ to a norm-isomorphism $\varphi_\sigma: F_\sigma \to V$. We set $F_\sigma' = \varphi_\sigma(F_\sigma)$ and note as before that our conditions are satisfied.

Case 3: σ is odd. Choose $\nu' = \nu'(\sigma)$ to be the smallest ordinal $\nu' < \aleph$ such that $t_{\nu'}' \notin F_{\sigma-1}'$ and set $F_\sigma' = \tilde{F}_{\sigma-1}'(t_{\nu'}')^r$. Proceed as in Case 2 to get a norm-isomorphism $\psi: F_\sigma \to V$. Then set $F_\sigma = \psi(F_\sigma')$ and $\varphi_\sigma = \psi^{-1}$.

Thus we have proved the existence of F_λ, F_λ' and φ_λ for all $\lambda < \aleph$. Let $W = \bigcup_{\lambda<\aleph} F_\lambda$. We claim that $B \subseteq W$. Indeed, suppose there exists $\mu < \aleph$ such that $t_\mu \notin W$. Then ν, as defined by (iii) (a) has domain, D, the even ordinals $< \aleph$, and image $\subseteq R = \{\alpha \mid \alpha < \mu\}$; but since ν is one-one we have $\aleph = \#D \leq \#R \leq \#\mu < \aleph$, a contradiction. Therefore $B \subseteq W$ and so V/W is algebraic. But W is relatively algebraically closed in V since the F_λ are, so we have $W = V$. Similarly, we prove that $V' = \bigcup_{\lambda<\aleph} F_\lambda'$. Then if we define $\varphi: V \to V'$ by $\varphi: F_\lambda = \varphi_\lambda$, φ is an ord-isomorphism, and the proof of Theorem A is complete.

REMARK. $\aleph = \dim (V/C) = \#V$, and, in particular, is independent of which maximal trivially valued subfield C is chosen, providing $G \neq 0$.

PROOF. Since C is infinite (char $C = 0$), it suffices to prove that $\dim V/C \geq$ Card C. Let $\gamma \in G - \{0\}$ such that $\gamma > 0$. Let $C((t))$ be a formal power series field over C valued such that ord $(t) = \gamma$ and C is trivially valued. Then by a one-sided version of the proof of Theorem 1 we get an ord-monomorphism: $C((t)) \to V$.

Hence it suffices to show $\dim C((t))/C \geq$ Card C. Since we are given $\aleph > \aleph_0$, we may assume $\#C > \aleph_0$. Thus $\#C = [C: \mathbf{Q}]$, so we need to show $\dim C((t))/C \geq [C: \mathbf{Q}]$. But if $\gamma_1, \ldots, \gamma_n \in C$ are $\mathbf{Q}$-linearly-independent, we assert that $\exp(\gamma_i t) \in C((t))$, $i = 1, \ldots, n$ are algebraically independent over C. Indeed, suppose there exists

$$H(X_1, \ldots, X_n) = \sum_{j=(j_1,\ldots,j_n)} a_j X_1^{j_1}, \ldots, X_n^{j_n}$$

in $C[X_1, \ldots, X_n]$ such that $0 = H(\exp(\gamma_1 t), \ldots, \exp(\gamma_n t)) = \sum_{j\in J} a_j \exp((j: \gamma)t)$ (using vector notation). Then we have

$$0 = \sum_{j\in J} a_j \frac{(j \cdot \gamma)^m}{m!}, \qquad m = 0, 1, 2 \ldots .$$

But the $(j \cdot \gamma)$ are distinct for distinct j, since the γ_i are $\mathbf{Q}$-linearly-independent.

Hence

$$\det\left((j\gamma)^m\right)_{j\in J;m=0,\ldots,(\#J)-1} \neq 0 \quad \text{(a Vandermonde determinant)},$$

so we have $a_j = 0$ for all j.

This remark allows us to restate Theorem A in the following form which contains the Isomorphism Theorem of the Lecture.

THEOREM B. *Let V and V' be Henselian, $\aleph_1$-pseudo-complete valued fields of cardinality $\aleph_1$, having cross-sections and isomorphic value groups and isomorphic residue class fields of characteristic zero. Then V is value-isomorphic to V'.*

STATE UNIVERSITY OF NEW YORK, STONY BROOK
LONG ISLAND, NEW YORK

HILBERT'S TENTH PROBLEM

JULIA ROBINSON[1]

Ju. V. Matijasevič [4] has recently shown that Hilbert's tenth problem is unsolvable—there is no general method of telling whether a (polynomial) diophantine equation has a solution. Hilbert had asked for such a method. However in his address at the 1900 International Congress when he proposed his list of problems for the twentieth century, Hilbert spoke of "the conviction (which every mathematician shares, but which no one has yet supported by a proof) that every definite mathematical problem must necessarily be susceptible of an exact settlement, either in the form of an actual answer to the question asked, or by the proof of the impossibility of its solution and therewith the necessary failure of all attempts".[2] So Hilbert would be satisfied with Matijasevič's solution.

Matijasevič actually proved the stronger theorem: *Every recursively enumerable (listable) relation is diophantine.* A relation $R(m_1, \ldots, m_j)$ is *diophantine* if there is a polynomial $P(x_1, \ldots, x_j, y_1, \ldots, y_k)$ with integer coefficients such that $R(m_1, \ldots, m_j)$ holds if and only if $P(m_1, \ldots, m_j, y_1, \ldots, y_k) = 0$ has a solution for $y_1, \ldots, y_k$ in natural numbers.

Since there are recursively enumerable sets which are not recursive (computable), Hilbert's problem is unsolvable. For suppose there is a general method for telling whether a diophantine equation has a solution. Let S be a recursively enumerable set which is not recursive. Choose P so that m belongs to S if and only if $P(m, y_1, \ldots, y_k) = 0$ has a solution for $y_1, \ldots, y_k$. For each m, we could then tell whether m belongs to S by using the assumed method to tell whether the

[1]My lectures at Stony Brook were an expository account of earlier work looking towards a negative solution of Hilbert's tenth problem. They are covered by the article *Diophantine decision problems* in the MAA volume, Studies in number theory (referred to here as Studies). The original paper submitted to these Proceedings was a short review written with the hope of stimulating number theorists to work on the question of the existence of a "roughly exponential" diophantine relation R (see I below). I learned with great delight of Matijasevič's ingenious answer in February, 1970. These notes are minor comments on the consequences of Matijasevič's theorem. There is also an appendix containing a simpler proof of one of the results in Studies.

[2]For an account of the circumstances of Hilbert's address, see Reid [5].

corresponding equation has a solution. But this is impossible, since S is not computable.

BACKGROUND. Matijasevič's proof is based on two earlier results:

I. (J. Robinson, 1952, [7]). The relation given by $r = s^t$ is diophantine if there is a diophantine relation $R(u, v)$ and an exponential function $E(u)$ such that $R(u, v)$ implies $v < E(u)$ but there is no polynomial for which $R(u, v)$ implies $v < P(u)$.

II. (M. Davis, H. Putnam, and J. Robinson, 1961, [2]). Every recursively enumerable relation is exponential diophantine. (A relation is *exponential diophantine* if it can be defined by a diophantine equation in which the variables may occur as exponents in the same way as a diophantine relation is defined by a polynomial diophantine equation.)

Matijasevič has found a diophantine relation $R(u, v)$ satisfying I. Namely,

III. (Matijasevič, 1970). The relation given by $v = \phi_{2u}$ is diophantine, where ϕ_k is the kth Fibonacci number.

Putting I and III together, we obtain that exponentiation is diophantine. Thus every exponential diophantine equation can be reduced to a polynomial diophantine equation in more variables. Hence Matijasevič's theorem follows by II. Proofs of I and II are in Studies. The entire proof uses only very elementary number theory—the binomial theorem, Chinese Remainder Theorem, solutions of Pell's equations, and Lucas' sequences.[3]

N. K. Kosovskiĭ, G. V. Čudnovskiĭ, Martin Davis, K. Schütte and Matijasevič have since given direct diophantine definitions of exponentiation based on Matijasevič's work.

The starting point for the proof of II was Gödel's theorem that every primitive recursive function is arithmetical. His proof showed more, namely, every recursively enumerable set S can be put in the form

$$(1) \qquad m \in S \leftrightarrow \mathbf{Q}_1 y_1 \ldots \mathbf{Q}_k y_k [P(m, y_1, \ldots, y_k) = 0]$$

where P is a polynomial with integer coefficients and each $\mathbf{Q}_i$ is either an existential quantifier or a bounded universal quantifier. Martin Davis [1] showed in 1950 that it was sufficient to have just one bounded universal quantifier and any number of existential quantifiers. This was essential for the proof of II. Indeed, the key lemma is that a bounded universal quantifier applied to a diophantine relation can be transformed into an exponential diophantine relation. Gödel's form is not enough, since we could not handle a bounded universal quantifier applied to an exponential diophantine relation. S. Kochen pointed out that now we can remove the bounded universal quantifiers one at a time, obtaining first an exponential diophantine relation and then converting it to a diophantine relation by Matijasevič's theorem. Hence §7 of Studies is no longer needed.

CONSEQUENCES. What are the practical consequences of Matijasevič's theorem to the business of solving diophantine equations? None at all. Because for any particular diophantine equation of reasonable size (the only ones we are interested in), we may still with ingenuity and perhaps new insight be able to tell if it has a solution. At least, Matijasevič has not destroyed our conviction that this is so. What we cannot do, is write out instructions once and for all and then turn over the testing of equations for solvability to a machine. We will have to add constantly to our tools as we take on new equations.

---

[3]Both the numbers a'_n in Studies and Matijasevič's $\psi_{m,n}$ are Lucas' sequences [3].

Still in a sense Hilbert's problem lives on. Hilbert asked for a method for telling whether an arbitrary diophantine equation has a solution and there is none. So now we ask for what classes of diophantine equations, is there a method, and seek to find it. For example, Alan Baker, these Proceedings, has made great advances on equations in two variables.

One of the corollaries to Matijasevič's theorem, which he mentions [**4**, p. 279] is the existence of a universal diophantine equation. This follows immediately by well-known properties of recursively enumerable sets. But since it is the most interesting purely number theoretical consequence, it seems worthwhile to give a direct proof. The standard argument is especially simple in this case. (See IV, p. 83, Studies for an intuitive argument.)

Number diophantine equations as on p. 79, Studies. Then, there is a polynomial $U(v, y_0, y_1, \ldots, y_k)$ with integer coefficients such that m belongs to the nth diophantine set D_n if and only if $U(n, m, y_1, \ldots, y_k) = 0$ has a solution for $y_1, \ldots, y_k$. Here $D_n = \{m$: the nth diophantine equation has a solution with $x_0 = m\}$. We call $U = 0$ a *universal diophantine equation*, since every diophantine set is obtained for some n. We will use the same notation as in §2 of Studies. Let $s_k = \text{Rem}\,(K(s), 1 + (1 + k)L(s))$ for all s and k. Notice that the relation given by $t = s_k$ is diophantine, since K, L, and $\text{Rem}\,(x, 1 + y)$ are. By Gödel's lemma, to every finite sequence of natural numbers $a, a', a'', \ldots, a^{(n)}$, there is an s such that $a^{(i)} = s_i$ for all $i \leqq n$. Number the terms built up from natural numbers and the variables $x_0, x_1, \ldots$ as follows:

$$\tau_{4n} = n, \qquad \tau_{4n+1} = x_n, \qquad \tau_{4n+2} = \tau_{Kn} + \tau_{Ln}, \qquad \tau_{4n+3} = \tau_{Kn} \cdot \tau_{Ln}.$$

Let the nth diophantine equation be $\tau_{Kn} = \tau_{Ln}$. Also let $\tau_n(s)$ be the value of τ_n when $s_0, s_1, \ldots$ are substituted for $x_0, x_1, \ldots$ respectively. We can then write

$$\begin{aligned} m \in D_n \leftrightarrow \bigvee_{s,t} \Big[\bigwedge_{k \leqq n} (t_{4k} = k \wedge t_{4k+1} = s_k \wedge t_{4k+2} \\ = t_{Kk} + t_{Lk} \wedge t_{4k+3} = t_{Kk} \cdot t_{Lk}) \wedge t_{Kn} = t_{Ln} \wedge m = s_0 \Big]. \end{aligned} \tag{2}$$

Notice that if t satisfies the expression in parentheses for all $k \leqq n$, then $t_j = \tau_j(s)$ for all $j \leqq 4n + 3$. Hence $t_{Kn} = t_{Ln}$ insures that $x_0 = s_0, x_1 = s_1, \ldots$ is a solution of the nth diophantine equation. On the other hand, if m belongs to D_n, then such t and s must exist by Gödel's lemma. Then the right side of (2) can be put in the Davis normal form, reduced to an exponential diophantine relation by the lemma on p. 103 of Studies, and then to a diophantine relation by Matijasevic's theorem. Clearly, each of these reductions can be carried out explicitly but the universal equation obtained would be very complicated.

If m is composite, there is a proof that m is composite consisting of one multiplication. Until now it was not known if there was a similar proof consisting of a bounded number of additions and multiplications to show that a prime p is prime. Let c be the number of additions and multiplications necessary to check that $U(n, m, y_1, \ldots, y_k) = 0$. Then if p is prime there is a proof that p is prime using just c additions and multiplications. The work of finding the right values of $y_1, y_2, \ldots$ to use in the computation is of course scratch work. The same c can be used for any recursively enumerable set.

OPEN PROBLEMS. I think the most exciting problem is to find some interesting bound on the number of variables needed in a universal diophantine equation.

A bound can be obtained by starting with the result of R. M. Robinson [8] that every recursively enumerable set can be put in the Davis form with just four inner existential quantifiers, reducing it to an exponential diophantine equation in more variables, and then to a diophantine equation in still more variables. A bound obtained this way would be too large to be interesting. We might define an interesting equation to be one written in the latin alphabet without subscripts.

Another natural problem is the decision problem for nontrivial solutions of homogeneous diophantine equations, or equivalently, for rational solutions of arbitrary diophantine equations. The only result in the negative direction is that the set of natural numbers is arithmetically definable in the field of rational numbers [6]. This implies that there is no general method of telling whether an arithmetical statement is true in the rational field.

APPENDIX. The result of Davis and Putnam that $m = \prod_{k=1}^{n} (a + bk)$ is exponential diophantine can be proved more simply than on pp. 98–101 of Studies. We may assume $b \neq 0$. Then $m = \prod_{k=1}^{n} (a + bk)$ if and only if there are t and x such that

$$(3) \quad t > (a + bn)^n, \qquad bx \equiv a \pmod{t}, \qquad \operatorname{Rem}\left(\binom{x+n}{n} n!b^n, t\right) = m.$$

If (3) holds, then

$$\binom{x+n}{n} n!b^n = (bx + b)(bx + 2b) \cdots (bx + nb)$$
$$\equiv (a + b)(a + 2b) \cdots (a + nb) \pmod{t}.$$

Since the product on the right is less than t, it must be the remainder m. On the other hand, suppose m, n, a, b satisfy the relation. Choose $t > (a + bn)^n$ and prime to b. Then there is an x such that $bx \equiv a \pmod{t}$. For this x and t, the third condition of (3) holds. Since all the relations of (3) are exponential diophantine (see pp. 98–100 of Studies), the result is obtained.

REFERENCES

1. Martin Davis, a) *On the theory of recursive unsolvability*, Doctoral Dissertation, Princeton University, Princeton, N.J., 1950; b) *Arithmetical problems and recursively enumerable predicates*, J. Symbolic Logic **18** (1953), 33–41. MR **14,** 1052.

2. Martin Davis, Hilary Putnam and Julia Robinson, *The decision problem for exponential diophantine equations*, Ann. of Math. (2) **74** (1961), 425–436. MR **24** #A3061.

3. E. A. Lucas, *Théorie des fonctions numérique simplement périodique*, Amer. J. Math. **1** (1878), 184–239, 289–321.

4. Ju. V. Matijasevič, *Enumerable sets are diophantine*, Dokl. Akad. Nauk SSSR **191** (1970), 279–282. = Soviet Math. Dokl. **11** (1970).

5. Constance Reid, *Hilbert. With an appreciation of Hilbert's mathematical work by H. Weyl*, Springer-Verlag, New York, 1970.

6. Julia Robinson, *Definability and decision problems in arithmetic*, J. Symbolic Logic **14** (1949), 98–114. MR **11,** 151.

7. ———, *Existential definability in arithmetic*, Trans. Amer. Math. Soc. **72** (1952), 437–449. MR **14,** 4.

8. R. M. Robinson, *Arithmetical representation of recursively enumerable sets*, J. Symbolic Logic **21** (1956), 162–186. MR **18,** 272.

UNIVERSITY OF CALIFORNIA
BERKELEY, CALIFORNIA

EFFECTIVE METHODS IN DIOPHANTINE PROBLEMS

A. BAKER

1. Introduction

These notes are intended to serve as a guide to the various effective results in the theory of numbers that have been obtained as a consequence of recent researches. Most of the theorems derive in some way from the author's papers on the logarithms of algebraic numbers and we shall begin with an account of this work. The reader will be referred to the original memoirs for proofs.

2. On the logarithms of algebraic numbers

At the International Congress of mathematicians held in Paris in 1900, Hilbert raised, as the seventh of his famous list of 23 problems, the question whether an irrational logarithm of an algebraic number to an algebraic base is transcendental. The question is capable of various alternative formulations; thus one can ask whether an irrational quotient of natural logarithms of algebraic numbers is transcendental, or whether α^β is transcendental for any algebraic number $\alpha \neq 0, 1$ and any algebraic irrational β. A special case relating to logarithms of rational numbers had been posed by Euler more than a century before but no apparent progress had been made towards its solution. Indeed Hilbert expressed the opinion that the resolution of the problem lay farther in the future than a proof of the Riemann hypothesis or Fermat's last theorem.

The first significant advance was made by Gelfond in 1929. Employing interpolation techniques of the kind that he had utilized previously in researches on integral integer-valued functions, Gelfond showed that the logarithm of an algebraic number to an algebraic base cannot be an imaginary quadratic irrational, that is, α^β is transcendental for any algebraic number $\alpha \neq 0, 1$ and any imaginary quadratic irrational β; in particular, one sees that $e^\pi = (-1)^{-i}$ is transcendental. The result was extended to real quadratic irrationals β by Kuzmin in 1930. But it was clear that direct appeal to an interpolation series for $e^{\beta z}$, on which the Gelfond-

Kuzmin work was based, was not appropriate for more general β, and further progress awaited a new idea. The search for the latter was concluded successfully by Gelfond and Schneider, independently, in 1934. The arguments they discovered were applicable for any irrational β and, though differing in detail, both depended on the construction of an auxiliary function that vanished at certain selected points. A similar technique had been used a few years earlier by Siegel in the course of his investigations on the Bessel functions. Herewith, Hilbert's seventh problem was finally solved.

The Gelfond-Schneider theorem shows that for any nonzero algebraic numbers $\alpha_1, \alpha_2, \beta_1, \beta_2$ with $\log \alpha_1, \log \alpha_2$ linearly independent over the rationals we have

$$\beta_1 \log \alpha_1 + \beta_2 \log \alpha_2 \neq 0.$$

It was natural to conjecture that an analogous theorem would hold for arbitrarily many logarithms of algebraic numbers and moreover it was soon realized that generalizations of this kind would have important consequences in number theory. But, for some thirty years, the problem of extension seemed resistant to attack. It was finally settled in 1966 [6], and the techniques devised for its solution have been the main instruments in establishing the various results described herein. The original theorem has been extended slightly to include also the case in which an additional nonzero algebraic number is present on the left and now reads as follows:

THEOREM 1 [7]. *If $\alpha_1, \ldots, \alpha_n$ denote nonzero algebraic numbers such that $\log \alpha_1, \ldots, \log \alpha_n$ are linearly independent over the rationals then $1, \log \alpha_1, \ldots, \log \alpha_n$ are linearly independent over the field of all algebraic numbers.*

Here $\log \alpha_1, \ldots, \log \alpha_n$ are any fixed determinations of the logarithms. The proof of the theorem depends on the construction of an auxiliary function of several complex variables which generalizes the original function of a single variable employed by Gelfond. The subsequent arguments, however, involve an extrapolation procedure that is special to the present context and for which there is no precise earlier counterpart. Quantitative extensions of Theorem 1 will be discussed in the next section and applications of the results to various branches of number theory will be the theme of §§4 to 6.

We record now a few immediate corollaries of Theorem 1.

THEOREM 2. *Any nonvanishing linear combination of logarithms of algebraic numbers with algebraic coefficients is transcendental.*

THEOREM 3. *$e^{\beta_0}\alpha_1^{\beta_1} \ldots \alpha_n^{\beta_n}$ is transcendental for any nonzero algebraic numbers $\alpha_1, \ldots, \alpha_n, \beta_0, \beta_1, \ldots, \beta_n$.*

THEOREM 4. *$\alpha_1^{\beta_1} \ldots \alpha_n^{\beta_n}$ is transcendental for any algebraic numbers $\alpha_1, \ldots, \alpha_n$, other than 0 or 1, and any algebraic numbers $\beta_1, \ldots, \beta_n$ with $1, \beta_1, \ldots, \beta_n$ linearly independent over the rationals.*

Particular cases of the above theorems show that $\pi + \log \alpha$ is transcendental for any algebraic number $\alpha \neq 0$, and $e^{\alpha\pi+\beta}$ is transcendental for any algebraic numbers α, β with $\beta \neq 0$. One might also mention an analogy with Lindemann's classical theorem; this asserts that

$$\beta_1 \exp \alpha_1 + \cdots + \beta_n \exp \alpha_n \neq 0$$

for any distinct algebraic numbers $\alpha_1, \ldots, \alpha_n$ and any nonzero algebraic numbers $\beta_1, \ldots, \beta_n$; Theorem 1 shows that the same holds with "exp" replaced by "log" provided that the logarithms are linearly independent over the rationals.

3. Lower bounds for linear forms

By the *height* of an algebraic number we shall mean the maximum of the absolute values of the relatively prime integer coefficients in its minimal defining polynomial. Soon after obtaining his solution to the seventh problem of Hilbert, Gelfond established an important refinement expressing a positive lower bound for a linear form in two logarithms; he proved that for any nonzero algebraic numbers α_1, α_2, β_1, β_2 with $\log \alpha_1$, $\log \alpha_2$ linearly independent over the rationals and any $\kappa > 5$ we have

$$|\beta_1 \log \alpha_1 + \beta_2 \log \alpha_2| > Ce^{-(\log H)^\kappa},$$

where H denotes the maximum of the heights of β_1, β_2, and $C > 0$ denotes a computable number depending only on $\log \alpha_1$, $\log \alpha_2$, and the degrees of β_1, β_2. Gelfond later improved the condition $\kappa > 5$ to $\kappa > 2$. Also he showed that, as a corollary to the Thue-Siegel theorem, about which we shall speak in §5, an inequality of the form

$$|b_1 \log \alpha_1 + \cdots + b_n \log \alpha_n| > Ce^{-\delta H}$$

holds, where $\delta > 0$ and $b_1, \ldots, b_n$ are rational integers, not all 0, with absolute values at most H; but here $C > 0$ could not be effectively computed.

After the demonstration of Theorem 1 it proved relatively easy to obtain extensions of Gelfond's inequalities relating to arbitrarily many logarithms of algebraic numbers, and indeed the following theorem was established [7]

THEOREM 5. *Let* $\alpha_1, \ldots, \alpha_n$ *denote nonzero algebraic numbers with* $\log \alpha_1, \ldots, \log \alpha_n$ *linearly independent over the rationals, and let* $\beta_0, \ldots, \beta_n$ *denote algebraic numbers, not all* 0, *with degrees and heights at most d and H respectively. Then, for any* $\kappa > n + 1$, *we have*

$$|\beta_0 + \beta_1 \log \alpha_1 + \cdots + \beta_n \log \alpha_n| > Ce^{-(\log H)^\kappa}$$

where $C > 0$ *denotes an effectively computable number depending only on n,* $\log \alpha_1, \ldots, \log \alpha_n$, *and d.*

Very recently the number on the right of the above inequality has been improved by Feldman to $C^{-1}H^{-\kappa}$, where $C > 0$, $\kappa > 0$ depend only on n, $\log \alpha_1, \ldots, \log \alpha_n$ and d [23], [24]. Estimates for C and κ have been explicitly calculated, but their values are large; the estimate for C takes the form $C' \exp[(\log A)^{\kappa'}]$ where A denotes the maximum of the heights of $\alpha_1, \ldots, \alpha_n$, κ' depends only on n, and C'

depends only on n, d and the degrees of $\alpha_1, \ldots, \alpha_n$. The value of C and, in particular, its dependence on A is of importance in applications; a more special result, but giving sharper estimates with respect to the parameters other than H, was recently established by the author:

THEOREM 6 [7]. *Suppose that $\alpha_1, \ldots, \alpha_n$ are $n \geq 2$ nonzero algebraic numbers and that the heights and degrees of $\alpha_1, \ldots, \alpha_n$ do not exceed integers A, d respectively, where $A \geq 4$, $d \geq 4$. Suppose further that $0 < \delta \leq 1$ and let $\log \alpha_1, \ldots, \log \alpha_n$ denote the principal values of the logarithms. If rational integers $b_1, \ldots, b_n$ exist with absolute values at most H such that*

$$0 < |b_1 \log \alpha_1 + \cdots + b_n \log \alpha_n| < e^{-\delta H}$$

then

$$H < (4^{n^2} \delta^{-1} d^{2n} \log A)^{(2n+1)^2}.$$

Theorem 6 will be the only transcendence result to which we shall refer directly in the sequel. It is useful for application to a wide class of Diophantine problems and yields estimates that will be found, in many cases, to be accessible to practical computation [17].

4. On imaginary quadratic fields with class number 1

In 1966 Stark [32] and the author [6], [12] showed independently how one could resolve the long-standing conjecture that the only imaginary quadratic fields $Q(\sqrt{-d})$ with class number 1 are those given by $d = 1, 2, 3, 7, 11, 19, 43, 67, 163$. Heilbronn and Linfoot had proved in 1934 that there could be at most ten such fields, and calculations had shown that the tenth field, if it existed, would satisfy $d > \exp(2 \cdot 2 \times 10^7)$. The work of Stark was motivated by an earlier paper of Heegner, and the work of the author was based on an idea of Gelfond and Linnik. We shall sketch briefly the latter method.

Let $-d < 0$ and $k > 0$ denote the discriminants of the quadratic fields $Q(\sqrt{-d})$ and $Q(\sqrt{k})$ respectively, and suppose that the corresponding class numbers are given by $h(d)$ and $h(k)$. Suppose also that $(d, k) = 1$ and let

$$\chi(n) = (k/n), \qquad \chi'(n) = (d/n)$$

denote the usual Kronecker symbols. Further let

$$f = f(x, y) = ax^2 + bxy + cy^2$$

run through a complete set of inequivalent quadratic forms with discriminant $-d$. From a well-known formula for $L(s, \chi)L(s, \chi\chi')$ with $s > 1$ we obtain, on taking limits as $s \rightarrow 1$ and applying classical results of Dirichlet,

$$h(k)h(kd) \log \epsilon = \tfrac{1}{12}\pi k\sqrt{d} \left(\sum_f \chi(a)a^{-1}\right) \prod_{p|k} (1 - p^{-2})$$
$$+ B_0 + \sum_f \sum_{r=-\infty;\, r\neq 0}^{\infty} B_r e^{\pi i r b/(ka)},$$

where ϵ denotes the fundamental unit in the field $Q(\sqrt{k})$, $h(kd)$ denotes the class number of the field $Q(\sqrt{(-kd)})$, $B_0 = -\log p\sum_f \chi(a)$ if k is the power of a prime p, $B_0 = 0$ otherwise and, for $r \neq 0$, we have

$$|B_r| \leq k|r|e^{-\pi|r|\sqrt{d}/(ak)}.$$

On choosing $k = 21$ or 33, so that $h(k) = 1$ and the units ϵ are given by $\alpha_1 = \frac{1}{2}(5 + \sqrt{21})$ and $\alpha_2 = 23 + 4\sqrt{33}$ respectively, we see that, if $h(d) = 1$,

$$|h(21d)\log\alpha_1 - \tfrac{32}{21}\pi\sqrt{d}| < e^{-\pi\sqrt{d}/100}, \qquad |h(33d)\log\alpha_2 - \tfrac{80}{33}\pi\sqrt{d}| < e^{-\pi\sqrt{d}/100}.$$

Since, in particular, $h(21d) < 4\sqrt{d}$, $h(33d) < 4\sqrt{d}$, it follows, on writing

$$\delta^{-1} = 14 \times 10^3, \qquad H = 140\sqrt{d}, \qquad b_1 = 35h(21d), \qquad b_2 = -22h(33d),$$

that the hypotheses of Theorem 6 are satisfied. We conclude that $H < 10^{250}$ whence $d < 10^{500}$. Note that, instead of Theorem 6, it would in principle have sufficed to have appealed to the earlier work of Gelfond and moreover, since $\pi = -2i\log i$, to have referred to the above formulae for just one value of k. A similar argument leads to the bound $d < 10^{500}$ for all imaginary quadratic fields $Q(\sqrt{-d})$ with class number 2, where d denotes a square-free positive integer with $d \not\equiv 3 \pmod 8$ [12].[1]

5. On the representation of integers by binary forms

We come now to the fundamental work begun by Thue in 1909 and subsequently developed by Siegel, Roth and others. Thue obtained a nontrivial inequality expressing a limit to the accuracy with which any algebraic number (not itself rational) can be approximated by rationals and thereby showed, in particular, that the Diophantine equation $f(x, y) = m$, where f denotes an irreducible binary form with integer coefficients and degree at least 3, possesses only a finite number of solutions in integers x, y. The work was much extended by Siegel first in 1921 when he strengthened the basic approximation inequality and then in 1929 when he applied his result, together with the Mordell-Weil theorem, to give a simple necessary and sufficient condition for any equation of the form $f(x, y) = 0$, where f denotes a polynomial with integer coefficients, to possess only a finite number of integer solutions. Siegel's work gave rise to many further developments; in particular, Mahler obtained far-reaching p-adic generalizations of the original theorems, and Roth succeeded in further improving the approximation inequality, establishing a result that is essentially best possible.

All the work which I have just described, however, suffers from one basic limitation, that of its noneffectiveness. The proofs depend on an assumption, made at the outset, that the Diophantine inequalities, or the corresponding equations, possess at least one solution in integers with large absolute values and the arguments provide no way of deciding whether or not such a solution exists. Thus although the Thue-Siegel theory supplies information on the number of solutions of the

[1] For another proof of the class number 1 result see P. Bundschuh and A. Hock [19]. An interesting application of Theorem 5 in a related context has been given by E. A. Anfert'eva and N. G. Čudakov [2].

equation $f(x, y) = 0$, it does not enable one to determine whether or not a particular equation of this type is soluble; of course, such a determination would amount to a solution of Hilbert's tenth problem for polynomials in two unknowns. For the special equation $f(x, y) = m$, where $f(x, 1)$ has at least one complex zero, another proof of the finiteness of the number of solutions was given by Skolem in 1935 by means of a p-adic argument very different from the original, but here the work depends on the compactness property of the p-adic integers and so is again noneffective.

As a consequence of Theorem 6 one can now give a new and effective proof of Thue's result on the representation of integers by binary forms.

THEOREM 7 [8]. *If f denotes an irreducible binary form with degree $n \geq 3$ and with integer coefficients then, for any positive integer m, the equation $f(x, y) = m$ has only a finite number of solutions in integers x, y, and these can be effectively determined.*

The method of proof leads in fact to an explicit bound for the size of all the solutions; assuming that $\mathcal{H}$ is some number exceeding the maximum of the absolute values of the coefficients of f we have

$$\max(|x|, |y|) < \exp\{(n\mathcal{H})^{(10n)^5} + (\log m)^{2n+2}\}.$$

In view of the mean-value theorem, this corresponds to an effective result on the approximation of algebraic numbers by rationals; indeed one can show that for any algebraic number α with degree $n \geq 3$ and for any $\kappa > n$ there exists an effectively computable number $c = c(\alpha, \kappa) > 0$ such that

$$|\alpha - p/q| > cq^{-n} \exp[(\log q)^{1/\kappa}]$$

for all integers p, q $(q > 0)$. Some slightly stronger quantitative results in this direction have been established for certain fractional powers of rationals [3], [4], [5] but here the work depends on particular properties of Gauss' hypergeometric function and is therefore of a special nature.

As regards the proof of Theorem 7, we assume, without loss of generality, that the coefficient of x^n in $f(x, y)$ is ± 1 and we denote the zeros of $f(x, 1)$ by $\alpha_1, \ldots, \alpha_n$, where it is assumed that $\alpha^{(1)}, \ldots, \alpha^{(s)}$ only are real and $\alpha^{(s+1)}, \ldots, \alpha^{(s+t)}$ are the complex conjugates of $\alpha^{(s+t+1)}, \ldots, \alpha^{(n)}$; thus it is implied that $n = s + 2t$. The algebraic number field generated by $\alpha = \alpha^{(1)}$ over the rationals will be denoted by K, and $\theta^{(1)}, \ldots, \theta^{(n)}$ will represent the conjugates of any element θ of K corresponding to the conjugates $\alpha^{(1)}, \ldots, \alpha^{(n)}$ of α. $C_1, C_2, \ldots$ will denote numbers greater than 1 which can be specified explicitly in terms of m, n and the coefficients of f. Finally we denote by $\eta_1, \ldots, \eta_r$ a set of $r = s + t - 1$ units in K such that

$$|\log |\eta_i^{(j)}|| < C_1 \qquad (1 \leq i, j \leq r)$$

and such that also the determinant Δ of order r with $\log |\eta_i^{(j)}|$ in the ith row and jth column satisfies $|\Delta| > C_2^{-1}$.

Suppose now that x, y are rational integers satisfying $f(x, y) = m$ and put $\beta = x - \alpha y$. Clearly β is an algebraic integer in K, and we have $|\beta^{(1)} \ldots \beta^{(n)}| = m$.

Further it is easily seen that an associate γ of β can be determined such that

$$\left|\log |\gamma^{(j)}|\right| < C_3 \qquad (1 \le j \le n).$$

Writing $\gamma = \beta\eta_1^{b_1} \dots \eta_r^{b_r}$ and $H = \max |b_j|$ we deduce from the equations

$$\log |\gamma^{(j)}/\beta^{(j)}| = b_1 \log |\eta_1^{(j)}| + \cdots + b_r \log |\eta_r^{(j)}| \qquad (1 \le j \le r)$$

that the maximum of the absolute values of the numbers on the left must exceed $C_4^{-1}H$, whence

$$\log |\beta^{(l)}| \le -(C_4^{-1}H - C_3)/(n-1)$$

for some l. In particular we have $|\beta^{(l)}| \le C_5$ and so $|\beta^{(k)}| \ge C_6^{-1}$ for some $k \ne l$. Since $n \ge 3$, there exists a superscript $j \ne k, l$, and we have the identity

$$(\alpha^{(k)} - \alpha^{(l)})\beta^{(j)} - (\alpha^{(j)} - \alpha^{(l)})\beta^{(k)} = (\alpha^{(k)} - \alpha^{(j)})\beta^{(l)}.$$

This gives $\alpha_1^{b_1} \dots \alpha_r^{b_r} - \alpha_{r+1} = \omega$, where

$$\alpha_s = \eta_s^{(k)}/\eta_s^{(j)} \qquad (1 \le s \le r),$$

$$\alpha_{r+1} = \frac{(\alpha^{(j)} - \alpha^{(l)})\gamma^{(k)}}{(\alpha^{(k)} - \alpha^{(l)})\gamma^{(j)}} \quad \text{and} \quad \omega = \frac{(\alpha^{(k)} - \alpha^{(j)})\beta^{(l)}\gamma^{(k)}}{(\alpha^{(k)} - \alpha^{(l)})\beta^{(k)}\gamma^{(j)}}.$$

Now by the choice of k and l we see that $0 < |\omega| < C_7 \exp(-H/C_8)$. Further, the degrees and heights of $\alpha_1, \dots, \alpha_{r+1}$ are bounded above by numbers depending only on f and m. Noting that $|e^z - 1| < \frac{1}{4}$ implies that $|z - ik\pi| < 4|e^z - 1|$ for some rational integer k, we easily obtain an inequality of the type considered in Theorem 6. Hence we conclude that $H < C_9$; this gives $|\beta^{(j)}| < C_{10}$ for each j and thus

$$\max(|x|, |y|) < C_{11} \max(|\beta^{(1)}|, |\beta^{(2)}|) < C_{12}.$$

6. The elliptic and hyperelliptic equations

Theorem 7 and its natural generalization to algebraic number fields can be used to solve effectively many other Diophantine equations in two unknowns. In particular it enables one to treat $y^2 = x^3 + k$ for any $k \ne 0$, an equation with a long and famous history in the theory of numbers [9], [27]. The work rests on the classical theory of the reduction of binary cubic forms, due mainly to Hermite, and the techniques used by Mordell in establishing the finiteness of the number of solutions of the equation. More generally, by means of the theory of the reduction of binary quartic forms one can prove:

THEOREM 8 [10]. *Let a ($\ne 0$), b, c, d denote rational integers with absolute values at most $\mathcal{H}$, and suppose that the cubic on the right of the equation*

$$y^2 = ax^3 + bx^2 + cx + d$$

has distinct zeros. Then all solutions in integers x, y satisfy

$$\max(|x|, |y|) < \exp\{(10^6\mathcal{H})^{10^6}\}.$$

Still more generally one can give bounds for all the solutions in integers x, y of equations of the form $y^m = f(x)$, where $m \geq 2$ and f denotes a polynomial with integer coefficients [11]. The work here, however, is based on a paper of Siegel and involves the theory of factorization in algebraic number fields; the bounds are therefore much larger than that specified in Theorem 8. As immediate consequences of the results one obtains inequalities of the type

$$|x^m - y^n| > c\,(\log\log x)^{1/n^2} \qquad (m, n \geq 3),$$

where $c = c(m, n) > 0$; in particular one can effectively solve the Catalan equation $x^m - y^n = 1$ for any given m, n.

We referred earlier to the celebrated Theorem of Siegel on the equation $f(x, y) = 0$. By means of appropriate extensions of the results described above, a new and effective proof of Siegel's theorem in the case of curves of genus 1 has recently been obtained.

THEOREM 9 [16]. *Let $F(x, y)$ be an absolutely irreducible polynomial with degree n and with integer coefficients having absolute values at most $\mathcal{H}$ such that the curve $F(x, y) = 0$ has genus 1. Then all integer solutions x, y of $F(x, y) = 0$ satisfy*

$$\max(|x|, |y|) < \exp\exp\exp\{(2\mathcal{H})^{10^{n^{10}}}\}.$$

The proof of the theorem involves a combination of some work of J. Coates [22] on the construction of rational functions on curves with prescribed poles, together with the techniques just mentioned for treating the equation $Y^2 = f(X)$. More precisely it is shown by means of the Riemann-Roch theorem that the integer solutions of $F(x, y) = 0$ can be related by a birational transformation to the solutions of an equation $Y^2 = f(X)$ as above, where now f denotes a cubic in X and where the coefficients and variables denote algebraic integers in a fixed field. From bounds for X, Y and their conjugates we immediately obtain the desired bounds for x, y.

In a recent series of papers [20], [21], Coates has generalized many of the theorems described above by employing analysis in the p-adic domain. In particular he has obtained explicit upper bounds for all integer solutions $x, y, j_1, \ldots, j_s$ of equations of the type

$$f(x, y) = mp_1^{j_1} \ldots p_s^{j_s} \quad \text{and} \quad y^2 = x^3 + kp_1^{j_1} \ldots p_s^{j_s},$$

where $p_1, \ldots, p_s$ denote fixed primes. The work involves, amongst other things, utilization of the Schnirelman line integral and the theory of S-units in algebraic number fields. As particular applications of his results, one can now give explicit lower bounds of the type $c(\log\log x)^{1/4}$ for the greatest prime factor of a binary form $f(x, y)$, and one can determine effectively all elliptic curves with a given conductor.

Several other extensions, applications and refinements of the theorems discussed here have been obtained by N. I. Feldman [25], [26], V. G. Sprindžuk [29], [30] and A. I. Vinogradov [31]. Recently Siegel [28] established some improved estimates for units in algebraic number fields which are likely to be of value in reducing the size of bounds. And, in 1967, Brumer [18] derived a natural p-adic analogue of Theorem 1 which, in conjunction with work of Ax [1], resolved a well-known problem of Leopoldt on the nonvanishing of the p-adic regulator of an abelian number field.

7. On the Weierstrass elliptic functions

Let $\wp(z)$ denote a Weierstrass $\wp$-function, let g_2, g_3 denote the usual invariants occurring in the equation

$$(\wp'(z))^2 = 4(\wp(z))^3 - g_2\wp(z) - g_3$$

and let ω, ω' denote any pair of fundamental periods of $\wp(z)$. Seigel proved in 1932 that if g_2, g_3 are algebraic then at least one of ω, ω' is transcendental; hence both are transcendental if $\wp(z)$ admits complex multiplication. Seigel's work was much improved by Schneider in 1937; Schneider showed that if g_2, g_3 are algebraic then any period of $\wp(z)$ is transcendental and moreover, the quotient ω/ω' is transcendental except in the case of complex multiplication. Furthermore Schneider proved that if $\zeta(z)$ is the corresponding Weierstrass ζ-function, given by $\wp(z) = -\zeta'(z)$, and if $\eta = 2\zeta(\frac{1}{2}\omega)$ then any linear combination of ω, η with algebraic coefficients, not both 0, is transcendental.

By means of techniques similar to those used in the proof of Theorem 1 these results can now be generalized as follows. Let $\wp_1(z)$, $\wp_2(z)$ be Weierstrass $\wp$-functions (possibly with $\wp_1 = \wp_2$) for which the invariants g_2, g_3 are algebraic and let $\zeta_1(z)$, $\zeta_2(z)$ be the associated Weierstrass ζ-functions. Further let ω_1, ω_1' and ω_2, ω_2' be any pairs of fundamental periods of $\wp_1(z)$, $\wp_2(z)$ respectively, and put $\eta_1 = 2\zeta(\frac{1}{2}\omega_1)$, $\eta_2 = 2\zeta(\frac{1}{2}\omega_2)$. We have

THEOREM 10 [13], [14]. *Any nonvanishing linear combination of* ω_1, ω_2, η_1, η_2 *with algebraic coefficients is transcendental.*

It will be recalled that ω_1, ω_2 and η_1, η_2 can be expressed as elliptic integrals of the first and second kinds respectively and so one sees, for instance, that the theorem establishes the transcendence of the sum of the circumferences of two ellipses with algebraic axes-lengths. Also, by an appropriate refinement of Theorem 10, one can obtain an upper estimate for the values assumed by a $\wp$-function with algebraic invariants at an algebraic point. In particular, for any positive integer n, we have $|\wp(n)| < C\exp[(\log n)^\kappa]$ for some absolute constant $\kappa > 0$ and some $C > 0$ depending only on g_2, g_3.[2] The proof of Theorem 10 utilizes results on the division values of the elliptic functions.

8. Concluding remarks

The three main problems left open by the work discussed here are

(i) To determine effectively all imaginary quadratic fields with a given class number ≥ 2.

---

[2]An account of this work is given in a paper submitted to the American J. Math [15]; it is easily seen that $|\wp(n)| > Cn$ for some $C > 0$ and infinitely many n.

(ii) To find an effective algorithm for determining all the integer points on any curve of genus ≥ 2.

(iii) To establish, under suitable conditions, the algebraic independence of the logarithms of algebraic numbers.

The resolution of these problems would represent a considerable advance in our knowledge.

REFERENCES

1. James Ax, *On the units of an algebraic number field*, Illinois J. Math. **9** (1965), 584–589. MR **31** #5858.

2. E. A. Anfert'eva and N. G. Čudakov, *The minima of a normed function in imaginary quadratic fields*, Dokl. Akad. Nauk SSSR **183** (1968), 255–256 = Soviet Math. Dokl. **9** (1968), 1342–1344; erratum, ibid. **187** (1969). MR **39** #5472.

3. A. Baker, *Rational approximations to certain algebraic numbers*, Proc. London Math. Soc. (3) **4** (1964), 385–398. MR **28** #5029.

4. ———, *Rational approximations to* $\sqrt[3]{2}$ *and other algebraic numbers*, Quart. J. Math. Oxford Ser. (2) **15** (1964), 375–383. MR **30** #1977.

5. ———, *Simultaneous rational approximations to certain algebraic numbers*, Proc. Cambridge Philos. Soc. **63** (1967), 693–702. MR **35** #4167.

6. ———, *Linear forms in the logarithms of algebraic numbers*, Mathematika **13** (1966), 204–216. MR **36** #3732.

7. ———, *Linear forms in the logarithms of algebraic numbers.* II, III, IV, Mathematika **14** (1967), 102–107, 220–228; ibid. **15** (1968), 204–216. MR **36** #3732.

8. ———, *Contributions to the theory of Diophantine equations.* I: *On the representation of integers by binary forms*, Philos. Trans. Roy. Soc. London A **263** (1967/68), 173–291. MR **37** #4005.

9. ———, *Contributions to the theory of Diophantine equations.* II: *The Diophantine equation* $y^2 = x^3 + k$, Philos. Trans. Roy. Soc. London Ser. A **263** (1967/68), 193–208. MR **37** #4006.

10. ———, *The Diophantine equation* $y^2 = ax^3 + bx^2 + cx + d$, J. London Math. Soc. **43** (1968), 1–9. MR **38** #111.

11. ———, *Bounds for the solutions of the hyperelliptic equation*, Proc. Cambridge Philos. Soc. **65** (1969), 439–444. MR **38** #3226.

12. ———, *A remark on the class number of quadratic fields*, Bull. London Math. Soc. **1** (1969), 98–102. MR **39** #2723.

13. ———, *On the quasi-periods of the Weierstrass* ζ*-function*, Nachr. Akad. Wiss. Göttingen Math.-Phys. Kl. II **1969,** 145–157.

14. ———, *On the periods of the Weierstrass* $\wp$*-function*, Proc. Sympos. Math. (Rome), vol. IV (to appear).

15. ———, *An estimate for the* $\wp$*-function at an algebraic point*, Amer. J. Math. (to appear).

16. A. Baker and J. Coates, *Integer points on curves of genus* 1, Proc. Cambridge Philos. Soc. **67** (1970), 595–602.

17. A. Baker and H. Davenport, *The equations* $3x^2 - 2 = y^2$ *and* $8x^2 - 7 = z^2$, Quart. J. Math. Oxford Ser. (2) **20** (1969), 129–137.

18. Armand Bruner, *On the units of algebraic number fields*, Mathematika **14** (1967), 121–124. MR **36** #3746.

19. P. Bundschuh and A. Hock, *Bestimmung aller imaginärquadratischen Zahlkörper der Klassenzahl Eins mit Hilfe eines Satzes von Baker*, Math. Z. **111** (1969), 191–204.

20. J. Coates, *An effective p-adic analogue of a theorem of Thue*, Acta Arith. **15** (1968/69), 279–305. MR **39** #4095.

21. ———, *An effective p-adic analogue of a theorem of Thue.* II: *On the greatest prime factor of a binary form.* III: *The Diophantine equation* $y^2 = x^3 + k$, Acta Arith. **16** (1970), 399–412, 425–435.

22. ———, *Construction of rational functions on a curve*, Proc. Cambridge Philos. Soc. **68** (1970), 105–123.

23. N. I. Feld'man, *Estimate for a linear form of logarithms of algebraic numbers*, Mat. Sb. **76** (**118**) (1968), 304–319 = Math. USSR Sb. **5** (1968), 291–307. MR **37** #4025.

24. ———, *An improvement of the estimate of a linear form in the logarithms of algebraic numbers*, Mat. Sb. **77** (**119**) (1968), 423–436 = Math. USSR Sb. **6** (1968), 393–406. MR **38** #1059.

25. ———, *An inequality for a linear form in the logarithms of algebraic numbers*, Mat. Zametki **6** (1969), 681–689.

26. ———, *Refinement of two effective inequalities of A. Baker*, Mat. Zametki **6** (1969), 767–769.

27. L. J. Mordell, *A chapter in the theory of numbers*, Cambridge Univ. Press, Cambridge; Macmillan, New York, 1947. MR **8**, 502.

28. C. L. Siegel, *Abschätzung von Einheiten*, Nachr. Akad. Wiss. Göttingen Math.-Phys. Kl. II **1969**, 71–86.

29. V. G. Sprindžuk, *Concerning Baker's theorem on linear forms in logarithms*, Dokl. Akad. Nauk BSSR **11** (1967), 767–769. MR **36** #1396.

30. ———, *Effectivization in certain problems of Diophantine approximation theory*, Dokl. Akad. Nauk BSSR **12** (1968), 293–297. MR **37** #6247.

31. V. G. Sprindžuk and A. I. Vinogradov, *The representation of numbers by binary forms*, Mat. Zametki **3** (1968), 293–297. MR **37** #151.

32. H. M. Stark, *A complete determination of the complex quadratic fields of class-number one*, Michigan Math. J. **14** (1967), 1–27. MR **36** #5102.

TRINITY COLLEGE
CAMBRIDGE, ENGLAND

ON SCHANUEL'S CONJECTURES AND SKOLEM'S METHOD

JAMES AX

Concerning the transcendentality properties of the exponential function, Schanuel has made a conjecture which embodies all its known transcendality properties such as the theorems of Lindemann and Baker as well as a whole collection of special conjectures including the algebraic independence of π and e and the unsolved problems 1, 7, and 8 on page 138 of Schneider's *Einführung in die Transzendenten Zahlen.*

The conjecture runs as follows.

(S) Let $y_1, \ldots, y_n \in \mathbf{C}$ be linearly independent over $\mathbf{Q}$ then

$$\dim_{\mathbf{Q}} \mathbf{Q}(y_1, \ldots, y_n, e^{y_1}, \ldots, e^{y_n}) \geq n.$$

Here $\dim_E F$ for any extension of fields F/E denotes the cardinality of a maximal E-algebraically independent subset of F.

Schanuel has also made the analogous conjectures for power series fields and differential fields. Our main purpose here is to prove these analogues and to indicate their connection with diophantine problems via the methods of Skolem [1] and Chabauty [2].

Let C denote a field containing $\mathbf{Q}$.

THEOREM 1 (SCHANUEL'S POWER SERIES CONJECTURE). *Let $y_1, \ldots, y_n \in tC[[t]]$ be linearly independent over $\mathbf{Q}$. Then*

$$\dim_{C(t)} C(t)(y_1, \ldots, y_n, \exp y_1, \ldots, \exp y_n) \geq n.$$

An alternate version states

THEOREM 1′. *Let $y_1, \ldots, y_n \in tC[[t]]$ be linearly independent over $\mathbf{Q}$. Then*

$$\dim_C C(y_1, \ldots, y_n, \exp y_1, \ldots, \exp y_n) \geq n + 1.$$

DEFINITION. A differential field over C is an extension F/C and a derivation D of F (into F) such that for all $c \in C$, $Dc = 0$.

EXAMPLE. $F = C((t))$, $D = d/dt$.

THEOREM 2 (SCHANUEL'S DIFFERENTIAL FIELD CONJECTURE). *Let F be a differential field over C. Let $y_1, \ldots, y_n, z_1, \ldots, z_n \in F$ be such that $Dy_1, \ldots, Dy_n$ are linearly independent over $\mathbf{Q}$ and $Dy_i = Dz_i/z_i$ for $i = 1, \ldots, n$. Then*

$$\dim_C C(y_1, \ldots, y_n, z_1, \ldots, z_n) \geq n + 1.$$

Theorem 2 represents a generalization of Theorem 1. Schanuel and his student Dale Brownawell proved certain cases of Theorem 2, e.g.: if $n \leq 2$; if $\dim_C C(y_1, \ldots, y_n) = 1$ or n; or if $\dim_C C(z_1, \ldots, z_n) = n$. Robert Risch in his work on elementary functions obtained a result equivalent to the case where for each i, $1 \leq i < n$ at most one of y_{i+1}, z_{i+1} is algebraic over

$$C(y_1, \ldots, y_i, z_1, \ldots, z_i)$$

as well as some additional related results.

In connection with Skolem's method for proving the finiteness of the set of $x \in \mathbf{Z}^l$ such that

$$(*) \qquad F(x) = c$$

for $c \in \mathbf{Z} - 0$ and certain forms $F \in \mathbf{Z}[X_1, \ldots, X_l]$, Borevič and Šafarevič [3] raised the problem of proving

(BS) Let $y_1, \ldots, y_n \in tC[[t]]$ be such that $n \geq 2$ and

$$\operatorname{rank}_C (y_1, \ldots, y_n) + \operatorname{rank}_C (\exp y_1, \ldots, \exp y_n) \leq n.$$

Then there exist distinct i and j for which $y_i = y_j$.

Here the $\operatorname{rank}_C (S)$ denotes the cardinality of a maximal C-linearly independent subset of S.

We can show by means of examples the falsity of (BS). On the other hand, Theorem 1 contains as a special case the following results

THEOREM 3. *Let $y_1, \ldots, y_n \in tC[[t]]$ be such that*

$$\operatorname{rank}_C (y_1, \ldots, y_n) + \operatorname{rank}_C (\exp y_1, \ldots, \exp y_n) \leq n.$$

Then $y_1, \ldots, y_n$ are linearly dependent over $\mathbf{Q}$.

Under the same hypotheses as in Theorem 3, (BS) asserts the existence of a very special $\mathbf{Q}$-linear dependency $y_i = y_j$; this is important for the application. As a matter of fact Skolem's results on the equation $(*)$ follow from his establishing (BS) in the special cases where $n \leq 5$ or where $\operatorname{rank}_C (\exp y_1, \ldots, \exp y_n) \leq 2$. It is therefore of interest to find hypotheses on the y_i of the type occurring in Theorem 3 which guarantee the equality $y_i = y_j$.

The following result at least has the merit of implying the results of Skolem.

The following is a result in this direction.

THEOREM 4. *Let $n \geq 2$ and $0 = y_0, y_1, \ldots, y_{n-1} \in tC[[t]]$ be such that*
(α) $\exp y_1, \ldots, \exp y_s$ *are C-algebraically independent;*
(β) $y_{s+1}, \ldots, y_{n-1}$ *are C-linearly independent;*
(γ) $\mathrm{rank}_C(y_0, \ldots, y_{n-1}) + \mathrm{rank}_C(\exp y_0, \ldots, \exp y_{n-1}) \leq n$. *Then there exist distinct i and j for which $y_i = y_j$.*

Our proof of Theorem 4 which appears in [5] is a direct calculation rather than an application of Theorem 1. However for the applications we note that the results of Skolem are contained in those of Chabauty and that these in turn are consequences of Theorem 2.

The main idea of the proof of Theorem 2 is to pass to the dual problem about differential 1-forms.

Dualization

If A is a commutative ring and if B is a commutative A-algebra then $\Omega_{B/A}$ denotes the B-module of relative differentials [4, Chapter 3, §1, pp. 279–280]. We denote the canonical A-derivation of B into $\Omega_{B/A}$ by $d = d_A = d_A^B \colon B \to \Omega_{B/A}$.

Let F/C be an extension of fields. Let $y_1, \ldots, y_n, z_1, \ldots, z_n \in F$. Set

$$\omega_i = dy_i - (1/z_i)\, dz_i \in \Omega_{F/C}.$$

THEOREM $\hat{2}$. *Assume*
(i) $\sum_{i=1}^{n} F\omega_i \cap dF = \{0\}$ *in* $\Omega_{F/C}$;
(ii) $dy_1, \ldots, dy_n$ *linearly independent over* $\mathbf{Q}$.
Then $\dim_C C(y_1, \ldots, y_n, z_1, \ldots, z_n) \geq n + 1$.

PROOF THAT THEOREM $\hat{2} \Rightarrow$ THEOREM 2. Since the theorems have the same conclusion, it suffices to show that the hypotheses of Theorem 2 imply (i) and (ii). We assume these hypotheses and (as we may) $C = \{f \in F \mid Df = 0\}$.

By the universal property of the C-derivation

$$F \xrightarrow{d} \Omega_{F/C}$$

there exists an F-linear map v such that

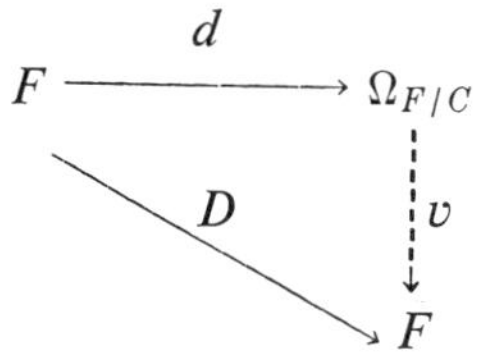

commutes.

Then for all $f \in F$, $Df = v(df)$. Hence

$$v(\omega_i) = v(dy_i - (1/z_i)dz_i) = Dy_i - (1/z_i)Dz_i = 0$$

for $i = 1, \ldots, n$.

Thus $v(\sum_{i=1}^n F\omega_i) = \{0\}$. Therefore if $f \in F$ and $df \in \sum_{i=1}^n F\omega_i$ then $v(df) = Df = 0$. It follows that $f \in C$ which implies $df = 0$. We have established (i). Since the $Dy_i = v(dy_i)$ are **Q**-linearly independent, (ii) is immediate.

Proof of Theorem $\hat{2}$. Assume the theorem false, i.e. that (i) and (ii) hold but that $\dim_C F' \leq n$, where $F' = C(y_1, \ldots, y_n, z_1, \ldots, z_n)$. Then it follows that the F'-space $\Omega_{F'/C} \hookrightarrow \Omega_{F/C}$ has F'-rank at most n. Thus $\omega_1, \ldots, \omega_n$ and dy_1 are F'-linearly dependent. By (i) dy_1 is not an F'-linear combination of the ω_ν and by (ii) $dy_1 \neq 0$. It follows that the ω_ν are F'-linearly dependent. By changing notation we can assume $f_1, \ldots, f_m \in F'$ and

$$\sum_{\mu=1}^m f_\mu \omega_\mu = 0 \tag{1}$$

is an F'-linear relation of shortest length. We may further assume $f_1 = 1$. We now show that all the $f_\mu \in C$. We may assume $m \geq 2$. As in the classical case, the canonical derivation $d: F' \to \Omega_{F'/C}$ extends to an anti-derivation (again denoted by d) of the exterior F'-algebra $\bigwedge \Omega_{F'/C}$ built on $\Omega_{F'/C}$. Moreover a simple computation shows that for all $y, z \in F'$, dy and dz/z are in the kernel of d. Thus (1) yields

$$0 = \sum_{\mu=1}^m df_\mu \wedge \omega_\mu = \sum_{\mu=2}^m df_\mu \wedge \omega_\mu. \tag{2}$$

Wedging (2) with $\bigwedge_{\mu=3}^m \omega_\mu$ (or leaving it alone if $m = 2$) yields

$$df_2 \wedge \bigwedge_{\mu=2}^m \omega_\mu = 0. \tag{3}$$

By the minimality of (1), the ω_μ, $\mu = 2, \ldots, m$ are F'-linearly independent and so (3) implies that $df_2 \in \sum_{\mu=2}^m F'\omega_\mu$. By (ii) we conclude $df_2 = 0$, i.e. $f_2 \in C$, since C, being the kernel of D, is relatively algebraically closed in F'. In similar fashion we have $f_\mu \in C$, $\mu = 1, \ldots, m$. This shows that the ω_μ are C-linearly dependent. Let A be the **Z**-submodule of $\Omega_{F'/C}$ generated by the ω_μ, $\mu = 1, \ldots, m$. Then the C-linear dependency of the ω_μ implies there exists $c_1, \ldots, c_l \in C$, linearly independent over **Q** and $\alpha_\lambda \in A$, $\lambda = 1, \ldots, l$ such that

$$\sum_{\lambda=1}^l c_\lambda \alpha_\lambda = 0 \tag{4}$$

in $\Omega_{F'/C}$ where the α_λ are nontrivial **Z**-linear combinations of the ω_μ.

Let E be a subfield of F' containing C and algebraically closed in F' such that $\dim_E F' = 1$. Then applying the canonical F'-linear map

$$\Omega_{F'/C} \xrightarrow{\tau} \Omega_{F'/E}$$

we obtain from (4) that

$$\sum_{\lambda=1}^l c_\lambda \tau(\alpha_\lambda) = 0. \tag{5}$$

Now if $\alpha \in A$, there exist $a_1, \ldots, a_m \in Z$ such that

$$\alpha = \sum_{\mu=1}^{m} a_\mu \omega_\mu = \sum_{\mu=1}^{m} a_\mu (dy_\mu - dz_\mu/z_\mu)$$
$$= dy' - dz'/z'$$

where $y' = \sum_{\mu=1}^{m} a_\mu y_\mu$, $z' = \prod_{\mu=1}^{m} z_\mu^{a_\mu} \in F'$. Thus, we can set

$$\alpha_\lambda = dy'_\lambda - dz'_\lambda/z'_\lambda$$

and obtain from (5)

$$\sum_{\lambda=1}^{l} c_\lambda (d_E y'_\lambda - d_E z'_\lambda / z'_\lambda) = 0. \tag{6}$$

F' is an algebraic function field in one variable over E. If $\mathbf{p}$ is an E-place of F' we have that $\operatorname{res}_{\mathbf{p}}$ is E-linear and $\operatorname{res}_{\mathbf{p}} d_E y = 0$ and $\operatorname{res}_{\mathbf{p}} d_E z/z = \operatorname{ord}_{\mathbf{p}} z$ where $\operatorname{ord}_{\mathbf{p}} z \in \mathbf{Z}$ is the order of the zero or pole of z at $\mathbf{p}$. Thus (6) implies

$$\sum_{\lambda=1}^{l} c_\lambda \operatorname{ord}_{\mathbf{p}} z'_\lambda = 0 \quad \text{for all } \mathbf{p}. \tag{7}$$

Since the c_λ are $\mathbf{Q}$-linearly independent, we conclude

$$\operatorname{ord}_{\mathbf{p}} z'_\lambda = 0 \quad \text{for all } \mathbf{p} \text{ and } \lambda,$$

i.e. $z'_\lambda \in E$ for all λ. Since this holds for every E relatively algebraically closed in F' with $E \supseteq C$ and $\dim_E F' = 1$, we conclude $z'_\lambda \in C$ for all λ. Now if $\alpha_1 = \sum_{\mu=1}^{m} b_\mu \omega_\mu$ then the $b_\mu \in \mathbf{Z}$ are not all zero and $z'_1 = \prod_{\mu=1}^{m} z_\mu^{b_\mu}$. We have

$$\sum_{\mu=1}^{m} b_\mu \omega_\mu = \sum_{\mu=1}^{m} b_\mu (dy_\mu - dz_\mu/z_\mu) = d \sum_{\mu=1}^{m} b_\mu y_\mu$$

exact and so by (i), zero. But then $0 = \sum_{\mu=1}^{m} b_\mu \, dy_\mu$ implies that the dy_μ are $\mathbf{Q}$-linearly dependent. This contradiction to (ii) completes the proof.

By a variation of this proof which appears in [5], it is possible to establish certain stronger results including the following.

THEOREM 1″. *Let $y_1, \ldots, y_n \in C[[t_1, \ldots, t_r]]$ be power series without constant term, $\mathbf{Q}$-linearly independent. Then*

$$\dim_C C(y_1, \ldots, y_n, \exp y_1, \ldots, \exp y_n) \geq n + \operatorname{rank}\left(\frac{\partial y_\nu}{\partial t_\rho}\right)_{\substack{\nu=1,\ldots,n;\\ \rho=1,\ldots,r}}.$$

On the methods of Chabauty and Skolem

By a p-adic method [3: Chapter 4, §6] due to Skolem [1] the problem of proving the finiteness of the number of solutions of certain diophantine equations is reduced to consideration of the algebraic relations satisfied by the exponential function. Skolem's results [1] on these relations are contained in those of Chabauty [2] and these in turn follow from Theorem 1″ as we show next.

Chabauty's results. Let C be an algebraically closed field (containing $\mathbf{Q}$) and complete with respect to a nondiscrete absolute value.

Let $b_{\mu\nu} \in C$ and $q_\nu \in C^*$ for $\mu = 1, \ldots, m$, $\nu = 1, \ldots, n$. Then, following Chabauty [2: p. 144], we say that the local analytic subvariety M of C^n at $q = (q_1, \ldots, q_n)$ defined by

$$(*) \qquad \sum_{\nu=1}^{n} b_{\gamma\nu} \log (x_\nu/q_\nu) = 0, \qquad \gamma = 1, \ldots, c$$

is a *μ-variety*. If we can choose the $b_{\mu\nu}$ to be in $\mathbf{Z}$ we shall call M an algebraic μ-variety for in this case M is the local analytic variety at q defined by the algebraic variety with defining equations

$$\prod_{\nu=1}^{n} (x_\nu/q_\nu)^{b_{\gamma\nu}} = 1, \qquad \gamma = 1, \ldots, c.$$

The following result is a restatement of [2: Lemmas 2.1, 2.2, 2.3].

THEOREM (CHABAUTY). *Let M be a μ-variety at q and W be an algebraic variety containing q. Then for each component I of $W \cap M$ there exists an algebraic μ-variety A such that $A \supseteq I$ and we have $a \leq m + w - i$ where* $\dim A = a$, $\dim I = i$, $\dim M = m$ *and* $\dim W = w$.

PROOF. We can assume $q_\nu = 1$ for $\nu = 1, \ldots, n$ by applying to C^n the map

$$(x_1, \ldots, x_n) \to (x_1/q_1, \ldots, x_n/q_n).$$

Let I be an irreducible component of $M \cap W$ of dimension i which we can assume to be positive. Then we can parameterize I at q; i.e. we can find $z_1, \ldots, z_n \in C[[t_1, \ldots, t_i]]$ convergent in a polydisk D about 0 in C^i such that $z_\nu(0) = 1$ for $\nu = 1, \ldots, n$ and such that for all $c \in I$ sufficiently close to q there exists $\tau \in D$ with $z(\tau) = (z_1(\tau), \ldots, z_n(\tau)) = c$. This implies that

$$\operatorname{rank}\left(\frac{\partial z_\nu}{\partial t_j}(\tau)\right)_{\substack{\nu=1,\ldots,n;\\ j=1,\ldots,i}} = i \quad \text{for some } \tau \in D$$

and hence that $\operatorname{rank}(\partial z_\nu/\partial t_j) = i$. Set $y_\nu = \log z_\nu$ for $\nu = 1, \ldots, n$; these y_ν are power series without constant terms. Let a be the $\mathbf{Z}$-rank of $y_1, \ldots, y_n$, say $y_1, \ldots, y_a$ are $\mathbf{Z}$-independent. We have

$$\operatorname{rank}_C \{y_1, \ldots, y_n\} \leq \dim M = m$$

and

$$\dim_C C(z_1, \ldots, z_n) \leq \dim W = w.$$

Thus $\dim_C C(y_1, \ldots, y_n, \exp y_1, \ldots, \exp y_n) \leq m + w$. But by Theorem 1″,

$$\begin{aligned}&\dim_C C(y_1, \ldots, y_n, \exp y_1, \ldots, \exp y_n)\\ &\quad = \dim_C C(y_1, \ldots, y_a, \exp y_1, \ldots, \exp y_a) \geq a + \operatorname{rank}(\partial y_\alpha/\partial t_j)_{\substack{\alpha=1,\ldots,a;\\ j=1,\ldots,i}}.\end{aligned}$$

Since

$$\frac{\partial z_\alpha}{\partial t_j} = z_\alpha \frac{\partial y_\alpha}{\partial t_j}, \quad \text{we have} \quad \operatorname{rank}\left(\frac{\partial y_\alpha}{\partial t_j}\right)_{\substack{\alpha=1,\ldots,a;\\ j=1,\ldots,i}} = \operatorname{rank}\left(\frac{\partial z_\alpha}{\partial t_j}\right)_{\substack{\alpha=1,\ldots,a;\\ j=1,\ldots,i}}$$

$$= \operatorname{rank}\left(\frac{\partial z_\nu}{\partial t_j}\right)_{\substack{\nu=1,\ldots,n;\\ j=1,\ldots,i}} = i.$$

Thus $m + w \geq a + i$. I is contained in the algebraic μ-variety A of dimension a defined by the system $(*)$ where

$$(b_{\gamma\nu})_{\nu=1,\ldots,n}, \qquad \gamma = 1, \ldots, c = n - a$$

is a basis for the set of $(b_1, \ldots, b_n) \in \mathbf{Z}^n$ such that

$$\sum_{\nu=1}^{n} b_\nu y_\nu = 0.$$

This completes the proof.

Finally, let us indicate some counterexamples to (BS). Let N be a nonnegative integer. Set $n = N(N+1)/2$ and

$$P_1, \ldots, P_n = \{a \log (1 - t) + b \log (1 + t) \mid a + b < N\}.$$

Then $\operatorname{rank}_C (P_1, \ldots, P_n) \leq 2$. Also

$$\operatorname{rank}_C (\exp P_1, \ldots, \exp P_n) = \operatorname{rank}_C ((1 - t)^a(1 + t)^b\colon a + b < N)) \leq N.$$

Hence for $N(N+1)/2 > N + 2$ we get counterexamples to (BS), the smallest corresponding value $n = 6$ coinciding with the first unproven case of (BS).

References

1. T. Skolem, *Einige Sätze über p-adische Potenzreihen mit Anwendung auf gewisse exponentielle Gleichungen*, Math. Ann. **111** (1935), 399–424.

2. C. Chabauty, *Sur les équations diophantiennes liées aux unités d'un corps de nombres algébrique fini*, Ann. Mat. Pura Appl. **17** (1938), 217–268.

3. Z. Borevič and I. Šafarevič, *Number theory*, "Nauka", Moscow, 1964; English transl., Pure and Appl. Math., vol. 20, Academic Press, New York, 1966. MR **30** #1080; MR **33** #4001.

4. D. Mumford, *Introduction to algebraic geometry*, Lecture Notes, Harvard University, Cambridge, Mass., 1967.

5. J. Ax, *On Schanuel's conjectures* (to appear in Ann. of Math.).

State University of New York, Stony Brook
Long Island, New York

ON APPROXIMATIONS OF ALGEBRAIC NUMBERS BY ALGEBRAIC NUMBERS OF BOUNDED DEGREE

EDUARD A. WIRSING

0. Summary

Let α be an algebraic and t a natural number. Denote by $\| \ldots \|$ the height of a polynomial or an algebraic number.

Suppose there are infinitely many algebraic numbers β of degree $\leq t$ such that $|\alpha - \beta| \leq \|\beta\|^{-\phi}$. Then $\phi \leq 2t$. Similarly, if there are infinitely many polynomials Q of degree $\leq t$ with integral rational coefficients such that

$$0 \neq |Q(\alpha)| \leq \|Q\|^{1-\phi}$$

then $\phi \leq 2t$.

If $t = 1$, this is the Thue-Siegel-Roth theorem. For $t > 2$, apparently, the results are not best possible (it should be $\phi \leq t + 1$).[1]

There are also results on simultaneous approximations

$$0 \neq |Q(\alpha_\nu)| \leq \|Q\|^{1-\phi_\nu} \qquad (\nu = 1, 2, \ldots, n).$$

These yield in particular a lower estimate of the resultant Res (P, Q) where P is fixed and Q varies (Theorem 5).

As in the Thue-Siegel-Roth Theorem the estimates are noneffective.

The progress as compared to Roth [4] is due, to a good extent, to probabilistic considerations used in the construction of the approximating polynomial.

As a contribution to real analysis the paper contains a number of inequalities which seem to be new (see §3). All other results are listed in §2.

1. Introduction

1.1. *About the results*

1.1.1. The famous theorem of Thue-Siegel-Roth states that for any real algebraic α and any $\epsilon > 0$ there are at most finitely many rational numbers p/q such that

$$|\alpha - p/q| \leq q^{-2-\epsilon}.$$

[1] Added in proof. This has been proved meanwhile by W. Schmidt.

The exponent 2 is the best possible (unless α is rational).

When instead of the rational p/q we approximate by algebraic numbers β of some fixed (or more generally bounded) degree greater than one new difficulties arise. Only if β is restricted to some particular number field of finite degree is the generalisation fairly straightforward. For this situation Leveque [2] proves that there are at most finitely many solutions of $|\alpha - \beta| \leq \|\beta\|^{-2-\epsilon}$.

1.1.2. In this paper we shall treat the general case, where β is allowed to run through all algebraics of degree $\leq t$, say. We shall see (Theorem 1) that

$$|\alpha - \beta| \leq \|\beta\|^{-\phi}, \qquad \deg \beta \leq t$$

has at most finitely many solutions if $\phi > 2t$.

For $t = 1$ this is Roth's Theorem and the exponent therefore best possible. For $t > 1$ however it seems very likely that the assertion is true already if $\phi > t + 1$. This, if true, would be best possible. Actually for $t = 2$ Theorem 1 has been proved in its best possible form, i.e. for $\phi > 3$, by W. Schmidt [5].

Though Theorem 1 apparently is not best possible it is, as far as we know, the first estimate of its kind where the condition on ϕ does not involve the degree of α. For earlier results see C. L. Siegel [9], [10], Ramachandra [3].

1.1.3. Closely related is the question how small for $Q \in \mathbf{Z}[x]$ of degree $\leq t$ the value $Q(\alpha)$ can be. We shall prove (Theorem 2), that

$$0 \neq |Q(\alpha)| < \|Q\|^{1-\phi}$$

has at most finitely many such solutions if $\phi > 2t$. Theorem 1 is an easy consequence of Theorem 2 since any zero β of Q that lies close to α causes $Q(\alpha)$ to be small. The opposite is not true, since for $Q(\alpha)$ to be small the reason may be several zeros of Q that lie fairly close to α rather than a single very close one.

1.1.4. If we consider Theorem 2 as a statement about simultaneous approximations of α by all zeros β of a polynomial Q, our method yields further information (Theorem 3) without additional labour. Number the zeros of Q according to their distance from α:

$$|\alpha - \beta^{(1)}| \leq |\alpha - \beta^{(2)}| \leq \cdots \leq |\alpha - \beta^{(t)}|$$

and let t_1 be the index for which

$$|\alpha - \beta^{(t_1)}| < 1 \leq |\alpha - \beta^{(t_1+1)}|.$$

Now form

$$\Gamma_\alpha(Q) := \prod_{\tau=1}^{t_1} |\alpha - \beta^{(\tau)}|^{2\tau-1}.$$

Then, if α is algebraic and $\phi > 2t$, there are at most finitely many Q with $\Gamma_\alpha(Q) \leq \|Q\|^{-\phi}$.

Since Theorem 3 shall be proved for general Q, while in the central part of the proof Q is required to be irreducible we become involved with estimating $\Gamma_\alpha(Q_1 \cdot Q_2)$ by $\Gamma_\alpha(Q_1)$ and $\Gamma_\alpha(Q_2)$. In this connection an additive notation is more convenient. We introduce a symmetric function $M(a_1, \ldots, a_t)$ of nonnegative real variables a_τ by setting

$$M(a_1, \ldots, a_t) := \sum_{\tau=1}^{t} (2\tau - 1)a_\tau \quad \text{if } a_1 \geq a_2 \geq \cdots.$$

If we employ the usual notation

$$\log^+ x = \max(\log x, 0)$$

we obtain

$$\log \Gamma_\alpha^{-1}(Q) = M(\log^+ |\alpha - \beta^{(\tau)}|^{-1} \mid \tau = 1, \ldots, t),$$

where a specific enumeration of the β is no longer required.

The inequalities that we need for the function M are developed in §3. The most surprising one is probably the one given below as (3.8)

$$M(a_1 + b_1, \ldots, a_n + b_n)^{1/2} \leq M(a_1, \ldots, a_n)^{1/2} + M(b_1, \ldots, b_n)^{1/2}.$$

Note that the a_i and b_j can be in any order. This is not a proper convexity theorem since $M(a_1, \ldots, a_n)$ is positive homogeneous of the first degree instead of the second.

1.1.5. If $\mathfrak{K}$ is some number field of finite degree it can be shown that in Theorem 1 the degree t of the β can be replaced by the relative degree t' of β over $\mathfrak{K}$ (Theorem 1.1). If we consider the $\beta \in \mathfrak{K}$ only, then $t' = 1$ and we obtain Leveque's theorem. It should be noted, that the height of β is always understood relative to $\mathbf{Q}$; we do not introduce a height relative to $\mathfrak{K}$.

Theorem 2 has a corresponding version, Theorem 2.1: If we admit such polynomials Q only that split over $\mathfrak{K}$ into factors of degree $\leq t'$ and $\phi > 2t'$ then

$$0 \neq |Q(\alpha)| \leq \|Q\|^{1-\phi}$$

has at most finitely many solutions Q.

1.1.6. If α is not real, it is natural to expect that the limiting exponent in our theorems can be taken smaller than in the general case because if the β are counted in the order of their height a positive fraction of all β's are real. Consequently the β are much denser on the real axis than they are in the plane. For such a sharpening of Theorem 2, say, one must try to utilize the fact that if $Q(\alpha_1)$ is small then $Q(\alpha_2)$ with $\alpha_2 = \bar{\alpha}_1$ is as small. We can indeed do this, though at the cost of a further complication of the proof. Once this set-up is given however it is found quite irrelevant whether $\alpha_2 = \bar{\alpha}_1$ or not. Furthermore we can deal with more than two distinct algebraic numbers $\alpha_1, \ldots, \alpha_n$. If here $\phi > 2t/n$ then there are at most finitely many $Q \in \mathbf{Z}[z]$ of degree t such that

$$0 \neq |Q(\alpha_\nu)| \leq \|Q\|^{1-\phi} \quad \text{for } \nu = 1, 2, \ldots, n.$$

The particular case $\alpha_2 = \bar{\alpha}_1 \neq \alpha_1$ is formulated as Theorem 2.2. We may consider the still more general system

$$0 \neq |Q(\alpha_\nu)| < \|Q\|^{1-\phi_\nu} \qquad (\nu = 1, \ldots, n). \tag{1.1}$$

One might expect that $\sum_\nu \phi_\nu > 2t$ suffices to prevent more than finitely many solutions. This we cannot prove, but it will be seen that the stronger condition:

$$n^2\left(\sum_{\nu=1}^{n} \phi_\nu^{-1}\right)^{-1} > 2t \tag{1.2}$$

is sufficient.

Let us call the left hand side here the harmonic sum of $\phi_1, \ldots, \phi_n$ since it is n times the harmonic means. If all ϕ_ν are equal the harmonic sum equals the usual sum, otherwise the harmonic sum is smaller.

If (1.2) is a sufficient condition it is clear that the corresponding condition for any one subset of the ϕ_ν will also suffice. It is important to note this since the harmonic sum of a subset can well be larger than the one of the full set. Therefore we define

$$L(\phi_1, \ldots, \phi_n) := \max_{r=1}^{n} \max_{\nu_1<\ldots<\nu_r} r^2\left(\sum_{\rho=1}^{r} \phi_{\nu_\rho}^{-1}\right)^{-1},$$

that is the maximal harmonic subsum of the ϕ_ν, and obtain the theorem: If $L(\phi_1, \ldots, \phi_n) > 2t$, then (1.1) has at most finitely many solutions $Q \in \mathbf{Z}[x]$ of degree t. The final formulation that we give as Theorem 4 in §2 is obtained by replacing in L the bounds ϕ_ν by the numbers to be bounded i.e. by

$$-\log \frac{|Q(\alpha_\nu)|}{\|Q\|} \Big/ \log \|Q\|.$$

The introduction of L into our formulation serves much more than just to combine several conditions into one. In the course of the proof, more specifically in the reduction from general Q to irreducible ones, we shall need nontrivial properties of the function L that are not true for the harmonic sum itself. The most striking one, given in Lemma 7 is

$$L(a_1 + b_1, \ldots, a_n + b_n)^{1/2} \leq L(a_1, \ldots, a_n)^{1/2} + L(b_1, \ldots, b_n)^{1/2}.$$

1.1.7. As a special case of Theorem 4 we have mentioned Theorem 2.2. Another application arises from considering the resultant of a fixed polynomial P and variable Q of degree t. The zeros of P play the part of the above α_ν. Our result (Theorem 5) is that for a $P \in \mathbf{Z}[x]$ of degree s, say, without multiple zeros and for

$$\phi > 2t\left(1 + \frac{1}{3} + \cdots + \frac{1}{2t-1}\right)$$

there are at most finitely many Q such that

$$\operatorname{Res}(P, Q) \leq \|Q\|^{s-\phi}.$$

The unpleasant factor $1 + \frac{1}{3} + \cdots + 1/(2t-1)$ reflects the weakness of Theorem 4 when the ϕ_ν are not equal. This factor could be dropped if L in Theorem 4 could be replaced by $\sum \phi_\nu$.

1.2. *About the proofs*

1.2.1. Roth's proof of his theorem [4], which incorporates ideas of Thue [11],

Siegel [9], [10], Dyson [1] and Schneider [7], [8], shall be summarized here as far as it is needed for an understanding of our generalization.

Let s be the degree of the algebraic number α.

Assume that several approximations

$$\left|\alpha - \frac{p_\kappa}{q_\kappa}\right| \leq q_\kappa^{-\phi} \qquad (\kappa = 1, \ldots, k) \tag{1.3}$$

are given, where $\phi > 2$.

In a first step one constructs a polynomial $P \in \mathbf{Z}[x_1, \ldots, x_k]$, $P \neq 0$, which at the point $(\alpha, \ldots, \alpha)$ vanishes to some high order as compared to its degrees, while its coefficients are not too large. To be more precise, denoting

$$D_{(i)} := D_{(i_1,\ldots,i_k)} := \prod_{\kappa=1}^{k} \frac{\partial^{i_\kappa}}{i_\kappa!\, \partial x_\kappa^{i_\kappa}},$$

one wants to have

$$D_{(i)}P(\alpha, \ldots, \alpha) = 0 \quad \text{for all } (i) \in \mathfrak{M}, \tag{1.4}$$

where $\mathfrak{M}$ is a suitable set of lattice points $(i_1, \ldots, i_k)$, and

$$\|P\| \leq C(\alpha)^{\Sigma_\kappa r_\kappa}, \tag{1.5}$$

where r_κ is the degree of P with respect to x_κ.

By (1.4) the unknown coefficients of P are subject to one homogeneous linear condition over $\mathbf{Q}(\alpha)$ for each $(i) \in \mathfrak{M}$, or equivalently to s such conditions over $\mathbf{Q}$ for each $(i) \in \mathfrak{M}$. If

$$|\mathfrak{M}| \leq \frac{1}{2s} \prod_\kappa (r_\kappa + 1), \tag{1.6}$$

then the number of conditions is at most half the number of unknowns. The pidgeon-hole principle provides a solution in integers which also fulfills (1.5).

In a second step one tries to exploit (1.3, 4, 5) by way of Taylor-expansion of P at $(\alpha, \ldots, \alpha)$ in order to show that $P(p_1/q_1, \ldots, p_k/q_k)$ is small. So small in fact, that the integer $q_1^{r_1} \ldots q_k^{r_k} P(p_1/q_1, \ldots, p_k/q_k)$ must be zero. In order to obtain an effective estimate by Taylor-expansion, one has to make

$$q_1^{r_1} \sim q_2^{r_2} \sim \cdots \sim q_k^{r_k}$$

and to specify

$$\mathfrak{M} := \left\{(i_1, \ldots, i_k);\ 0 \leq i_\kappa \leq r_\kappa,\ \sum_\kappa \frac{i_\kappa}{r_\kappa} \leq \vartheta\right\} \tag{1.7}$$

where ϑ is taken as large as possible without violating (1.6). Now the crucial point is that here $\vartheta = (\frac{1}{2} - \epsilon)k$ is an admissible value, provided $k \geq k_0(\epsilon)$ (Schneider [7]). Schneider's result can also be deduced from the law of large numbers.

If, as we assume, $\phi > 2$ the method furnishes an $\epsilon > 0$ such that

$$D_{(j)}P\left(\frac{p_1}{q_1}, \ldots, \frac{p_k}{q_k}\right) = 0 \quad \text{for} \quad \frac{j_1}{r_1} + \cdots + \frac{j_k}{r_k} \leq \epsilon k. \tag{1.8}$$

In the last step Roth shows that under certain additional conditions (q_κ large and very different) (1.8) contradicts (1.5), thus proving his theorem.

We need not dwell on the details of this step since the generalization that we need in this paper has been developed by Leveque [2].

1.2.2. If one copies this procedure starting with

$$(1.9) \qquad |\alpha - \beta_\kappa| \le \|\beta_\kappa\|^{-\phi} \qquad (\kappa = 1, \ldots, k)$$

where β_κ are algebraics of degree t one arrives easily at the corresponding estimate of $P(\beta_1, \ldots, \beta_k)$. But in order to continue apparently one has to change to the norm of this number. In general there will be t^k conjugates. With $\mathfrak{M}$ determined by (1.7) we have a good estimate at $(\beta_1, \ldots, \beta_k)$ but we cannot expect a nontrivial estimate at the other points

$$(\beta_1^{(\tau_1)}, \ldots, \beta_k^{(\tau_k)}) \qquad (\text{let } \beta_\kappa = \beta_\kappa^{(1)}).$$

Therefore in this situation k must be kept fixed and Schneider's lemma cannot be used. Ramachandra [3] shows about how much can be proved this way. His result is $\phi \le kt^{k-1}(s/k!)^{1/k}$ for all $k = 1, 2, \ldots$.

1.2.3. We cannot assume, in general, any nontrivial estimate for $|\alpha - \beta_\kappa^{(\tau)}|$ if $\tau \ne 1$. Therefore, the fewer of the τ_κ ($\kappa = 1, \ldots, k$) that have value one, the smaller are the chances to do anything nontrivial about $P(\beta_1^{(\tau_1)}, \ldots, \beta_k^{(\tau_k)})$, even if we change the specification (1.7) of $\mathfrak{M}$. Since we want an effective estimate of

$$\prod_{\tau_1, \ldots, \tau_k = 1, \ldots, t} P(\beta_1^{(\tau_1)}, \ldots, \beta_k^{(\tau_k)})$$

for large k it is essential, however, to give nontrivial upper bounds for a positive fraction of all factors. Now, by the law of large numbers for almost all vectors (τ) about k/t of the components τ_κ are one. We can and must concentrate, therefore, on estimating the $P(\beta_\kappa^{(\tau_\kappa)})$ with

$$(1.10) \qquad \sum_{\kappa, \tau_\kappa = 1} 1 \sim \frac{k}{t}.$$

It is probably optimal, though we cannot prove it, to choose $\mathfrak{M}$ in such a way that all $P(\beta_\kappa^{(\tau_\kappa)})$ with (1.10) can be equally well estimated. This idea leads to setting

$$(1.11) \quad \mathfrak{M} := \left\{(i);\, 0 \le i_\kappa \le r_\kappa,\, \sum_\kappa{}' \frac{i_\kappa}{r_\kappa} \le \vartheta \text{ for at least one } m\text{-term subsum } \sum{}' \right\}$$

with $m = [(1 - \epsilon)k/t]$ and ϑ as large as condition (1.6) allows. Probabilistic considerations show that

$$\vartheta = (1 - \epsilon)k/2t^2$$

is an admissible value, $(1 + \epsilon)k/2t^2$ is not.

1.2.4. The numbers $P(\beta_\kappa^{(\tau_\kappa)})$ need not all be conjugates, as we have assumed so far and as is used in the proof sketched above. However, if we have infinitely many solutions of (1.9) then there is a normal extension $\mathfrak{N}$ of $\mathbf{Q}$ of finite degree and an infinite selection of the β_κ such that all these β_κ have the same degree t_1, say, over $\mathfrak{N}$ and take their conjugates with respect to $\mathfrak{N}$ independently (Lemma 18 in §5.3). In this situation we can work similarly as above.

1.2.5. It is very convenient to have the conjugates $\beta_\kappa^{(\tau_\kappa)}$ bounded. This can be achieved by a linear transformation and, if necessary, by selecting the β_κ further. We do this step in §5.5. The method seems to be due to Sprindžuk [12].

1.2.6. If information on the approximation of α by the conjugates $\beta_\kappa^{(\tau_\kappa)}$, $\tau_\kappa \neq 1$, is given, there is no new difficulty to exploit it by our method. Thus we arrive at the Theorems 2 and 3. The construction of the set $\mathfrak{M}$ must be refined a little further.

1.2.7. Theorem 4 requires for its proof a polynomial P that vanishes at all points $(\alpha_{\nu_1}, \ldots, \alpha_{\nu_k})$, where $\nu_1, \ldots, \nu_k$ are arbitrary, $1 \leq \nu_\kappa \leq n$. Thus, if $\beta_\kappa^{(\tau_\kappa)}$ approximates any one of the α_ν, this can be used to estimate $P(\beta_\kappa^{(\tau_\kappa)})$. The original set-up does not seem to give such a polynomial easily, so we work differently instead. $P(x_1, \ldots, x_k)$ is written down in the form of an interpolation series as a linear combination of power products of factors $(x_\kappa - \alpha_\nu)$ with coefficients in $\mathbf{Q}(\alpha_1, \ldots, \alpha_n)$. Thereby the zeros of the required order are immediately built in. For the unknowns, the rational components of the coefficients, we have again linear conditions. They come from the requirement that P as a polynomial in $x_1, \ldots, x_k$ must have rational coefficients.

1.2.8. Theorems 2, 3, 4, which deal with estimates of $Q(\alpha)$ are proved for irreducible polynomials Q in the first instance. For the reduction of the general case to that with irreducible polynomials an important tool is of course the approximate multiplicativity of the height

$$C_1(n) \prod_i \|Q_i\| \leq \left\| \prod_i Q_i \right\| \leq C_2(n) \prod_i \|Q_i\|$$

($n = \deg \prod Q_i$), the essential part of which, namely

$$\prod_{\tau=1}^{t} (1 + |\beta^{(\tau)}|) \leq 6^t \left\| \prod_{\tau=1}^{t} (x - \beta^{(\tau)}) \right\|,$$

is due to Siegel [9].

Other tools that we need for this purpose are the inequalities on M and L in §3.

2. Formulation of the results

2.1. *Notations*

$\mathbf{Q}$: the set of rational numbers.

$\mathbf{Z}$: the set of rational integers.

The letters α, β are reserved for algebraic numbers, the α (or α_ν) being fixed and the β (β_κ, $\beta_{\nu\kappa}$) approximating the α.

P, Q denote polynomials, usually $P(\alpha) = 0$, $Q(\beta) = 0$.

$$D_{(i)} := D_{(i_1,\ldots,i_k)} := \prod_{\kappa=1}^{k} \frac{\partial^{i_\kappa}}{i_\kappa!\, \partial x_\kappa^{i_\kappa}}.$$

C stands for positive constants. We do not always distinguish different constants. Dependences may be shown as in $C(\alpha)$, $C(s, t)$.

$\log^+ x := \max(0, \log x)$.

M, L are the symmetric functions of nonnegative variables that in the region

$$a_1 \geq \cdots \geq a_n \geq 0$$

are given by

$$M(a_\nu \mid \nu = 1, \ldots, n) = \sum_{\nu=1}^{n} (2\nu - 1)a_\nu,$$

$$L(a_\nu \mid \nu = 1, \ldots, n) = \max_{r=1}^{n} r^2 \left(\sum_{\nu=1}^{r} \frac{1}{a_\nu}\right)^{-1}.$$

2.2. *The theorems*

THEOREM 1. *Let α be an algebraic, t a natural number. Then for any real $\phi > 2t$ there are at most finitely many algebraic numbers β of degree $\leq t$ such that*

$$|\alpha - \beta| \leq \|\beta\|^{-\phi}.$$

THEOREM 1.1. *Let α be an algebraic, t' a natural number and $\mathfrak{K}$ a numberfield of finite degree over* **Q**. *If $\phi > 2t'$ then there are at most finitely many numbers β of degree $\leq t'$ relative to $\mathfrak{K}$ such that $|\alpha - \beta| \leq \|\beta\|^{-\phi}$.*

THEOREM 2. *Let α be an algebraic and t a natural number. If $\phi > 2t$ then there are at most finitely many $Q \in \mathbf{Z}[x]$ of degree $\leq t$ such that*

$$0 \neq |Q(\alpha)| \leq \|Q\|^{1-\phi}.$$

THEOREM 2.1. *Let α be an algebraic, t and t' natural numbers and $\mathfrak{K}$ a number field of finite degree. If $\phi > 2t'$ then there are at most finitely many polynomials $Q \in \mathbf{Z}[x]$ of degree $\leq t$ that split over $\mathfrak{K}$ into factors of degree $\leq t'$ such that*

$$0 \neq |Q(\alpha)| \leq \|Q\|^{1-\phi}.$$

THEOREM 2.2. *Let α be a nonreal algebraic and t a natural number. If $\phi > t$ then there are at most finitely many $Q \in \mathbf{Z}[x]$ of degree $\leq t$ such that*

$$0 \neq |Q(\alpha)| \leq \|Q\|^{1-\phi}.$$

THEOREM 3. *Let α be an algebraic and t a natural number. If $\phi > 2t$ then there are at most finitely many $Q \in \mathbf{Z}[x]$,*

$$Q(x) = b_0 \prod_{\tau=1}^{t} (x - \beta^{(\tau)}),$$

such that $Q(\alpha) \neq 0$ and

$$M(\log^+ |\alpha - \beta^{(\tau)}|^{-1} \mid \tau = 1, \ldots, t) > \phi \log \|Q\|.$$

THEOREM 4. *Let $\alpha_1, \ldots, \alpha_n$ be distinct algebraic numbers and t a natural number. If $\phi > 2t$ then there are at most finitely many $Q \in \mathbf{Z}[x]$ of degree $\leq t$*

such that

$$L\left(\log^+ \frac{\|Q\|}{|Q(\alpha_\nu)|} \,\middle|\, \nu = 1, \ldots, n\right) \geq \phi \log \|Q\|$$

and

$$Q(\alpha_\nu) \neq 0 \quad \textit{for } \nu = 1, 2, \ldots, n.$$

THEOREM 5. *Let* $P \in \mathbf{Z}[x]$, $\deg P = s$ *and* t *a natural number. Suppose that* P *has no multiple zeros. If*

$$\phi > 2t\left(1 + \frac{1}{3} + \cdots + \frac{1}{2t-1}\right)$$

then there are at most finitely many $Q \in \mathbf{Z}[x]$ *of degree* $\leq t$ *such that*

$$0 \neq |\mathrm{Res}\,(P, Q)| \leq \|Q\|^{s-\phi}.$$

3. Some inequalities

3.1. We consider for the beginning

$$N(a; b) := N(a_1, \ldots, a_m; b_1, \ldots, b_n) := \sum_{i,j} \min\,(a_i, b_j)$$

together with the function M already mentioned:

$$\begin{aligned} M(a) &= M(a_1, \ldots, a_m) = M(a_i \mid i = 1, \ldots, m) \\ &:= N(a; a) = \sum_{i,j} \min\,(a_i, a_j). \end{aligned}$$

Obviously M is a symmetric function and N is symmetric in the a as well as in the b. If $a_1 \geq a_2 \geq \cdots$ one easily sees that

$$M(a) = \sum_i (2i - 1)a_i.$$

In any case we therefore have

$$M(a) = \sum_i (2i - 1)a_{\pi_i}, \tag{3.1}$$

where π is such a permutation that

$$a_{\pi_1} \geq a_{\pi_2} \geq \cdots \geq a_{\pi_m}. \tag{3.2}$$

For any other permutation the right hand side of (3.1) takes larger values, as follows from the easy and well-known theorem that $\sum a_{\pi_i} b_i$ with $b_1 < b_2 < \cdots$ becomes minimal under the condition (3.2) only. A handy representation for M and N is given by

$$N(a; b) = \int_0^\infty A(x)B(x)\,dx, \qquad M(a) = \int_0^\infty A^2(x)\,dx, \tag{3.3}$$

where

$$A(x) = \sum_{i, a_i \geq x} 1, \qquad B(x) = \sum_{j, b_j \geq x} 1.$$

The proof is immediate:

$$\int_0^\infty AB\,dx = \int_0^\infty \sum_{\substack{i,j \\ a_i \geq x, b_j \geq x}} 1\,dx = \sum_{i,j} \int_{\substack{0 \leq x \leq a_i; \\ 0 \leq x \leq b_j}} dx$$
$$= \sum_{i,j} \min(a_i, b_j).$$

Lemma 1.

$$N(a; b) \leq M(a)^{1/2} M(b)^{1/2}, \tag{3.4}$$

$$M(a_1, \ldots, a_m, b_1, \ldots, b_n)^{1/2} \leq M(a)^{1/2} + M(b)^{1/2}. \tag{3.5}$$

Formula (3.5) *generalizes immediately to subdivisions into more subsets.*

Proof. By the Schwarz inequality

$$N(a; b) = \int_0^\infty AB\,dx \leq \left(\int_0^\infty A^2\,dx \int_0^\infty B^2\,dx\right)^{1/2}$$
$$= (M(a)M(b))^{1/2},$$

and by the Minkowski inequality

$$\left(\int_0^\infty (A + B)^2\,dx\right)^{1/2} \leq \left(\int_0^\infty A^2\,dx\right)^{1/2} + \left(\int_0^\infty B^2\,dx\right)^{1/2},$$

which is (3.5). (3.5) can also be deduced from (3.4).

Lemma 2. *With any set of coefficients* $\gamma_i > 0$ *we have*

$$M(a_{ij} \mid i, j) \leq \left(\sum_i \gamma_i\right) \sum_i \frac{1}{\gamma_i} M(a_{ij} \mid j).$$

Proof. Lemma 1 and the Schwarz inequality give

$$M(a_{ij} \mid i, j) \leq \left(\sum_i M(a_{ij} \mid j)^{1/2}\right)^2$$
$$= \left(\sum_i \gamma_i^{1/2} \frac{1}{\gamma_i^{1/2}} M(a_{ij} \mid j)^{1/2}\right)^2$$
$$\leq \left(\sum_i \gamma_i\right)\left(\sum_i \frac{1}{\gamma_i} M(a_{ij} \mid j)\right).$$

LEMMA 3. *With* $n = \sum_{i=1}^{m} n_i$ *we have*

$$M(a_{ij} \mid i = 1, \dots, m; j = 1, \dots, n_i) \le M(a_{1j} \mid j) + 2n \sum_{i=2}^{m} M(a_{ij} \mid j).$$

PROOF. Take

$$A_i(x) := \sum_{j, a_{ij} \ge x} 1$$

then $A_i(x) \le n_i$ and

$$\begin{aligned} \left(\sum_1^m A_i(x)\right)^2 &= A_1^2(x) + \left(2A_1(x) + \sum_{i=2}^m A_i(x)\right) \sum_{i=2}^m A_i(x) \\ &\le A_1^2(x) + \left(2n_1 + \sum_2^m n_i\right) \sum_2^m A_i(x), \qquad (3.6) \\ \left(\sum_1^m A_i(x)\right)^2 &\le A_1^2(x) + 2n \sum_2^m A_i^2(x). \end{aligned}$$

Integrating this inequality gives the result. For a later application we also integrate (3.6) in the particular case $m = 2$, $n_1 = n_2$. This leads to

$$M(a_{ij} \mid i = 1, 2; j = 1, \dots, n_1) \le M(a_{1j} \mid j) + 3n_1 \sum_j a_{2j}. \qquad (3.7)$$

LEMMA 4.

$$M\left(\sum_{j=1}^{n_i} a_{ij} \;\middle|\; i = 1, \dots, m\right) \le M(a_{ij} \mid i = 1, \dots, m; j = 1, \dots, n_i).$$

PROOF. The right hand side can be transformed into the left hand side by a number of steps each of which replaces a pair of variables, u and u', say, by $u + u'$ and 0. It is enough, therefore, to study a single such step. Out of $u_1, \dots, u_n$ we want to add u_r and u_s. Without loss of generality let $u_1 \ge u_2 \ge \cdots \ge u_n$ and $r < s$. Then

$$\begin{aligned} M(u_i \mid i) &= \sum_{i \ne r,s} (2i - 1)u_i + (2r - 1)u_r + (2s - 1)u_s \\ &\ge \sum_{i \ne r,s} (2i - 1)u_i + (2r - 1)(u_r + u_s) + (2s - 1) \cdot 0 \\ &= \sum_{i=1}^{n} (2i - 1)v_i, \end{aligned}$$

where $v_i = u_i$ for $i \ne r, s$; $v_r = u_r + u_s$, $v_s = 0$. By an earlier remark we have

$$\sum_{i=1}^{n} (2i - 1)v_i \ge M(v_i \mid i), \qquad M(u_i \mid i) \ge M(v_i \mid i).$$

By a combination of Lemmata 1 and 4 one obtains

$$M(a_i + b_i \mid i)^{1/2} \le M(a_i \mid i)^{1/2} + M(b_i \mid i)^{1/2} \qquad (3.8)$$

which, on the surface, resembles the Minkowski inequality. One must not forget, however, that M is (positive) homogeneous of the first degree only.

LEMMA 5. *If in* $(b_1, \ldots, b_n)$ *each number occurs at least* h *times as often as in* $(a_1, \ldots, a_m)$ *then* $M(b) \geq h^2 M(a)$.

The proof of this rather trivial statement goes from $B(x) \geq hA(x)$ via (3.3).

3.2. We turn to considering the function L which if

$$a_1 \geq a_2 \geq \cdots \geq 0$$

is defined by

$$L(a_i \mid i = 1, \ldots, n) := \max_{r=1}^{n} r^2 \left(\sum_{i=1}^{r} \frac{1}{a_i} \right)^{-1} \tag{3.9}$$

and which is symmetrical besides.

LEMMA 6. *Allowing* $\gamma_i \geq 0$ *only we have*

$$\max_{\Sigma\gamma_i=1} M(\gamma_i a_i \mid i = 1, \ldots, n) = L(a_i \mid i = 1, \ldots, n).$$

PROOF. If any $a_i = 0$ the lemma reduces to the case $n - 1$. Since the case $n = 1$ is trivial we can assume that all $a_i > 0$. The domain

$$\gamma_i \geq 0, \qquad \sum \gamma_i = 1$$

is split into convex polyhedra $H(\pi)$ by

$$\gamma_{\pi_1} a_{\pi_1} \geq \gamma_{\pi_2} a_{\pi_2} \geq \cdots \geq \gamma_{\pi_n} a_{\pi_n} \tag{3.10}$$

where π runs through all permutations of $\{1, \ldots, n\}$. In each $H(\pi)$ the function

$$M(\gamma_i a_i \mid i) = \sum_i (2i - 1)\gamma_{\pi_i} a_{\pi_i}$$

is linear in the γ_i. The maximum in $H(\pi)$ is taken at an extremal point of $H(\pi)$, that is a point $(\gamma_1, \ldots, \gamma_n)$ that fulfills n linearly independent ones of the boundary equations

$$\gamma_1 = 0, \ldots, \qquad \gamma_n = 0, \qquad \sum_i \gamma_i = 1, \tag{3.11}$$

$$\gamma_{\pi_1} a_{\pi_1} = \gamma_{\pi_2} a_{\pi_2}, \ldots, \gamma_{\pi_{n-1}} a_{\pi_{n-1}} = \gamma_{\pi_n} a_{\pi_n}. \tag{3.12}$$

Consider an extremal point with exactly r of the γ_i nonvanishing. In view of (3.10) these are $\gamma_{\pi_1}, \ldots, \gamma_{\pi_r}$, while $\gamma_{\pi_{r+1}} = \cdots = \gamma_{\pi_n} = 0$. Thus of (3.11) $n - r + 1$ equations hold. Of (3.12) the equations $\gamma_{\pi_{r+1}} a_{\pi_{r+1}} = \cdots = \gamma_{\pi_n} a_{\pi_n}$ depend on the former ones and $\gamma_{\pi_r} a_{\pi_r} = \gamma_{\pi_{r+1}} a_{\pi_{r+1}}$ does not hold. So we have to use all the other equations of (3.12) to make the number of n linearly independent conditions full:

$$\gamma_{\pi_1} a_{\pi_1} = \cdots = \gamma_{\pi_r} a_{\pi_r}.$$

Call the common value λ. Then

$$1 = \sum_i \gamma_i = \sum_{i=1}^{r} \frac{\lambda}{a_{\pi_i}}, \qquad \lambda = \left(\sum_{i=1}^{r} \frac{1}{a_{\pi_i}}\right)^{-1},$$

$$\begin{aligned} M(\gamma_i a_i \mid i = 1, \ldots, n) &= M(\lambda \mid i = 1, \ldots, r) \\ &= r^2 \lambda \\ &= r^2 \left(\sum_{i=1}^{r} \frac{1}{a_{\pi_i}}\right)^{-1}. \end{aligned}$$

The maximum of these values, taken over all π and r is, by definition, $L(a_1, \ldots, a_n)$. By Lemma 6 one can deduce inequalities for L from the known inequalities of M.

LEMMA 7.

(3.13) $$L(a_i + b_i \mid i)^{1/2} \leq L(a_i \mid i)^{1/2} + L(b_i \mid i)^{1/2},$$

(3.14) $$L\left(\sum_{j=1}^{n} a_{ij} \mid i = 1, \ldots, m\right) \leq \left(\sum_j \gamma_j\right) \sum_j \frac{1}{\gamma_j} L(a_{ij} \mid i).$$

PROOF. By (3.8)

$$\begin{aligned} L(a + b)^{1/2} &= \max_{\Sigma\gamma_i = 1} M(\gamma a + \gamma b)^{1/2} \\ &\leq \max_{\gamma} (M(\gamma a)^{1/2} + M(\gamma b)^{1/2}) \\ &\leq \max_{\gamma} M(\gamma a)^{1/2} + \max_{\gamma} M(\gamma b)^{1/2} \\ &= L(a)^{1/2} + L(b)^{1/2}. \end{aligned}$$

In the same way (3.14) is obtained from a combination of Lemmata 2 and 4.

It should be noted that (3.13) with the harmonic mean value instead of L does not hold. A counterexample is $(a) = (1, 0)$ and $(b) = (0, 1)$.

Applying our argument to Lemma 1 (3.5) one obtains

$$L(a_1, \ldots, a_m, b_1, \ldots, b_n) \leq L(a_1, \ldots, a_m) + L(b_1, \ldots, b_n).$$

But this can be seen without difficulties from the definition, since the corresponding inequality holds already without the "max" in (3.9).

LEMMA 8.

$$L(a_i + b_i \mid i = 1, \ldots, n) \leq L(a_i \mid i) + 3n \max_i b_i.$$

PROOF. We combine Lemma 4 and (3.7).

$$\begin{aligned} L(a+b) &= \max_{\gamma} M(\gamma a + \gamma b) \\ &\leq \max_{\gamma} \Big(M(\gamma a) + 3n \sum_i \gamma_i b_i \Big) \\ &\leq \max_{\gamma} M(\gamma a) + 3n \max_{\gamma} \big(\sum \gamma_i b_i \big) \\ &= L(a) + 3n \max_i b_i. \end{aligned}$$

LEMMA 9.

$$\max_{L(a) \leq u} \sum_{i=1}^{m} a_i = u \sum_{i=1}^{m} \frac{1}{2i-1}.$$

PROOF. By Lemma 6 we have

$$M(\gamma_i a_i \mid i) \leq L(a) \sum_i \gamma_i$$

whatever the $\gamma_i \geq 0$. Assuming $a_1 \geq a_2 \geq \cdots$ and picking

$$\gamma_i = \frac{1}{2i-1}$$

gives

$$\begin{aligned} \sum a_i &= \sum (2i-1) \frac{a_i}{2i-1} = M(\gamma a) \\ &\leq L(a) \sum \gamma_i \leq u \sum_i \frac{1}{2i-1}. \end{aligned}$$

On the other hand, if we take $a_i = u/(2i-1)$, then

$$\sum_{i=1}^{r} \frac{1}{a_i} = \frac{1}{u} \sum_{1}^{r} (2i-1) = \frac{r^2}{u}.$$

Hence $L(a) = u$, while

$$\sum_i a_i = u \sum_i \frac{1}{2i-1}.$$

4. Proof of the Main Lemma

4.1. *Construction of the approximating polynomial*

LEMMA 10. *Let $\alpha_1, \ldots, \alpha_n$ be algebraic numbers and $s \geq 2$ the degree of the field*

$$\mathfrak{K} := \mathbf{Q}(\alpha_1, \ldots, \alpha_n).$$

Let

$$k, r_1, \ldots, r_k \in \mathbf{N}$$

be arbitrary and $\mathfrak{M}$ *a set of k-tuples* $(i) = (i_1, \ldots, i_k)$ *with*

$$0 \leq i_\kappa \leq r_\kappa, \qquad i_\kappa \in \mathbf{Z}$$

such that

$$|\mathfrak{M}| \leq \frac{1}{2s} \prod_{\kappa=1}^{k} (r_\kappa + 1). \tag{4.1}$$

Further let $p_i(x)$, $i = 0, 1, \ldots$ *be a sequence of polynomials of one variable and of the form*

$$p_i(x) = \prod_{\nu=1}^{n} (x - \alpha_\nu)^{j_\nu(i)}$$

where

$$\deg p_i = \sum_\nu j_\nu(i) = i. \tag{4.2}$$

Then there are numbers

$$h_{(i)} \in \mathfrak{K} \qquad (0 \leq i_\kappa \leq r_\kappa)$$

such that

$$\begin{aligned} |h_{(i)}| &\leq C_6^{\sum_{\kappa=1}^k r_\kappa} && \text{for all } (i), \\ h_{(i)} &= 0 && \text{for } (i) \in \mathfrak{M}, \end{aligned} \tag{4.3}$$

and such that for

$$P(x_1, \ldots, x_k) := \sum_{(i)} h_{(i)} p_{i_1}(x_1) \ldots p_{i_k}(x_\kappa) \text{ we have}$$

$$P \in \mathbf{Z}[x_1, \ldots, x_k], \qquad P \neq 0. \tag{4.4}$$

Furthermore

$$\|P\| \leq C_7^{\sum_{\kappa=1}^k r_\kappa}. \tag{4.5}$$

The constants C_6 *and* C_7 *depend on* $\alpha_1, \ldots, \alpha_n$ *only.*

PROOF. We choose some integral primitive element ϑ of $\mathfrak{K}$. By the basis $\{\vartheta^\sigma; \sigma = 0, \ldots, s-1\}$ we represent the α_ν:

$$\alpha_\nu = \frac{1}{d} \sum_{\sigma=0}^{s-1} a_{\nu\sigma} \vartheta^\sigma, \qquad d \in \mathbf{N}, a_{\nu\sigma} \in \mathbf{Z}.$$

For the $h_{(i)}$ we put

$$h_{(i)} = d^{\sum_\kappa i_\kappa} \sum_{\sigma=0}^{s-1} h_{(i)\sigma} \vartheta^\sigma, \qquad h_{(i)\sigma} \in \mathbf{Z}.$$

The polynomial P also splits over the basis $\{\vartheta^\sigma\}$:

$$P(x) = \sum_{\sigma=0}^{s-1} P_\sigma(x)\vartheta^\sigma.$$

Condition (4.4) apparently means

$$P_\sigma = 0 \quad \text{for } \sigma = 1, \ldots, s-1.$$

As is well known and can easily be seen by induction we have

$$\vartheta^r = \sum_{\sigma=0}^{s-1} d_{r\sigma}\vartheta^\sigma \quad \text{for all } r \in \mathbf{N} \text{ with } d_{r\sigma} \in \mathbf{Z},$$
$$|d_{r\sigma}| \leq C_1^r, \qquad C_1 := 1 + \|\vartheta\|.$$

Similarly for each product γ_r of r factors α_ν one obtains a representation

$$\gamma_r = \frac{1}{d^r}\sum_{\sigma=0}^{s-1} g_\sigma\vartheta^\sigma, \qquad g_\sigma \in \mathbf{Z}$$

with

$$|g_\sigma| \leq C_2^r, \qquad C_2 := s^2C_1^{2s}\max_{\nu,\sigma}|a_{\nu\sigma}|.$$

Multiplication of $p_{i_1}(x_1)\ldots p_{i_k}(x_k)$ produces a polynomial with coefficients that are sums of not more than $2^{\Sigma i_\kappa}$ terms $\pm\gamma_r$, $r \leq \sum i_\kappa$. Splitting $p_{i_1}\ldots p_{i_k}$ into components with respect to our basis therefore yields

$$p_{i_1}(x_1)\ldots p_{i_k}(x_k) = d^{-\Sigma_\kappa i_\kappa}\sum_{\sigma=0}^{s-1}\vartheta^\sigma\Omega_{(i)\sigma}(x_1, \ldots, x_k)$$

with

$$\Omega_{(i)\sigma} \in \mathbf{Z}[x_1, \ldots, x_k],$$
$$\|\Omega_{(i)\sigma}\| \leq C_3^{\Sigma i_\kappa} \leq C_3^{\Sigma r_\kappa}, \qquad C_3 := 2dC_2.$$

From

$$P(x_1, \ldots, x_k) = \sum_{(i)}\Big(\sum_\sigma h_{(i)\sigma}\vartheta^\sigma\Big)\Big(\sum_\tau \Omega_{(i)\tau}(x)\vartheta^\tau\Big)$$
$$= \sum_{(i),\sigma,\tau,\rho} h_{(i)\sigma}\Omega_{(i)\tau}(x)\, d_{\sigma+\tau,\rho}\vartheta^\rho$$

we infer

$$P_\rho(x) = \sum_{(i),\sigma} h_{(i)\sigma}\sum_\tau d_{\sigma+\tau,\rho}\Omega_{(i)\tau}(x).$$

This shows that the condition (4.4), i.e. $P_1 = \cdots = P_{s-1} = 0$ represents

$$M := (s-1)\prod_\kappa (r_\kappa + 1) \tag{4.6}$$

homogeneous linear equations for the $h_{(i)\sigma}$. Since $h_{(i)\sigma} = 0$ for $(i) \in \mathfrak{M}$ the number of unknowns is

$$(4.7)\qquad \begin{aligned} N &:= s\left(\prod_\kappa (r_\kappa + 1) - |\mathfrak{M}|\right) \\ &\geq s\left(1 - \frac{1}{2s}\right)\prod_\kappa (r_\kappa + 1). \end{aligned}$$

The moduli of the coefficients in the linear forms are bounded by

$$A := sC_1^s C_3^{\Sigma r_\kappa} \leq C_4^{\Sigma r_\kappa}.$$

As is well known (see e.g. Schneider [6, Hilfssatz 29]) such a system has a nontrivial solution in rational integers $h_{(i)\sigma}$ with

$$|h_{(i)\sigma}| \leq (NA)^{M/(N-M)} + 2$$

provided $M < N$. In fact, by (4.6) and (4.7),

$$\frac{N - M}{M} \geq \frac{(s - \frac{1}{2}) - (s - 1)}{s - 1} = \frac{1}{2(s - 1)} \geq \frac{1}{2s}$$

and

$$\begin{aligned} NA &\leq s\prod_\kappa (r_\kappa + 1)C_4^{\Sigma r_\kappa} \\ &\leq (s \cdot 2 \cdot C_4)^{\Sigma r_\kappa} \\ &=: C_5^{\Sigma r_\kappa}. \end{aligned}$$

Therefore we find nontrivial $h_{(i)\sigma}$ with

$$\begin{aligned} |h_{(i)\sigma}| &\leq C_5^{2s\Sigma r_\kappa} + 2, \\ |h_{(i)}| &\leq \sum_d i_\kappa \sum_0^{s-1} |h_{(i)\sigma}|\, |\vartheta|^\sigma \leq C_6^{\Sigma r_\kappa}, \end{aligned}$$

which is (4.3).

By construction $p_i(x)$ has degree i. Hence, the products $p_{i_1}(x_1) \ldots p_{i_k}(x_k)$ are linearly independent, and the fact that not all $h_{(i)}$ vanish implies that P is not the zero polynomial. From (4.3) one obtains (4.5) easily by expanding all $p_{i_1}(x_1) \ldots p_{i_k}(x_k)$ into power products of $x_1, \ldots, x_k$,

$$C_7 := 2\left(1 + \max_\nu |\alpha_\nu|\right) C_6$$

is a possible choice.

4.2. *Two versions of the law of large numbers.* The k-dimensional cube

$$\mathfrak{W}_k := \{(i);\, i_\kappa \in \mathbf{Z},\, 0 \leq i_\kappa \leq r_\kappa\}$$

with the measure

$$\mu\mathfrak{A} := \frac{|\mathfrak{A}|}{\prod_\kappa (r_\kappa + 1)}$$

shall be considered a probability space. We are interested in the stochastic variables

$$\xi_\kappa := \frac{i_\kappa}{r_\kappa} \qquad (\kappa = 1, \ldots, k).$$

Apparently these are independent.

Expectation and dispersion are denoted by E, D respectively.

LEMMA 11. *Let* $0 < \epsilon < 1$,

$$k \geq 12/\epsilon^4, \tag{4.8}$$

$$r_\kappa \geq 3/\epsilon \quad \textit{for } \kappa = 1, \ldots, k. \tag{4.9}$$

Then there is a set $\mathfrak{M} \subset \mathfrak{W}_k$ *with*

$$\mu\mathfrak{M} \leq \epsilon \qquad \left(|\mathfrak{M}| \leq \epsilon \prod_\kappa (r_\kappa + 1)\right) \tag{4.10}$$

such that for all $(i) \notin \mathfrak{M}$ *and all* $x \in [0, 1]$

$$\left| \sum_{\kappa, \xi_\kappa \leq x} 1 - kx \right| \leq \epsilon k. \tag{4.11}$$

For the proof write

$$\chi_\kappa(x) := 1 \quad \text{if } \xi_\kappa \leq x,$$
$$:= 0 \quad \text{if } \xi_\kappa > x,$$

then

$$s(x) := \sum_{\kappa=1}^{k} \chi_\kappa(x) = \sum_{\kappa, \xi_\kappa \leq x} 1.$$

Apparently

$$E\chi_\kappa(x) = \frac{[3xr_\kappa + 1]}{r_\kappa + 1},$$

$$|E\chi_\kappa(x) - x| = \left| \frac{1 - x - xr_\kappa + [xr_\kappa]}{r_\kappa + 1} \right|$$

$$\leq \frac{1}{r_\kappa + 1} \leq \frac{\epsilon}{3},$$

$$|Es(x) - kx| \leq k\left(\frac{\epsilon}{3}\right).$$

Since χ_κ takes values 0 and 1 only, we have $D^2\chi_\kappa(x) \leq \frac{1}{4}$ and since the χ_κ are independent

$$D^2s(x) = \sum_\kappa D^2\chi_\kappa(x) \leq \frac{k}{4}.$$

Now the Tschebyscheff inequality gives

$$\mu\left\{(i);\ |s(x) - Es(x)| \geq \frac{\epsilon}{3}k\right\} \leq \frac{9}{\epsilon^2k^2} D^2s(x) \leq \frac{9}{4\epsilon^2 k}.$$

If we set

$$\mathfrak{M} := \bigcup_{\nu=0}^{n} \left\{(i);\ \left| s\left(\frac{\nu}{n}\right) - Es\left(\frac{\nu}{n}\right)\right| > \frac{\epsilon}{3}k\right\}$$

where n is chosen according to $3/\epsilon < n \leq 4/\epsilon$ then by (4.8)

$$\mu\mathfrak{M} \leq (n+1)\frac{9}{4\epsilon^2 k} \leq \frac{5}{\epsilon}\frac{9}{4\epsilon^2 k} \leq \frac{5 \cdot 9}{4 \cdot 12}\epsilon < \epsilon.$$

On the other hand for $(i) \notin \mathfrak{M}$ and $x \in [0, 1]$,

$$(\nu - 1)/n \leq x \leq \nu/n,$$

say,

$$\begin{aligned} s(x) &\leq s\left(\frac{\nu}{n}\right) \leq Es\left(\frac{\nu}{n}\right) + \frac{\epsilon}{3}k \\ &\leq k\left(\frac{\nu}{n} + \frac{2\epsilon}{3}\right) \leq k\left(x + \frac{1}{n} + \frac{2\epsilon}{3}\right) \\ &\leq k(x+\epsilon). \end{aligned}$$

Similarly $s(x) \geq k(x - \epsilon)$, hence (4.11).

LEMMA 12. *Let $\mathfrak{M}$ be the set of Lemma 11 and let $\eta_1, \ldots, \eta_k$ be such a permutation of $\xi_1, \ldots, \xi_k$ that $\eta_1 \leq \eta_2 \leq \cdots \leq \eta_k$. Then for all $(i) \notin \mathfrak{M}$ and for all $\kappa = 1, 2, \ldots, k$*

$$|\eta_\kappa - \kappa/k| \leq \epsilon. \tag{4.12}$$

PROOF. By definition of s and of the η_κ we have

$$s(\eta_\kappa -) < \kappa \leq s(\eta_\kappa).$$

From (4.11) we see

$$s(\eta_\kappa) \leq k(\eta_\kappa + \epsilon), \qquad s(\eta_\kappa -) \geq k(\eta_\kappa - \epsilon).$$

Taken together this gives (4.12).

LEMMA 13. *We consider the cube $\mathfrak{B}_k$ of k-tuples $(\tau_1, \ldots, \tau_k)$ with $\tau_\kappa = 1, 2, \ldots, t$. Let $0 < \epsilon < 1$. Then there is a set $\mathfrak{T} \subset \mathfrak{B}_k$ such that*

$$|\mathfrak{T}| \leq t^k/(\epsilon^2 k) \tag{4.13}$$

and such that for all $(\tau) \notin \mathfrak{S}$ and $\sigma = 1, \ldots, t$

$$\left|\sum_{\kappa, \tau_\kappa = \sigma} 1 - \frac{k}{t}\right| \leq \epsilon k. \tag{4.14}$$

The proof goes as that of Lemma 11, using

$$\begin{aligned}\chi_\kappa(\sigma) &:= 1 \quad \text{if } \tau_\kappa = \sigma,\\ &:= 0 \quad \text{if } \tau_\kappa \neq \sigma,\\ s(\sigma) &:= \sum_\kappa \chi_\kappa(\sigma),\\ E\chi_\kappa(\sigma) &= \frac{1}{t}, \qquad D^2\chi_\kappa(\sigma) = \frac{1}{t}\left(1 - \frac{1}{t}\right) < \frac{1}{t},\\ Es(\sigma) &= \frac{k}{t}, \qquad D^2 s(\sigma) < \frac{k}{t}.\end{aligned}$$

For

$$\mathfrak{T} := \bigcup_{\sigma=1}^{t} \left\{(\tau); \left|s(\sigma) - \frac{k}{t}\right| > \epsilon k\right\}$$

the Tschebyscheff inequality gives

$$\mu\mathfrak{T} \leq 1/\epsilon^2 k.$$

4.3. *Estimates of* $D_{(l)}P(\beta)$

Lemma 14. *Let* P *be the polynomial of Lemma* 10 *and* $\beta_1, \ldots, \beta_k$ *numbers such that*

$$|\alpha_\nu - \beta_\kappa| \leq 1 \quad \textit{for all } \nu = 1, \ldots, n \textit{ and } \kappa = 1, \ldots, k.$$

Then for any $(l_1, \ldots, l_k)$ *we have*

$$|D_{(l)}P(\beta_\kappa)| \leq C_8^{\Sigma r_\kappa} \qquad (C_8 = 4C_6).$$

Proof. Differentiating the product $p_i(x) = \prod_\nu (x - \alpha_\nu)^{j v_\nu - u_\nu}$ produces

$$\frac{1}{l!}\frac{d^l}{dx^l} p_i(x) = \sum_{\substack{u_1,\ldots,u_n;\\ u_1+\ldots+u_n=l}} \prod_\nu \binom{j_\nu}{u_\nu} (x - \alpha_\nu)^{j_\nu - u_\nu}. \tag{4.15}$$

By $|\alpha_\nu - \beta_\kappa| \leq 1$ this yields

$$\begin{aligned}\left|\frac{1}{l!}\frac{d^l}{dx^l} p_i(\beta_\kappa)\right| &\leq \sum_{\substack{u_1,\ldots,u_n;\\ u_1+\ldots+u_n=l}} \prod_\nu \binom{j_\nu}{u_\nu}\\ &\leq \prod_\nu \sum_{u_\nu=0}^{j_\nu} \binom{j_\nu}{u_\nu}\\ &= 2^{\Sigma j_\nu} = 2^i.\end{aligned}$$

Thereby

$$\begin{aligned}|D_{(1)}P(\beta_1, \ldots, \beta_k)| &\leq \sum_{(i)} |h_{(i)}| \prod_\kappa \left|\frac{1}{l_\kappa!}\left(\frac{d}{dx_\kappa}\right)^{l_\kappa} p_{i_\kappa}(\beta_\kappa)\right|\\ &\leq C_6^{\Sigma r_\kappa} \sum_{(i)} 2^{\Sigma i_\kappa}\\ &\leq (2C_6)^{\Sigma r_\kappa} \prod_\kappa (r_\kappa + 1)\\ &\leq (4C_6)^{\Sigma r_\kappa}.\end{aligned}$$

Before we estimate $D_{(l)}P(\beta_\kappa)$ in a more detailed way we specialize the $p_i(x)$ that are used in the construction of P. We attach weights γ_ν to the α_ν such that

$$\gamma_\nu \geq 0, \qquad \sum_\nu \gamma_\nu = 1 \tag{4.16}$$

and put $j_\nu(i) = [\gamma_\nu i]$ for $\nu \leq n-1$, $j_n(i) = i - \sum_{\nu=1}^{n-1} j_\nu(i)$. Then we have

$$j_\nu(i) \geq i\gamma_\nu - 1 \quad \text{for } \nu = 1, \ldots, n. \tag{4.17}$$

Next we make the assumption

$$\epsilon \leq 1/2s. \tag{4.18}$$

Then the set $\mathfrak{M}$ constructed in Lemma 11 is a possible choice for Lemma 10, since (4.10) and (4.18) imply (4.1).

LEMMA 15. *Assume* (4.8), (4.9), (4.16)–(4.18) *and* $\epsilon \leq 1/t$. *Let* $\mathfrak{M}$ *and* $\mathfrak{T}$ *be the sets of Lemmata* 11 *and* 13, P *the polynomial of Lemma* 10. *Further let* $\beta_\kappa^{(\tau)}$ $(\kappa = 1, \ldots, k, \tau = 1, \ldots, t)$ *be complex numbers that approximate the* α_ν *according to*

$$|\beta_\kappa^{(\tau)} - \alpha_\nu| \leq q_\kappa^{-\phi_\nu(\tau)} \quad \textit{for all } \nu, \kappa, \tau, \tag{4.19}$$

with $q_\kappa \in \mathbf{N}$ *and* $\phi_\nu(\tau) \geq 0$ (*later on the* $\beta_\kappa^{(\tau)}$ $(\tau = 1, \ldots, t)$ *will be conjugates and* q_κ *their heights*).

Assume further that

$$q_\kappa^{r_\kappa} > q_1^{r_1} \qquad (\kappa = 2, \ldots, k) \tag{4.20}$$

and that for each τ *there is at most one* ν *with* $\phi_\nu(\tau) > 0$.

Then for any (l) *with*

$$\sum_\kappa \frac{l_\kappa}{r_\kappa} \leq \epsilon k \tag{4.21}$$

and any $(\tau) \notin \mathfrak{T}$ *we have*

$$|D_{(l)}P(\beta_1^{(\tau_1)}, \ldots, \beta_k^{(\tau_k)})| \leq C_8^{\Sigma r_\kappa} q_1^{kr_1(4\epsilon t\Phi - M/2t^2)}$$

where

$$\Phi := \max_{\nu,\tau} \phi_\nu(\tau),$$

$$M := M\left(\sum_\nu \gamma_\nu \phi_\nu(\tau) \,\middle|\, \tau = 1, \ldots, t\right).$$

PROOF. For an abbreviation put $\psi(\tau) = \sum_\nu \gamma_\nu \phi_\nu(\tau)$. Apparently $\psi(\tau) \leq \sum_\nu \gamma_\nu \Phi = \Phi$. If $\beta_\kappa^{(\tau_\kappa)}$ is substituted for x into (4.15) from (4.19) and (4.20) it follows that

$$\left|\frac{1}{l!}\frac{d^l}{dx^l}p_i(\beta_\kappa^{(\tau)})\right| \leq \sum_{\substack{u_1+\dots+u_n=l;\\ u_\nu\leq j_\nu}} \prod_\nu \binom{j_\nu}{u_\nu} |\beta_\kappa^{(\tau)}-\alpha_\nu|^{j_\nu-u_\nu}$$
$$\leq \sum_{\substack{u_1+\dots+u_n=l;\\ u_\nu\leq j_\nu}} \prod_\nu \binom{j_\nu}{u_\nu} q_\kappa^{(u_\nu-j_\nu)\phi_\nu(\tau)}$$
$$\leq \sum_{u_1+\dots+u_n=l} q_1^{(r_1/r_\kappa)\Sigma_\nu(u_\nu-j_\nu)\phi_\nu(\tau)} \prod_\nu \binom{j_\nu}{u_\nu}.$$

Since at most one $\phi_\nu(\tau)$ does not vanish and because of (4.17) the expression becomes

$$\leq q_1^{(r_1/r_\kappa)(l+1)\Phi-(r_1/r_\kappa)i\Sigma_\nu\gamma_\nu\phi_\nu(\tau)} \sum_{u_1,\dots,u_n}\prod_\nu \binom{j_\nu}{u_\nu}$$
$$= 2^i q_1^{(r_1/r_\kappa)((l+1)\Phi-i\psi(\tau))}.$$

If we take $i = i_\kappa$, $l = l_\kappa$, $\tau = \tau_\kappa$ and multiply, we obtain

$$\left|D_{(l)}\prod_\kappa p_{i_\kappa}(x_\kappa)\Big/_{x_\kappa=\beta_\kappa}(\tau_\kappa)\right| \leq 2^{\Sigma i_\kappa} q_1^{r_1(\Sigma_\kappa((2l_\kappa/r_\kappa)-(i_\kappa/r_\kappa)\Psi(\tau_\kappa))}$$
$$\leq 2^{\Sigma r_\kappa} q_1^{2\epsilon k r_1\Phi - r_1\Sigma_\kappa(i_\kappa/r_\kappa)\psi(\tau_\kappa)}, \tag{4.22}$$

where in the last step (4.21) has been applied.

Because of $h_{(i)} = 0$ for $(i) \in \mathfrak{M}$ we shall need (4.22) for $(i) \notin \mathfrak{M}$ only. So in the following treatment of $\sum(i_\kappa/r_\kappa)\psi(\tau_\kappa)$ we can use this assumption.

A suitable permutation π of the indices transforms the $\xi_\kappa = i_\kappa/r_\kappa$ into $\eta_\kappa = \xi_{\pi(\kappa)}$, such that $\eta_1 \leq \eta_2 \leq \cdots \leq \eta_k$. By Lemma 12 we find

$$\sum_\kappa \frac{i_\kappa}{r_\kappa}\psi(\tau_\kappa) = \sum_\kappa \eta_\kappa\psi(\tau_{\pi(\kappa)})$$
$$\geq \sum_\kappa\left(\frac{\kappa}{k}-\epsilon\right)\psi(\tau_{\pi(\kappa)})$$
$$\geq \sum_\kappa \frac{2\kappa-1}{2k}\psi(\tau_{\pi(\kappa)}) - \epsilon k\Phi.$$

According to the remark following (3.2) this is

$$\geq \frac{1}{2k}M(\psi(\tau_\kappa)\mid \kappa = 1,\dots,k) - \epsilon k\Phi.$$

Next we use the assumption $(\tau) \notin \mathfrak{T}$. By Lemma 13, τ_κ as $\kappa = 1,\dots,k$, takes each value $1,\dots,t$ at least m times with some

$$m \geq k(1/t - \epsilon) \geq 0.$$

Hence by Lemma 5

$$M(\psi(\tau_\kappa)\mid \kappa = 1,\dots,k) \geq m^2 M(\psi(\tau)\mid \tau = 1,\dots,t)$$
$$= m^2 M.$$

Taken together

$$\begin{aligned}\sum_{\kappa} \frac{i_\kappa}{r_\kappa} \psi(\tau_\kappa) &\geq \frac{m^2}{2k} M - \epsilon k \Phi \\ &\geq \frac{k}{2}\left(\frac{1}{t} - \epsilon\right)^2 M - \epsilon k \Phi \\ &\geq \frac{k}{2}\left(\frac{1}{t^2} - \frac{2\epsilon}{t}\right) M - \epsilon k \Phi.\end{aligned}$$

Since $M \leq t^2 \max \psi(\tau) \leq t^2 \Phi$ the ϵ-term can be simplified,

$$\begin{aligned}\sum_{\kappa} \frac{i_\kappa}{r_\kappa} \psi(\tau_\kappa) &\geq \frac{k}{2t^2} M - \epsilon t k \Phi - \epsilon k \Phi \\ &\geq \frac{k}{2t^2} M - 2\epsilon t k \Phi.\end{aligned}$$

We insert this estimate into (4.22), multiply with $h_{(i)}$ and sum over (i). Thus we obtain

$$\begin{aligned}|D_{(l)} P(\beta_1^{(\tau_1)}, \dots, \beta_k^{(\tau_k)})| &\leq (2C_6)^{\Sigma r_\kappa} q_1^{k r_1 (2\epsilon\Phi + 2\epsilon t \Phi - M/2t^2)} \sum_{(i)} 1 \\ &\leq (4C_6)^{\Sigma r_\kappa} q_1^{k r_1 (4\epsilon t \Phi - M/2t^2)}.\end{aligned}$$

This is our proposition.

4.4. *Roth's lemma.* The following is Leveque's generalization of Roth's lemma. While quoting from [2] we translate it into our notation.

LEMMA 16. *Assume*

$$P \in \mathbf{Z}[x_1, \dots, x_k],\ P \not\equiv 0,$$
$$\operatorname{grad}_\kappa P \leq r_\kappa \qquad (\kappa = 1, \dots, k).$$

Let $\beta_1, \dots, \beta_k$ *be algebraic numbers and* $q_\kappa = \|\beta_\kappa\|$. *Let*

$$D_{(l)} P(\beta_1, \dots, \beta_k) = 0 \quad \textit{for all } (l) \textit{ with } \sum_{\kappa} \frac{l_\kappa}{r_\kappa} \leq 10^k\, \delta^{(1/2)^k}. \tag{4.23}$$

Concerning k, δ, r_κ, q_κ *it is assumed that*

$$0 < \delta < \frac{1}{4k2^k}, \tag{4.24}$$

$$r_k > 10/\delta, \tag{4.25}$$

$$r_\kappa > \frac{1}{\delta} r_{\kappa+1} \quad \textit{for } \kappa = k-1, \dots, 1, \tag{4.26}$$

$$\log q_1 > \frac{6k^2}{\delta}, \tag{4.27}$$

$$q_\kappa^{r_\kappa} > q_1^{r_1} \quad \textit{for } \kappa = 2, \ldots, k. \tag{4.28}$$

Then we have

$$\|P\| \geq q_1^{\delta r_1}. \tag{4.29}$$

4.5. THE MAIN LEMMA. *Let $\alpha_1, \ldots, \alpha_n$ be distinct algebraic numbers and $Q_\kappa \in \mathbf{Z}[x]$ infinitely many irreducible polynomials of degree $t = t_1 t_2$. Assume that the zeros of Q_κ can be enumerated as $\beta_\kappa^{(\rho,\tau)}$ ($\rho = 1, \ldots, t_2$; $\tau = 1, \ldots, t_1$) in such a way that the isomorphisms of $\mathbf{Q}(\beta_1^{(1,1)}, \ldots, \beta_k^{(1,1)})$ are all the substitutions*

$$(\beta_1^{(1,1)}, \ldots, \beta_k^{(1,1)}) \mapsto (\beta_1^{(\rho,\tau_1)}, \ldots, \beta_k^{(\rho,\tau_k)})$$

with arbitrary $\rho = 1, \ldots, t_2$; $\tau_1 = 1, \ldots, t_1; \ldots; \tau_k = 1, \ldots, t_1$.

Assume further that the β approximate the α according to

$$|\alpha_\nu - \beta_\kappa^{(\rho,\tau)}| \leq \|Q_\kappa\|^{-\phi_\nu(\rho,\tau)} \qquad (\textit{for all } \nu, \kappa, \rho, \tau) \tag{4.30}$$

with nonnegative $\phi_\nu(\rho, \tau)$.

Then for arbitrary weights γ_ν with

$$\gamma_\nu \geq 0 \ (\nu = 1, \ldots, n), \qquad \sum_{\nu=1}^{n} \gamma_\nu = 1, \tag{4.16}$$

we have

$$\sum_{\rho=1}^{t_2} M\left(\sum_{\nu=1}^{n} \gamma_\nu \phi_\nu(\rho, \tau) \,\middle|\, \tau = 1, \ldots, t_1\right) \leq 2t_1. \tag{4.31}$$

PROOF. As before let $\|Q_\kappa\| = q_\kappa$ and let $b_{0\kappa}$ denote the main coefficient of Q_κ. Since the Q_κ are distinct, we have $q_\kappa \to \infty$. If $\nu \neq \mu$ then $\phi_\nu(\rho, \tau)\phi_\mu(\rho, \tau) = 0$. For if both these numbers were positive, then

$$\begin{aligned} |\alpha_\nu - \alpha_\mu| &\leq |\alpha_\nu - \beta_\kappa^{(\rho,\tau)}| + |\alpha_\mu - \beta_\kappa^{(\rho,\tau)}| \\ &\leq q_\kappa^{-\phi_\nu(\rho,\tau)} + q_\kappa^{-\phi_\mu(\rho,\tau)} \\ &\to 0 \quad \text{as } \kappa \to \infty, \end{aligned}$$

and $\alpha_\nu = \alpha_\mu$ contrary to our assumption.

We choose the parameters ϵ, k, $r_1, \ldots, r_k$ so as to satisfy the conditions (4.8), (4.9), (4.18), (4.20) and $\epsilon \leq 1/t$. We construct P as in Lemma 10, specializing the $j_\nu(i)$ by (4.17) and using the $\mathfrak{M}$ supplied by Lemma 11. Now Lemmata 13 and 15 can be applied with t_1 and $\beta^{(\rho,\tau)}$ instead of t and $\beta^{(\tau)}$. Thus, provided

$$\sum_\kappa \frac{l_\kappa}{r_\kappa} \leq \epsilon k$$

and $(\tau) \notin \mathfrak{T}$, we obtain

(4.32) $$|D_{(l)}P(\beta_\kappa^{(\rho,\tau_\kappa)})| \leq C_8^{\Sigma r_\kappa} q_1^{kr_1(4\epsilon t_1\Phi - M_\rho/2t_1^2)}$$

with

$$\Phi := \max_{\nu,\rho,\tau} \phi_\nu(\rho, \tau)$$

and

$$M_\rho := M\left(\sum_\nu \gamma_\nu\phi_\nu(\rho, \tau) \mid \tau = 1, \ldots, t_1\right).$$

For $(\tau) \in \mathfrak{T}$ we have

(4.33) $$|D_{(l)}P(\beta_\kappa^{(\rho,\tau_\kappa)})| \leq C_8^{\Sigma r_\kappa}$$

by Lemma 14.

Now we restrict the parameters further by the assumption

(4.34) $$q_\kappa^{r_\kappa - 1} \leq q_1^{r_1},$$

which, in connection with (4.20) determines $r_2, \ldots, r_k$ as functions of $r_1, q_1, \ldots, q_k$. Because of (4.9) we have $r_\kappa \leq (1 + \epsilon)(r_\kappa - 1)$,

(4.35) $$q_\kappa^{r_\kappa} \leq q_\kappa^{(1+\epsilon)(r_\kappa-1)} \leq q_1^{(1+\epsilon)r_1}.$$

From now on we assume

(4.36) $$q_1 \leq q_2 \leq \cdots \leq q_k,$$

consequently $r_1 \geq r_2 \geq \cdots \geq r_k$, and in particular $\sum r_\kappa \leq kr_1$.

The number

$$Z := \left(\prod_\kappa b_{0\kappa}^{r_\kappa t_1^{k-1}}\right) \prod_{\rho=1}^{t_2} \prod_{\tau_1,\ldots,\tau_k=1,\ldots,t_1} D_{(l)}P(\beta_\kappa^{(\rho,\tau_\kappa)})$$

is rational since the right-hand product goes over all conjugates of an algebraic number. Furthermore Z is integral for no $\beta_\kappa^{(\rho,\tau)}$ occurs with an exponent greater than $r_\kappa t_1^{k-1}$ and it appears multiplied with $b_{0\kappa}$ to just this exponent.

The first product is easily estimated:

$$\begin{aligned}\prod_\kappa b_{0\kappa}^{r_\kappa t_1^{k-1}} &\leq \prod_\kappa q_\kappa^{r_\kappa t_1^{k-1}} \\ &\leq \prod_\kappa q_1^{r_1 t_1^{k-1}(1+\epsilon)} \\ &\leq q_1^{kr_1 t_1^{k-1}(1+\epsilon)}.\end{aligned}$$

This result combines with (4.32) and (4.33) into

$$|Z| \leq C_8^{kr_1 t_1^k t_2} q_1^{kr_1 t_1^{k-1}(1+\epsilon) + kr_1(t_1^k - |\mathfrak{T}|)\Sigma_\rho(4\epsilon t_1\Phi - M_\rho/2t_1^2)}.$$

By Lemma 13 and (4.8)

$$|\mathfrak{T}| \leq \frac{t_1^k}{\epsilon^2 k} \leq \epsilon t_1^k$$

hence

$$\begin{aligned} |Z| &\leq (C_8^{t_1^2 t_2} q_1^{(1+\epsilon)t_1 + (\epsilon - 1)\Sigma_\rho M_\rho/2 + 4\epsilon t_1^3 t_2 \Phi})^{k r_1 t_1^{k-2}} \\ &\leq (C_8^{t_1^2 t_2} q_1^{t_1 - (1/2)\Sigma_\rho M_\rho} q_1^{\epsilon(t_1 + 4 t_1^3 t_2 \Phi + \Sigma_\rho M_\rho)})^{k r_1 t_1^{k-2}}. \end{aligned}$$

If we assume now what we want to disprove, that is $\sum_\rho M_\rho > 2t_1$, then ϵ can be taken small enough to make the exponent of q_1 inside the parentheses less than $-\epsilon$, and therefore

$$|Z| \leq (C_8^{t_1^2 t_2} q_1^{-\epsilon})^{k r_1 t_1^{k-2}}.$$

This $\epsilon > 0$ depends on nothing but t_1, t_2 and the $\phi_\nu(\rho, \tau)$.

Since by assumption we have infinitely many polynomials Q_κ there is no obstacle to taking

$$q_1 > C_8^{t_1^2 t_2/\epsilon}. \tag{4.37}$$

This makes $|Z| < 1$. But $Z \in \mathbf{Z}$. Hence $Z = 0$.

Now, apart from factors $b_{0\kappa}$, which are nonzero, Z is the norm of $D_{(l)}P(\beta_\kappa^{(1,1)})$. Therefore we have proved

$$D_{(l)}P(\beta_\kappa^{(1,1)}) = 0 \quad \textit{for all } (l) \textit{ with } \sum_\kappa \frac{l_\kappa}{r_\kappa} \leq \epsilon k. \tag{4.38}$$

To (4.38) we shall apply Roth's lemma (Lemma 16). The numbers k and ϵ have already been fixed, while $r_1, q_1, \ldots, q_k$ are unrestricted so far, apart from (4.9), (4.36), (4.37).

We choose $\delta > 0$ small enough to fulfill (4.24) and $10^k \delta^{(1/2)^k} \leq \epsilon k$. Then, by (4.38), (4.23) holds.

From our infinite store of polynomials Q we choose a Q_1 with q_1 sufficiently large to satisfy (4.37), (4.27) as well as

$$q_1 > C_7^{k/\delta} \tag{4.39}$$

and the other Q_κ successively such that

$$\log q_{\kappa+1} \geq \frac{2}{\delta} \log q_\kappa \quad \text{for } \kappa = 1, 2, \ldots, k-1. \tag{4.40}$$

The last parameter at our disposal, the degree r_1, is taken so large that the other r_κ, which depend on it by (4.20), (4.28), (4.34), satisfy (4.9) and (4.25).

From (4.28), (4.35), (4.40) we infer

$$\frac{r_{\kappa+1}}{r_\kappa} < (1+\epsilon) \frac{\log q_\kappa}{\log q_{\kappa+1}} \leq (1+\epsilon) \frac{\delta}{2} \leq \delta,$$

that is (4.26).

All conditions of Lemma 16 have been shown. But the conclusion (4.29) cannot hold because of (4.5) and (4.39), which give

$$\|P\| \leq C_7^{\Sigma r_\kappa} \leq C_7^{k r_1} < q_1^{\delta r_1}.$$

This contradiction proves the Main Lemma.

5. Deduction of the theorems from the Main Lemma

5.1. *Branching off towards Theorems 4, 3 and 2.1.* We note that each sum $\sum_\nu \gamma_\nu \phi_\nu(\rho, \tau)$ contains one nonzero element at most. Since adding a number of new variables, each with value zero, does not change the value of M we can rewrite (4.31) as

$$\sum_{\rho=1}^{t_2} M(\gamma_\nu \phi_\nu(\rho, \tau) \mid \nu = 1, \dots, n; \tau = 1, \dots, t_1) \leq 2t_1. \tag{5.1}$$

Let $n = 1$. In this case we write $\phi(\rho, \tau)$ for $\phi_1(\rho, \tau)$ and have $\gamma_1 = 1$. Since $\sum_\tau \phi(\rho, \tau) \leq M(\phi(\rho, \tau) \mid \tau)$ we obtain from (4.31)

$$\sum_{\rho,\tau} \phi(\rho, \tau) \leq 2t_1. \tag{5.2}$$

Another implication of (4.31) is found if we use Lemma 2 ($\phi(\rho, \tau)$ as a_{ij}, all $\gamma_i = 1$):

$$M(\phi(\rho, \tau) \mid \rho, \tau) \leq 2t_1 t_2 = 2t. \tag{5.3}$$

Now let n be arbitrary. Using Lemma 2 first and Lemma 4 then we see

$$\begin{aligned} t_2 \sum_\rho M(\gamma_\nu \phi_\nu(\rho, \tau) \mid \tau, \nu) &\geq M(\gamma_\nu \phi_\nu(\rho, \tau) \mid \rho, \tau, \nu) \\ &\geq M\left(\gamma_\nu \sum_{\rho,\tau} \phi_\nu(\rho, \tau) \mid \nu\right). \end{aligned}$$

Therefore

$$M\left(\gamma_\nu \sum_{\rho,\tau} \phi_\nu(\rho, \tau) \mid \nu = 1, \dots, n\right) \leq 2t.$$

Since this is true for all $\gamma_\nu \geq 0$ with $\sum \gamma_\nu = 1$, we can take the maximum with respect to the γ_ν. By Lemma 6 this yields

$$L\left(\sum_{\rho,\tau} \phi_\nu(\rho, \tau) \mid \nu = 1, \dots, n\right) \leq 2t. \tag{5.4}$$

Formulae (5.2)-(5.4), which we have proved under the conditions stated in the Main Lemma, contain the kernel of the Theorems 2.1, 3 and 4 respectively.

5.2 *Making the $\phi_\nu(\rho, \tau)$ variable.* We are still considering algebraic numbers $\alpha_1, \dots, \alpha_n$ and a sequence of polynomials Q_κ as described in the Main Lemma. Assuming $q_\kappa > 1$ let us write

$$|\alpha_\nu - \beta_\kappa^{(\rho,\tau)}| = q_\kappa^{-\omega_{\nu\kappa}(\rho,\tau)}$$

(in the case $n = 1$ the index ν shall be omitted).

LEMMA 17. *Under the conditions of the Main Lemma, but with* (4.30) *replaced by*

$$(5.5) \qquad \omega_{\nu\kappa}(\rho, \tau) \geq 0 \qquad (\textit{that is } |\alpha_\nu - \beta_\kappa^{(\rho,\tau)}| \leq 1) \quad \textit{for all } \nu, \kappa, \rho, \tau$$

we have for $n = 1$

$$(5.6) \qquad \overline{\lim_{\kappa \to \infty}} \sum_{\rho,\tau} \omega_\kappa(\rho, \tau) \leq 2t_1,$$

$$(5.7) \qquad \overline{\lim_{\kappa \to \infty}} M(\omega_\kappa(\rho, \tau) \mid \rho, \tau) \leq 2t,$$

and for arbitrary n

$$(5.8) \qquad \overline{\lim_{\kappa \to \infty}} L\Big(\sum_{\rho,\tau} \omega_{\nu\kappa}(\rho, \tau) \mid \nu = 1, \ldots, n\Big) \leq 2t.$$

PROOF. The Main Lemma obviously implies, as does the well-known estimate

$$|\alpha - \beta| \geq \frac{C(s, t)}{\|\alpha\|^t \|\beta\|^s} \qquad (s = \deg \alpha, \; t = \deg \beta) \qquad \text{if } \alpha \neq \beta,$$

that the $\omega_{\nu\kappa}(\rho, \tau)$ are bounded as functions of κ.

We can therefore pick a part sequence for which the upper limits in (5.2)-(5.4), become proper limits and for which moreover all the limits

$$(5.9) \qquad \lim_{\kappa} \omega_{\nu\kappa}(\rho, \tau) =: \phi_\nu(\rho, \tau)$$

exist. We change notation so that now Q_κ are the elements of this part sequence. If $\epsilon > 0$ and κ is sufficiently large we have, by (5.9) if $\phi_\nu(\rho, \tau) > 0$ and by (5.5) otherwise,

$$\omega_{\nu\kappa}(\rho, \tau) \geq (1 - \epsilon)\phi_\nu(\rho, \tau),$$

that is

$$|\alpha_\nu - \beta_\kappa^{(\rho,\tau)}| \leq q_\kappa^{-(1-\epsilon)\phi_\nu(\rho,\tau)}.$$

Thereby (5.2, 3, 4) are obtained with $\phi_\nu(\rho, \tau)$ replaced by $(1 - \epsilon)\phi_\nu(\rho, \tau)$, and, letting $\epsilon \to 0$, in their original form. Finally, using (5.9) again, we see

$$\lim_{\kappa} \sum_{\rho,\tau} \omega_\kappa(\rho, \tau) = \sum_{\rho,\tau} \phi(\rho, \tau) \leq 2t_1,$$
$$\lim_{\kappa} M(\omega_\kappa(\rho, \tau) \mid \rho, \tau) = M(\phi(\rho, \tau) \mid \rho, \tau) \leq 2t,$$
$$\lim_{\kappa} L\Big(\sum_{\rho,\tau} \omega_{\nu\kappa}(\rho, \tau) \mid \nu\Big) = L\Big(\sum_{\rho,\tau} \phi_\nu(\rho, \tau) \mid \nu\Big) \leq 2t.$$

The continuity of L, which we need here, is easily seen from Lemma 8.

5.3. *Eliminating the conditions on the conjugates.*

LEMMA 18. *Let* $\mathfrak{F}$ *be an infinite set of irreducible polynomials* $Q \in \mathbf{Z}[x]$ *of degree* t. *Then there is a decomposition* $t = t_1 t_2$ *and an infinite sequence of* $Q_\kappa \in \mathfrak{F}$ *that fulfill the first condition of the Main Lemma, i.e. the zeros of the* Q_κ *can be* *enumerated as* $\beta_\kappa^{(\rho,\tau)}$ $(\rho = 1, \dots, t_2; \tau = 1, \dots, t_1)$ *in such a way that for all* k *the isomorphisms* λ *of* $\mathbf{Q}(\beta_1^{(1,1)}, \dots, \beta_k^{(1,1)})$ *are given by all the substitutions*

$$(\beta_1^{(1,1)}, \dots, \beta_k^{(1,1)}) \mapsto (\beta_1^{(\rho,\tau_1)}, \dots, \beta_k^{(\rho,\tau_k)}),$$

where $\rho = 1, \dots, t_2; \tau_1 = 1, \dots, t_1; \dots; \tau_k = 1, \dots, t_1$ *are taken independently.*

PROOF. If $Q \in \mathbf{Z}[x]$ is irreducible and $\mathfrak{N}$ a normal extension of $\mathbf{Q}$ then Q decomposes over $\mathfrak{N}$ into irreducible factors of equal degree. We pick a normal extension $\mathfrak{N}$ of $\mathbf{Q}$ of finite degree such that

(a) infinitely many $Q \in \mathfrak{F}$ decompose over $\mathfrak{N}$ into irreducible factors of some degree t_1 (maybe $= t$), and

(b) this t_1 is minimal, i.e. for no $\mathfrak{N}'$ we have the situation (a) with some $t_1' < t_1$.

Let $\mathfrak{F}_1$ be the collection of all $Q \in \mathfrak{F}$ which factorize over $\mathfrak{N}$ in the indicated way:

$$Q = \prod_{\rho=1}^{t_2} Q^{(\rho)}, \qquad \deg Q^{(\rho)} = t_1, \qquad t_1 t_2 = t.$$

The enumeration of the factors of each Q is arbitrary but shall not be changed.

The $Q^{(\rho)}$ are conjugates with respect to $\mathfrak{N}$. Therefore each automorphism σ of $\mathfrak{N}$ will permute the $Q^{(\rho)}$:

$$\sigma Q^{(\rho)} = Q^{(\pi\rho)}, \qquad \pi\colon \{1, 2, \dots, t_1\} \to \{1, 2, \dots, t_1\}. \tag{5.10}$$

In the first instance the π associated with σ will depend on Q, but since there are only finitely many permutations π and automorphisms σ we can select such an infinite $\mathfrak{F}_2 \subset \mathfrak{F}_1$ that the map $\sigma \mapsto \pi$ given by (5.10) does not depend on $Q \in \mathfrak{F}_2$.

From $\mathfrak{F}_2$ we pick Q_1 arbitrarily. Let $Q_1, \dots, Q_k$ be chosen. Let $\mathfrak{N}'$ be the splitting field of $Q_1, \dots, Q_k$ over $\mathfrak{N}$. Because of (b) for almost all $Q \in \mathfrak{F}_2$ the irreducible factors $Q^{(\rho)}$ over $\mathfrak{N}$ stay irreducible over $\mathfrak{N}'$. Any such Q can be chosen as Q_{k+1}.

The factors of Q_κ are $Q_\kappa^{(\rho)}$ and the zeros of $Q_\kappa^{(\rho)}$ shall be denoted by $\beta_\kappa^{(\rho,\tau)}$ $(\tau = 1, \dots, t_1)$ in any order. By construction the degree of $\beta_k^{(1,1)}$ over $\mathbf{Q}(\beta_1^{(1,1)}, \dots, \beta_{k-1}^{(1,1)})$ is at least t_1. Hence, by induction,

$$[\mathbf{Q}(\beta_1^{(1,1)}, \dots, \beta_k^{(1,1)}) : \mathbf{Q}] \geq t_1^k t_2. \tag{5.11}$$

Any isomorphism λ of $\mathbf{Q}(\beta_1^{(1,1)}, \dots, \beta_k^{(1,1)})$ must take each $\beta_\kappa^{(1,1)}$ into one of its conjugates, $\beta_\kappa^{(\rho_\kappa,\tau_\kappa)}$, say. λ can be extended to an isomorphism λ' of $\mathfrak{N}(\beta_1^{(1,1)}, \dots, \beta_k^{(1,1)})$, which by construction permutes the factors of all Q_κ in some manner not depending on κ. Therefore all the ρ_κ must be equal,

$$\lambda(\beta_\kappa^{(1,1)}) = \beta_\kappa^{(\rho,\tau_\kappa)}.$$

On the other hand all these substitutions give isomorphisms since (5.11) requests at least $t_1^k t_2$ different isomorphisms.

The purpose of Lemma 18 is to free Lemma 17 of the condition on the isomorphisms. We consider any infinite set of polynomials

$$Q_\kappa(x) = b_{0\kappa} \prod_{\tau=1}^{t} (x - \beta_\kappa^{(\tau)}) \in \mathbf{Z}[x]$$

and define $\omega_{\nu\kappa}(\tau)$ by

$$|\alpha_\nu - \beta_\kappa^{(\tau)}| = q_\kappa^{-\omega_{\nu\kappa}(\tau)}.$$

Then Lemma 17 evolves into the following statements.

If

$$\text{(5.12)} \qquad \text{all } Q_\kappa \text{ are irreducible,}$$

$$\text{(5.13)} \qquad |\alpha_\nu - \beta_\kappa^{(\tau)}| \leq 1 \quad \text{for all } \nu, \kappa, \tau,$$

then

$$\text{(5.14)} \qquad \overline{\lim_{\kappa\to\infty}} L\Big(\sum_\tau \omega_{\nu\kappa}(\tau) \mid \nu\Big) \leq 2t,$$

and in the case $n = 1$

$$\text{(5.15)} \qquad \overline{\lim_{\kappa\to\infty}} M(\omega_\kappa(\tau) \mid \tau) \leq 2t.$$

If in the case $n = 1$ we have any field $\mathfrak{K}$ of finite degree over $\mathbf{Q}$ such that all Q_κ decompose over $\mathfrak{K}$ into factors of degree $\leq t'$ then

$$\text{(5.16)} \qquad \overline{\lim_{\kappa\to\infty}} \sum_\tau \omega_\kappa(\tau) \leq 2t'.$$

The proofs are immediate. Concerning (5.6)–(5.16) we note that $\mathfrak{N}$ can be taken large enough to contain $\mathfrak{K}$, whence $t_1 \leq t'$.

5.4. *Waiving irreducibility*. We shall drop the index κ, simply considering now all polynomials Q with the required properties as $\|Q\| =: q \to \infty$. Let Q be any polynomial and Q_λ ($\lambda = 1, \ldots, l$) its irreducible factors, $\deg Q_\lambda =: t_\lambda$, $\sum t_\lambda = t$, $\|Q_\lambda\| =: q_\lambda$. We designate the zeros in two ways, as $\beta^{(\tau)}$ $(1 \leq \tau \leq t)$ on the one hand and as $\beta_\lambda^{(\tau)}$ $(1 \leq \lambda \leq l,\ 1 \leq \tau \leq t_\lambda)$ on the other. From the preceding section we know that, under the conditions stated there, for any $\epsilon > 0$ and a suitable $R_0 = R_0(\epsilon, t)$ we have

$$\text{(5.17)} \qquad \begin{aligned} L\Big(\sum_{\tau=1}^{t_\lambda} \log|\alpha_\nu - \beta_\lambda^{(\tau)}|^{-1} \,\Big|\, \nu = 1, \ldots, n\Big) &\leq (2+\epsilon)t_\lambda \log q_\lambda, \\ M(\log|\alpha - \beta_\lambda^{(\tau)}|^{-1} \mid \tau = 1, \ldots, t_\lambda) &\leq (2+\epsilon)t_\lambda \log q_\lambda, \\ \sum_{\tau=1}^{t_\lambda} \log|\alpha - \beta_\lambda^{(\tau)}|^{-1} &\leq (2+\epsilon)t' \log q_\lambda \end{aligned}$$

respectively provided $q_\lambda \geq R_0$.

For the finitely many polynomials Q_λ with degree $\leq t$ and height $q_\lambda < R_0$ we can use some bound like

$$\text{(5.18)} \qquad \begin{aligned} \sum_{\tau=1}^{t_\lambda} \log|\alpha_\nu - \beta_\lambda^{(\tau)}|^{-1} &\leq H, \\ M(\log|\alpha_\nu - \beta_\lambda^{(\tau)}|^{-1} \mid \tau = 1, \ldots, t_\lambda) &\leq H, \end{aligned}$$

barring only those Q_λ that have some zero α_ν.

Of the approximate multiplicativity of the height we need the one inequality

$$\sum_{\lambda=1}^{l} \log q_\lambda \leq \log q + C. \tag{5.19}$$

In the following we use the abbreviations

$$\sum_\lambda{}' := \sum_{\lambda=1,\ldots,l;\, q_\lambda \geq R_0}, \qquad \sum_\lambda{}'' := \sum_{\lambda=1,\ldots,l;\, q_\lambda < R_0}.$$

We apply Lemmata 8 and 7 to (5.17), (5.18) and use (5.19) in the end:

$$\begin{aligned}
L\Big(\sum_{\tau=1}^{l} \log|\alpha_\nu - \beta^{(\tau)}|^{-1} \Big| \nu = 1, \ldots, n\Big) \\
&= L\Big(\sum_{\lambda=1}^{l} \sum_{\tau=1}^{t_\lambda} \log|\alpha_\nu - \beta_\lambda^{(\tau)}|^{-1} \Big| \nu = 1, \ldots, n\Big) \\
&\leq L\Big(\sum_\lambda{}' \sum_{\tau=1}^{t_\lambda} \log|\alpha_\nu - \beta_\lambda^{(\tau)}|^{-1} \Big| \nu = 1, \ldots, n\Big) + 3n \sum_\lambda{}'' H \\
&\leq \Big(\sum_\lambda{}' \log q_\lambda\Big) \sum_\lambda{}' \frac{1}{\log q_\lambda} L\Big(\sum_{\tau=1}^{t_\lambda} \log|\alpha_\nu - \beta_\lambda^{(\tau)}|^{-1} \Big| \nu\Big) + 3ntH \\
&\leq (\log q + C)(2 + \epsilon) \sum_\lambda{}' t_\lambda + 3ntH \\
&\leq (\log q + C)(2 + \epsilon)t + 3ntH.
\end{aligned}$$

Hence

$$\varlimsup_{q\to\infty} \frac{1}{\log q} L\Big(\sum_{\tau=1}^{t} \log|\alpha_\nu - \beta^{(\tau)}|^{-1} \Big| \nu = 1, \ldots, n\Big) \leq 2t, \tag{5.20}$$

if we consider all $Q \in \mathbf{Z}[x]$ with degree t and property (5.13) and with $Q(\alpha_\nu) \neq 0$ for all ν.

Concerning (5.15) we work similarly with Lemmata 3 and 2:

$$\begin{aligned}
M(\log|\alpha - \beta^{(\tau)}|^{-1} \mid \tau = 1, \ldots, t) \\
&= M(\log|\alpha - \beta_\lambda^{(\tau)}|^{-1} \mid \lambda = 1, \ldots, l;\, \tau = 1, \ldots, t_\lambda) \\
&\leq M(\log|\alpha - \beta_\lambda^{(\tau)}|^{-1} \mid \lambda, q_\lambda \geq R_0;\, \tau = 1, \ldots, t_\lambda) \\
&\quad + 2t \sum_\lambda{}'' M(\log|\alpha - \beta_\lambda^{(\tau)}|^{-1} \mid \tau) \\
&\leq \Big(\sum_\lambda{}' \log q_\lambda\Big) \sum_\lambda{}' \frac{1}{\log q_\lambda} M(\log|\alpha - \beta_\lambda^{(\tau)}|^{-1} \mid \tau) + 2t^2 H \\
&\leq (\log q + C)(2 + \epsilon)t + 2t^2 H
\end{aligned}$$

and obtain

$$\varlimsup_{q\to\infty} \frac{1}{\log q} M(\log|\alpha - \beta^{(\tau)}|^{-1} \mid \tau = 1, \ldots, t) \leq 2t, \tag{5.21}$$

where Q runs through the polynomials over $\mathbf{Z}$ with degree t, property (5.13) and $Q(\alpha) \neq 0$.

The corresponding result from (5.16) is

$$\overline{\lim_{q\to\infty}} \frac{1}{\log q} \sum_{\tau=1}^{t} \log |\alpha - \beta^{(\tau)}|^{-1} \leq 2t', \tag{5.22}$$

where Q are now the polynomials that split over some fixed extension $\mathfrak{K}$ of $\mathbf{Q}$ of finite degree into factors of degree $\leq t'$ and which have property (5.13) and do not vanish at α.

The proof of this statement is straightforward and is left to the reader.

5.5. *Allowing distant zeros.* The only step left in the proof of Theorems 4, 3, and 2.1 is to remove condition (5.13). Let $Q \in \mathbf{Z}[x]$ and

$$Q(x) = b_0 \prod_{\tau=1}^{t} (x - \beta^{(\tau)}).$$

Since not every one of the disks

$$|z - m| < \tfrac{1}{2}, \qquad m = 0, 1, \ldots, n + t,$$

can contain one of the points $\alpha_1, \ldots, \alpha_n, \beta^{(1)}, \ldots, \beta^{(t)}$, we can choose m such that

$$|\alpha_\nu - m| \geq \tfrac{1}{2} \qquad (\nu = 1, \ldots, n),$$
$$|\beta^{(\tau)} - m| \geq \tfrac{1}{2} \qquad (\tau = 1, \ldots, t).$$

Now consider the new numbers

$$\alpha_{\nu *} := \frac{1}{4(\alpha_\nu - m)}, \qquad \beta_*^{(\tau)} := \frac{1}{4(\beta^{(\tau)} - m)}.$$

The $\beta_*^{(\tau)}$ are the zeros of

$$Q_*(x) := (4x)^t Q\left(m + \frac{1}{4x}\right) = b_{0*} \prod_{\tau} (x - \beta_*^{(\tau)}).$$

By construction

$$|\alpha_{\nu *}| \leq \tfrac{1}{2}, \qquad |\beta_*^{(\tau)}| \leq \tfrac{1}{2}, \qquad |\alpha_{\nu *} - \beta_*^{(\tau)}| \leq 1 \tag{5.23}$$

and

$$|\alpha_{\nu *} - \beta_*^{(\tau)}| = \frac{|\alpha_\nu - \beta^{(\tau)}|}{4|\alpha_\nu - m|\,|\beta^{(\tau)} - m|} \leq |\alpha_\nu - \beta^{(\tau)}|,$$

therefore

$$\log^+ |\alpha_\nu - \beta^{(\tau)}|^{-1} \leq \log |\alpha_{\nu *} - \beta_*^{(\tau)}|^{-1}.$$

An easy calculation gives

$$C^{-1}\|Q\| \leq \|Q_*\| \leq C\|Q\| \qquad (C = C(n + t)),$$

whence

$$\log \|Q_*\| = \log \|Q\| + O(1).$$

Now we can apply (5.21) to α_*, Q_*, $\beta_*^{(\tau)}$. With (5.23) we have (5.13). The fact that m and therefore α_* depend on Q does not create difficulties since there are only $n + t + 1$ possible values of m and correspondingly of α_*. We may subdivide the Q into as many classes and apply (5.21) to each class separately. Hence Theorem 3:

For $Q \in \mathbf{Z}[x]$, $Q(\alpha) \neq 0$, $\deg Q = t$ we have

$$\overline{\lim_{\|Q\| \to \infty}} \frac{1}{\log \|Q\|} M(\log^+ |\alpha - \beta^{(\tau)}|^{-1} \mid \tau = 1, \ldots, t) \leq 2t.$$

Concerning (5.20) and (5.22) we note first that

$$Q_*(\alpha_{\nu *}) = (4\alpha_{\nu *})^t Q(\alpha_\nu),$$
$$|Q_*(\alpha_{\nu *})| \leq C|Q(\alpha_\nu)| \qquad (C = 2^t).$$

Next we estimate the symmetric functions of the $\beta_*^{(\tau)}$

$$|b_{\nu *}/b_{0 *}| \leq \prod_\tau (1 + |\beta_*^{(\tau)}|) \leq (3/2)^t$$

and obtain $\|Q_*\| \leq C|b_{0*}|$.

Taken together these lines imply

$$\text{(5.24)} \qquad \frac{\|Q\|}{|Q(\alpha_\nu)|} \leq C \frac{|b_{0*}|}{|Q_*(\alpha_{\nu *})|} = C \prod_{\tau=1}^{t} |\alpha_{\nu *} - \beta_*^{(\tau)}|^{-1}.$$

Now let $Q(\alpha) \neq 0$ and assume that over some field $\mathfrak{K}$ the Q of degree t split into factors of degree $\leq t'$. Then (5.22) applies to α_* and Q_* and we obtain

$$\overline{\lim_{\|Q\| \to \infty}} \frac{1}{\log \|Q\|} \log \frac{\|Q\|}{|Q(\alpha)|} \leq 2t',$$
$$\overline{\lim_{\|Q\| \to \infty}} \frac{1}{\log \|Q\|} \log |Q(\alpha)|^{-1} \leq 2t' - 1,$$

which is Theorem 2.1.

Without loss of generality in (5.24) $C \geq 1$. Taking (5.23) into account we see

$$\log^+ \frac{\|Q\|}{|Q(\alpha_\nu)|} \leq \log C + \sum_\tau \log |\alpha_{\nu *} - \beta_*^{(\tau)}|^{-1}.$$

Aided by Lemma 8 we apply (5.20):

$$L\left(\log^+ \frac{\|Q\|}{|Q(\alpha_\nu)|} \middle| \nu = 1, \ldots, n\right)$$
$$\leq L\left(\sum_\tau \log |\alpha_{\nu *} - \beta_*^{(\tau)}|^{-1} \middle| \nu = 1, \ldots, n\right) + 3n \log C,$$

$$\overline{\lim_{\|Q\| \to \infty}} \frac{1}{\log \|Q\|} L\left(\log^+ \frac{\|Q\|}{|Q(\alpha_\nu)|} \middle| \nu = 1, \ldots, n\right) \leq 2t,$$

provided $Q \in \mathbf{Z}[x]$, $\deg Q = t$, $Q(\alpha_\nu) \neq 0$ $(\nu = 1, \ldots, n)$. This is Theorem 4.

5.6. *Proof of Theorem* 5. Let $P, Q \in \mathbf{Z}[x]$, $\deg P = s$, $\deg Q = t$, all zeros $\alpha^{(\sigma)}$ of P distinct and $Q(\alpha^{(\sigma)}) \neq 0$ for all $\sigma = 1, \ldots, s$. We consider P fixed and Q variable. Let 2δ be the minimal distance of any two $\alpha^{(\sigma)}$. At most t of the disks

$$|z - \alpha^{(\sigma)}| < \delta, \qquad \sigma = 1, \ldots, s$$

can contain a zero $\beta^{(\tau)}$ of Q. Let these be the ones with centers $\alpha^{(\sigma)}$, $\sigma = 1, \ldots, t'$, $t' \leq t$. We have

$$\begin{aligned}
|\mathrm{Res}\,(P, Q)| &= |a_0|^t \prod_{\sigma=1}^{s} |Q(\alpha^{(\sigma)})| \\
&\geq \prod_{\sigma=1}^{s} |Q(\alpha^{(\sigma)})| \\
&= \|Q\|^s \prod_{\sigma=1}^{t'} \frac{|Q(\alpha^{(\sigma)})|}{\|Q\|} \prod_{\sigma=t'+1}^{s} \frac{|Q(\alpha^{(\sigma)})|}{\|Q\|}.
\end{aligned}$$

The first product is estimated by Theorem 4 and Lemma 9:

$$L\left(\log^+ \frac{\|Q\|}{|Q(\alpha^{(\sigma)})|} \,\middle|\, \sigma = 1, \ldots, t'\right) \leq (2 + \epsilon)t \log \|Q\|,$$

provided $\|Q\|$ is large enough. Therefore

$$\begin{aligned}
\sum_{\sigma=1}^{t'} \log \frac{\|Q\|}{|Q(\alpha^{(\sigma)})|} &\leq \sum_{\sigma=1}^{t'} \log^+ \frac{\|Q\|}{|Q(\alpha^{(\sigma)})|} \\
&\leq \left(1 + \frac{1}{3} + \cdots + \frac{1}{2t' - 1}\right)(2 + \epsilon)t \log \|Q\| \\
&\leq (2 + \epsilon)t\left(1 + \frac{1}{3} + \cdots + \frac{1}{2t - 1}\right) \log \|Q\|.
\end{aligned}$$

For the $\alpha^{(\sigma)}$ with $\sigma > t'$ we conclude

$$\begin{aligned}
\|Q\| &\leq |b_0| \prod (1 + |\beta^{(\tau)}|), \\
|Q(\alpha^{(\sigma)})| &= |b_0| \prod |\alpha^{(\sigma)} - \beta^{(\tau)}|, \\
\frac{\|Q\|}{|Q(\alpha^{(\sigma)})|} &\leq \prod_\tau \frac{1 + |\beta^{(\tau)}|}{|\alpha^{(\sigma)} - \beta^{(\tau)}|} \\
&\leq \prod_\tau \frac{1 + |\alpha^{(\sigma)}| + |\alpha^{(\sigma)} - \beta^{(\tau)}|}{|\alpha^{(\sigma)} - \beta^{(\tau)}|} \\
&\leq \left(1 + \frac{1 + |\alpha^{(\sigma)}|}{\delta}\right)^t \\
&\leq C \quad \text{for } \sigma = t' + 1, \ldots, s.
\end{aligned}$$

Altogether we have

$$|\mathrm{Res}\,(P, Q)| \geq C^{-s}\|Q\|^{s-(2+\epsilon)t(1+1/3+\cdots+1/(2t-1))}.$$

This is Theorem 5.

5.7. *The remaining theorems.* Theorem 2.2 is an immediate corollary of Theorem 4. We simply take $\alpha_1 := \alpha$, $\alpha_2 := \bar{\alpha}$. Theorem 2 is the special case $t' = t$ of Theorem 2.1. Theorem 1.1 could be proved by a simplified version of the proof of Theorem 2.1. Once Theorem 2.1 is proved it is much simpler however to derive Theorem 1.1 from there. The degree of the numbers β is bounded by $t' \cdot [\mathfrak{K}: \mathbf{Q}]$. Let Q be the minimal polynomial of β and $\deg \beta =: t$. There is no loss of generality in considering β with $|\alpha - \beta| \leq 1$ only. Then we have

$$\begin{aligned} Q(\alpha) &= \sum_{\tau=1}^{t} (\alpha - \beta)^{\tau} \frac{1}{\tau!} Q^{(\tau)}(\beta), \\ |Q(\alpha)| &\leq |\alpha - \beta| \sum_{\tau=1}^{t} |Q^{(\tau)}(\beta)| \\ &\leq |\alpha - \beta| \, \|Q\| C \qquad (C = C(\alpha, t)), \\ |\alpha - \beta| &\geq C^{-1} |Q(\alpha)| / \|Q\|. \end{aligned}$$

If we assume, as we may, that $\mathfrak{K}$ is normal then all zeros of Q are of degree $\leq t'$ over $\mathfrak{K}$. Hence Theorem 2.1 applies and gives

$$\begin{aligned} |Q(\alpha)| &\geq \|Q\|^{1-2t'-\epsilon} \qquad (\|Q\| = \|\beta\| \text{ large}), \\ |\alpha - \beta| &\geq C^{-1} \|Q\|^{-2t'-\epsilon} \\ &\geq \|Q\|^{-2t'-2\epsilon} \\ &= \|\beta\|^{-2t'-2\epsilon}. \end{aligned}$$

Theorem 1 is the case $\mathfrak{K} = \mathbf{Q}$ of Theorem 1.1.

REFERENCES

1. F. J. Dyson, *The approximation to algebraic numbers by rationals*, Acta Math. **79** (1947), 225–240. MR **9,** 412.

2. W. J. Leveque, *Topics in number theory.* Vol. II, Addison-Wesley, Reading, Mass., 1956. MR **18,** 283.

3. K. Ramachandra, *Approximation of algebraic numbers*, Nachr. Akad. Wiss. Göttingen Math.-Phys. Kl. II **1966,** 45–52. MR **36** #112.

4. K. F. Roth, *Rational approximations to algebraic numbers*, Mathematika **2** (1955), 1–20. MR **17,** 242.

5. W. M. Schmidt, *On simultaneous approximations of two algebraic numbers by rationals*, Acta Math. **119** (1967), 27–50. MR **36** #6357.

6. Th. Schneider, *Einführung in die transzendenten Zahlen*, Springer-Verlag, Berlin, 1957. MR **19,** 252.

7. ———, *Über die Approximation algebraischer Zahlen*, J. Reine Angew. Math. **175** (1936).

8. ———, *Über eine DYSONsche Verschärfung des SIEGEL-THUEschen Satzes*, Arch. Math. **1** (1949), 288–295. MR **10,** 592.

9. C. L. Siegel, *Approximation algebraischer Zahlen*, Math. Z. **10** (1921), 173–213.

10. ———, *Über Näherungswerte algebraischer Zahlen*, Math. Ann. **84** (1921), 80–99.

11. A. Thue, *Über Annäherungswerte algebraischer Zahlen*, J. Reine Angew. Math. **135** (1909), 284–305.

12. V. G. Sprindžuk, *On some general problems of approximating numbers by algebraic numbers*, Litovsk. Mat. Sb. **2** (1962) no. 1, 129–145. (Russian) MR **28** #1165.

MATHEMATISCHES INSTITUT DER UNIVERSITÄT MARBURG
MARBURG DER LAHN, FEDERAL REPUBLIC OF GERMANY

LECTURES ON TRANSCENDENTAL NUMBERS

KURT MAHLER

I. In this introductory lecture I shall collect certain properties of transcendental numbers which are of interest in themselves and may suggest further work.

We shall be concerned only with real or complex numbers, but analogous theories can be developed for p-adic numbers and for formal power series, say with coefficients in a finite field.

The number ξ is called algebraic if there is at least one polynomial

$$(1) \qquad a(n) = a_0 + a_1 x + \cdots + a_m x^m, \qquad a_m \neq 0,$$

with integral coefficients such that $a(\xi) = 0$, and it is called transcendental if no such polynomial exists.

That there are transcendental numbers was first proved by Liouville in 1844, and the transcendency of e was established by Hermite in 1873. Since then much progress has been made and still is being made, and I shall in the following lectures report on some of this work. However, let us begin with a general necessary and sufficient, condition for transcendency. For this purpose, it is convenient to use the notations

$$L(a) = |a_0| + |a_1| + \cdots + |a_m|, \qquad \Lambda(a) = 2^m L(a).$$

The use of $L(a)$, the length of a, is advantageous because this function has much simpler properties than the height of a. Thus

$$L(a \mp b) \leq L(a) + L(b), \qquad L(ab) \leq L(a)L(b),$$

and if a and b are of the degrees m and n, respectively,

$$L(ab) \geq 2^{-(m+n)} L(a)L(b),$$

that is, $\Lambda(ab) \geq L(a)L(b)$. Roth's theorem establishes a necessary, but *not* sufficient, condition for transcendence. A necessary and sufficient condition is given by the following theorem:

THEOREM. *The number ξ is transcendental if and only if there exist*
(i) *an infinite sequence of distinct polynomials*

$$\{a_1(x), a_2(x), a_3(x), \ldots\}$$

with integral coefficients, and
(ii) *an infinite sequence of positive numbers*

$$\{\omega_1, \omega_2, \omega_3, \ldots\}$$

tending to ∞, such that

$$0 < |a_r(\xi)| < \Lambda(a_r)^{-\omega_r} \qquad (r = 1, 2, 3, \ldots).$$

Thus transcendental numbers, but not algebraic ones, can be approximated very closely by algebraic numbers distinct from them.

I shall not deal with the old classification of transcendental numbers into S, T, and U-numbers, but would like to mention a new classification which may possibly become useful.

If ξ is any real or complex number and t is a positive integer, let $\Sigma = \Sigma(\xi \mid t)$ be the set of all polynomials of arbitrary degree n, $a(z) = a_0 + \cdots + a_n z^n$, with integer coefficients such that

$$a(\xi) \neq 0, \qquad \Lambda(a) = 2^n L(a) \leq t,$$

and then put

$$\Omega(\xi \mid t) = \inf_{a(z) \in \Sigma} |a(\xi)|,$$

so that $0 \leq \Omega(\xi \mid t) \leq 1$, and $\Omega(\xi \mid t)$ is a decreasing function of t. On putting

$$\omega(\xi \mid t) = \log \{1/\Omega(\omega \mid t)\},$$

we obtain a nondecreasing function of t, with the following properties:

(1) $\omega(\xi \mid t) = O(\log t)$ if ξ is algebraic;

(2) $\omega(\xi \mid t) > c(\log t)^2$ if ξ is transcendental ($c > 0$ depends only on ξ);

(3) if ξ and η are two transcendental numbers which are algebraically dependent over $\mathbf{Q}$, then there exist constants $c_1 > 0$, $c_2 > 0$, $\gamma_1 > 0$, $\gamma_2 > 0$, $t_0 > 0$, such that for all $t \geq t_0$

$$(*) \qquad \omega(\xi \mid t^{c_1}) \geq \gamma_1 \omega(\eta \mid t) \quad \text{and} \quad \omega(\eta \mid t^{c_2}) \geq \gamma_2 \omega(\xi \mid t).$$

We may distribute transcendental numbers ξ into classes according to the order of magnitude for $t \to \infty$ of $\omega(\xi \mid t)$. Then algebraically dependent numbers fall into the same class provided that functions satisfying (*) are put into the same class.

The most interesting classes of numbers for which transcendency has been proved are given as the values of suitable analytic functions. These functions in many cases are defined as power series with integral or rational or algebraic coefficients. Since the time of Weierstrass, many mathematicians have posed conjectures on values of such functions at algebraic points, e.g. that they cannot always be algebraic numbers. Surprisingly, most of these conjectures turned out to be wrong, and mathematicians like Häckel, Faber, Hurwitz, Gelfond, Lekkerkerker have obtained results as follows:

(1) There are entire transcendental functions

$$f(z) = \sum_{h=0}^{\infty} f_h z^h$$

with rational coefficients f_h such that, for every algebraic α, all values

$$f(\alpha), f'(\alpha), f''(\alpha), \ldots$$

are algebraic.

(2) There exist transcendental power series

$$f(z) = \sum_{h=0}^{\infty} f_h z^h$$

with integral coefficients f_h, which converge for $|z| < 1$, such that, for every algebraic α satisfying $|\alpha| < 1$, all values

$$f(\alpha), f'(\alpha), f''(\alpha), \ldots$$

are algebraic.

(3) There exists a transcendental power series

$$f(z) = \sum_{h=0}^{\infty} f_h z^h$$

with rational coefficients f_h which converges at least for $|z| < \rho$ and is here algebraic for algebraic z and transcendental for transcendental z.

(4) Let

$$f(z) = \sum_{h=0}^{\infty} f_h z^h$$

be a power series with real coefficients which represents an entire transcendental function, say with exactly the zeros $\{\zeta_1, \zeta_2, \zeta_3, \ldots\}$. Then there exists also an entire transcendental function

$$F(z) = \sum_{h=0}^{\infty} F_h z^h$$

with *rational* coefficients and exactly the same zeros

$$\{\zeta_1, \zeta_2, \zeta_3, \ldots\}.$$

(5) Let $(\alpha_1, \beta_1), \ldots, (\alpha_n, \beta_n)$ be finitely many pairs of real or complex numbers,

as follows:

(a) $0 < |\alpha_k| < 1$ (for $k = 1, 2, \ldots, n$).

(b) If α_k is real for any k, so is β_k.

(c) If α_k is not real, there is an α_l, $l \neq k$, such that

$$\alpha_l = \bar{\alpha}_k, \qquad \beta_l = \bar{\beta}_k.$$

Then there exists a power series

$$f(z) = \sum_{k=0}^{\infty} f_h x^h$$

with bounded integral coefficients f_s such that

$$f(\alpha_k) = \beta_k \qquad (k = 1, 2, \ldots, n).$$

We may choose for the a_k conjugate algebraic numbers. The result shows then that $f(z)$ may be algebraic in one of these points and transcendental in the conjugate algebraic points.

All these theorems make quite clear that for general power series with rational or integral coefficients *no general assertions on transcendency can be made* with respect to their values at algebraic points. Such values will sometimes be algebraic and sometimes transcendental.

One has succeeded in proving the transcendence of function values $f(\alpha)$, α algebraic, mainly in the case where $f(z)$ satisfies one or more functional equations. Thus Hermite's proof of the transcendence of e is based on the pair of functional equations

$$\frac{d}{dz} e^z = e^z, \qquad e^{z+w} = e^z e^w.$$

Siegel's proof of the transcendency of $J_0(\alpha)$ uses the linear differential equation for $J_0(z)$, and Shidlovski's more general results apply to the solutions of systems of linear differential equations.

Let us mention at this point several unsolved problems. They are all in some way connected with the problem of the digits of a transcendental decimal fraction.

(I). Does there exist a transcendental power series

$$f(z) = \sum_{h=0}^{\infty} f_h z^h$$

with *bounded* integral coefficients which is algebraic in all algebraic points $z = \alpha$ where $|\alpha| < 1$?

If the condition of boundedness is dropped, we found that such series do exist. I conjecture that for transcendental power series with bounded integral coefficients f_h the sequence $\{\alpha_1, \alpha_2, \alpha_3, \ldots\}$ of algebraic points $z = \alpha_k$ for which $f(\alpha_k)$ also is algebraic always satisfies $\lim_{k\to\infty} |\alpha_k| = 1$. Such points have thus no limit point in the interior of the unique circle. A simple example is

$$f(z) = \prod_{n=0}^{\infty} (1 - 2z^{2^n}).$$

If my conjecture is wrong, $f(z)$ may be algebraic in all points $z = 1/g$, $g = 2, 3, \ldots$. It would then follow that, for sufficiently large g, the g-adic fraction

$$\sum_{h=0}^{\infty} f_h g^{-h}$$

is algebraic. Since $|f_h|$ is bounded, we could add a multiple of the rational number

$$\sum_{h=0}^{\infty} g^{-h} = \frac{g}{g-1},$$

and would then get a g-adic series where all coefficients are digits $0, 1, \ldots, c$ with $c < g - 1$. There would thus be algebraic irrationals, the g-adic series of which would not contain all digits $0, 1, \ldots, g - 1$.

In the case $g = 3$, my conjecture takes the form

(II) Cantor's set of all triadic series

$$\sum_{n=0}^{\infty} \frac{a_n}{3^n}, \quad \text{where } a_n = 0 \text{ or } = 2,$$

does not contain any irrational algebraic number. (It is obvious that there are infinitely many rational numbers in Cantor's set.)

II. As a preparation to the deep results by Siegel and Shidlovski, I shall today discuss some simpler results of mine which appeared in 1929 and 1930 in three papers in Mathematische Annalen and Mathematische Zeitschrift.

The problem to be discussed is under which additional conditions analytic functions defined, say, by convergent power series

$$f(z) = \sum_{h=0}^{\infty} f_h z^h$$

can for algebraic z inside the circle of convergence assume algebraic values.

If $f(z)$ is an algebraic function of z, it is not difficult to prove that

(i) $f(z)$ is algebraic at all regular algebraic points z if all the Taylor coefficients f_h are algebraic, but

(ii) there are at most finitely many algebraic points z for which $f(z)$ is algebraic if at least one coefficient f_h is transcendental, and these points can be determined.

We exclude now algebraic functions and impose on $f(z)$ the

1st restriction. $f(z)$ is a transcendental function of z, and, in the hope of simpler results, also

2nd restriction. The Taylor coefficients f_h of $f(z)$ are algebraic numbers, say they lie in a finite algebraic number field K.

Even if these two restrictions are combined, it is not possible to make general assertions on the function values of $f(z)$ at algebraic points. For we saw already that there exist even transcendental entire functions with *rational* coefficients f_h for which $f(z)$ is algebraic for all algebraic z. Thus still further restrictions have to be imposed on $f(z)$.

These additional restrictions on $f(z)$ usually take the form of one or more functional equations, in particular of differential equations. By way of example, Hermite's proof of the transcendency of e and Lindemann's proof of the transcendency of π are both based on the pair of functional equations

$$\frac{d}{dz}e^z = e^z \quad \text{and} \quad e^{z+z'} = e^z e^{z'}.$$

In the work of Siegel and Shidlovski, an analogous role is played by a system of linear differential equations.

Let us begin with the simpler case where the additional condition takes the following form.

3rd restriction. Let $\rho \geq 2$ be a fixed positive integer, and let m be an integer satisfying $1 \leq m < \rho$; let further

$$a_l(z),\, b_l(z) \qquad (l = 0, 1, \ldots, m)$$

be polynomials in z with algebraic coefficients where $a_m(z)$ and $b_m(z)$ do not both vanish identically. Further let $f(z)$ satisfy the functional equation

$$(1) \qquad f(z^\rho) = \frac{\sum_{l=0}^{m} a_l(z) f(z)^l}{\sum_{l=0}^{m} b_l(z) f(z)^l}.$$

This class of functional equation has interest in itself, but not much seems to be known about it. It can be generalized; thus one might consider the more general kind of functional equation $P(z, f(z), f(z^\rho)) = 0$ where P is a polynomial in its arguments, at most of degree $< \rho$ in both $f(z)$ and $f(z^\rho)$. When P is of degree $\geq \rho$ in $f(z)$ and $f(z^\rho)$, difficulties arise which I have not so far overcome. This is regrettable because the transformation equations of the modular function $f(z) = j(\log z/2\pi i)$ are exactly of this kind.

Let $f(z)$ satisfy our three restrictions, and let not only the Taylor coefficients f_h, but also the coefficients of the polynomials $a_l(z)$ and $b_l(z)$ lie in the finite number field K, and so let z_0 and $f(z_0)$. The problem to solve is for which values of z_0 this can be the case. Naturally K can always be replaced by a larger algebraic number field; the hypothesis just made is therefore a natural one when both z_0 and $f(z_0)$ have algebraic values.

If $z = 0$, $f(z) = f_0$ certainly is algebraic; we exclude this trivial case and assume that $0 < |z_0| < 1$. Then $z_0^{\rho^n}$ lies for sufficiently large n in the circle of convergence of $f(z)$, and hence the functional equation (1) enables us to obtain the value of $f(z_0)$ from the series, possibly after solving an algebraic equation.

In fact, on applying (1) successively to

$$z_0^\rho,\, z_0^{\rho^2}, \ldots, z_0^{\rho^n}$$

and eliminating $f(z_0^\rho), f(z_0^{\rho^2}), \ldots, f(z_0^{\rho^{n-1}})$ from the equations so obtained, we evidently obtain a relation of the form

$$(2) \qquad f(z_0^{\rho^n}) = \frac{\sum_{l=0}^{m^n} a_l^{(n)}(z_0) f(z_0)^l}{\sum_{l=0}^{m^n} b_l^{(n)}(z_0) f(z_0)^l}$$

where

$$a_l^{(n)}(z) \quad \text{and} \quad b_l^{(n)}(z) \qquad (l = 0, 1, \ldots, m^n)$$

are certain $2(m^n + 1)$ polynomials in z which have again coefficients in K, and without loss of generality *integral* coefficients. From the known value of $f(z_0^{\rho^n})$ (known from the power series), the value of $f(z_0)$ is obtained by solving (2) for $f(z_0)$. This will only then become impossible when the right-hand side becomes indeterminate because the polynomials in u,

$$\sum_{l=0}^{m^n} a_l^{(n)}(z) u^l \quad \text{and} \quad \sum_{l=0}^{m^n} b_l^{(n)}(z) u^l,$$

have a common zero $u = u_0$. A detailed discussion shows that this can happen only if z_0 satisfies one of the equations

$$\Delta(z^{\rho^k}) = 0 \qquad (k = 0, 1, 2, \ldots)$$

where $\Delta(z)$ is the resultant of

$$\sum_{l=0}^{m} a_l(z) u^l \quad \text{and} \quad \sum_{l=0}^{m} b_l(z) u^l$$

with respect to u.

Such values of z_0 may indeed lead to algebraic values of $f(z_0)$ as can be seen in simple examples. We exclude this difficulty by imposing the

4th restriction. For every integer $h \geq 0$, z_0 satisfies $\Delta(z_0^{\rho\cdot}) \neq 0$.

I would like to add that in the two special cases

$$f(z^\rho) = \frac{\sum_{l=0}^{m} a_l(z) f(z)^l}{b_0(z)} \qquad (b_0(z) \not\equiv 0)$$

and

$$f(z^\rho) = \frac{a_0(z)}{\sum_{l=0}^{m} b_l(z) f(z)^l} \qquad (a_0(z) \not\equiv 0)$$

the resultant $\Delta(z)$ is to be defined by

$$\Delta(z) = a_m(z) b_0(z) \quad \text{and} \quad \Delta(z) = a_0(z) b_m(z),$$

respectively.

The four restrictions are sufficient to settle the problem of transcendency $f(z)$.

THEOREM. *If the function $f(z)$ and the number z_0 satisfy the four restrictions and if $0 < |z_0| < 1$, then z_0 and $f(z_0)$ cannot both lie in K, and therefore at least one of these two numbers is transcendental.*

The proof runs as follows. Denote by p a large positive integer. One can then construct $p + 1$ polynomials not all identically zero,

$$A_0(z), A_1(z), \dots, A_p(z),$$

of degree at most p with integral coefficients in K such that in the new power series

$$E_p(z) = \sum_{l=0}^{p} A_l(z) f(z)^l = \sum_{h=0}^{\infty} B_h z^h, \quad \text{say,} \tag{3}$$

all coefficients B_h with $h \leq (p + 1)^2 - 2$ are zero. For we have $(p + 1)^2$ coefficients of the $A_l(z)$ at our disposal and need satisfy only the $(p + 1)^2 - 1$ homogeneous linear equations

$$B_0 = B_1 = \cdots = B_{(p+1)^2-2} = 0$$

for these coefficients where these linear equations have coefficients in K.

By the 1st restriction, $E_p(z)$ is not identically zero; there is thus a suffix h_0 satisfying $h_0 \geq (p + 1)^2 - 1 > p^2$ such that

$$B_{h_0} \neq 0. \tag{4}$$

Let now n be a large positive integer, and let

$$E_p^{(n)}(z) = E_p(z^{\rho^n}) \left\{ \sum_{l=0}^{m^n} b_l^{(n)}(z) f(z)^l \right\}^p .$$

By the formula (2), we can also write

$$E_p^{(n)}(z) = \sum_{l=0}^{km^n} B_l^{(n)}(z) f(z)^l \tag{5}$$

where the $B_l^{(n)}(z)$ are again polynomials with integral coefficients in K. One can easily obtain majorants for these polynomials and for $E_p^{(n)}(z)$. The hypothesis of the 4th restriction shows that, for the given z_0,

$$\sum_{l=0}^{m^n} b_l^{(n)}(z_0) f(z_0)^l \neq 0.$$

Further, for large n,

$$E_p(z_0^{\rho^n}) \sim B_{h_0} z_0^{h_0 \rho^n}, \quad \text{hence } E_p^{(n)}(z_0) \neq 0.$$

With the usual taking of the norm it follows then that

$$0 < |E_p^{(n)}(z_0)| \leq \exp(-c_1 p^2 \rho^n), \tag{6}$$

while

$$(7) \qquad |E_p^{(n)}(z_0)| > \exp(-c_2 p \rho^n).$$

Here $c_1 > 0$ and $c_2 > 0$ depend on z_0, but not on p and n. From (6) and (7), a contradiction arises as soon as p and n are sufficiently large. This proves the theorem.

By way of example, the two functions

$$f_1(z) = \prod_{n=0}^{\infty} (1 + z^{2^n}) \quad \text{and} \quad f_2(z) = \prod_{u=0}^{\infty} (1 - z^{2^n})$$

have power series convergent for $|z| < 1$, and they satisfy the functional equations

$$f_1(z^2) = \frac{f(z)}{1+z} \quad \text{and} \quad f_2(z^2) = \frac{f(z)}{1-z},$$

respectively. Further the resultants become

$$\Delta(z) = 1 + z \quad \text{and} \quad \Delta(z) = 1 - z,$$

respectively, and hence, for all n,

$$\Delta(z^{2^n}) \neq 0 \quad \text{if} \quad 0 < |z| < 1.$$

Hence if

$$0 < |z_0| < 1 \quad \text{and} \quad z_0 \text{ is algebraic},$$

then $f_k(z_0)$ is transcendental provided $f_k(z)$ is a transcendental function. But it is easily proved that $f_1(z) \equiv 1/(1 - z)$ is an algebraic function.

The second function $f_2(z)$, however, *is* transcendental and in fact cannot be continued beyond $|z| = 1$.

Much more, and for more general classes of functions, can be proved. Thus if, e.g.

$$f(z) = \sum_{n=0}^{\infty} \frac{z^{\rho^n}}{1 - z^{\rho^n}},$$

all derivatives

$$f^{(k)}(z) \qquad (k = 0, 1, 2, \ldots)$$

are easily proved to satisfy a simple system of functional equations similar to the one studied. One can also show that $f(z)$ does not satisfy any algebraic differential equation, and that the Taylor coefficients of $f(z)$ are rational integers. From this it can again be deduced that, if

$$(8) \qquad 0 < |z_0| < 1, \quad \text{and } z_0 \text{ is an algebraic number},$$

then, for every m, the $m + 1$ function values

$$(9) \qquad f(z_0), f'(z_0), \ldots, f^{(m)}(z_0)$$

are algebraically independent over **Q**.

Perhaps even more interesting is the analogous result for

$$f(z) = \sum_{h=0}^{\infty} [h\omega]z^m$$

where $\omega > 0$ is a real quadratic irrationality, and [ ] denotes the integral part. Again, if (8) holds, the function values (9) are algebraically independent.

I have discussed the functions of today's lecture because somewhat similar ideas play a role in Shidlovski's work.

III. In the remaining three lectures, I shall discuss the beautiful results obtained by Shidlovski by generalizing Siegel's ideas of 1929.

These results are concerned with entire functions satisfying linear differential equations with rational functions as coefficients. It is convenient to consider instead systems of linear differential equations

$$Q^*\colon w_h' = \sum_{k=1}^{m} q_{hk}w_k + q_{h0} \qquad (h = 1, 2, \ldots, m),$$

where the coefficients

$$q_{hk}, q_{h0}$$

are arbitrary rational functions of z. We shall also have to deal with the corresponding homogeneous system

$$Q\colon w_h' = \sum_{k=1}^{m} q_{hk}w_k \qquad (h = 1, 2, \ldots, m).$$

While there are no further restrictions on the coefficients q_{hk}, q_{h0} of Q^* and Q, the theory of Siegel and Shidlovski is specialized by restrictions on the solution vectors

$$\mathbf{w} = \begin{bmatrix} w_1 \\ \vdots \\ w_m \end{bmatrix}$$

of these systems. Not only will only that case be considered in which all the components

$$w_h = \sum_{l=0}^{\infty} w_{hl}z^l \qquad (h = 1, 2, \ldots, m)$$

are *entire* functions, but these entire functions will be restricted to a very special class, the so-called E-functions of Siegel.

These are defined as follows:

Let K be a number field of finite degree N over $\mathbf{Q}$. If $\alpha \in K$, denote as usual by

$$\overline{|\alpha|} = \max(|\alpha|, |\alpha'|, \ldots, |\alpha^{(N-1)}|)$$

the maximum of the absolute values of the conjugates of α relative to $\mathbf{Q}$.

The series

$$f(z) = \sum_{l=0}^{\infty} f_l \frac{z^l}{l!}$$

is now a Siegel E-function over K if the following conditions are satisfied:

(1) All $f_l \in K$.

(2) $\overline{|f_l|} = O(l^{\epsilon l})$ for all l and all $\epsilon > 0$.

(3) There exists for each $l \geq 0$ a positive rational integer $d_l = O(l^{\epsilon l})$ for all l and $\epsilon > 0$ such that

$$d_l f_k = \text{algebraic integer for } k = 0, 1, \ldots, l.$$

The E-functions so defined are entire functions, possibly polynomials. If $\gamma \in K$, $f(\gamma z)$ also is an E-function. Further the set of all E-functions forms a ring which is moreover closed under differentiation and under the integration $\int_0^z \ldots dz$.

The E-functions are so important because of the following lemma by Siegel.

LEMMA. *Let*

$$f_h(z) = \sum_{l=0}^{\infty} f_{hl} \frac{z^l}{l!} \qquad (h = 1, 2, \ldots, m)$$

be finitely many E-functions, say over K; let $0 < \phi < 1$; *and let n be any positive integer. Then there exist n polynomials*

$$p_h(z) = \sum_{l=0}^{\infty} G_{hl} z^l \qquad (h = 1, 2, \ldots, n)$$

with integral coefficients in K not all zero where

$$\max_{h,l} \overline{|G_{hl}|} = O(n^{(1+\epsilon)n})$$

for all $\epsilon > 0$, *while, on putting*

$$p = mn - [\phi n] - 1 \quad \textit{and} \quad \sum_{h=1}^{m} p_h(z) f_h(z) = \sum_{l=0}^{\infty} a_l \frac{z^l}{l!},$$

all coefficients $a_0 = a_1 = \cdots = a_{p-1} = 0$, *and* $\overline{|a_l|} = n^n O(l^{\epsilon l})$ *for all* $l \geq p$ *and* $\epsilon > 0$.

Let now in particular

$$\mathbf{f}(z) = \begin{bmatrix} f_1(z) \\ \vdots \\ f_m(z) \end{bmatrix}$$

be a solution of the homogeneous system

$$Q\colon w_h' = \sum_{k=1}^{m} q_{hk} w_k \qquad (h = 1, \ldots, m).$$

Denote by $\kappa(z)$ the polynomial with leading coefficient 1 which is the least common denominator of all the q_{hk}. As can be shown easily, since the series $f_h(z)$ have Taylor coefficients in K, the same is without loss of generality true for the coefficients of the q_{hk} and thus of κ.

We put now

$$\lambda_1\{\mathbf{w}(z)\} = \sum_{k=1}^{m} p_{1k}(z)w_k(z),$$

where

$$p_{1k}(z) = p_k(z) \qquad (k = 1, \ldots, m)$$

and deduce from λ_1 infinitely many further linear forms

$$\lambda_h\{\mathbf{w}(z)\} = \sum_{k=1}^{m} p_{hk}(z)w_k(z)$$

where

$$\lambda_{h+1} = \kappa \frac{d}{dz} \lambda_h.$$

Here $\mathbf{w}(z)$ denotes a general solution of Q, and during the differentiation w_h' is replaced by its expression from Q so that

$$p_{h+1,k} = \kappa p_{hk}' + \sum_{j=1}^{m} p_{hj}\kappa q_{jk} \qquad (h = 1, 2, \ldots, k = 1, 2, \ldots, m).$$

It is clear that also the p_{hk} are polynomials in $K[z]$, and the lemma leads to simple estimates for these coefficients and their conjugates, and also for the functions

$$\lambda_h\{\mathbf{f}(z)\} = \sum_{k=1}^{m} p_{hk}(z)f_k(z) \qquad (k = 1, 2, \ldots).$$

Siegel proved in special cases, and Shidlovski under very general conditions, that the determinant

$$P(z) = \begin{vmatrix} p_{11}(z) & \cdots & p_{1m}(z) \\ \vdots & & \vdots \\ p_{m1}(z) & \cdots & p_{mm}(z) \end{vmatrix}$$

is not identically zero. I shall discuss this fundamental question in detail; but let us for the present just assume that

(H) $$P(z) \not\equiv 0.$$

Denote by $\alpha \in K$ an algebraic number such that

(A) $$\alpha \neq 0, \qquad \kappa(\alpha) \neq 0;$$

the second condition means that $z = a$ is not a singular point of $\mathbf{Q}$. It can be deduced easily from (H), and was first done by Siegel in a special case, that there exist m suffixes $h_1, \ldots, h_m$ satisfying

$$1 \leq h_1 < h_2 < \cdots < h_m \leq [\phi n] + n_0,$$

where n_0 is a constant integer independent of n, such that by (H)

$$\text{(K)} \qquad \begin{vmatrix} p_{h_1 1}(\alpha) & \cdots & p_{h_1 m}(\alpha) \\ \vdots & & \vdots \\ p_{h_m 1}(\alpha) & \cdots & p_{h_m m}(\alpha) \end{vmatrix} \neq 0.$$

One of Shidlovski's conditions for (H) is that

$$f_1(z), \ldots, f_m(z)$$

are linearly independent over $C(z)$ and hence also over $K(z)$. Let us on the other hand assume that not more than $r < m$ of the function values

$$f_1(\alpha), \ldots, f_m(\alpha)$$

are linearly independent over K. There exists then an $(m - r) \times m$ matrix of rank $m - r$,

$$\begin{bmatrix} s_{11} & \cdots & s_{1m} \\ \vdots & & \vdots \\ s_{m-r,1} & \cdots & s_{m-r,m} \end{bmatrix}$$

with elements in K such that

$$\text{(1)} \qquad s_{h1}f_1(\alpha) + \cdots + s_{hm}f_m(\alpha) = 0 \qquad (h = 1 \ldots m - r);$$

and we can select r suffixes $h_1, \ldots, h_m$, say, $j_1, \ldots, j_r$, such that

$$S \equiv \begin{vmatrix} p_{j_1 1}(\alpha) & \cdots & p_{j_1 m}(\alpha) \\ \cdot & & \cdot \\ p_{j_r 1}(\alpha) & \cdots & p_{j_r m}(\alpha) \\ \cdot & & \cdot \\ s_{11} & \cdots & s_{1m} \\ \cdot & & \cdot \\ s_{n-r,1} & \cdots & s_{m-r,m} \end{vmatrix} \neq 0.$$

The equations (1) together with

$$\text{(2)} \qquad \lambda_h(\mathbf{f}(\alpha)) = p_{h1}(\alpha)f_1(\alpha) + \cdots + p_{hm}(\alpha)f_m(\alpha) \qquad (h = j_1, \ldots, j_r)$$

lead therefore to

$$\text{(S)} \qquad Sf_k(\alpha) = \sum_{i=1}^{r} S_{ik}\lambda_{j_i}(\mathbf{f}(\alpha)) \qquad (k = 1, \ldots, m)$$

where the S_{ik} are the cofactors of S (row i, column k).

One proceeds now in (S) to take the absolute values of the conjugates on both sides, and assumes that n is large and $\epsilon > 0$ small.

The lemma leads easily to the estimates

$$\overline{|p_{hk}(\alpha)|} = O(n^{(1+\phi+\epsilon)n}) \quad \text{for } h = j_1, \ldots, j_r,$$

and

$$|\lambda_h(\mathbf{f}(\alpha))| = O(n^{-(m-1-7\phi)n}) \quad \text{for } h = j_1, \ldots, j_m.$$

Therefore

$$\boxed{S} = O(n^{(1+\phi+\epsilon)nr}), \qquad \boxed{S_{ik}} = O(n^{(1+\phi+\epsilon)n(r-1)}).$$

Also $S \neq 0$ lies in K, and there is a positive integer

$$T = O(e^{cn}) \qquad (c > 0 \text{ const.})$$

such that ST is an algebraic integer and thus $|\text{norm}\,(ST)| \geq 1$. This implies by the estimate for S a lower bound for $|S|$ which may be written as

$$|S|^{-1} = O(n^{(1+2\phi)nr(N/\sigma-1)}).$$

Here N is the degree of the field K, and

$$\begin{aligned} \sigma &= 1 \quad \text{if } K \text{ is a real field,} \\ &= 2 \quad \text{if } K \text{ is an imaginary field.} \end{aligned}$$

For in the second case two of the conjugates of S have equal absolute value.

Finally, by (S), and since we can choose k such that $f_k(\alpha) \neq 0$ (otherwise $f(z) \equiv 0$),

$$1 = O(n^{(1+2\phi)nr(N/\sigma-1)})O(n^{(1+\phi+\epsilon)n(r-1)})O(n^{-(m-1-7\phi)n}).$$

Here we make $n \to \infty$. The sum of the exponents of n is necessarily ≥ 0, hence

$$(1 + 2\phi)r\{(N/\sigma) - 1\} + (1 + \phi + \epsilon)(r - 1) - (m - 1 - 7\phi) \geq 0.$$

Now m, r, and N/σ are fixed, and both ϕ and ϵ are arbitrarily small. Thus in the limit

$$r\{(N/\sigma) - 1\} + (r - 1) - (m - 1) \geq 0,$$

which means that $r \geq \sigma m/N$. Thus if m of the functions $f_1(z), \ldots, f_m(z)$ are linearly independent over $C(z)$, then at least $\sigma m/N$ of the function values $f_1(\alpha), \ldots, f_m(\alpha)$ are linearly independent over K and hence also over $\mathbf{Q}$.

Here $\sigma = N$ if $K = \mathbf{Q}$, or if K is an imaginary quadratic field.

The result just obtained is in this generality due to Shidlovski. He has extended it in the following way:

(I) Let not all m functions $f_h(z)$ be linearly independent over $f(z)$, but say only $\rho(z) \leq m$, and let similarly $\rho(\alpha)$ denote the maximum number of function values $f_h(\alpha)$ that are linearly independent over K or $\mathbf{Q}$. Then

$$\text{(I)} \qquad \rho(\alpha) \geq \sigma\rho(z)/N.$$

Thus again $\rho(\alpha) = \rho(z)$ if $K = \mathbf{Q}$, or if K is an imaginary quadratic field; for $\rho(\alpha)$ cannot be larger than $\rho(z)$.

(II) The results so far deal with linear independence of the components of $\mathbf{f}(z)$ or $\mathbf{f}(\alpha)$. The factor σ/N on the right hand side of (I) depends only on the field K. It is this fact which will allow us to deduce from (I) the following final theorems of Shidlovski.

1ST THEOREM. *Let* $\mathbf{f}(z)$ *be a solution in terms of E-functions of*

$$Q^*\colon w_h' = q_{h0} + \sum_{k=1}^{m} q_{hk}w_k \qquad (h = 1, 2, \ldots, m).$$

Let α *be an algebraic number* $\neq 0$ *which is a regular point of* Q^*. *Here as many of the components*

$$f_1(z), f_2(z), \ldots, f_m(z)$$

of $\mathbf{f}(z)$ *are algebraically independent over* $\mathbf{C}(z)$ *as there are components*

$$f_1(\alpha), f_2(\alpha), \ldots, f_m(\alpha)$$

of $\mathbf{f}(\alpha)$ *that are algebraically independent over* $\mathbf{Q}$.

2ND THEOREM. *Let* $\mathbf{f}(z)$ *be a solution in terms of E-functions of*

$$Q\colon w_h' = \sum_{k=1}^{m} q_{hk}w_k \qquad (h = 1, 2 \ldots m).$$

Then for α *as above as many of the function ratios*

$$f_1(z) \mid f_2(z)\colon \ldots \colon f_m(z)$$

are algebraically independent over $\mathbf{C}(z)$ *as there are function value ratios*

$$f_1(\alpha)\colon f_2(\alpha)\colon \ldots\colon f_m(\alpha)$$

that are algebraically independent over $\mathbf{Q}$.

These two general theorems have many important specializations, and I hope to find the time in my last lecture to say a little about it.

IV. This fourth lecture is to deal with the discussion of the determinant P of the last lecture, and like the next lecture, depends essentially on the work of Shidlovski. However, I shall for the present slightly generalize his method because this will bring out the basic ideas in a clearer way.

Let K be any field of characteristic 0, c any constant in K, and $K\langle z - c\rangle$ the field of formal series

$$f = \sum_{l=\lambda}^{\infty} f_l(z - c)^l, \qquad f_l \in K,$$

where λ is any integer. If $f_\lambda \neq 0$, we put

$$\operatorname{ord} f = \operatorname{ord}_c f = \lambda.$$

We consider a fixed system of formal differential equations

$$Q\colon w_h' = \sum_{k=1}^{m} q_{hk}w_k \qquad (h = 1, 2 \ldots m)$$

where the q_{hk} lie in $K(z)$. We shall be concerned in particular with the solutions $\mathbf{f}$ of Q that have components $f_1, \ldots, f_m$ in $K\langle z - c\rangle$.

Denote by V_Q the set of all such solutions; then V_Q is a vector space over K. A basic result asserts that if $f_1, \ldots, f_M$ are finitely many solutions in V_Q which are linearly independent over K, they are also linearly independent over $K\langle z - c\rangle$. Hence the dimension of V_Q over K, M say, satisfies $0 \leq M \leq m$.

In the special case when the least common denominator κ of all the q_{hk} is such that $\kappa(c) \neq 0$, i.e. when c is not a pole of any q_{hk}, always $M = m$.

For the deeper study of Q one introduces also a vector space over $K(z)$. Let for the present $p_1, \ldots, p_m$ be m rational functions in $K(z)$. One can then form the linear space of all forms

$$\lambda = \lambda(\mathbf{w}) = p_1 w_1 + \cdots + p_m w_m$$

where $\mathbf{w}$ denotes any solution in $K\langle z - c\rangle$ of Q.

Denote by Λ not this whole linear space, but any linear subspace; thus λ_1, $\lambda_2 \in \Lambda$ implies $\lambda_1 \mp \lambda_2 \in \Lambda$, and $r\lambda \in \Lambda$ if $r \in K(z)$.

We can differentiate linear forms $\lambda(\omega)$,

$$\frac{d}{dz}\lambda(w) = \frac{d}{dz}\sum_{k=1}^{m} p_k w_k = \sum_{k=1}^{m} (p_k' w_k + p_k w_k'),$$

and hence by Q,

$$\kappa \frac{d}{dz}\lambda(w) = \sum_{h=1}^{m} p_h^* w_h$$

where

$$p_h^* = \kappa\left(p_h' + \sum_{j=1}^{m} p_j q_{jh}\right).$$

On putting $D = \kappa d/dz$, $D\lambda$ is then a linear form of the same type as λ. In particular, if the p_h are polynomials, so are the p_h^*.

The following definition is now basic:

DEFINITION. The vector space Λ is said to be closed under D if $\lambda \in \Lambda$ implies $D\lambda \in \Lambda$.

THEOREM A. *Let V_Q be of dimension M over K and Λ of dimension n over $K(z)$ where $M > n$; let further Λ be closed under D. Then there exists a basis*

$$\mathbf{w}_1, \ldots, \mathbf{w}_M$$

of V_Q over K such that

$$\lambda(\mathbf{w}_1) = \cdots = \lambda(\mathbf{w}_{M-n}) = 0 \quad \text{for all } \lambda \in \Lambda.$$

I come now to the main lemma of Shidlovski. Let

$$p_{11} = p_1, \ldots, p_{1m} = p_m$$

be any polynomials in $K[z]$, and let

$$\lambda_1(\mathbf{w}) = p_{11}w_1 + \cdots + p_{1m}w_m;$$

let further, as in yesterday's lecture,

$$p_{h+1,k} = \kappa p_{hk}^* + \sum_{j=1}^{m} p_{hj}\kappa q_{jk}.$$

Then all the p_{hk} are polynomials, and the linear forms

$$\lambda_h(\mathbf{w}) = p_{h1}w_1 + \cdots + p_{hm}w_m \qquad (\mathbf{w} \in V_Q)$$

satisfy the recursive relations $\lambda_{h+1}(\mathbf{w}) = D\lambda_h(\mathbf{w})$. It is clear that the definition of the p_{hk} and λ_h is independent of c; hence we are allowed to assume that $\kappa(c) \neq 0$ so that $z = c$ is a regular point of Q. This means that V_Q has the dimension $M = m$ and that for every solution of Q,

$$\mathbf{w} = \begin{bmatrix} w_1 \\ \vdots \\ w_m \end{bmatrix}$$

with components in $K\langle z - c\rangle$, $\text{ord}_c\, w_h \geq 0$.

Having fixed λ_1 and hence λ_h for $h = 1, 2, 3, \ldots$, let Λ be the vector space over $K(z)$ spanned by these vectors, and let μ be the dimension of Λ over $K(z)$. Since $D\lambda_h = \lambda_{h+1}$, Λ is closed under D. If $\mu = m$,

$$\lambda_1, \lambda_2, \ldots, \lambda_m$$

are linearly independent forms, and hence

$$\begin{vmatrix} p_{11} & \cdots & p_{1m} \\ \vdots & & \vdots \\ p_{m1} & \cdots & p_{mm} \end{vmatrix} \neq 0.$$

For the present let this easy case be excluded so that $1 \leq \mu \leq m - 1$. Since $\lambda_1, \ldots, \lambda_\mu$ are linearly independent over $K(z)$, the matrix

$$p^* = \begin{bmatrix} p_{11} & \cdots & p_{1m} \\ \vdots & & \vdots \\ p_{\mu 1} & \cdots & p_{\mu m} \end{bmatrix}$$

has the rank μ. Therefore, without loss of generality, the minor

$$P = \begin{vmatrix} p_{11} & \cdots & p_{1\mu} \\ \vdots & & \vdots \\ p_{\mu 1} & \cdots & p_{\mu\mu} \end{vmatrix}$$

does not vanish identically. Thus the first μ columns of p^* are linearly independent over $K(z)$, and the other columns are linearly dependent on them. This means that there exist rational functions e_{ij} in $K(z)$ such that

$$p_{hj} = \sum_{i=1}^{\mu} p_{hi} e_{ij} \qquad (1 \leq h \leq \mu,\ \mu + 1 \leq j \leq m).$$

These e_{ij} naturally are unique since $P \not\equiv 0$, and they depend only on the p_{hk} and q_{hk} and not on c.

Since Λ is closed under D and evidently has the dimension

$$n = \mu \quad \text{where} \quad \mu < m = M,$$

it follows from Theorem A that there exists a basis

$$\mathbf{w}_1, \ldots, \mathbf{w}_m$$

of V_Q such that

(*) $$\lambda_h(\mathbf{w}_k) = 0 \quad \text{if } 1 \leq h \leq \mu;\ 1 \leq k \leq m - \mu.$$

Let in explicit form

$$\mathbf{w}_k = \begin{bmatrix} w_{1k} \\ \vdots \\ w_{mk} \end{bmatrix}.$$

Then

$$\begin{aligned} \lambda_h(\mathbf{w}_k) &= \sum_{i=1}^{m} p_{hi} w_{ik} = \sum_{i=1}^{\mu} + \sum_{i=\mu+1}^{m} \\ &= \sum_{i=1}^{\mu} p_{hi} \left(w_{ik} + \sum_{j=\mu+1}^{m} e_{ij} w_{jk} \right) = \sum_{i=1}^{\mu} p_{hi} W_{ik}, \end{aligned}$$

where we have put

$$W_{ik} = w_{ik} + \sum_{j=\mu+1}^{m} e_{ij} w_{jk}.$$

By (*) we have now for each $k = 1, 2, \ldots, m - \mu$

(**) $$\lambda_h(\mathbf{w}_k) = \sum_{i=1}^{\mu} p_{hi} W_{ik} = 0 \qquad (h = 1, 2, \ldots, \mu).$$

Since $P \not\equiv 0$, this requires that

$$W_{ik} = 0 \qquad (i = 1, 2, \ldots, \mu;\ k = 1, 2, \ldots, m - \mu),$$

for (**) is a system of μ homogeneous equations for μ unknowns.

We have thus the result that

(1) $$W_{ik} \equiv w_{ik} + \sum_{j=\mu+1}^{m} e_{ij} w_{jk} = 0 \qquad (1 \leq i \leq \mu,\ 1 \leq k \leq m - \mu).$$

It is then not difficult to deduce that the matrix of order $m - \mu$

$$w^{(0)} = \begin{bmatrix} w_{\mu+1,1} & \cdots & w_{\mu+1,m-\mu} \\ \vdots & & \vdots \\ w_{m1} & \cdots & w_{m,m-\mu} \end{bmatrix}$$

is nonsingular,

$$\det w^{(0)} \neq 0.$$

For the full solution matrix

$$w = \begin{bmatrix} w_{11} & \cdots & w_{1m} \\ \vdots & & \vdots \\ w_{m1} & \cdots & w_{mm} \end{bmatrix}$$

certainly is regular, and if $\det w^{(0)}$ were $= 0$, one could use the identities (1) to deduce that also $\det w = 0$.

Put $\Omega = \det w^{(0)}$ so that $\Omega \neq 0$. One can solve the equations (1) for the rational functions e_{ij} in the form

$$(2) \qquad e_{ij} = -\frac{\Omega_{ij}}{\Omega} \qquad (1 \leq i \leq \mu;\ \mu + 1 \leq j \leq m)$$

where Ω_{ij} is obtained from Ω on replacing the row

$$w_{j1}, \ldots, w_{j,m-\mu}$$

by the new row

$$w_{i1}, \ldots, w_{i,m-\mu}.$$

The formulae (2) lead to deeper results on these rational functions e_{ij}.

For this purpose, let us vary the coefficients

$$(3) \qquad p_{11}, \ldots, p_{1m}$$

of λ_1 in all ways such that $1 \leq \mu \leq m - 1$. Naturally for each choice of the coefficients (3) we can expect different e_{ij} and also different bases

$$\mathbf{w}_1, \ldots, \mathbf{w}_m$$

of V_Q. Thus the determinants Ω, Ω_{ij} in (2) will vary.

However, if $\mathbf{w}_1^0, \ldots, \mathbf{w}_m^0$ is any one basis of V_Q chosen once for all, the most general basis $\mathbf{w}_1, \ldots, \mathbf{w}_m$ has the form

$$\mathbf{w}_h = \sum_{k=1}^{m} a_{hk}\mathbf{w}_k^0 \qquad (h = 1, \ldots, m)$$

where

$$(a_{hk}) \qquad (h, k = 1, 2, \ldots, m)$$

is an arbitrary nonsingular matrix with elements in K. Thus we arrive at the following results for the e_{ij}.

Form the matrix of the vectors

$$\mathbf{w}_1^0, \ldots, \mathbf{w}_m^0$$

and denote by

$$\phi_1, \phi_2, \ldots, \phi_s$$

the set which consists of all the elements of this matrix, all its minors of order 2, of order 3, etc., and finally of its determinant. Then clearly Ω and Ω_{ij} can be written as

$$\Omega = c_1\phi_1 + \cdots + c_s\phi_s$$
$$-\Omega_{ij} = c_{ij1}\phi_1 + \cdots + c_{ijs}\phi_s$$

where the c's are certain elements in K. Then

$$e_{ij} = \frac{c_{ij1}\phi_1 + \cdots + c_{ijs}\phi_s}{c_1\phi_1 + \cdots + c_s\phi_s} \tag{4}$$

where the e_{ij} are rational functions while the ϕ by their definition lie in $K\langle z - c\rangle$.

If we change now the p_{hk} so that $\mu < m$ remains fixed, only the constant coefficients c in (4), but not the ϕ's are changed. From this it can easily be deduced that

THEOREM B. *While the rational functions e_{ij} may vary with the changes of the polynomials p_{hk}, the degrees of their numerators and their denominators remain bounded.*

We come finally to the consideration of the determinant

$$\begin{vmatrix} p_{11} & \cdots & p_{1m} \\ \vdots & & \vdots \\ p_{m1} & \cdots & p_{mm} \end{vmatrix}.$$

Let us assume that Q has a solution

$$\mathbf{f} = \begin{bmatrix} f_1 \\ \vdots \\ f_m \end{bmatrix}$$

with components in $K\langle z - c\rangle$ which are linearly independent over $K(z)$.

Denote by ∂p the degree of a polynomial p, by $\operatorname{ord}_c w$ the order of any element w in $K\langle z - c\rangle$. For a polynomial obviously

$$\operatorname{ord}_c p \le \partial p \quad \text{if } p \not\equiv 0.$$

Let us now assume that the p_{hk} are such that $1 \le \mu \le m - 1$, that further X and Y are two integers such that

$$\partial p_{1k} \le X \quad \text{and} \quad \operatorname{ord}_c \lambda_1(\mathbf{f}) \ge Y.$$

We have found already that

$$P = \begin{vmatrix} p_{11} & \cdots & p_{1\mu} \\ \vdots & & \vdots \\ p_{\mu 1} & \cdots & p_{\mu\mu} \end{vmatrix} \not\equiv 0.$$

It is not difficult to deduce from the recursive formulae for the p_{hk} that $\partial P \le \mu X + \mu(\mu - 1)/2C_1$ where C_1 depends only on the q_{hk}. Hence also

$$\operatorname{ord}_c P \le \mu X + \frac{\mu(\mu - 1)}{2} C_1.$$

Let us put

$$\text{(O)} \qquad F_i = f_i + \sum_{j=\mu+1}^{m} e_{ij} f_j \qquad (i = 1 \dots \mu),$$

so that certainly all $F_i \not\equiv 0$ because $f_1, \dots, f_m$ are linearly independent over $K(z)$. It is easily verified that

$$\lambda_h(\mathbf{f}) = \sum_{i=1}^{\mu} p_{hi} F_i \qquad (h = 1 \dots \mu)$$

and since $P \not\equiv 0$, for all i

$$(+) \qquad PF_i = \sum_{h=1}^{\mu} P_{ih} \lambda_h(\mathbf{f}) \qquad (1 \leq i \leq \mu).$$

The P_{ih} are cofactors of the polynomials in P, hence are themselves polynomials, and so $\text{ord}_c P_{ih} \geq 0$. The e_{ij}, as we saw, have numerators and denominators of bounded degrees. Let ϵ be their common denominator which is also of bounded degree.

From (O),

$$\epsilon F_i = \epsilon f_i + \sum_{j=\mu+1}^{m} (\epsilon e_{ij}) f_j,$$

a formula from which it can be deduced that

$$\max_{1 \leq i \leq \mu} (\text{ord}_c F_i) \quad \text{is bounded.}$$

It can then easily be proved from (+) that

$$\text{ord}_c P \geq Y - (\mu - 1) - C_2$$

where also C_2 is an integer independent of the p_{hk}.

Since on the other hand

$$\text{ord}_c P \leq \mu X + \frac{\mu(\mu - 1)}{2} C_1,$$

we deduce that for $\mu \leq m - 1$

$$\begin{aligned} Y - (m - 1)X \leq Y - \mu X &\leq \frac{\mu(\mu - 1)}{2} C_1 + (\mu - 1) + C_2 \\ &\leq \frac{m(m - 1)}{2} C_1 + (m - 1) + C_2, = C \quad \text{say.} \end{aligned}$$

Hence, conversely, if

$$\text{(S)} \qquad Y - (m - 1)X > C,$$

then we cannot have $1 \leq \mu \leq m - 1$ and therefore

$$\mu = m,$$

thus

$$\begin{vmatrix} p_{11} & \cdots & p_{1m} \\ \vdots & & \vdots \\ p_{m1} & \cdots & p_{mm} \end{vmatrix} \neq 0$$

(S) is Shidlovski's main lemma. In the application, we had

$$X = \max \partial p_{1k} \leq n - 1, \qquad Y = \operatorname{ord} \lambda_1(f) \geq mn - [\phi n] - 1,$$

where $0 < \phi < 1$; hence

$$\begin{aligned} Y - (m-1)X &\geq mn - [\phi n] - 1 - (m-1)(n-1) \\ &\geq (1-\phi)n - \text{const.}, \end{aligned}$$

and so (S) can certainly be applied as soon as n is sufficiently large.

V. Let again K be a finite number field,

$$Q\colon w_h' = \sum_{k=1}^{m} q_{hk} w_k \qquad (h = 1, 2, \ldots, m),$$

a system of homogeneous linear differential equations with coefficients $q_{hk} \in K(z)$, and κ the least common denominator of these coefficients. Let further α be any algebraic number satisfying $\alpha \neq 0$, $\kappa(\alpha) \neq 0$, and let

$$\mathbf{f}(z) = \begin{bmatrix} f_1(z) \\ \vdots \\ f_m(z) \end{bmatrix}$$

be a solution of Q with components that are Siegel E-functions. We saw that Shidlovski proved the following result.

THEOREM 1. *Denote by $\rho(z)$ the maximum number of components of $\mathbf{f}(z)$ that are linearly independent over $K(z)$, by $\rho(\alpha)$ the maximum number of components of $\mathbf{f}(\alpha)$ that are linearly independent over K. Then*

$$\rho(\alpha) \geq \frac{\sigma}{N}\,\rho(z).$$

Here N is the degree of K over $\mathbf{Q}$, and $\sigma = 1$ if K is real, $\sigma = 2$ if K is imaginary.

From Theorem 1 we shall deduce two general results on algebraic independence.

Denote by L and L^* any two fields of characteristic zero such that $L \subset L^*$, and by $x_1, \ldots, x_n$ any finite number of elements of L^*. These elements are called algebraically $\begin{Bmatrix} H\text{-dependent} \\ H\text{-independent} \end{Bmatrix}$ over L if there $\begin{Bmatrix} \text{exists} \\ \text{does not exist} \end{Bmatrix}$ a homogeneous polynomial with coefficients in L,

$$P_H(X_1, \ldots, X_n) \not\equiv 0,$$

such that

$$P_H(x_1, \ldots, x_n) = 0.$$

They are similarly called

$$\text{algebraically} \begin{Bmatrix}\text{dependent} \\ \text{independent}\end{Bmatrix} \text{ over } L \text{ if there } \begin{Bmatrix}\text{exists} \\ \text{does not exist}\end{Bmatrix}$$

a polynomial with coefficients in L,

$$P(X_1, \ldots, X_n) \not\equiv 0,$$

such that

$$P(x_1, \ldots, x_n) = 0.$$

If $x_1, \ldots, x_n$ are H-independent, evidently $x_n \neq 0$, and

$$y_1 = \frac{x_1}{x_n}, \ldots, y_{n-1} = \frac{x_{n-1}}{x_n}$$

are independent, and vice versa.

We denote by $d_H = d_H(x_1, \ldots, x_n)$ the maximum number of the $x_1, \ldots, x_n$ that are H-independent, by $d = d(x_1, \ldots, x_n)$ the maximum number that are independent, and we put

$$D_H = d_H - 1.$$

Consider now the set $V(t)$ of all the homogeneous polynomials

$$P_H(X_1, \ldots, X_n) = \sum_{\substack{h_1 \geq 0 \\ h_1 + \cdots + h_n = t}} \cdots \sum_{h_n \geq 0} P^{(H)}_{h_1 \ldots h_n} X_1^{h_1} \ldots X_n^{h_n}$$

with coefficients in L, of exact degree t, and the subset $S(t)$ of all such polynomials for which

$$P_H(x_1, \ldots, x_n) = 0.$$

Evidently $V(t)$ is a linear vector space over L of dimension

$$v(t) = \binom{n + t - 1}{n - 1},$$

and $S(t)$ is a subspace, say of dimension $s(t)$. The difference $h(t) = v(t) - s(t)$ gives the number of linearly independent homogeneous linear equations with coefficients in L which the coefficients of $P_H \in S(t)$ must satisfy.

As a special case of a much more general theorem by Hilbert of 1890, it can be proved that

$$h(t) = h_0\binom{t}{D_H} + h_1\binom{t}{D_{H^{-1}}} + \cdots + h_{D_H} \quad \text{for } t \geq t_0,$$

where $h_0 > 0, h_1, \ldots, h_d$ are certain constant integers. Thus at $t \to \infty$,

$$h(t) \sim ct^{D_H}$$

where $c > 0$ is a certain constant.

We return now to the study of the solutions $\mathbf{f}(z)$ of Q where as before $f_1(z), \ldots,$ $f_m(z)$ are E-functions. As before let $\alpha \neq 0$, $\alpha \in K$, $\kappa(\alpha) \neq 0$. In difference from the previous notation denote by $D_H(z) + 1$ the maximum number of the functions

$$f_1(z), \ldots, f_m(z)$$

that are algebraically H-independent over $K(z)$, by $D_H(\alpha) + 1$ the maximum number of function values

$$f_1(\alpha), \ldots, f_m(\alpha)$$

that are algebraically H-independent over K.

If t is any positive integer, let $V_z(t)$ and $V_\alpha(t)$ be the sets of all H-polynomials of exact degree t with coefficients in $K(z)$, and K, respectively, and $S_z(t)$ and $S_\alpha(t)$ the subsets of the polynomials in these sets for which

$$\text{(z)} \qquad P_H(f_1(z), \ldots, f_m(z)) = 0$$

and

$$(\alpha) \qquad P_H(f_1(\alpha), \ldots, f_m(\alpha)) = 0,$$

respectively. Let similarly

$$v_z(t),\ v_\alpha(t), \qquad s_z(t),\ s_\alpha(t)$$

be the dimensions of these vector spaces, and

$$h_z(t) = v_z(t) - s_z(t), \qquad h_\alpha(t) = v_\alpha(t) - s_\alpha(t)$$

the corresponding dimensions. By Hilbert's theorem there exist then positive constants c_z and c_α such that, as $t \to \infty$,

$$\text{(A)} \qquad h_z(t) \sim c_z t^{D_H(z)}, \qquad h_\alpha(t) \sim c_\alpha t^{D_H(\alpha)}.$$

We finally derive relations between $h_z(t)$ and $h_\alpha(t)$. For this purpose let

$$\mathbf{w} = \begin{bmatrix} w_1 \\ \vdots \\ w_m \end{bmatrix}$$

be the general solution of

$$Q\colon w_h' = \sum_{k=1}^{m} q_{hk} w_k \qquad (h = 1, 2, \ldots, m).$$

With each set of integers $h_1, \ldots, h_m$ satisfying

$$h_1 \geq 0, \ldots, h_m \geq 0, \qquad h_1 + \cdots + h_m = t$$

we associate the two products

$$W_{(h)} = W_{h_1 \ldots h_m} = w_1^{h_1} \ldots w_m^{h_m}$$

and

$$F_{(h)}(z) = F_{h_1 \dots h_m}(z) = f_1^{h_1}(z) \dots f_m^{h_m}(z).$$

The equations (z) have now on their left-hand side linear forms in the

$$\tau = v_z(t) = \binom{t+m-1}{m-1}$$

products $F_{(h)}(z)$, and the equations (α) have similarly linear forms in the τ products $F_{(h)}(\alpha)$. In either case there are $h_z(t)$ and $h_\alpha(t)$ such linearly independent homogeneous linear forms.

The original system Q for $\mathbf{w}$ and $\mathbf{f}$ implies an analogous system for the vectors $\mathbf{W}_{(h)}$ and $\mathbf{F}_{(h)}(e)$. Thus

$$W'_{(h)} = W_{(h)} \sum_{j=1}^{m} h_j w_j^{-1} w'_j = W_{(h)} \sum_{j=1}^{m} h_j w_j^{-1} \sum_{k=1}^{m} q_{jk} w_k,$$

and this is equivalent to a new system

$$Q(t): \quad W'_{(h)} = \sum_{(k)} q_{(h)(k)} W_{(k)},$$

where the coefficients $q_{(h)(k)}$ are linear forms in the q_{hk} with numerical integral coefficients. Thus also the $q_{(h)(k)}$ have κ as denominator, and α, by $\kappa(\alpha) \neq 0$, is a regular point also for $Q(t)$.

A particular solution for $Q(t)$ is the vector $\mathbf{F}(z)$ with the components $F_{(h)}(z)$ which evidently are Siegel E-functions. Denote by $\rho_z(t)$ and $\rho_\alpha(t)$ the maximal number of components of $\mathbf{F}(z)$ and $\mathbf{F}(\alpha)$ that are linearly independent over $K(z)$ and K, respectively. By Shidlovski's first result,

$$\text{(B)} \qquad \rho_\alpha(t) \geq \frac{\sigma}{N} \rho_z(t).$$

We assert now that

$$\rho_z(t) = h_z(t), \qquad \rho_\alpha(t) = h_\alpha(t),$$

i.e. these ranks are simply the Hilbert functions. The proofs being the same, it suffices to prove the first relation. There are $\tau = v_z(t)$ components of $\mathbf{F}(z)$, and these satisfy $s_z(t)$ linearly independent homogeneous linear equations. Hence the number of linearly independent components of $\mathbf{F}(z)$ is indeed

$$v_z(t) - s_z(t) = h_z(t).$$

Thus, by (A) and (B),

$$c_\alpha t^{D_H(\alpha)} \sim h_\alpha(t) \geq \frac{\sigma}{N} h_z(t) \sim \frac{\sigma}{N} c_z t^{D_H(z)}.$$

Allow here $t \to \infty$. Then it follows that $D_H(\alpha) \geq D_H(z)$. In fact,

$$\text{(C)} \qquad D_H(\alpha) = D_H(z).$$

For assume that $D_H(\alpha) > D_H(z)$. We can then without loss of generality assume that

$$f_1(\alpha), \ldots, f_\delta(\alpha), \quad \text{where } \delta = D_H(\alpha) + 1,$$

are algebraically H-independent over K. On the other hand,

$$f_1(z), \ldots, f_\delta(z)$$

are certainly algebraically H-dependent over K. Thus there is an H-polynomial $P_H(X_1, \ldots, H, z) \not\equiv 0$ such that

$$P_H(f_1(z), \ldots, f_\sigma(z), z) = 0$$

identically in z. The coefficients of this polynomial are polynomials in $K(z)$, and we can assume that these polynomials are relatively prime. But then

$$P_H(X_1, \ldots, X_\delta, \alpha)$$

is not identically zero, and

$$P_H(f_1(\alpha), \ldots, f_\delta(\alpha), \alpha) = 0$$

is a nontrivial homogeneous algebraic equation for $f_1(\alpha), \ldots, f_m(\alpha)$, which is impossible.

The relation (C) is equivalent to the

First Main Theorem by Shidlovski. *Let* $\mathbf{f}(z)$ *be a solution of* Q *in E-functions, and let* α *be an algebraic number such that*

$$\alpha \neq 0, \qquad (\alpha) \neq 0.$$

Then the number of components of $\mathbf{f}(z)$ *that are algebraically H-independent over* $K(z)$ *is equal to the number of components of* $\mathbf{f}(\alpha)$ *that are algebraically H-independent over* K.

It is not difficult to show that in this independence $K(z)$ may be replaced by $\mathbf{C}(z)$ and K by $\mathbf{Q}$.

From this first result we can immediately deduce a perhaps even more striking result.

Second Main Theorem by Shidlovski. *Let* $\mathbf{f}(z)$ *be a solution of the inhomogeneous equations*

$$Q^*: w_h' = q_{h0} + \sum_{k=1}^{m} q_{hk} w_k \qquad (h = 1, 2, \ldots, m)$$

in terms of E-functions, and let again $\alpha \neq 0$ *be a regular algebraic point so that* $\kappa(\alpha) \neq 0$. *Then the number of components of* $\mathbf{f}(z)$ *that are algebraically independent over* $\mathbf{C}(z)$ *is equal to the number of components of* $\mathbf{f}(\alpha)$ *that are algebraically independent over* $\mathbf{Q}$.

For put $w_0 \equiv 1, f_0(z) \equiv 1$ and consider the two vectors with the components

$$w_0, w_1, \ldots, w_m \quad \text{and} \quad f_0(z), f_1(z), \ldots, f_m(z).$$

Both vectors satisfy the homogeneous equations

$$w_0' = 0, \qquad w_h' = q_{h0}w_0 + \sum_{k=1}^{m} q_{hk}w_k \qquad (h = 1, 2, \ldots, m).$$

The result is thus an immediate consequence of the First Main Theorem.

Siegel, Shidlovski, and Shidlovski's students like Oleinikov, have applied the main theorems to special E-functions and obtained many striking results. Thus Siegel was the first to show that, for algebraic $\alpha \neq 0$, $J_0(\alpha)$ and $J_0'(\alpha)$ are algebraically independent over **Q**.

That the transcendency of e and π is contained in our results is obvious. For consider Q: $w' = w$ with the solution $f(z) = e^z$ which is an E-function and is moreover transcendental. Hence e^α, for algebraic $\alpha \neq 0$, is a transcendental number by the Second Main Theorem.

Of the many other consequences I mention only two. Firstly, any finite number of the integrals

$$\int_0^1 e^{-z} (\log z)^n \, dr \qquad (n = 0, 1, 2, \ldots)$$

are algebraically independent over **Q**. Secondly, the very complicated number

$$\frac{\pi}{2} \frac{Y_0(\alpha)}{J_0(\alpha)} - \left(\gamma + \log \frac{\alpha}{2}\right), \qquad \alpha \neq 0 \text{ algebraic}$$

is transcendental. Here γ is Euler's constant.

Early in 1967 I thought I had a proof of the transcendency of γ itself. I made a mistake.

Ohio State University
Columbus, Ohio 43210

MAHLER'S *T*-NUMBERS

WOLFGANG M. SCHMIDT[1]

1. Introduction

The real numbers are usually defined as the limits of certain sequences of rationals. Hence it is natural to classify real numbers by their approximation properties by rational or, more generally, by algebraic numbers. Such a classification was given by Mahler [5]. The algebraic numbers are called *A*-numbers, and the transcendental numbers are divided into three classes, called *S*-numbers, *T*-numbers, and *U*-numbers. The existence of *S*-numbers and of *U*-numbers is relatively easy to see, and the existence of *T*-numbers was recently shown by the author [6]. In the present lectures I intend to refine and at the same time to simplify the construction in [6] and to make some further remarks on Mahler's classification.

Throughout, the symbol β_n will denote algebraic numbers of precise degree n. The height $H(\beta)$ of an algebraic number is defined in the standard way.

Let ξ be a real number. Put

$$\chi_n(H, \xi) = \min_{H(\beta_n) \leqq H;\, \beta_n \neq \xi} |\xi - \beta_n|. \tag{1}$$

Thus $\chi_n(H, \xi)$ is a measure of how well ξ may be approximated by algebraic numbers β_n of degree n with $\beta_n \neq \xi$ and with $H(\beta_n) \leqq H$. Write $\chi_n = \chi_n(\xi)$ for the supremum of the numbers χ such that

$$\chi_n(H, \xi) < H^{-\chi} \tag{2}$$

for arbitrarily large values of H. Finally put

$$\kappa_n = \kappa_n(\xi) = \max(\chi_1, \chi_2, \dots, \chi_n). \tag{3}$$

---

[1]Written with partial support from NSF-GP-9581.

(Remark. In the notation of Koksma [14], we have $\kappa_n = \omega_n^* + 1$.)

It is easy to see that if ξ is algebraic of degree m, then $\kappa_n \leqq m$ $(n = 1, 2, \ldots)$. On the other hand it was shown by Wirsing [8], that for transcendental ξ one has $\kappa_n \geqq (n + 3)/2$.

Suppose ξ and ξ' are transcendental numbers which are algebraically dependent. Then they satisfy a relation of the type $P(\xi, \xi') = 0$ where $P(x, y)$ is a polynomial with integer coefficients, not identically zero, and irreducible over the rationals. Suppose $P(x, y)$ has degree r in x and degree s in y. As we shall show in the next section, one can easily combine known estimates to obtain $\kappa_n \leqq 2r(\kappa'_{ns} - 1)$ $(n = 1, 2, \ldots)$, where $\kappa_n = \kappa_n(\xi)$, $\kappa'_n = \kappa_n(\xi')$ $(n = 1, 2, \ldots)$. We also shall give a direct proof of

(4) $$\kappa_n \leqq r\kappa'_{ns} \qquad (n = 1, 2, \ldots).$$

Let $K = (\kappa_1, \kappa_2, \ldots)$ and $K' = (\kappa'_1, \kappa'_2, \ldots)$ be increasing sequences of positive numbers. Call K and K' *equivalent* and write $K \sim K'$, if there is a positive integer t such that

(5) $$\kappa_n \leqq t\kappa'_{nt} \quad \text{and} \quad \kappa'_n \leqq t\kappa_{nt} \qquad (n = 1, 2, \ldots).$$

Given such a sequence $K = (\kappa_1, \kappa_2, \ldots)$, let $\mathbf{C}(K)$ be the class of all real numbers ξ with $K(\xi) = (\kappa_1(\xi), \kappa_2(\xi), \ldots) \sim K$. Then *every real number falls into precisely one class* $\mathbf{C}(K)$, *and algebraically dependent transcendental numbers fall into the same class. There are continuum many distinct classes* $\mathbf{C}(K)$.

Let $K_A = (1, 1, \ldots)$. The equivalence class of K_A consists precisely of sequences $K = (\kappa_1, \kappa_2, \ldots)$ with bounded κ_i. Thus by what we said above, $\mathbf{C}(K_A)$ consists precisely of the algebraic numbers. Mahler calls them A-numbers.

By the result of Wirsing, $\mathbf{C}(K)$ is empty if $K = (\kappa_1, \kappa_2, \ldots)$ is not bounded and if not $\kappa_n \gg n$. Hence we may now assume that $\kappa_n \gg n$.

Put $K_S = (1, 2, 3, \ldots)$. The equivalence class of K_S consists precisely of sequences $K = (\kappa_1, \kappa_2, \ldots)$ with $n \ll \kappa_n \ll n$. Numbers in the class $\mathbf{C}(K_S)$ are called S-numbers.

Write $K_U = (\infty, \infty, \ldots)$. The equivalence class of K_U consists of all sequences $K = (\kappa_1, \kappa_2, \ldots)$ with $\kappa_n = \infty$ for $n > n_0(K)$. The numbers in $\mathbf{C}(K_U)$ are the U-numbers.

The T-numbers are those which are neither A-, nor S-, nor U-numbers. More briefly, the four classes A, S, T, U may be characterized as follows:

$$A\colon \kappa_n \ll 1,$$
$$S\colon \kappa_n \ll 1,\ \kappa_n \ll n,$$
$$T\colon \kappa_n \ll n,\ \kappa_n \text{ finite},$$
$$U\colon \kappa_n = \infty \text{ for large } n.$$

This is precisely the definition given by Koksma [4], which he showed to be equivalent with the classification Mahler [5] had given earlier. From what we said above it is clear that algebraically dependent transcendental numbers lie in the same class.

The class of T-numbers can be split up into smaller classes as follows. Write $\alpha(\xi)$ for the infimum of the numbers α with $\kappa_n(\xi) \ll n^\alpha$. For any α in $1 \leqq \alpha \leqq \infty$ let $T(\alpha)$ consist of all T-numbers with $\alpha(\xi) = \alpha$. By (4), algebraically dependent T-numbers fall into the same class $T(\alpha)$. We now split up the U-numbers: Write $\beta(\xi)$ for the infimum of the numbers β such that for every n, but with a constant in $\ll$ which may depend on n,

$$\chi_n(H, \xi) \ll e^{-H^\beta} \quad \text{as } H \to \infty.$$

For any β in $0 \leqq \beta \leqq \infty$, let $U(\beta)$ be the class of U-numbers ξ with $\beta(\xi) = \beta$. It will be clear from the argument in the next section that algebraically dependent U-numbers fall into the same class $U(\beta)$.

Our main result is as follows.

THEOREM. *Let $\alpha_1, \alpha_2, \ldots$ be nonzero algebraic numbers and $\varphi_1(x), \varphi_2(x), \ldots$ positive functions defined for positive x which have*

$$\varphi_n(x) = o(1/x) \quad \text{as } x \to \infty \qquad (n = 1, 2, \ldots). \tag{6}$$

(This estimate need not hold uniformly in n.) Let $\chi_1, \chi_2, \ldots$ be real numbers with

$$\chi_n > 2n^3 + 2n^2 + 3n + 1. \qquad (n = 1, 2, \ldots).[2] \tag{7}$$

There is a number ξ with the following properties.

(a) *For every integer i there are infinitely many distinct nonzero numbers $\gamma_{i1}, \gamma_{i2}, \ldots$ such that γ_{it}/α_i is rational $(t = 1, 2, \ldots)$ and*

$$\tfrac{1}{2}\varphi_i(H(\gamma_{it})) < |\xi - \gamma_{it}| < \varphi_i(H(\gamma_{it})). \tag{8}$$

(b) *For every positive integer n there is a positive constant λ_n such that for every algebraic number β_n of degree n which is distinct from the numbers γ_{it} $(i, t = 1, 2, \ldots)$, one has*

$$|\xi - \beta_n| \geqq \lambda_n H(\beta_n)^{-\chi_n}. \tag{9}$$

COROLLARY 1. *Suppose $\chi_1, \chi_2, \ldots$ satisfy* (2). *There is a number ξ with*

$$\chi_n(\xi) = \chi_n \qquad (n = 1, 2, \ldots). \tag{10}$$

PROOF. Apply the theorem with α_n of degree n and $\varphi_n(x) = x^{-\chi_n}$ $(n = 1, 2, \ldots)$.

COROLLARY 2. *Suppose we have $0 < \kappa_1 \leqq \kappa_2 \leqq \ldots$ and $\kappa_n > 2n^3 + 2n^2 + 3n + 1$. There is a number ξ with*

$$\kappa_n(\xi) = \kappa_n \qquad (n = 1, 2, \ldots). \tag{11}$$

PROOF. Apply Corollary 1 with $\chi_n = \kappa_n$.

---

[2]The result is probably still true with the cubic polynomial on the right replaced by n. This polynomial just happens to come out with the present method, and no importance is attached to it.

COROLLARY 3. *Suppose the sequence $K = (\kappa_1, \kappa_2, \ldots)$ is nondecreasing and $\kappa_n \gg n^3$. Then $\mathbf{C}(K)$ is nonempty. In particular, $T(\alpha)$ is nonempty for every α in $3 \leqq \alpha \leqq \infty$.*

PROOF. Our hypotheses imply the existence of a sequence $K' = (\kappa_1', \kappa_2', \ldots)$ with $K' \sim K$ and $\kappa_n' > 2n^3 + 2n^2 + 3n + 1$. By Corollary 2 there is a ξ with $\kappa_n(\xi) = \kappa_n'$ $(n = 1, 2, \ldots)$. Thus $\mathbf{C}(K) = \mathbf{C}(K')$ is nonempty. Let K_α be the sequence $\kappa_n = n^\alpha$ $(n = 1, 2, \ldots)$ if $3 \leqq \alpha < \infty$ and the sequence $\kappa_n = e^n$ $(n = 1, 2, \ldots)$ if $\alpha = \infty$. Since $\mathbf{C}(K_\alpha)$ is nonempty and is contained in $T(\alpha)$, the set $T(\alpha)$ is nonempty.

I am unable to show that $T(\alpha)$ is nonempty for $1 \leqq \alpha < 3$.

COROLLARY 4. *The classes $U(\beta)$ are nonempty.*

PROOF. Apply the theorem with α_n of degree n and with

$$\begin{aligned} \varphi_n(x) &= e^{-(\log x)^2} \quad \text{if } \beta = 0, \\ &= e^{-x^\beta} \quad \text{if } 0 < \beta < \infty, \qquad (n = 1, 2, \ldots). \\ &= e^{-e^x} \quad \text{if } \beta = \infty \end{aligned}$$

Let F be a real algebraic number field. A real number ξ will be called *F-Liouville* if for every $\omega > 0$ there are infinitely many numbers γ of F with

$$|\xi - \gamma| < H(\gamma)^{-\omega}.$$

According to this definition the ordinary Liouville numbers are the $\mathbf{Q}$-Liouville numbers where $\mathbf{Q}$ is the rational field. It is clear that if $F_1 \subset F_2$ and if ξ is F_1-Liouville, then ξ is F_2-Liouville.

COROLLARY 5. *Let $\mathbf{A}$ be a collection of real algebraic number fields such that $F_1 \subset F_2$ and $F_1 \in \mathbf{A}$ implies $F_2 \in \mathbf{A}$. There is a number ξ which is F-Liouville precisely for the fields F of $\mathbf{A}$.*

PROOF. If $\mathbf{A}$ is empty, any number ξ which is not an U-number will do. If $\mathbf{A}$ is nonempty, it contains countably many fields $F_1, F_2, \ldots$. In each field F_n pick a number α_n such that $F_n = \mathbf{Q}(\alpha_n)$. Put $\varphi_n(x) = e^{-x}$ and $\chi_n = 9n^3$ $(n = 1, 2, \ldots)$. The theorem may now be applied. It is clear that the number ξ of the theorem is F_n-Liouville for $n = 1, 2, \ldots$. Now let F be a real number field which is not in $\mathbf{A}$, and let d be its degree. A number β of F is distinct from all the numbers γ_{it} since $\mathbf{Q}(\gamma_{it}) = \mathbf{Q}(\alpha_i) = F_i$ and since $F_i \not\subset F$. Every number β of F has some degree $j \leqq d$, whence by (9),

$$|\xi - \beta| \geqq \min(\lambda_1 H(\beta)^{-\chi_1}, \ldots, \lambda_d H(\beta)^{-\chi_d}).$$

Hence ξ is not F-Liouville.

Like the proof in [6], our proof of the main theorem will heavily depend on the following result of Wirsing.

THEOREM[3] (WIRSING). *Suppose α is algebraic, n a positive integer and $\lambda > 2n$. There are only finitely many numbers β_n with*

$$|\alpha - \beta_n| < H(\beta)^{-\lambda}.$$

Hence there is a constant $c(\alpha, n, \lambda) > 0$ such that

$$|\alpha - \beta_n| > c(\alpha, n, \lambda)H(\beta)^{-\lambda}$$

for every β_n which is distinct from α. Like its parent, namely Roth's Theorem, Wirsing's Theorem is "ineffective", and it is not possible to compute $c(\alpha, n, \lambda)$.

We shall give an explicit construction of T-numbers, but it will depend on the constants $c(\alpha, n, \lambda)$.

2. The inequality (4)

Let ω_n and ω_n^* ($n = 1, 2, \ldots$) be the constants introduced by Mahler [5] and Koksma [4], respectively. It is known that if ξ, ξ' are transcendental and if they satisfy a polynomial equation of the type discussed in the paragraph preceding (4), then

$$\omega_n(\xi) \leqq r - 1 + r\omega_{ns}(\xi').$$

(See, e.g. LeVeque [2, vol. II], or Schneider [7, Chapter III].) According to Wirsing [8], we have for every transcendental η,

$$\tfrac{1}{2}(\omega_n(\eta) + 1) \leqq \omega_n^*(\eta) \leqq \omega_n(\eta).$$

Since as we remarked above, $\kappa_n(\xi) = \omega_n^*(\xi) + 1$, we obtain

$$\begin{aligned} \kappa_n = \kappa_n(\xi) = \omega_n^*(\xi) + 1 \leqq \omega_n(\xi) + 1 \leqq r + r\omega_{ns}(\xi') \\ \leqq 2r\omega_{ns}^*(\xi') = 2r(\kappa_{ns}(\xi') - 1) = 2r(\kappa_{ns}' - 1). \end{aligned}$$

To prove (4) we recall the following facts.

LEMMA 1. *Let β be algebraic of degree n, and let*

$$Q(x) = b_n x^n + \cdots + b_0 = b_n(x - \beta)(x - \beta^{(2)}) \ldots (x - \beta^{(n)})$$

be the defining polynomial of β, with coprime coefficients $b_n, \ldots, b_0$. Then if $\nu_1, \ldots, \nu_t$ are distinct numbers among $1, \ldots, n$, the number

$$b_n \beta^{(\nu_1)} \ldots \beta^{(\nu_t)} \tag{12}$$

(a) *is an algebraic integer*

and

(b) *it has absolute value less than $c(n)H(\beta)$.*

---

[3]See Wirsing's paper in the present set of notes.

Proof. The assertion (a) is Hilfssatz 17 on page 77 in [7]. The assertion (b) may be found, e.g., in [1] or [3].

Now let $P(x, y)$ be the irreducible polynomial with $P(\xi, \xi') = 0$. The partial derivatives of P do not vanish at (ξ, ξ'). (Namely, suppose to the contrary that $(\partial/\partial y)P$ vanishes at (ξ, ξ'). By considering $P(\xi, y)$ as a polynomial in y with coefficients in $\mathbf{Q}(\xi)$ one sees that $P(\xi, y) = P_1^*(\xi, y)P_2^*(\xi, y)$ where P_1^*, P_2^* are polynomials of positive degree in y, with coefficients in $\mathbf{Q}(\xi)$. Since ξ is transcendental, we have in fact $P(x, y) = P_1^*(x, y)P_2^*(x, y)$. By Gauss' Lemma one can find *polynomials* $P_1(x, y)$, $P_2(x, y)$ in x, y of positive degree in y and with $P(x, y) = P_1(x, y)P_2(x, y)$. This contradicts the irreducibility of $P(x, y)$.)

By the implicit function theorem there is a neighborhood $\mathbf{N}$ of ξ and a one to one and differentiable function $y(x)$ defined on $\mathbf{N}$ such that $y(\xi) = \xi'$ and $P(x, y(x)) \equiv 0$.

Suppose β_n is the number with

$$\chi_n(H, \xi) = |\xi - \beta_n|.$$

Let $Q(x) = a_n(x - \beta) \dots (x - \beta^{(n)})$ be the defining polynomial of β_n, and set $\beta' = y(\beta_n)$. Then

$$|\xi' - \beta'| = |y(\xi) - y(\beta_n)| \ll |\xi - \beta_n| = \chi_n(H, \xi).$$

The number β' is a root of the polynomial $P(\beta_n, y)$ in y, hence a root of the polynomial

$$R(y) = a_n^r P(\beta_n, y)P(\beta_n^{(2)}, y) \dots P(\beta_n^{(n)}, y).$$

$P(\beta_n, y)$ is not identically zero, since $P(x, y)$ would otherwise be divisible by the defining polynomial of β_n. Thus $R(y)$ is not identically zero, and it has rational coefficients. Hence β' is algebraic of a degree less than or equal to ns. The coefficients of $R(y)$ are linear combinations with rational integer coefficients of terms of the type

$$a_n^r \beta_n^{(1)i_1} \dots \beta_n^{(n)i_n} \quad \text{with} \quad 0 \leqq i_j \leqq r \qquad (1 \leqq j \leqq n).$$

By Lemma 1a, $R(y)$ has integral, hence rational integer coefficients. By Lemma 1b, the coefficients of $R(y)$ have absolute values $\ll H^r$. The defining polynomial of β' is a factor of $R(y)$, and using Lemma 1b again it is clear that its coefficients also are $\ll H^r$. Hence $H(\beta') \ll H^r$, say $H(\beta') \leqq cH^r$. Thus

$$\min (\chi_1(cH^r, \xi'), \dots, \chi_{ns}(cH^r, \xi')) \leqq |\xi' - \beta'| \ll \chi_n(H, \xi).$$

For every $\epsilon > 0$ one has $\chi_n(H, \xi) < H^{-(\kappa_n - \epsilon)}$ for some arbitrarily large values of H, and this implies that $\kappa_{ns}(\xi') \geqq (\kappa_n(\xi) - \epsilon)/r$, whence $r\kappa_{ns}' \geqq \kappa_n$.

3. A proposition which implies the theorem

Proposition. *Suppose $\alpha_1, \alpha_2, \dots$ and $\varphi_1(x), \varphi_2(x), \dots$ and $\chi_1, \chi_2, \dots$ satisfy the hypotheses of the theorem. There is a sequence of positive numbers $\lambda_1, \lambda_2, \dots$ and a sequence of nonzero numbers $\gamma_1, \gamma_2, \dots$ with γ_n/α_n rational $(n = 1, 2, \dots)$ and with the following properties.*

(I_i) $5^{i-1}\varphi_{i-1}(H(\gamma_{i-1})) > 5^i\varphi_i(H(\gamma_i)) \qquad (i = 2, 3, \dots)$.

(II$_i$) *The number γ_i lies in the interval $\mathbf{I}_{i-1}$ defined by*

$$\gamma_{i-1} + \tfrac{1}{2}\varphi_{i-1}(H(\gamma_{i-1})) < x < \gamma_{i-1} + \tfrac{3}{4}\varphi_{i-1}(H(\gamma_{i-1})) \qquad (i = 2, 3, \ldots).$$

(III$_i$) $|\gamma_i - \beta_n| \geqq \lambda_n H(\beta_n)^{-\chi_n}$
for every number β_n of degree $n \leqq i$ which is distinct from $\gamma_1, \ldots, \gamma_i$ $(i = 1, 2, \ldots)$.

This proposition implies the theorem: For $i < j$ we have

$$\begin{aligned}\gamma_i + \tfrac{1}{2}\varphi_i(H(\gamma_i)) < \gamma_j &< \gamma_i + \tfrac{3}{4}(\varphi(H(\gamma_i)) + \varphi_{i+1}(H(\gamma_{i+1})) + \cdots \\ &\qquad + \varphi_{j-1}(H(\gamma_{j-1}))) \\ &< \gamma_i + \tfrac{3}{4}\varphi_i(H(\gamma_i))(1 + 1/5 + 1/25 + \cdots) \\ &= \gamma_i + (15/16)\varphi_i(H(\gamma_i)).\end{aligned}$$

Hence $\gamma_1, \gamma_2, \ldots$ increase to a limit ξ, and

$$\gamma_i + \tfrac{1}{2}\varphi_i(H(\gamma_i)) < \xi < \gamma_i + \varphi_i(H(\gamma_i)) \qquad (i = 1, 2, \ldots).$$

If β_n of degree n is distinct from $\gamma_1, \gamma_2, \ldots,$ then

$$|\xi - \beta_n| = \lim_{i\to\infty} |\gamma_i - \beta_n| \geqq \lambda_n H(\beta_n)^{-\chi_n}.$$

The two inequalities above are almost the same as (8) and (9). But now we have only one number γ_i corresponding to each α_i, while in the theorem there was a sequence of distinct numbers $\gamma_{i1}, \gamma_{i2}, \ldots$. However, every pair $(\alpha, \varphi(x))$ with α a nonzero algebraic number and $\varphi(x)$ a function with $\varphi(x) = o(1/x)$ may occur infinitely often among $(\alpha_1, \varphi_1(x)), (\alpha_2, \varphi_2(x)), \ldots$. Since the sequence $\gamma_1, \gamma_2, \ldots$ is strictly increasing, the numbers γ_i corresponding to $(\alpha, \varphi(x))$ are all distinct.

Let $\epsilon_1, \epsilon_2, \ldots$ be a sequence of numbers satisfying $0 < \epsilon_i < 1$ and

(13) $$\chi_n > 2n^3 + 2n^2 + 3n + 1 + 20n^3\epsilon_n \qquad (n = 1, 2, \ldots).$$

Write $\mathbf{J}_i$ for the subset of $\mathbf{I}_i$ consisting of numbers x of $\mathbf{I}_i$ which have

(14) $$|x - \beta_n| \geqq 2\lambda_n H(\beta_n)^{-\chi_n}$$

for all numbers β_n of degree $n \leqq i$ which are distinct from $\gamma_1, \ldots, \gamma_i$ and which satisfy

(15$_i$) $$\varphi_i(H(\gamma_i)) \geqq \lambda_n H(\beta_n)^{-\chi_n}.$$

Let $\mathbf{J}_i'$ be the larger set defined as $\mathbf{J}_i$ but with β_n in (14) distinct not only from $\gamma_1, \ldots, \gamma_i$ but also from x.

In order to carry out our inductive construction of sequences $\lambda_1, \lambda_2, \ldots$ and $\gamma_1, \gamma_2, \ldots$ satisfying (I), (II), (III), we shall impose four additional conditions.

(IV$_i$) $$\gamma_i \in \mathbf{J}_{i-1}' \qquad (i = 2, 3, \ldots),$$

i.e. $\gamma_i \in \mathbf{I}_{i-1}$ and

(16) $$|\gamma_i - \beta_n| \geqq 2\lambda_n H(\beta_n)^{-\chi_n}$$

for all numbers β_n of degree $n \leqq i-1$ which are distinct from $\gamma_1, \ldots, \gamma_i$ and which satisfy

$$(15_{i-1}) \qquad \varphi_{i-1}(H(\gamma_{i-1})) \geqq \lambda_n H(\beta_n)^{-\chi_n}.$$

(V_i) The inequality (16) is true for all numbers $\beta_n \neq \gamma_i$ of degree $n = i$. $(i = 1, 2, \ldots)$.

(VI_i) Writing $\gamma_i = \alpha_i(a_i/b_i)$ where the rational a_i/b_i is in its lowest terms and has positive denominator b_i, we have

$$|\gamma_i - \beta_n| \geqq 1/b_i$$

provided $n \leqq i$ and provided

$$(17) \qquad H(\beta_n) \leqq b_i^{1/(n+1+\epsilon_n)} \qquad (i = 1, 2, \ldots).$$

(VII_i) The measure $\mu(\mathbf{J}_i)$ of $\mathbf{J}_i$ satisfies

$$\mu(\mathbf{J}_i) \geqq \tfrac{1}{2}\mu(\mathbf{I}_i).$$

4. The inductive construction

We shall inductively construct

$$\gamma_1, \lambda_1, \gamma_2, \lambda_2, \ldots .$$

The number γ_1 will lie in the interval $1 < \gamma_1 < 2$. By (I) and (II) there is then a constant C such that all the numbers γ_i will lie in the interval

$$(18) \qquad 1 < \gamma_i < C.$$

Our inductive construction will consist of two steps (A_i) and (B_i).

(A_i). Here we either assume that $i = 1$ or we assume that $i > 1$ and $\gamma_1, \lambda_1, \ldots, \gamma_{i-1}, \lambda_{i-1}$ have been constructed and have the desired properties. If $i = 1$ we shall show that for every sufficiently large prime b_1 it is possible to construct γ_1 of the type $\gamma_1 = \alpha_1(a_1/b_1)$, g.c.d. $(a_1, b_1) = 1$, $1 < \gamma_1 < 2$ and to construct λ_1 in $0 < \lambda_1 < 1$ such that (III_1), (V_1), (VI_1) holds. (There is no condition (I_1), (II_1), or (IV_1).) If $i > 1$ we shall show that for every sufficiently large prime b_i it is possible to construct $\gamma_i = \alpha_i(a_i/b_i)$ with coprime a_i, b_i and λ_i with $0 < \lambda_i < 1$ satisfying (I_i) – (VI_i).

(B_i). Our assumptions here are the same as for (A_i). We shall show there is a constant ρ_i depending perhaps on

$$(19) \quad \alpha_1, \alpha_2, \ldots ; \varphi_1, \varphi_2, \ldots ; \chi_1, \chi_2, \ldots ; \epsilon_1, \epsilon_2, \ldots ; \gamma_1, \lambda_1, \ldots, \gamma_{i-1}, \lambda_{i-1},$$

but independent of λ_i such that the numbers γ_i, λ_i constructed in (A_i) will automatically satisfy (VII_i) provided

$$(20) \qquad b_i > \rho_i.$$

In what follows, the constants implied by $\ll$ may depend on the elements in (19), but they will be independent of a_i, b_i and of numbers β_n occurring in our inequalities. Assuming that $\gamma_i = \alpha_i(a_i/b_i)$ with coprime a_i, b_i satisfies (I_i) and (II_i) we shall have (18) and hence

$$(21) \qquad 1 \ll |a_i/b_i| \ll 1,$$

whence if $k = k(i)$ denotes the degree of α_i,

$$(22) \qquad b_i^k \ll H(\gamma_i) \ll b_i^k.$$

For β_n of degree $n \leqq i$ we shall have

$$(23) \qquad H\left(\frac{b_i}{a_i}\beta_n\right) \ll b_i^n H(\beta_n).$$

If $n \leqq i$ and if β_n is distinct from γ_i, then it follows from Wirsing's Theorem that

$$|\gamma_i - \beta_n| = \left|\frac{a_i}{b_i}\right| \left|\alpha_i - \frac{b_i}{a_i}\beta_n\right| \gg \left|\alpha_i - \frac{b_i}{a_i}\beta_n\right|$$
$$\gg H\left(\frac{b_i}{a_i}\beta_n\right)^{-2n-\epsilon_n} \gg b_i^{-2n^2-n\epsilon_n} H(\beta_n)^{-2n-\epsilon_n}.$$

Hence if b_i is sufficiently large, we have

$$(24) \qquad |\gamma_i - \beta_n| > b_i^{-2n^2-2n\epsilon_n} H(\beta_n)^{-2n-\epsilon_n}.$$

5. Step (A_1)

For every large prime b_1 there are $\gg b_1$ numbers $\gamma_1 = \alpha_1(a_1/b_1)$ with coprime a_1, b_1 and with $1 < \gamma_1 < 2$. These numbers have mutual distances not less than $|\alpha_1|/b_1$. On the other hand there are only $o(b_1)$ rationals β_1 satisfying (17) with $n = i = 1$. Hence among our numbers $\alpha_1(a_1/b_1)$ there will be some which have a distance greater than $1/b_1$ from all the rationals β_1 with (17). Any such number $\gamma_1 = \alpha_1(a_1/b_1)$ will satisfy (VI_1).

By the case $n = 1$ of Wirsing's Theorem (this case is in fact Roth's Theorem), there is a number λ_1 in $0 < \lambda_1 < 1$ such that (III_1) and (V_1) hold.

6. Step (A_i) **where** $i > 1$

In view of (22), the condition (I_i) will be clearly satisfied if $\gamma_i = \alpha_i(a_i/b_i)$ with coprime a_i, b_i and if b_i is large. Now let b_i be a large prime. We want to find numbers a_i such that $\gamma_i = \alpha_i(a_i/b_i)$ satisfies (IV_i).

Suppose $n < i$ and suppose β_n satisfies (15_{i-1}). By (24), we have

$$|\gamma_i - \beta_n| \geqq 2\lambda_n H(\beta_n)^{-\chi_n}, \quad \text{i.e. (16)},$$

provided $\beta_n \neq \gamma_i$, provided b_i is large and provided

$$(25) \qquad H(\beta_n) > (2\lambda_n b_i^{2n^2+2n\epsilon_n})^{1/(\chi_n-2n-\epsilon_n)}.$$

Thus we have to be concerned only about numbers β_n which violate (25). By (13), their number is $o(b_i)$.

Thus the set of numbers x which have

$$|x - \beta_n| < 2\lambda_n H(\beta_n)^{-\chi_n}$$

for some β_n, $n < i$, which satisfies (15_{i-1}) and violates (25) is the union of $o(b_i)$ intervals. Hence the set $\mathbf{L}_{i-1}$ consisting of numbers x in $\mathbf{I}_{i-1}$ which satisfy (14) for all β_n, $n < i$, which are distinct from $\gamma_1, \ldots, \gamma_{i-1}$ and which satisfy (15_{i-1}) and violate (25) is the union of $o(b_i)$ intervals. Conceivably it could be empty. But $\mathbf{L}_{i-1}$ contains $\mathbf{J}_{i-1}$, and hence by (VII_{i-1}) we have $\mu(\mathbf{L}_{i-1}) \geqq \frac{1}{2}\mu(\mathbf{I}_{i-1}) \gg 1$. Since $\mu(\mathbf{L}_{i-1}) \gg 1$ and since $\mathbf{L}_{i-1}$ consists of $o(b_i)$ intervals, it contains $\gg b_i$ numbers $\gamma_i = \alpha_i(a_i/b_i)$ with coprime a_i, b_i.

Every such number γ_i satisfies (16) provided β_n is distinct from γ_i and satisfies (15_{i-1}) and violates (25). By what we said above it will also satisfy (16) provided β_n satisfies (15_{i-1}) and (25). Hence $\gamma_i \in \mathbf{J}_{i-1}$. We have seen:

For large primes b_i there are $\gg$ numbers $\gamma_i = \alpha_i(a_i/b_i)$ which satisfy (I_i), (IV_i), *and hence* (II_i).

The number of numbers β_n with $n \leqq i$ and satisfying (17) is $o(b_i)$. Hence there are numbers $\gamma_i = \alpha_i(a_i/b_i)$ as above which have distance greater than $1/b_i$ from any β_n with (17). Such a number will satisfy (I_i), (II_i), (IV_i) and (VI_i).

We are going to show that such a number satisfies the condition (III_i) as far as numbers β_n of degree $n < i$ are concerned. In view of (IV_i) we may assume that (15_{i-1}) does not hold, i.e. that $\varphi_{i-1}(H(\gamma_{i-1})) < \lambda_n H(\beta_n)^{-\chi_n}$. One either has

$$\varphi_t(H(\gamma_t)) < \lambda_n H(\beta_n)^{-\chi_n} \quad \text{with } t = n,$$

or there is a t in $n < t < i$ with

$$\varphi_t(H(\gamma_t)) < \lambda_n H(\beta_n)^{-\chi_n} \leqq \varphi_{t-1}(H(\gamma_{t-1})).$$

In either case we have

$$|\gamma_i - \beta_n| \geqq |\gamma_t - \beta_n| - |\gamma_i - \gamma_t| \geqq |\gamma_t - \beta_n| - \varphi_t(H(\gamma_t)).$$

Hence if β_n is distinct from $\gamma_1, \ldots, \gamma_i$ and hence is in particular distinct from γ_t, then we have by (V_t) if $t = n$ and by (IV_t) if $n < t < i$ (note that (15_{t-1}) holds),

$$|\gamma_i - \beta_n| \geqq 2\lambda_n H(\beta_n)^{-\chi_n} - \varphi_t(H(\gamma_t)) > \lambda_n H(\beta_n)^{-\chi_n}.$$

Thus (III_i) is indeed true for β_n with $n < i$.

To complete the step (A_i) it therefore remains to satisfy (III_i) for numbers β_n of degree $n = i$ and to satisfy (V_i). It is clear that the special case of (III_i) with $n = i$ is implied by (V_i). By Wirsing's Theorem and since $\chi_n > 2n$, we can choose λ_i with $0 < \lambda_i < 1$ such that (V_i) is in fact true.

7. Step (B_i)

Put

$$\psi_i = \varphi_i(H(\gamma_i)), \qquad \Psi_i = 1/\psi_i. \tag{26}$$

By (6) and (22) we have

$$\psi_i = o(1/b_i) \quad \text{as } b_i \to \infty. \tag{27}$$

In particular we see that ψ_i tends to zero and Ψ_i to infinity as $b_i \to \infty$. (The subscript i is fixed here.)

Suppose that $\gamma_i = \alpha_i(a_i/b_i)$ and λ_i with $0 < \lambda_i < 1$ satisfy (I_i)–(VI_i), or that $i = 1$ and they satisfy (III_1), (V_1) and (VI_1).

LEMMA 2. *Suppose β_n of degree $n \leqq i$ is distinct from γ_i and it satisfies* (15_i), *and suppose x lies in $\mathbf{I}_i$. Then* (14) *holds provided b_i is large (depending on* (19) *but independent of λ_i) and provided β_n satisfies* (17).

PROOF. By (VI_i), (26), (27) and (15_i) we have

$$\begin{aligned}|x - \beta_n| &\geqq |\gamma_i - \beta_n| - |\gamma_i - x| \\ &\geqq 1/b_i - \psi_i \geqq 2\psi_i \geqq 2\lambda_n H(\beta_n)^{-\chi_n}\end{aligned}$$

if b_i is large.

LEMMA 3. *Lemma* 2 *remains valid if the inequality* (17) *is replaced by*

$$H(\beta_n) \leqq \Psi_i^{1/(\chi_n - n - 1 - \epsilon_n)}. \tag{28}$$

PROOF. The conditions (15_i) and (28) together are equivalent with

$$(\lambda_n \Psi_i)^{1/\chi_n} \leqq H(\beta_n) \leqq \Psi_i^{1/(\chi_n - n - 1 - \epsilon_n)}. \tag{29}$$

We have to show that every x in $\mathbf{I}_i$ satisfies (14). By virtue of Lemma 2 we may assume that the right-hand side of (28) exceeds the right-hand side of (17), i.e. that

$$b_i^{\chi_n - n - 1 - \epsilon_n} < \Psi_i^{n + 1 + \epsilon_n}. \tag{30}$$

Suppose that x lies in $\mathbf{I}_i$ and that $n \leqq i$. We have

$$|x - \beta_n| \geqq |\gamma_i - \beta_n| - |\gamma_i - x| \geqq |\gamma_i - \beta_n| - \psi_i.$$

Furthermore, by (24), we have

$$|\gamma_i - \beta_n| > b_i^{-2n^2 - 2n\epsilon_n} H(\beta_n)^{-2n - \epsilon_n}$$

provided b_i is large. Thus (14) does hold if b_i is large and if

$$b_i^{-2n^2 - 2n\epsilon_n} H(\beta_n)^{-2n - \epsilon_n} \geqq 2 \max (\psi_i, 2\lambda_n H(\beta_n)^{-\chi_n}),$$

that is, if

$$(4\lambda_n b_i^{2n^2 + 2n\epsilon_n})^{1/(\chi_n - 2n - \epsilon_n)} \leqq H(\beta_n) \leqq (\tfrac{1}{2}\Psi_i b_i^{-2n^2 - 2n\epsilon_n})^{1/(2n + \epsilon_n)}. \tag{31}$$

It will suffice to verify that the interval (31) contains the interval (29) provided (30) holds and b_i is large. In what follows, we shall assume that (30) is satisfied. Then the right-hand side of (31) is

$$\gg \Psi_i^{(1 - (2n^2 + 2n\epsilon_n)(n + 1 + \epsilon_n)/(\chi_n - n - 1 - \epsilon_n))/(2n + \epsilon_n)}.$$

The exponent here is

$$((\chi_n - n - 1 - \epsilon_n) - (2n^2 + 2n\epsilon_n)(n + 1 + \epsilon_n))/((\chi_n - n - 1 - \epsilon_n)(2n + \epsilon_n)),$$

and is greater than $1/(\chi_n - n - 1 - \epsilon_n)$ by (13). Thus if b_i and hence Ψ_i is sufficiently large, the right-hand side of (31) will exceed the right-hand side of (29).

Again by (30), the left-hand side of (31) is

$$\ll \lambda_n^{1/\chi_n} \Psi_i^{(2n^2 + 2n\epsilon_n)(n + 1 + \epsilon_n)/((\chi_n - 2n - \epsilon_n)(\chi_n - n - 1 - \epsilon_n))}.$$

This estimate holds uniformly for all numbers λ_n in $0 < \lambda_n < 1$. By (13) we have

$$(\chi_n - 2n - \epsilon_n)(\chi_n - n - 1 - \epsilon_n) > \chi_n(2n^3 + 2n^2 + 18n^3\epsilon_n) > \chi_n(2n^2 + 2n\epsilon_n)(n + 1 + \epsilon_n).$$

Hence the exponent of Ψ_i is less than $1/\chi_n$, and if b_i and hence Ψ_i is sufficiently large, the left-hand side of (31) is less than the left-hand side of (29).

This finishes the proof of Lemma 2. Step (B_i) is now completed as follows. The set $\mathbf{J}_i$ contains the complement in $\mathbf{I}_i$ of the union of the intervals $E(\beta_n)$ defined by

$$|x - \beta_n| < 2\lambda_n H(\beta_n)^{-\chi_n},$$

where $n \leqq i$ and where $\beta_n \neq \gamma_i$ and where (15_i) holds. By Lemma 3, the intervals with (15_i) and (28) have empty intersection with $\mathbf{I}_i$. We therefore may restrict ourselves to intervals $E(\beta_n)$ with

$$H(\beta_n) > \Psi_i^{1/(\chi_n - n - 1 - \epsilon_n)}. \tag{32}$$

Now $E(\beta_n)$ has length $4\lambda_n H(\beta_n)^{-\chi_n} < 4H(\beta_n)^{-\chi_n}$, and the sum of the lengths of the intervals $E(\beta_n)$ with $H(\beta_n) = H$ is $\ll H^{n-\chi_n}$. The sum of the lengths of all intervals $E(\beta_n)$ with (32) is

$$\ll \Psi_i^{-(\chi_n - n - 1)/(\chi_n - n - 1 - \epsilon_n)} = O(\Psi_i^{-1}) = O(\varphi_i(H(\gamma_i)).$$

Thus for large b_i, the complement of $\mathbf{J}_i$ in $\mathbf{I}_i$ has measure less than $\frac{1}{2}\mu(\mathbf{I}_i) = \frac{1}{8}\varphi_i(H(\gamma_i))$, and $\mathbf{J}_i$ itself has measure $\mu(\mathbf{J}_i) \geqq \frac{1}{2}\mu(\mathbf{I}_i)$.

This finishes step (B_i) and hence our proof of the theorem.

8. Remark

In a paper to appear elsewhere A. Baker and I plan to discuss the Hausdorff dimension of certain sets defined in terms of the constants κ_n.

References

1. N. I. Feldmann, *The approximation of certain transcendental numbers.* I: *The approximation of logarithms of algebraic numbers*, Izv. Akad. Nauk SSSR Ser. Mat. **15** (1951), 53–74; English transl., Amer. Math. Soc. Transl. (2) **59** (1966), 224–245. MR **12,** 595.

2. W. J. LeVeque, *Topics in number theory*, Addison-Wesley, Reading, Mass., 1956. MR **18,** 283.

3. F. Kasch and B. Volkmann, *Zur Mahlerschen Vermutung über S-Zahlen*, Math. Ann. **136** (1958), 442–453. MR **21** #1297.

4. J. F. Koksma, *Über die Mahlersche Klasseneinteilung der transzendenten Zahlen und ide Approximation komplexer Zahlen durch algebraische Zahlen*, Monatsh. Math. Phys. **48** (1939), 176–189. MR **1,** 137.

5. K. Mahler, *Zur Approximation der Exponentialfunktion und des Logarithms.* I, J. Reine Angew. Math. **166** (1932), 118–136.

6. W. M. Schmidt, *T-numbers do exist*, Rendiconti convegno di Teoria dei numeri, Roma 1968.

7. Th. Schneider, *Einführung in die transzendenten Zahlen*, Springer-Verlag, Berlin and New York, 1957. MR **19,** 252.

8. E. Wirsing, *Approximation mit algebraischen Zahlen beschränkten Grades*, J. Reine angew. Math. **206** (1960), 67–77. MR **26** #79.

University of Colorado
Boulder, Colorado 80302

SELBERG'S SIEVE WITH WEIGHTS

H.-E. RICHERT

1. The sieve problem

The purpose of this paper is to present some general results on Selberg's sieve with certain weights together with some number-theoretical applications. A general description of the method introduced by A. Selberg [41], [42], [43], [44] has been given by several authors (cf. Ankeny-Onishi [1], Halberstam-Roth [18]). We shall follow the presentation given in a forthcoming joint book with H. Halberstam [17].

The sieve method deals with estimates for the number of elements in a (finite) sequence $\mathcal{A}$ that are not divisible by any prime number of a set $\mathcal{P}$ of primes. To be precise: we consider a finite sequence

$$\mathcal{A} = \{a: \dots\}$$

of (not necessarily distinct and not necessarily positive) integers. The problems we deal with require that $\mathcal{A}$ will depend on certain parameters that may vary. For our purposes the most important of these parameters are those which occur in the basic information we have to have about $\mathcal{A}$: the number of elements in

$$\mathcal{A}_d := \{a: a \in \mathcal{A}, a \equiv 0 \bmod d\},$$

i.e. the number of elements in $\mathcal{A}$ that are divisible by squarefree d, which we denote by $|\mathcal{A}_d|$. For $d = 1$ this is the number $|\mathcal{A}|$ of elements of $\mathcal{A}$. However, here, as for general d, it suffices and is, in fact, in most cases even more convenient to work with a sufficiently close approximation $X(> 1)$ to $|\mathcal{A}|$ (f.e. $X = \text{li}\,x$ for $|\mathcal{A}| = \pi(x)$); the remainder will be denoted by $R(X, 1)$:

$$R(X, 1) := |\mathcal{A}| - X. \tag{1.1}$$

Next, $(d = p)$, for each prime p we choose an $\omega(p)$ such that $(\omega(p)/p)X$ approximates $|\mathcal{A}_p|$, and the remainder is called $R(X, p)$:

$$R(X, p) := |\mathcal{A}_p| - \frac{\omega(p)}{p} X. \tag{1.2}$$

Finally, with these choices of X and $\omega(p)$ we define

$$\omega(1) := 1, \qquad \omega(d) := \prod_{p|d} \omega(p) \quad \text{for } \mu(d) \neq 0,$$

so that $\omega(d)$ is a multiplicative function and, in accordance with (1.1) and (1.2), we then introduce

$$R(X, d) := |\mathcal{A}_d| - \frac{\omega(d)}{d} X \quad \text{for } \mu(d) \neq 0.$$

Obviously, these choices can be made in various ways, but, as one expects, it will turn out that the smaller $|R(X, d)|$ is (at any rate on average), the better our results will be.

We give two simple examples in order to illustrate this concept.

EXAMPLE 1.1. Let F be a polynomial of degree g with integer coefficients and consider the sequence $\mathcal{A} = \{F(n): x - y < n \leq x\}$, where x and y are real numbers satisfying $1 < y \leq x$. Denoting by $\rho(d) = \rho_F(d)$ the number of solutions of the congruence

$$F(m) \equiv 0 \bmod d \tag{1.3}$$

we find

$$|\mathcal{A}_d| = |\{n: x - y < n \leq x, F(n) \equiv 0 \bmod d\}| = \rho(d)(y/d + \theta), \qquad |\theta| \leq 1.$$

Therefore, here a reasonably good approximation is provided by the choice of $X = y$, $\omega(d) = \rho(d)$, and we then infer that $|R(X, d)| \leq \omega(d)$. $\rho(d)$ is a multiplicative function, and by Lagrange's theorem

$$\rho(p) \leq g \quad \text{if } \rho(p) < p, \tag{1.4}$$

i.e. always, if F has no fixed prime divisor. In the general case we have very little knowledge about the individual numbers $\rho(p)$. However, on the average it follows from the prime ideal theorem that for some k (which is the number of irreducible components of F) as $x \to \infty$

$$\sum_{p<x} \frac{\rho(p)}{p} \log p = k \log x + O_F(1),$$

where O_F indicates that the O-constant may depend on the polynomial F.

EXAMPLE 1.2. Let F and ρ be defined as before and consider the sequence $\mathcal{A} = \{F(p): p \leq x\}$. Here

$$\begin{aligned} |\mathcal{A}_d| &= |\{p: p \leq x, F(p) \equiv 0 \bmod d\}| \\ &= \sum_{\substack{l=1 \\ F(l) \equiv 0 \bmod d}}^{d} |\{p: p \leq x, p \equiv l \bmod d\}|. \end{aligned}$$

The sum over l has $\rho(d)$ terms, and each term with $(l, d) > 1$ is at most 1; thus

$$(1.5)\quad |\mathfrak{A}_d| - \frac{\rho_1(d)}{\varphi(d)}\operatorname{li}x = \sum_{\substack{l=1\\(l,d)=1\\F(l)\equiv 0 \bmod d}}^{d} \left\{\pi(x; d, l) - \frac{\operatorname{li}x}{\varphi(d)}\right\} + \vartheta\rho(d), \qquad 0 \le \vartheta \le 1,$$

where $\rho_1(d)$ denotes the number of solutions of (1.3) that are coprime with d, i.e. of $F(l) \equiv 0 \bmod d$, $(l, d) = 1$. $\rho_1(d)$ is a multiplicative function, and we find that

$$\begin{aligned}\rho_1(p) &= \rho(p) && \text{if } p \nmid F(0),\\ &= \rho(p) - 1 && \text{if } p \mid F(0);\end{aligned}$$

also, by (1.4),

$$\rho_1(d) \le \rho(d) \le g^{\nu(d)} \quad \text{for } \mu(d) \ne 0, \quad \text{if } \rho(p) < p \quad \text{for all } p \mid d,$$

($\nu(d)$ denotes the number of prime divisors of d). In view of (1.5) here an appropriate choice will be $X = \operatorname{li}x\ (> 1)$, $\omega(d) = (\rho_1(d)/\varphi(d))d$, which yields $|R(X, d)| \le \rho(d)(E(x, d) + 1)$, where

$$E(x, d) := \max_{l;(l,d)=1} |\pi(x; d, l) - \operatorname{li}x/\varphi(d)|.$$

This time $|R(X, d)|$ is not as small as in the preceding example, however, it is still small on the average in the sense expressed by Bombieri's prime number theorem [8]; (a slightly weaker form was proved by A. I. Vinogradov [46]): "For any positive constant U there exists a positive constant $C = C(U)$ such that

$$(1.6)\qquad \sum_{d \le \sqrt{x}/\log^C x} E(x, d) = O_U\left(\frac{x}{\log^U x}\right)."$$

Incidentally, a corresponding theorem holds for the sequence p_1p_2 (and for many more sequences, cf. [4]) as a consequence of the Barban-Davenport-Halberstam Theorem [3], [15] for the sequence of primes.

On the other hand we consider a set

$$\mathcal{P} = \{p: \ldots\}$$

of primes, and we denote its complement with respect to the set of all primes by $\overline{\mathcal{P}}$. $\mathcal{P}$ will usually be infinite, and often it simply consists of all primes not dividing some integer K, in which case we may write

$$(1.7)\qquad \mathcal{P} = \{p: p \nmid K\}.$$

Furthermore, truncations of $\mathcal{P}$, i.e. sets of the type $\{p: p \in \mathcal{P}, p < z\}$, where z is a real number satisfying $z \ge 2$, will occur, and in this connection we introduce

$$P(z) := \prod_{p<z;p\in\mathcal{P}} p.$$

Using these notations we may say that a sieve problem is an investigation of the number of elements left in a sequence $\mathcal{A}$ when sifted by a truncation (at z) of a set $\mathcal{P}$, that is, for a triplet $\mathcal{A}$, $\mathcal{P}$, z, an estimation of the "sifting function"

$$S(\mathcal{A}; \mathcal{P}, z) := |\{a: a \in \mathcal{A}, (a, P(z)) = 1\}|,$$

which is clearly the number of members of $\mathcal{A}$ that are not divisible by any prime $< z$ of $\mathcal{P}$. In fact, in order to deal also with lower estimates we shall need to study the more general sifting function

$$S(\mathcal{A}_q; \mathcal{P}, z) := |\{a: a \in \mathcal{A}_q, (a, P(z)) = 1\}|,$$

where q is an arbitrary squarefree number satisfying

$$(1.8) \qquad (q, P(z)) = 1, \qquad (q, \overline{\mathcal{P}}) = 1,$$

instead of $S(\mathcal{A}; \mathcal{P}, z)$, which corresponds to the special case $q = 1$, only.

In order to deal with this problem, we shall use the arithmetic function γ which is defined by

$$\gamma(p) := \omega(p) \quad \text{for } p \in \mathcal{P}, \qquad = 0 \quad \text{for } p \in \overline{\mathcal{P}},$$

and extended to the set of all squarefree numbers by

$$\gamma(1) := 1, \qquad \gamma(d) := \prod_{p|d} \gamma(p) \quad \text{for } \mu(d) \neq 0,$$

so that γ is a multiplicative function, and γ depends on both $\mathcal{A}$ and $\mathcal{P}$. Correspondingly we introduce

$$(1.9) \qquad \eta(X, d) := \big||\mathcal{A}_d| - (\gamma(d)/d)X\big| \quad \text{for } \mu(d) \neq 0,$$

and find

$$\eta(X, d) = |R(X, d)| \quad \text{for } \mu(d) \neq 0, \quad \text{if } (d, \overline{\mathcal{P}}) = 1;$$

we have used the notation $(d, \overline{\mathcal{P}}) = 1$ here to indicate that d and $\mathcal{P}$ are coprime, namely that none of the prime divisors of d belongs to $\overline{\mathcal{P}}$ (so that if $\mathcal{P}$ is of the type (1.7), $(d, \overline{\mathcal{P}}) = 1$ means $(d, K) = 1$). Finally, we form with this new function γ the product

$$\Gamma(z) := \prod_{p<z} \left(1 - \frac{\gamma(p)}{p}\right).$$

A direct use of Möbius' function yields immediately the Eratosthenes-Legendre formula

THEOREM 1.

$$S(\mathcal{A}_q; \mathcal{P}, z) = \frac{\gamma(q)}{q} X\Gamma(z) + \theta \sum_{d|P(z)} \eta(X, qd), \qquad |\theta| \leq 1.$$

This holds without any further condition, however, it is of use only if z is very small in comparison with X, and this fact may be considered as the starting point for what is now called a sieve method.

Unless some further dependence is stated (as by O_λ), the O-constants of our theorems depend at most on the constants A_i, α, κ, as far as they occur in the conditions imposed there.

2. Selberg's upper estimate

The only condition we need to impose in order to obtain a first upper sieve estimate is

$$(\mathrm{A}_1) \qquad 0 \leq \gamma(p)/p \leq 1 - 1/A_1 \quad \text{for all } p,$$

and for some suitable constant A_1 (≥ 1). As a consequence, the multiplicative function

$$g(d) := \gamma(d) \Big/ \left(d \prod_{p|d} \left(1 - \frac{\gamma(p)}{p} \right) \right), \qquad \mu(d) \neq 0,$$

is well defined, and with g we form

$$G(x, z) := \sum_{d \leq x ; d | P(z)} g(d).$$

The well-known starting point for Selberg's method is the inequality

$$S(\mathcal{A}_q; \mathcal{P}, z) \leq \sum_{a \in \mathcal{A}_q} \Big(\sum_{d|a ; d|P(z)} \lambda_d \Big)^2 = \sum_{d_i | P(z) ; i=1,2} \lambda_{d_1} \lambda_{d_2} |\mathcal{A}_{qD}|, \qquad D = [d_1, d_2],$$

which is true for any set of real numbers λ_d satisfying $\lambda_1 = 1$. An approximation to $|\mathcal{A}_{qD}|$ is $(\gamma(qD)/qD)X$. By (1.9) and (1.8) we find

$$\begin{aligned} S(\mathcal{A}_q; \mathcal{P}, z) &\leq \frac{\gamma(q)}{q} X \sum_{d_i | P(z) ; i=1,2} \lambda_{d_1} \lambda_{d_2} \frac{\gamma(D)}{D} + \sum_{d_i | P(z) ; i=1,2} |\lambda_{d_1} \lambda_{d_2}| \eta(X, qD) \\ &= \frac{\gamma(q)}{q} X \sum\nolimits_1 + \sum\nolimits_2, \end{aligned}$$

say, and according to Selberg's idea the λ_d's are chosen in such a way that subject to

$$\lambda_1 = 1, \qquad \lambda_d = 0 \quad \text{for } d > \xi \ (> 1)$$

the sum $\sum_1$ is minimized (the general minimum problem seems to be very hard to solve). This leads to the choice of

$$\lambda_d = \frac{\mu(d)}{\prod_{p|d} (1 - \gamma(p)/p)} \frac{G_d(\xi/d, z)}{G(\xi, z)},$$

where

$$G_d(x, z) := \sum_{n \leq x ; n | P(z) ; (n,d)=1} g(n),$$

yielding

$$\sum\nolimits_1 = \frac{1}{G(\xi, z)}.$$

$\sum_2$ can be simplified by noticing that (cf. [28, p. 210] and [20, p. 223]) $|\lambda_d| \leq 1$, which is a consequence of

$$\begin{aligned} G(\xi, z) &= \sum_{t|d} \mu^2(t)g(t)G_d\left(\frac{\xi}{t}, z\right) \geq \sum_{t|d} \mu^2(t)g(t)G_d\left(\frac{\xi}{d}, z\right) \\ &= G_d\left(\frac{\xi}{d}, z\right) \Big/ \left(\prod_{p|d}\left(1 - \frac{\gamma(p)}{p}\right)\right). \end{aligned} \tag{2.1}$$

In view of $|\{d_1, d_2: [d_1, d_2] = d\}| = 3^{\nu(d)}$ for $\mu(d) \neq 0$, it follows that

$$\begin{aligned} \sum\nolimits_2 &\leq \sum_{d_i \leq \xi; d_i|P(z); i=1,2} \eta(X, qD) \leq \sum_{d \leq \xi^2; d|P(z)} 3^{\nu(d)}\eta(X, qd) \\ &\leq \sum_{d \leq \xi^2; (d, \overline{\mathcal{P}})=1} \mu^2(d)3^{\nu(d)}\eta(X, qd) \end{aligned} \tag{2.2}$$

and putting these estimates together we have

THEOREM 2. (A_1):[1]

$$S(\mathcal{A}_q; \mathcal{P}, z) \leq \frac{\gamma(q)}{q}\frac{X}{G(\xi, z)} + \sum_{d \leq \xi^2; d|P(z)} 3^{\nu(d)}\eta(X, qd) \quad \textit{for any } \xi > 1.$$

Often we find that

$$\eta(X, d) \leq \gamma(d) \quad \text{for } \mu(d) \neq 0, \qquad (d, \overline{\mathcal{P}}) = 1, \tag{η}$$

is satisfied. Under this condition one obtains ($q = 1$)

(2.3)

$$\sum_{d \leq \xi^2; d|P(z)} 3^{\nu(d)}\eta(X, d) \leq \sum_{d \leq \xi^2; d|P(z)} 3^{\nu(d)}\gamma(d) \leq \xi^2 \prod_{p<z}\left(1 + \frac{3\gamma(p)}{p}\right) \leq \frac{\xi^2}{\Gamma^3(z)}.$$

3. Special cases

Under special circumstances the function $G(\xi, z)$ can be readily estimated. If

$$\gamma(p) = 1 \quad \text{for } p \in \mathcal{P}$$

we obtain

$$G(\xi, z) = \sum_{d \leq \xi; d|P(z)} \frac{1}{\varphi(d)}. \tag{3.1}$$

---

[1]This way of writing is used to express "If the condition (A_1) is satisfied."

In this case it was noticed by Bombieri (cf. [10]) that also the large sieve gives a result that is similar to Selberg's upper estimate. We state the large sieve in the following form [11].

"Let M and X (> 0) be integers, and let $\mathcal{S} = \{s: \ldots\}$, $M < s \leq M + X$, be a set of integers. Put

$$T(x) = \sum_{s \in \mathcal{S}} e^{2\pi i x s}.$$

Then

$$\sum_{d \leq \xi} \sum_{l=1\,;(l,d)=1}^{d} \left| T\left(\frac{l}{d}\right) \right|^2 \leq (X + c\xi^2)|\mathcal{S}|, \tag{3.2}$$

where c is some numerical constant."

We shall give here a simple proof for Bombieri's observation. Putting $\mathcal{S} = \{a: a \in \mathcal{A}, (a, P(z)) = 1\}$, $\mathcal{D} = \{d: d \mid P(z)\}$ we have $|\mathcal{S}| = S(\mathcal{A}; \mathcal{P}, z)$ and $(\mathcal{S}, \mathcal{D}) = 1$, i.e. $(s, d) = 1$ for all $s \in \mathcal{S}$ and all $d \in \mathcal{D}$. Starting from the well-known formula

$$\mu(d) = \sum_{l=1\,;(l,d)=1}^{d} \exp\left(\frac{2\pi i l}{d} s\right)$$

it follows by summation over s that

$$\mu(d)|\mathcal{S}| = \sum_{l=1\,;(l,d)=1}^{d} T\left(\frac{l}{d}\right),$$

and Cauchy's inequality yields

$$|\mathcal{S}|^2 \leq \varphi(d) \sum_{l=1\,;(l,d)=1}^{d} \left| T\left(\frac{l}{d}\right) \right|^2.$$

Dividing by $\varphi(d)$ and summing over $d \leq \xi$, $d \in \mathcal{D}$ we obtain by (3.2)

$$S(\mathcal{A}; \mathcal{P}, z) \leq \frac{X + c\xi^2}{G(\xi, z)},$$

which is essentially Selberg's upper estimate.

The corresponding result for a general $\gamma(p)$ was obtained by Montgomery [31]. The precise connection between Selberg's sieve and the large sieve is still unknown.

Turning back to Selberg's sieve now, let us assume that $\mathcal{P}$ is of the type (1.7). Then, in view of (3.1), and as another consequence of the identity given in (2.1), we find

$$\begin{aligned} G(\xi, z) &= \sum_{d \leq \xi\,;(d,K)=1} \frac{\mu^2(d)}{\varphi(d)} \geq \prod_{p \mid K} \left(1 - \frac{1}{p}\right) \sum_{d \leq \xi} \frac{\mu^2(d)}{\varphi(d)} \\ &> \prod_{p \mid K} \left(1 - \frac{1}{p}\right) \log \xi \quad \text{for } \xi < z; \end{aligned}$$

also, the estimate (2.2) can slightly be improved. This case contains the following result of the Brun-Titchmarsh type

EXAMPLE 3.1. [28]

$$\pi(x; k, l) - \pi(x - y; k, l) < \frac{2y}{\varphi(k) \log (y/k)} \left(1 + \frac{8}{\log (y/k)}\right)$$

$$\text{for } 1 \leq k < y \leq x, (k, l) = 1.$$

Bombieri has shown that such an estimate can also be obtained by a large sieve application (unpublished). It is known that any improvement of the constant 2 in this estimate would have important consequences.

Another case which belongs here is

$$\gamma(p) = p/(p - 1) \quad \text{for } p \in \mathcal{P},$$

and as an example we choose

$$\mathcal{A} = \{ap + b: p \leq x, p \equiv l \bmod k\}, \qquad \mathcal{P} = \{p: p \nmid kab\}.$$

Then, making use of (1.6), Theorem 2 yields

EXAMPLE 3.2.

$$|\{p: p \leq x, p \equiv l \bmod k, ap + b = p'\}|$$

$$\leq 8 \prod_{p>2} \left(1 - \frac{1}{(p-1)^2}\right) \prod_{2<p|kab} \frac{p-1}{p-2} \frac{x}{\varphi(k) \log^2 x} \left\{1 + O_A\left(\frac{\log \log x}{\log x}\right)\right\},$$

as $x \to \infty$, uniformly in k, l, a, b, for

$$1 \leq k \leq \log^A x, \qquad (k, l) = 1, \qquad ab \neq 0, \qquad (a, b) = 1, \qquad 2 \mid ab.$$

This example includes an upper estimate for the prime-twin problem ($a = k = 1$), and has been used by Bombieri and Davenport [9] in their proof of

$$\liminf_{k \to \infty} \frac{p_{k+1} - p_k}{\log p_k} \leq \frac{2 + \sqrt{3}}{8},$$

which constitutes a considerable improvement of the former estimate $\leq 15/16$ given by Ricci [34].

4. *O*-estimates

If we do not want to confine ourselves to some special sequence as in §3, we have to need some general condition on γ. In case that no estimate with some numerical constant is required the condition

$$(\mathrm{A}_0) \qquad \gamma(p) \leq A_0 \quad \text{for all } p,$$

suffices. Then Theorem 2, when combined with (2.3) yields the very convenient

THEOREM 4.1. (A_0), (A_1), (η):

$$S(\mathcal{A}; \mathcal{P}, z) = O_\lambda\left(X \prod_{p<z} \left(1 - \frac{\gamma(p)}{p}\right)\right) \quad \textit{if } z \leq X^\lambda, \textit{ for any } \lambda > 0.$$

A more general result is

THEOREM 4.2. (A_0), (A_1):

$$S(\mathfrak{A}; \mathcal{P}, z) \leq O(X\Gamma(z)) + \sum_{d \leq z^2 ;(d,\overline{\mathcal{P}})=1} \mu^2(d)3^{\nu(d)}\eta(X, d).$$

We illustrate the possible applications of this theorem by an estimate for the number of primes $p \leq N$ for which $F(p)$ is not divisible by any prime of $\mathcal{P}$.

EXAMPLE 4. Let F be a polynomial with integral coefficients. Then for any set $\mathcal{P}$

$$|\{p: p \leq N, (F(p), \mathcal{P}) = 1\}| \ll \prod_{p<N ;p\in\mathcal{P}} \left(1 - \frac{\rho(p)}{p}\right) \prod_{p<N ;p\in\mathcal{P} ;p|F(0)} (1 - 1/p)^{-1} \frac{N}{\log N},$$

where $\rho(p)$ is defined as in (1.3) and the $\ll$-constant depends only on the degree of F.

Often, the set $\mathcal{P}$ used here is of the type

$$\mathcal{P}_{l,k} := \{p: p \equiv l \bmod k\} \qquad (k, l) = 1,$$

or a union of such sets, or is some similar set, because for sets of the type (1.7) we have more precise estimates. Thus, a typical special case of Example 4 is obtained by choosing $F(n) = N - n$, $\mathcal{P} = \mathcal{P}_{l,k}$.

5. More precise estimates

Although Theorems 4.1 and 4.2 were stated with the simple condition (A_0) they are valid even under the average condition

$$(A_2(\kappa)) \qquad \sum_{w\leq p<z} \frac{\gamma(p)}{p} \log p \leq \kappa \log \frac{z}{w} + A_2 \quad \text{for } z \geq w \geq 2.$$

More precise results can be obtained by introducing also a complementary condition from below, namely that with suitable constants κ (> 0), A_2 (≥ 1) and L (≥ 1)

$(A_2(\kappa; L))$

$$-L \leq \sum_{w\leq p<z} \frac{\gamma(p)}{p} \log p - \kappa \log \frac{z}{w} \leq A_2 \quad \text{for } z \geq w \geq 2.$$

Clearly, this is only one of the various ways one can state a condition expressing that $\gamma(p)$ equals κ on average; κ is an important parameter which may be called the dimension of the sieve. There are a number of different methods in order to deal with the difficult function $G(\xi, z)$ in Selberg's estimate (see f.e. Prachar [32], Klimov [21], Levin [25]); most effective is Wirsing's method [49], [50], [51] which reduces this problem to an integral equation. Under our conditions (A_1) and $(A_2(\kappa; L))$ it follows that

$$\frac{1}{G(\xi, z)} \leq \Gamma(z)e^{\gamma_0\kappa}\kappa! \left(\frac{\log z}{\log \xi}\right)^\kappa \left\{1 + O\left(\frac{L}{\log \xi}\right)\right\} \quad \text{for } z \geq \xi,$$

where γ_0 denotes Euler's constant, and

$$\Gamma(z) = \prod_p \frac{1 - \gamma(p)/p}{(1 - 1/p)^\kappa} \frac{e^{-\gamma_0 \kappa}}{\log^\kappa z} \left\{1 + O\left(\frac{L}{\log z}\right)\right\}.$$

Here, the infinite product is convergent, and a crude but explicit estimate in terms of our constants is

$$(5.1) \qquad \prod_p \frac{1 - \gamma(p)/p}{(1 - 1/p)^\kappa} \geq \exp\{-A_1 A_2(1 + \kappa + A_2)\} \qquad (>0).$$

Hence we obtain by Theorem 2, (2.3) and (2.2) for $q = 1$.

THEOREM 5.1. (A_1), $(A_2(\kappa; L))$, (η):

$$S(\mathfrak{A}; \mathfrak{P}, z) \leq \kappa! \prod_p \frac{(1 - \gamma(p)/p)}{(1 - 1/p)^\kappa} \frac{X}{\log^\kappa z} \left\{1 + O\left(\frac{\log\log 3X + L}{\log z}\right)\right\}$$

for $z \leq \sqrt{X}$.

THEOREM 5.2. (A_1), $(A_2(\kappa; L))$:

$$S(\mathfrak{A}; \mathfrak{P}, z) \leq X\Gamma(z)e^{\gamma_0\kappa}\kappa! \left\{1 + O\left(\frac{L}{\log z}\right)\right\} + \sum_{d \leq z^2; (d, \bar{\mathfrak{P}})=1} \mu^2(d) 3^{\nu(d)} \eta(X, d).$$

The following remarks may serve as an indication when to apply Theorem 5.1 and when Theorem 5.2. Although for practical reasons Theorems 5.1 and 5.2 have been put in a rather different form, their only essential distinction lies in the fact that Theorem 5.2 still contains a sum over remainder terms, whereas Theorem 5.1 does not, which amounts to stating that the additional condition (η) enabled us to keep that sum fairly small. Therefore, applying Theorem 5.1 is not only simpler, but an application of Theorem 5.2 should also be more cumbersome, because we may then assume that condition (η) is violated, so that the sum should be harder to estimate. There are problems which can be attacked in different ways: either by a sequence where (η) is satisfied or by another sequence where we do not have (η); sometimes the harder way leads to a better result. To be explicit: let us consider the general problem of finding an upper estimate for the number

$$(5.2) \qquad |\{p: x - y < p \leq x, p \equiv l \bmod k, F(p) = p'\}|,$$

i.e. the number of primes in the interval $x - y < p \leq x$ which are $\equiv l \bmod k$ such that at these primes the polynomial F is also a prime number (of course, we ought to impose some (natural) conditions as $1 < y \leq x$, $(k, l) = 1$, F has integer coefficients, a positive leading coefficient, irreducible etc.; they are, however, of no relevance here).

We have two possibilities of attacking this problem: either by considering the sequence

$$(5.3) \qquad \mathfrak{A} = \{nF(n): x - y < n \leq x, n \equiv l \bmod k\}$$

and apply a "double sieve", thereby estimating the numbers for which both n and $F(n)$ are primes, namely $n = p$ and $F(n) = p'$; or we could deal directly with the sequence

$$(5.4) \qquad \mathcal{A} = \{F(p): x - y < p \leq x, p \equiv l \bmod k\}$$

and estimating by a sifting procedure the numbers where $F(p) = p'$. Therefore, we may say that in the second case we have "linearized" our problem in comparison with the first procedure. Technically, the linearizing procedure was only made possible by the deep result of Bombieri's (1.6) and its predecessors. Here, i.e. for (5.4), our additional condition (η) is not satisfied, so that we must apply Theorem 5.2. The nonlinearized approach via (5.3) is much simpler because that condition is fulfilled and we can apply Theorem 5.1.

For (5.2), which covers many important problems, linearizing improves the constant in the estimate by a factor 2. On the other hand, the treatment of the more delicate remainder term in the linearized case necessitates certain restrictions: when linearizing we cannot obtain a nontrivial estimate for the interval $x - y \ldots x$ unless

$$y \geq x/\log^C x,$$

and we cannot deal with an arithmetic progression modulo k unless $k \leq \log^C x$ (in both cases with an arbitrary but fixed constant C).

Let us finally turn to the extension of problem (5.2) to g polynomials where (under appropriate conditions) we ask for an estimate of the number

$$|\{p: x - y < p \leq x, p \equiv l \bmod k, F_i(p) \text{ prime for } i = 1, \ldots, g\}|,$$

and of which (5.2) is the special case $g = 1$. Here again putting

$$F(n) = F_1(n) \cdots F_g(n)$$

we may either deal with $nF(n)$ by a $(g + 1)$-dimensional sieve or with $F(p)$ by a g-dimensional sieve, and the same remarks as before apply. However, whereas in the most important case $g = 1$ linearizing means an improvement in the estimate by a factor 2, for $g = 2$ it is the factor 3/2, whereas for $g \geq 3$ linearizing gives no improvement at all, and there is no point in choosing the more difficult and restrictive method (for $g = 3$ both procedures give the same constant, and for $g > 3$ we obtain a better constant if we do not linearize).

Let $F_1, \ldots, F_g$ be distinct irreducible polynomials with integral coefficients and leading coefficients positive. Put $F = F_1 \cdots F_g$ and let $\rho(p)$ $(< p)$ denote the number of solutions of $F(n) \equiv 0 \bmod p$. Then in papers of Bateman-Stemmler [7] and Bateman-Horn [5], [6], Schinzel's conjecture H [39], [37] was put into the following quantitative form

$$(\mathrm{H}) \quad |\{n: 1 \leq n \leq x, F_i(n) \text{ prime for } i = 1, \ldots, g\}| \sim \frac{1}{h_1 \cdots h_g} \prod_p \frac{1 - \rho(p)/p}{(1 - 1/p)^g} \frac{x}{\log^g x},$$

where h_i is the degree of F_i, $i = 1, \ldots, g$.

An application of Theorem 5.1 gives an upper estimate for the left of (H) that is asymptotically $2^g g! h_1 \cdots h_g$ times the conjectured value. This was proved for $g = 1$ by Wang [47], and in the general case by Batemann and Stemmler [7]; and it contains a large number of individual sieve results that were proved by several authors.

Conjecture (H) was extended by Schinzel [38] to the corresponding problem with

$$(\mathrm{H}_N) \qquad F(n)(N - G(n)).$$

Both conjectures cover some of the most prominent unproven conjectures in Number Theory, some of them were already formulated by Hardy and Littlewood [19], and they constitute unsolved problems except for the case of one polynomial of degree one, where (H) reduces to the prime number theorem. (H_N) is much more difficult to deal with than (H): we do not even have explicit estimates by a sieve method for the problem $N - G(p) = p'$ if the degree of G is > 1 (regardless of papers of Schwarz [40], Wang [48], Levin [26], Levin [27]).

As an application of Theorem 5.2 we mention

EXAMPLE 5.1. Let $F(n)$ ($\neq n$) be an irreducible polynomial with integral coefficients. Let $\rho(p)$ ($< p$) denote the number of solutions of $F(n) \equiv 0 \bmod p$, and suppose that $\rho(p) < p - 1$ if $p \nmid F(0)$. Let k, l be integers, let x be a real number, such that with some constant A, $1 \leq k \leq \log^A x$, $(k, l) = 1$. Then, as $x \to \infty$,

$$\begin{aligned} |\{p: p \leq x, p \equiv l \bmod k, F(p) = p'\}| \\ \leq 8 \prod_{p>2}\left(1 - \frac{1}{(p-1)^2}\right) \prod_{2<p\nmid F(0);p\nmid k}\left(1 - \frac{\rho(p)-1}{p-2}\right) \\ \times \prod_{2<p|F(0);p\nmid k}\left(1 - \frac{\rho(p)-2}{p-2}\right) \prod_{2<p|k} \frac{p-1}{p-2} \frac{x}{\varphi(k)\log^2 x} \\ \times \left\{1 + O_F\left(\frac{\log\log 3x}{\log x}\right)\right\}, \end{aligned}$$

where the O-constant is independent of x, k and l; it may, however, depend on the polynomial F, i.e. on its coefficients and on its degree, as well as on A.

A special case of Example 5.1 is an estimate that was proved by Bateman and Stemmler [7], but since Theorem 5.2 uses linearization our constant here is 8 instead of 16; accordingly, Theorem 4 of [7] can be improved by a factor 2.

EXAMPLE 5.2. With an absolute O-constant we have

$$\begin{aligned} |\{p: p \leq x, p^2 + p + 1 = p'\}| \\ \leq 8 \prod_{p>2}\left(1 - \frac{1}{(p-1)^2}\right) \prod_{p>3}\left(1 - \frac{\chi(p)}{p-2}\right) \frac{x}{\log^2 x}\left\{1 + O\left(\frac{\log\log 3x}{\log x}\right)\right\}, \end{aligned}$$

where $\chi(p) = 1$ or -1 according as $p \equiv 1$ or $-1 \bmod 3$.

6. The simplest lower estimate

All results we have mentioned so far were consequences of Selberg's upper estimate given in Theorem 2. A natural and, for most of the problems we have in mind, even more interesting question is that for an estimation of our sifting function

$S(\mathcal{A}_q; \mathcal{P}, z)$ from below. The simplest way of obtaining such a lower sieve estimate is based on the observation that for any permissible choice of our parameters the following identity holds true

$$S(\mathcal{A}_q; \mathcal{P}, z) = |\mathcal{A}_q| - \sum_{p<z;p\in\mathcal{P}} S(\mathcal{A}_{qp}; \mathcal{P}, p); \tag{6.1}$$

this is actually a formula of mathematical logic. It is clear that an approximation for $|\mathcal{A}_q|$ and upper estimates for each of the functions $S(\mathcal{A}_{qp}; \mathcal{P}, p)$ on the right of (6.1) yield a lower estimate for $S(\mathcal{A}_q; \mathcal{P}, z)$.

Subject to certain conditions, but what is most important, for any dimension κ (> 0), the resulting lower estimates were worked out by Ankeny and Onishi [1]. The most delicate part that occurs in such an estimate consists of a function that satisfies a difference-differential equation, and [1] contains also a detailed discussion of these functions for every $\kappa > 0$.

Without any additional aid, the indicated method of estimating from below by means of (6.1) does not give very satisfactory results, especially when compared with the upper estimates we have described before; it turns out that both are rather far apart unless z is small in comparison with X. However, on the other hand, here z does not have to be so small as is required for a nontrivial application of Theorem 1, and in fact, it results in a considerable extension of Theorem 1, sometimes called Fundamentallemma (cf. [24], [2]).

We mention a result which can be obtained in this way.

THEOREM 6.1. (A_0), (A_1): *For* $\xi \geq z$

$$S(\mathcal{A}_q; \mathcal{P}, z) = \frac{\gamma(q)}{q} X\Gamma(z)\{1 + O(\exp\{-\tau \log \tau + (2 + \log A_0)\tau\})\}$$
$$+ \theta \sum_{d\leq \xi^2;d|P(z)} 3^{\nu(d)}\eta(X, qd), \qquad \tau = \frac{\log \xi}{\log z}, \qquad |\theta| \leq 1.$$

A particularly simple version that follows from Theorem 6.1 is

THEOREM 6.2. (A_0), (A_1), (η): *For* $X \geq z$

$$S(\mathcal{A}; \mathcal{P}, z) = X\Gamma(z)\{1 + O(\exp\{-\tfrac{1}{2}\log X/\log z\})\}.$$

7. Buchstab's method of iteration

An important method of improving upper and lower sieve estimates step by step was developed by Buchstab (cf. [12]) (already in connection with Brun's sieve method). The underlying idea is the following one. Taking two values for z in (6.1) and forming the difference we obtain

$$S(\mathcal{A}_q; \mathcal{P}, z) = S(\mathcal{A}_q; \mathcal{P}, z_1) - \sum_{z_1\leq p<z;p\in\mathcal{P}} S(\mathcal{A}_{qp}; \mathcal{P}, p) \quad \text{for } z \geq z_1 \geq 2. \tag{7.1}$$

Suppose now, we already had for this sifting function a pair of estimates, i.e. one estimate from above and one from below. Then, using the upper (lower) estimate for the first term on the right of (7.1) and using the lower (upper) estimate for each term of the sum, we arrive at a new upper (lower) estimate for the left-hand side of (7.1). The surprising effect is that often the new pair of estimates thus obtained improves (at any rate for certain values of the parameters involved) the original pair.

Buchstab and several authors after him made extensive use of this method. Often one starts with an upper Selberg estimate and a comparatively crude but simple lower estimate taken from Brun's sieve.

In some of the more important cases the resulting estimating functions have been tabulated, and computers have been used in order to complete and to extend those tables (see [12], [48], [13], [14]).

If the dimension of the sieve is different from 1, the method we just described, when combined with certain combinatorial arguments (see §9 below), leads in each case to the best results that are known so far for lower sieve estimates. However, we have very little general knowledge about the case $\kappa \neq 1$. Here it lies completely in the dark what the best possible results could be, and each problem is best considered individually.

8. The linear sieve

If the dimension κ of the sieve equals 1 (we may then speak of the linear sieve), the general theory can be carried considerably further. However, as was already indicated, this is the only case where we possess such satisfactory results, and therefore from now on we shall confine ourselves to this linear case $\kappa = 1$.

Clearly, any problem where, apart from some finite number of primes, $\gamma(p)$ equals 1 or is 1 on average, as for example with irreducible polynomials, belongs here; also any "binary" problem that can be linearized in the sense of §5, or any problem of higher dimension where the sparseness of the set $\mathcal{P}$ produces a $\kappa = 1$ in our condition $(A_2(\kappa; L))$. Hence it follows that this case covers the most important problems that had been attacked by the sieve method.

We start by noticing that Buchstab's formula (7.1) can be iterated, that is for each or for some of the sifting functions in our sum there, we can apply the corresponding formula (7.1) again. This was already done by Viggo Brun in order to obtain lower estimates, and the main problem is rather to which of the functions this procedure should be applied and to which not. One can find certain criteria to decide this question, and it turns out that the following generalization of (7.1) is most suitable.

LEMMA 8. [20] *Let $2 \leq z_1 \leq z \leq \xi$ and set $\xi_j^2 = \xi^2/(p_1 \ldots p_j)$, $j = 1, 2, \ldots$. Then for any natural number n we have*

$$S(\mathcal{A}_q; \mathcal{P}, z) = S(\mathcal{A}_q; \mathcal{P}, z_1) + \sum_{1 \leq i \leq n-1} (-1)^i \sum_{\substack{z_1 \leq p_i < \ldots < p_1 < z \\ p_j < \xi_j, p_j \in \mathcal{P}, j=1,\ldots,i}} S(\mathcal{A}_{qp_1 \ldots p_i}; \mathcal{P}, z_1)$$

$$+ (-1)^n \sum_{\substack{z_1 \leq p_n < \ldots < p_1 < z \\ p_j < \xi_j, p_j \in \mathcal{P}, j=1,\ldots,n}} S(\mathcal{A}_{qp_1 \ldots p_n}; \mathcal{P}, p_n)$$

$$+ \sum_{1 \leq i \leq n} (-1)^i \sum_{\substack{z_1 \leq p_i < \ldots < p_1 < z \\ p_j < \xi_j, p_j \in \mathcal{P}, j=1,\ldots,i-1 \\ \xi_i \leq p_i < \xi_i^2, p_i \in \mathcal{P}}} S(\mathcal{A}_{qp_1 \ldots p_i}; \mathcal{P}, p_i).$$

The idea for an application of Lemma 8 is as follows. In order to find an estimate for $S(\mathcal{A}_q; \mathcal{P}, z)$ on the left, we have to choose in a suitable way z_1 (sufficiently small) and n (sufficiently large), leaving ξ ($\geq z$) arbitrary. Then the first term and the first two sums on the right have comparatively small z's (namely z_1 and p_n, respectively) and Theorem 6.1 yields a satisfactory result. In the last sum, depending on the sign, we either estimate (from above) by Selberg's upper estimate or (from below) by zero. It looks as if we lose something by this procedure, especially by the use of $S \geq 0$. However, the condition $\xi_i \leq p_i < \xi_i^2$ ensures that in the last sum we are always within a range where these estimates cannot be improved (see below). It also looks as if the resulting estimating functions become rather complicated. This is, however, not the case, because our function $\Gamma(z)$ also satisfies an identity of the type (6.1) on which Lemma 8 was based, namely

$$\Gamma(z) = 1 - \sum_{p<z;p\in\mathcal{P}} \frac{\gamma(p)}{p} \Gamma(p);$$

and when multiplied by the aforementioned estimating functions which satisfy a difference-differential equation, we still obtain a corresponding identity, apart from an error term of minor importance, so that we can, so to speak, imitate the identity for S by the leading terms of its estimate.

The resulting estimating functions can be defined by [20, p. 226],

$$F(u) = 2e^{\gamma_0}/u, \qquad f(u) = 0 \quad \text{for } 0 < u \leq 2, \tag{8.1}$$

and[2]

$$(uF(u))' = f(u-1), \qquad (uf(u))' = F(u-1) \quad \text{for } u > 2. \tag{8.2}$$

One finds, for instance, [20, 226–227]

$$F(u) = 2e^{\gamma_0}/u \quad \text{for } 0 < u \leq 3,$$
$$f(u) = (2e^{\gamma_0}/u) \log (u-1) \quad \text{for } 2 \leq u \leq 4,$$

$F(u)$ is monotonically decreasing towards 1,
$f(u)$ is monotonically increasing towards 1.

By the method described above one obtains with these functions, and only subject to the conditions (A_1) and ($A_2(1; L)$)

THEOREM 8.1. [20], [35], [17]. (A_1), ($A_2(1; L)$): *For any* $\xi \geq z$

$$S(\mathcal{A}_q; \mathcal{P}, z) \leq \frac{\gamma(q)}{q} X\Gamma(z) \left\{ F\left(\frac{\log \xi^2}{\log z}\right) + B_1 \frac{L}{(\log \xi)^{1/14}} \right\} + \sum_{d \leq \xi^2; d | P(z)} 3^{\nu(d)} \eta(X, qd),$$

$$S(\mathcal{A}_q; \mathcal{P}, z) \geq \frac{\gamma(q)}{q} X\Gamma(z) \left\{ f\left(\frac{\log \xi^2}{\log z}\right) - B_2 \frac{L}{(\log \xi)^{1/14}} \right\} - \sum_{d \leq \xi^2; d | P(z)} 3^{\nu(d)} \eta(X, qd),$$

---

[2] At the point $u = 2$ the right-hand derivative has to be taken.

where B_1 *and* B_2 *depend only on the constants* A_1 *and* A_2. *The result remains true for* $1 < \xi < z$ *but* $z \ll \xi^{\lambda}$, *in which case* B_1 *may also depend on* λ.

As soon as an estimate for the sum containing the remainder terms is known, we can choose ξ^2 in an appropriate way. This can be secured by our next condition: "There are constants α $(0 < \alpha \leq 1)$, A_3 (≥ 1) and A_5 (≥ 1) such that

$$(\mathrm{A}_3(\alpha)) \qquad \sum_{d \leq X^{\alpha}/\log^{A_3} X;\,(d,\overline{\mathcal{P}})=1} \mu^2(d) 3^{\nu(d)} \eta(X, d) \leq A_5 \frac{X}{\log^{15/14} X}."$$

Then with the choice of $\xi^2 = X^{\alpha}/\log^{A_3} X$ we obtain the following main result for the linear sieve[3]

THEOREM 8.2. [20], [16], [35], [17]. (A_1), $(\mathrm{A}_2(1; L))$, $(\mathrm{A}_3(\alpha))$: *For* $z \leq X$

$$S(\mathcal{A}; \mathcal{P}, z) \leq X\Gamma(z) \left\{ F\left(\alpha \frac{\log X}{\log z}\right) + B \frac{L}{(\log X)^{1/14}} \right\},$$

$$S(\mathcal{A}; \mathcal{P}, z) \geq X\Gamma(z) \left\{ f\left(\alpha \frac{\log X}{\log z}\right) - B \frac{L}{(\log X)^{1/14}} \right\}$$

where B depends at most on the constants α *and* A_i, $i = 1, 2, 3, 5$.

By means of $(\mathrm{A}_3(\alpha))$ Theorem 8.2 is an easy consequence of Theorem 8.1, but Theorem 8.1 depends entirely on the method we have described above and which is based on the identity in Lemma 8. This method exhausts Buchstab's technique of improving the estimates step by step, that is deriving from one pair of estimating functions F_ν, f_ν, say, a new pair $F_{\nu+1}, f_{\nu+1}$ (and at each step in the most suitable way), completely. In fact, one can show that for each $u > 0$ the limits

$$(8.3) \qquad \lim_{\nu \to \infty} F_\nu(u), \qquad \lim_{\nu \to \infty} f_\nu(u)$$

exist, and that they equal the functions that were defined by (8.1) and (8.2).

Suppose that $\alpha = 1$, and put $\log X/\log z = u$. Then for $0 < u \leq 2$, and disregarding the error terms, Theorem 8.2 coincides with Selberg's upper estimate of Theorem 2 and the trivial estimate $S \geq 0$, respectively. This is necessarily so, because by considering the two sets

$$\mathcal{A}^{(\nu)} := \{n\colon 1 \leq n \leq x,\ \Omega(n) \equiv \nu \bmod 2\}, \qquad \nu = 1, 2,$$

where $\Omega(n)$ denotes the total number of prime factors of n, and taking for $\mathcal{P}$ the set of all primes, A. Selberg [42], [43] has shown that those estimates cannot be improved upon, as for $0 < u \leq 2$ they correspond with the asymptotic formulae for $S(\mathcal{A}^{(\nu)}; \mathcal{P}, z)$, $\nu = 1, 2$, respectively. For $u > 2$ this is no longer true with Theorem 2 and $S \geq 0$; however, Theorem 8.2 continues to yield the correct leading terms in the asymptotic developments of $S(\mathcal{A}^{(\nu)}; \mathcal{P}, z)$, $\nu = 1, 2$. Therefore Theorem

---

[3]In his lectures (see pp. 311–351) A. Selberg has pointed out that this result was also obtained in unpublished work by Barkley Rosser by means of Brun's sieve.

8.2 is best possible in the sense that Selberg's examples prove that (for $\alpha = 1$) in both estimates and for every $u > 0$ we may have equality.

The two sets $\mathfrak{A}^{(\nu)}$ are also of great value when one tries to construct a more sophisticated combinatorial formula (in the sense of §9 below) in order to obtain a sieve estimate. Actually, they enable us to check whether in such a formula we necessarily lose something by using an estimate or not. More specific, and considering for example an upper estimate, one does not necessarily lose anything by an application of Buchstab's formula (7.1), because using F for the first term and, because of the opposite sign, f for each term inside the sum, could still produce the right asymptotic, namely for $\mathfrak{A}^{(1)}$, where the first term on the right is then also formed with $\mathfrak{A}^{(1)}$ whereas each term inside our sum contains an $\mathfrak{A}^{(2)}$.

After these remarks it is clear that it would be highly desirable to have a result like Theorem 8.2 also for other dimensions $\kappa \neq 1$. However, although Lemma 8 is always true, the process of iteration, indicated by (8.3), has to converge, and it is this we were able to prove only in case that $\kappa = 1$. It seems possible that a corresponding result can be obtained for each κ with $0 < \kappa < 3e^{-1}$.

9. Selberg's sieve with weights

The most important results obtained by the sieve method, being consequences of lower estimates in the linear case, do not follow directly from Theorem 8.2 but rather by combining it with an additional idea which shall now be discussed. The means by which a lower estimate can be made more effective, originally entered the sieve method as a combinatorial device, first stated by Kuhn [22], [23] in connection with Brun's sieve. Several authors gave other ways of refining a lower sieve estimate, but all of these methods can be described by a sieve with weights. In most cases this means attaching the weight

$$1 - \sum_{p|a} \omega_p \tag{9.1}$$

with certain ω_p's to each $a \in \mathfrak{A}$.

Rényi [33] used weight functions in his well-known paper on Goldbach's problem. If P_r denotes a number having at most r prime factors, multiple prime factors being counted multiply, i.e. $\Omega(P_r) \leq r$ (such numbers are called almost-primes), Rényi proved the existence of a constant c such that $2N = p + P_c$ for all sufficiently large N. (Incidentally, it was in the course of this work that Rényi was the first to notice that the error term in the Brun/Selberg sieve can sometimes be handled effectively by means of Linnik's large sieve; the linearizing procedure of §5.)

A. Selberg suggested the use of general weights [42] and applied such functions in dealing with the twin-prime problem, where he showed for the first time that there exist infinitely many numbers P_2 such that $P_2 + 2 = P_3$ [44]. A generalization of this method has been used by Miech [29].

Ankeny and Onishi [1] have used beside Kuhn's weights, certain special weights when dealing with the problem of almost-primes in a product of linear forms.

Buchstab [13], [14] in his proof for the solvability of $2N = p + P_3$ formulated the procedure of choosing weights as a problem in linear programming, and was lead in this way to a very complicated set of constant weights.

The problem of finding optimal weights in Selberg's sieve seems to be very difficult since the problem of finding the minimum of a certain quadratic form, a basic feature in Selberg's method, is rather complicated as soon as the weights are no longer 1.

We shall confine ourselves here to a description of a certain extension of the Ankeny-Onishi method of constructing weighting functions, given in [35], which turns out to be rather effective, and avoids the much more complicated weights of the recent Buchstab method [14]. Using the abbreviations

$$X^{1/v} = z, \qquad X^{1/u} = y,$$

we put in (9.1) $\omega_p = 0$ unless $z \leq p < y$ and $p \in \mathcal{P}$ in which case we use

$$\omega_p = \lambda(1 - \log p/\log y).$$

Also in order that we count the total number of prime factors and not merely the number of distinct ones, we exclude the numbers which are divisible by the square of a prime $\geq z$, and for this we introduce one more condition of minor importance, namely[4]

$$(\mathrm{A}_4) \qquad |\mathcal{A}_{p^2}| \leq A_4\left(\frac{X \log X}{p^2} + 1\right), \qquad p \in \mathcal{P}.$$

Instead of $S(\mathcal{A}; \mathcal{P}, z)$ we now consider the sifting function with weights

$$W_\lambda(\mathcal{A}; \mathcal{P}, v, u) := \sum_{a\in\mathcal{A};(a,P(z))=1}' \left\{1 - \lambda \sum_{p|a;z\leq p<y;p\in\mathcal{P}} \left(1 - \frac{\log p}{\log y}\right)\right\}$$

where $\sum'$ indicates that summation is restricted to those a's for which

$$a \not\equiv 0 \bmod p^2 \quad \text{for } z \leq p < y, p \in \mathcal{P}.$$

The underlying idea when employing this function W is, roughly speaking, that the a's which have many prime divisors make the weight negative, and therefore from an estimate $W > 0$ we can draw the conclusion that there must be numbers having only few prime divisors.

For an estimate of W from below we use the lower estimate of Theorem 8.2 for the part originating from the 1, and condition (A_4) for an estimate of the number of a's that are not squarefree. The remaining part can be rewritten as

$$-\lambda \sum_{z\leq p<y;p\in\mathcal{P}} \left(1 - \frac{\log p}{\log y}\right) S(\mathcal{A}_p; \mathcal{P}, z),$$

and this in turn is estimated by Theorem 8.1 (each time with ξ^2/p instead of ξ^2, and $\xi^2 = X^\alpha/\log^{A_3} X$, with the constants of our condition $(\mathrm{A}_3(\alpha))$.

We thus obtain

THEOREM 9. (A_1), $(\mathrm{A}_2(1; L))$, $(\mathrm{A}_3(\alpha))$, (A_4): *For* $1/\alpha < u < v$, $0 \leq \lambda \leq A_6$ *we have*

$$W_\lambda(\mathcal{A}; \mathcal{P}, v, u) \geq X\Gamma(z)\{I(\alpha, \lambda, v, u) - bL/(\log X)^{1/14}\},$$

[4]This condition can be weakened to $A_4(X \log^{A_7} X/p^2 + 1)$ or even to some suitable average condition, without changing the results.

where

$$I(\alpha, \lambda, v, u) = f(\alpha v) - \lambda \int_u^v F\left(v\left(\alpha - \frac{1}{t}\right)\right)\left(1 - \frac{u}{t}\right)\frac{dt}{t},$$

and where b depends only on α, A_i, $i = 1, \ldots, 6$, *u and v.*

We remark that

$$I(\alpha, \lambda, v, u) = I(1, \lambda, \alpha v, \alpha u)$$

which means that it suffices to consider I for $\alpha = 1$ only, since if for $\alpha = 1, \lambda, v, u$ make $I > 0$, then the same is true for any α with the triplet $\lambda, v/\alpha, u/\alpha$. In this case, I is for $v \leq 4$ an elementary function.

LEMMA 9. *For* $1 < u < v$, $2 \leq v \leq 4$, $\lambda \geq 0$ *we have*

$$I(1, \lambda, v, u) = \frac{2e^{\gamma_0}}{v}\left\{\log(v-1) - \lambda u \log\frac{v}{u} + \lambda(u-1)\log\frac{v-1}{u-1}\right\}.$$

In our applications here and in §10 we have always used $v = 4/\alpha$ in order that Lemma 9 be applicable. The choice of $v > 4/\alpha$ would lead to rather cumbersome computations; on the other hand such a choice could possibly give further improvements of our results.

For integers h and N satisfying $0 < |h| \leq N$ we set

$$Z_3(N, h) = |\{p: p \leq N, p + h = P_3\}|, \quad G_3(N) = |\{p: p \leq N, N - p = P_3\}|.$$

For these functions Theorem 9 yields

EXAMPLE 9. There is an absolute constant N_0 such that, for $N \geq N_0$, we have

$$Z_3(N, h) \geq \frac{13}{3}\prod_{p>2}\left(1 - \frac{1}{(p-1)^2}\right)\prod_{2<p|h}\frac{p-1}{p-2}\frac{N}{\log^2 N} \quad \text{if } 2 \mid h,$$

in particular ($N \to \infty$), for any even number $h \neq 0$ the equation $p + h = P_3$ has infinitely many solutions; and

$$G_3(N) \geq \frac{13}{3}\prod_{p>2}\left(1 - \frac{1}{(p-1)^2}\right)\prod_{2<p|N}\frac{p-1}{p-2}\frac{N}{\log^2 N} \quad \text{if } 2 \mid N,$$

so that every sufficiently large even number N can be written as $N = p + P_3$.

The constant 13/3 actually enters the proof as $4 \log 3 - \epsilon$ for any $\epsilon > 0$.

Qualitatively these results were first proved by Buchstab [13] (see also [46], [14], [16], [45]).

10. Further applications

In many applications the following general consequence of Theorem 9 is more convenient, and it is also simpler to have ($A_2(1; L)$) replaced by

$$(A_2) \qquad -A_2 \log\log 3X \leq \sum_{w \leq p < z}\frac{\gamma(p)}{p}\log p - \log\frac{z}{w} \leq A_2 \quad \text{for } z \geq w \geq 2;$$

this amounts to requiring in $(A_2(1; L))$ that $L \leq A_2 \log \log 3X$.[5]

For each positive integer r we set

$$\Lambda_r = r + 1 - \log(4(1 + 3^{-r}))/\log 3$$

and remark that

$$r + 1 - \log 4/\log 3 \leq \Lambda_r \leq r + 1 - \log 3.6/\log 3 \quad \text{for } r \geq 2.$$

Then, with the choice of

$$v = 4/\alpha, \qquad u = (1 + 3^{-r})/\alpha, \qquad 1/\lambda = r + 1 - u\alpha(\Lambda_r - \delta),$$

Theorem 9 leads to

THEOREM 10. (A_1), (A_2), $(A_3(\alpha))$, (A_4): *Suppose that* $(a, \overline{\mathcal{P}}) = 1$ *for all* $a \in \mathcal{A}$. *Let* δ *be a real number satisfying* $0 < \delta \leq 2/3$, *and take the positive integer* r (≥ 2) *so large that*

$$|a| \leq X^{\alpha(\Lambda_r - \delta)} \quad \textit{for all } a \in \mathcal{A}.$$

Then

$$|\{P_r : P_r \in \mathcal{A}\}| \geq \frac{\delta}{\alpha} \prod_p \frac{1 - \gamma(p)/p}{1 - 1/p} \frac{X}{\log X} \quad \textit{for } X \geq X_0, \tag{10.1}$$

where X_0 *depends at most on* α, A_i, $i = 1, \ldots, 5$, r *and* δ. *Moreover, for the smallest prime factor* p_1, *say, of each* P_r *counted on the left, we have*

$$p_1(P_r) \geq X^{\alpha/4}.$$

By (5.1) the product in (10.1) satisfies

$$\prod_p \frac{1 - \gamma(p)/p}{1 - 1/p} \geq \exp\{-A_1 A_2(2 + A_2)\} \quad (> 0).$$

Theorem 10 can also be applied to yield the essence of Example 9. However, in Theorem 10 we have chosen our parameters in such a way that for various values of r a smallest possible r is admitted, whereas a slightly better result with respect to the constant is obtained by the choice we made for Example 9.

We mention the following applications of Theorem 10. For a more detailed discussion and further applications we refer to [35].

EXAMPLE 10.1. Let $(k, l) = 1$. Then there exists a number P_2 satisfying

$$P_2 \leq k^{2.2}, \qquad P_2 \equiv l \bmod k \quad \text{if } k \geq k_2;$$

and for each $r \geq 2$, there exists a number P_r satisfying

$$P_r \leq k^{1+1/(r-9/7)} P_r \equiv l \bmod k \quad \text{if } k \geq k_r;$$
$$x - x^{1/(r-2/7)} < P_r \leq x \quad \text{if } x \geq x_r.$$

---

[5]Actually, any function satisfying $o((\log X)^{1/14})$ instead of $\log \log 3X$ suffices.

With the choice of $\delta = 2/3$ and $r = g + 1$ Theorem 10 yields

EXAMPLE 10.2. Let F be an irreducible polynomial of degree g (≥ 1) with integral coefficients, and let $\rho(p)$ ($< p$) denote the number of solutions of

$$F(m) \equiv 0 \bmod p. \tag{10.2}$$

Then

$$|\{n: 1 \leq n \leq N, F(n) = P_{g+1}\}| \geq \frac{2}{3} \prod_p \frac{1 - \rho(p)/p}{1 - 1/p} \frac{N}{\log N} \quad \text{for } N \geq N_0 = N_0(F). \tag{10.3}$$

In particular (if F is independent of N), there are infinitely many n such that

$$F(n) = P_{g+1}, \tag{10.4}$$

i.e. such that $F(n)$ consists of at most $g + 1$ prime factors.

(10.4) was first stated by Buchstab [14] as a consequence of his new method; before it had only been known up to $g = 7$ (Levin [27]).

Inequality (10.3) should be compared with what is expected by Conjecture (H), namely

$$|\{n: 1 \leq n \leq N, F(n) = P_1\}| \sim \frac{1}{g} \prod_p \frac{1 - \rho(p)/p}{1 - 1/p} \frac{N}{\log N}.$$

There is a corresponding result for numbers represented by $F(p)$.

EXAMPLE 10.3. Let $F(n)$ ($\neq \pm n$) be an irreducible polynomial of degree g (≥ 1) with integral coefficients. Let $\rho(p)$ ($< p$) be defined as in (10.2), and suppose that

$$\rho(p) < p - 1 \quad \text{if } p \nmid F(0), p \leq g + 1. \tag{10.5}$$

Then

$$|\{p: p \leq N, F(p) = P_{2g+1}\}| \geq \frac{4}{3} \prod_{p \nmid F(0)} \frac{1 - \rho(p)/(p-1)}{1 - 1/p} \prod_{p \mid F(0)} \frac{1 - (\rho(p) - 1)/(p - 1)}{1 - 1/p} \frac{N}{\log^2 N}$$
$$\text{for } N \geq N_0 = N_0(F).$$

In particular (if F is independent of N), there are infinitely many primes p such that $F(p) = P_{2g+1}$, i.e. such that $F(p)$ consists of at most $2g + 1$ prime factors.

Since $F(n) = \pm n$ has been excluded, our conditions imply that $F(0) \neq 0$. (10.5) is also a natural condition, because from $\rho(p_0) = p_0 - 1$, and $p_0 \nmid F(0)$ it would follow that $p_0 \mid F(p)$ for all $p \neq p_0$.

For any integral valued polynomial F, using Brun's sieve, Miech [30] has proved the existence of infinitely many primes p such that $F(p) = P_{cg}$ where c is some sufficiently large constant; recently Rieger [36] has used Kuhn's method in order to prove this result for $p^2 - 2 = P_5$.

With respect to the Example 5.2 of Bateman and Stemmler, Example 10.3

yields

$$|\{p: p \leq N, p^2 + p + 1 = P_5\}| \geq \frac{8}{3} \prod_{p>2} \left(1 - \frac{1}{(p-1)^2}\right) \prod_{p>3} \left(1 - \frac{\chi(p)}{p-2}\right) \frac{N}{\log^2 N} \quad \text{for } N \geq N_0,$$

where $\chi(p) = \pm 1$ for $p \equiv \pm 1 \bmod 3$.

Finally, for the general linear case $F(n) = an + b$, it follows from Example 10.3.

EXAMPLE 10.4. Let a, b integers satisfying

$$ab \neq 0, \quad (a, b) = 1, \quad 2 \mid ab. \tag{10.6}$$

Then

$$|\{p: p \leq N, ap + b = P_3\}| \geq \frac{8}{3} \prod_{p>2} \left(1 - \frac{1}{(p-1)^2}\right) \prod_{2<p|ab} \frac{p-1}{p-2} \frac{N}{\log^2 N} \quad \text{for } N \geq N_0 = N_0(a, b);$$

in particular, under the conditions (10.6) there are infinitely many primes p such that $ap + b = P_3$.

REFERENCES

1. N. C. Ankeny and H. Onishi, *The general sieve*, Acta Arith. **10** (1964/65), 31–62. MR **29** #4740.

2. M. B. Barban, *On a theorem of I. P. Kubiljus*, Izv. Akad. Nauk UzSSR Ser. Fiz.-Mat. Nauk **1961,** no. 5, 3–9, 1963, no. 1, 82–83. (Russian) MR **25** #2051.

3. ———, *Analogues of the divisor problem of Titchmarsh*, Vestnik Leningrad. Univ. Ser. Mat. Meh. Astronom. **18** (1963), no. 4, 5–13. (Russian) MR **28** #57.

4. ———, *The "large sieve" method and its application to number theory*, Uspehi Mat. Nauk **21** (1966), no. 1 (127), 51–102. (Russian) MR **33** #7320.

5. P. T. Bateman and R. A. Horn, *A heuristic asymptotic formula concerning the distribution of prime numbers*, Math. Comp. **16** (1962), 363–367. MR **26** #6139.

6. ———, *Primes represented by irreducible polynomials in one variable*, Proc. Sympos. Pure Math., vol. 8, Amer. Math. Soc., Providence, R.I., 1965, pp. 119–132. MR **31** #1234.

7. P. T. Bateman and R. Stemmler, *Waring's problem for algebraic number fields and primes of the form* $(p^r - 1)/(p^d - 1)$, Illinois J. Math. **6** (1962), 142–156. MR **25** #2059.

8. E. Bombieri, *On the large sieve*, Mathematika **12** (1965), 201–225. MR **33** #5590.

9. E. Bombieri and H. Davenport, *Small differences between prime numbers*, Proc. Roy. Soc. Ser. A **293** (1966), 1–18. MR **33** #7314.

10. ———, *On the large sieve method*, Abh. Erinn. E. Landau (1968), 9–22.

11. ———, *Some inequalities involving trigonometrical polynomials*, Ann. Scuola Norm. Syp. Pisa (3) **23** (1969), 223–241.

12. A. A. Buchštab, *Sur la décomposition des nombres pairs en somme de deux composantes dont chacune est formée d'un nombre borné de facteurs premiers*, C.R. Acad. Sci. URSS **29** (1949), 544–548. MR **2,** 348.

13. ———, *New results in the investigation of the Goldbach-Euler problem and the problem of prime pairs*, Dokl. Akad. Nauk SSSR **162** (1965), 735–738 = Soviet Math. Dokl. **6** (1965), 729–732. MR **31** #2226.

14. ———, *Combinatorial strengthening of the sieve of Eratosthenes method*, Uspehi Mat. Nauk **22** (1967), no. 3 (135), 199–226. (Russian) MR **36** #1413.

15. H. Davenport and H. Halberstam, *Primes in arithmetic progressions*, Michigan Math. J. **13** (1966), 485–489. MR **34** #156.

16. H. Halberstam, W. B. Jurkat and H.-E. Richert, *Un nouveau résultat de la méthode du crible*, C.R. Acad. Sci. Paris Sér. A-B **264** (1967), A920–A923. MR **36** #6374.

17. H. Halberstam and H.-E. Richert, *Sieve methods*, Markham, Chicago, Ill. (to appear).

18. H. Halberstam and K. F. Roth, *Sequences*. Vol. I, Clarendon Press, Oxford, 1966. MR **35** #1565.

19. G. H. Hardy and J. E. Littlewood, *On the expression of a number as a sum of primes*, Acta Math. **44** (1923), 1–70.

20. W. B. Jurkat and H.-E. Richert, *An improvement of Selberg's sieve method*. I, Acta Arith. **11** (1965), 217–240. MR **34** #2540.

21. N. I. Klimov, *Combination of elementary and analytic methods in the theory of numbers*, Uspehi Mat. Nauk **13** (1958), no. 3 (81), 145–164. (Russian) MR **20** #3841.

22. P. Kuhn, *Zur Viggo Brun'schen Siebmethode*. I, Norske Vid. Selsk. Forh. Trondhjem **14** (1941), no. 39, 145–148. MR **8**, 503.

23. ———, *Neue Abschätzungen auf Grund der Viggo Brunschen Siebmethode*, Tolfte Skandinaviska Matematikerkongressen, Lund, 1953, pp. 160–168. MR **16**, 676.

24. J. Kubilius, *Probabilistic methods in the theory of numbers*, Gos. Izdat. Polit. Naučn. Lit. Litovsk. SSR, Vilna, 1962; English transl., Transl. Math. Monographs, vol. 11, Amer. Math. Soc., Providence, R.I., 1964. MR **28** #3956.

25. B. V. Levin, Trudy Tashkent Gos. Univ. **228** (1963), 56–68.

26. ———, *Distribution of "new primes" in polynomial sequences*, Mat. Sb. **61** (**103**) (1963), 389–407. (Russian) MR **30** #1991.

27. ———, *A one-dimensional sieve*, Acta Arith. **10** (1964/65), 387–397. (Russian) MR **31** #4774.

28. J. H. van Lint and H.-E. Richert, *On primes in arithmetic progressions*, Acta Arith. **11** (1965), 209–216. MR **32** #5613.

29. R. J. Miech, *Almost primes generated by a polynomial*, Acta Arith. **10** (1964/65), 9–30. MR **29** #1174.

30. ———, *Primes, polynomials and almost primes*, Acta Arith. **11** (1965), 35–56. MR **31** #3390.

31. H. L. Montgomery, *A note on the large sieve*, J. London Math. Soc. **43** (1968), 93–98. MR **37** #184.

32. K. Pracher, *Primzahlverteilung*, Springer-Verlag, Berlin and New York, 1957. MR **19**, 393.

33. A. Rényi, *On the representation of an even number as the sum of a prime and of an almost prime*, Izv. Akad. Nauk SSSR Ser. Mat. **12** (1948), 57–78; English transl., Amer. Math. Soc. Transl. (2) **19** (1962), 299–321. MR **9**, 413; MR **24** #A1264.

34. G. Ricci, *Sull'andamento della differenza di numeri primi consecutivi*, Rivista Mat. Univ. Parma **5** (1954), 3–54. MR **16**, 675.

35. H.-E. Richert, *Selberg's sieve with weights*, Mathematika **16** (1969), 1–22.

36. G. Rieger, *On polynomials and almost-primes*, Bull. Amer. Math. Soc. **75** (1969), 100–103. MR **38** #2104.

37. A. Schinzel, *Remarks on the paper "Sur certaines hypothèses concernant les nombres premiers*, Acta Arith. **7** (1961/62), 1–8. MR **24** #A70.

38. ———, *A remark on a paper of Bateman and Horn*, Math. Comp. **17** (1963), 445–447. MR **27** #3609.

39. A. Schinzel and W. Sierpiński, *Sur certaines hypothèses concernant les nombres premiers*, Acta Arith. **4** (1958), 185–208; erratum **5** (1959), 259. MR **21** #4936.

40. W. Schwarz, *Weitere, mit einer Methode von Erdös-Prachar erzielte Ergebnisse*, Math. Nachr. **23** (1961), 327–348. MR **25** #3004.

41. A. Selberg, *On an elementary method in the theory of primes*, Norske Vid. Selsk. Forh., Trondhjem **19** (1947), no. 18, 64–67. MR **9**, 271.

42. ———, *On elementary methods in prime number-theory and their limitations*, Den 11te Skandinaviske Matematikerkongress, Trondheim, 1949, pp. 13–22. MR **14**, 726.

43. ———, *The general sieve-method and its place in prime number theory*, Proc. Internat. Congress Math. (Cambridge, Mass., 1950) vol. 1, Amer. Math. Soc., Providence, R.I., 1952, pp. 286–292. MR **13**, 438.

44. ———, *Twin prime problem* (unpublished).

45. S. Uchiyama, *On the representation of large even integers as sums of a prime and an almost prime.* II, Proc. Japan Acad. **43** (1967), 567–571. MR **37** #179.

46. A. I. Vinogradov, *The density hypothesis for Dirichlet L-series*, Izv. Akad. Nauk SSSR Ser. Mat. **29** (1965), 903–934; correction, ibid., **30** (1966), 719–720; English transl., Amer. Math. Soc. Transl. (2) **82** (1969), 9–46. MR **33** #5579.

47. Y. Wang, *On some properties of integral valued polynomials*, Advance Math. **3** (1957), 416–423. (Chinese) MR **20** #4531.

48. ———, *On sieve methods and some of their applications*, Sci. Sinica **11** (1962), 1607–1624. MR **26** #3685.

49. E. Wirsing, *Über die Zahlen, daren Primteiler einer gegebenen Menge angehören*, Arch. Math. **7** (1956), 263–272. MR **18,** 642.

50. ———, *Das asymptotische Verhalten von Summen über multiplikative Funktionen*, Math. Ann. **143** (1961), 75–102. MR **24** #A1241.

51. ———, *Das asymptotische Verhalten von Summen über multiplikative Funktionen.* II, Acta Math. Acad. Sci. Hungar. **18** (1967), 411–467. MR **36** #6366.

MATHEMATISCHES INSTITUT DER UNIVERSITÄT MARBURG
MARBURG DER LAHN, FEDERAL REPUBLIC OF GERMANY

SIEVE METHODS

ATLE SELBERG

The main purpose of these lectures was to give a fuller exposition of the theory of the general sieve, which was already indicated to some extent in my lecture at the International Congress at Harvard in 1950 (though the most important results in §4 were proved in 1951). I hoped at that time that I might be able to find some practical procedure or algorithm for computation of the functions $T_k^+(\alpha)$ and $T_k^-(\alpha)$, or even perhaps to find some reasonably explicit formula for these functions. (I don't consider the expressions in Theorem 2 in §4 to be such!) I was, however, never able to do this, and partly for this and partly for other reasons I never came back to the subject in print. When I lectured on the sieve method, as I have done occasionally, this material always seemed rather awkward to handle, since it did not seem easy to develop a good system of notation and terminology, and so it always was simpler to concentrate on the special sieve procedures and applications to particular problems.

Since my Harvard lecture, nothing has appeared in the literature about the general sieve (the title of the paper [2] (see Bibliography at end) is quite misleading, it deals with a specific sieve procedure). Although quite a number of papers have appeared since then dealing with specific sieve procedures, these deal mostly with particular applications (and mostly contain little new in this direction too).

Of real interest are the papers of Ankeny [1], Ankeny-Onishi [2], and Jurkat-Richert [4]. Very important is also the unpublished work of Barkley Rosser, who first settled the problem of the 1-residue sieve, and thus anticipated the work of Jurkat-Richert by about a decade.

The part of §5 which deals with the Buchstab-Rosser sieve, refers to work done in the late 1950's after I had had the opportunity to see an unpublished manuscript by Rosser. Apart from that, the contents of these lectures (though not the notation nor the arrangement of the proofs, both of which have been revised for this occasion) date from the years 1946–51.

As I may never return to this subject, I have tried to make §4 complete enough that the missing details should be easy to fill in.

Since the "large sieve" is not really a sieve in the sense considered here, I found it best to leave it out entirely on this occasion.

1. Let S be a finite set of integers n, with each of which is associated a "weight" $w_n \geq 0$, let further P be a finite set of primes p and denote by (P) the set of all positive square free integers whose prime factors all belong to P. In the following we shall usually denote members of the set (P) by the letter d. By (n, P) we understand the largest member of (P) that divides n.

We shall be concerned with the problem of determining what bounds can be obtained for the quantity

$$M(S, P) = \sum_{(n,P)=1} w_n, \tag{1.1}$$

from information we may possess about all or some of the quantities

$$N_d = N_d(S) = \sum_{d|n} w_n, \tag{1.2}$$

where $d \in (P)$.

Möbius formula gives

$$M(S, P) = \sum_{d \in (P)} \mu(d) N_d, \tag{1.3}$$

where μ is the Möbius function.

We assume that our information about the quantities N_d is of the form

$$N_d^- \leq N_d \leq N_d^+, \tag{1.4}$$

where we assume that the quantities N_d^- and N_d^+ are such that there exist sets S and nonnegative weights w_n, such that the inequalities (1.4) are satisfied. We consider the class of all such S and wish to determine lower and upper bounds for $M(S, P)$.

From (1.3) we have clearly

$$M(S, P) \leq \sum_{d \in (P)} \mu(d) N_d^{\operatorname{sgn} \mu(d)}, \tag{1.5}$$

and

$$M(S, P) \geq \sum_{d \in (P)} \mu(d) N_d^{\operatorname{sgn} -\mu(d)}. \tag{1.5'}$$

These bounds, however, are usually much too crude to be of any use or interest, because of the large number of terms involved if the set P is large. To counteract this, we choose to replace the $\mu(d)$ in (1.3) by some other set of numbers λ_d defined for $d \in (P)$ in such a way that we get an inequality instead of an equation. We shall refer to such a set of λ_d by the symbol Λ or $\Lambda(P)$.

If Λ is such that for all $d_1 \in (P)$ we have

$$\sum_{d|d_1} \lambda_d \geq \sum_{d|d_1} \mu(d), \tag{1.6}$$

we use the notation Λ^+ or $\Lambda^+(P)$.

It is easily seen that for a $\Lambda^+(P)$ we have

$$(1.7)\qquad M(S, P) \leq \sum_{d \in (P)} \lambda_d N_d^{\operatorname{sgn} \lambda_d}.$$

Choosing now our $\Lambda^+(P)$ so as to make the right-hand side of (1.7) as small as possible, we get

$$(1.8)\qquad M(S, P) \leq M^+(S, P) = \min_{\Lambda^+(P)} \sum_{d \in (P)} \lambda_d N_d^{\operatorname{sgn} \lambda_d}.$$

Similarly, if we denote by $\Lambda^- = \Lambda^-(P)$ a set of λ_d such that

$$(1.9)\qquad \sum_{d \mid d_1} \lambda_d \leq \sum_{d \mid d_1} \mu(d),$$

for all $d_1 \in (P)$, then we have

$$(1.10)\qquad M(S, P) \geq M^-(S, P) = \max_{\Lambda^-(P)} \sum_{d \in (P)} \lambda_d N_d^{\operatorname{sgn} - \lambda_d}.$$

The bounds $M^+(S, P)$ and $M^-(S, P)$ defined by (1.8) and (1.10) are actually the best possible bounds for $M(S, P)$. That is to say, if the class of S satisfying (1.4) is nonempty, it contains sets S for which these bounds are attained.

This fact can be shown by relatively simple convexity considerations. If we consider, with P fixed, $M(S, P)$ and the quantities $N_d^-(S)$ and $N_d^+(S)$ as coordinates of a point in a euclidean space of sufficiently many dimensions, these points will fill a convex region. A convex region is completely determined by the linear inequalities one gets by considering the supporting planes. If we designate the positive direction on the axis for the coordinate $M(S, P)$ as "up," the inequalities (1.8) and (1.10) correspond respectively to the supporting planes lying above this convex region, and those lying below.

It is also seen that if some part of the information (1.4) is redundant (that is to say, for instance, it could be derived from the fact that $N_{d_1} \leq N_{d_2}^+$ if $d_2 \mid d_1$, or that $N_{d_1} \geq N_{d_2}^-$ if $d_1 \mid d_2$, or that $N_d \geq 0$ for all $d \in (P)$), then the minimum in (1.8) or the maximum in (1.10) is attained with a Λ^+ or a Λ^- such that the redundant $N_{d_1}^+$ or $N_{d_1}^-$ does not enter in the expression for $M^+(S, P)$ or $M^-(S, P)$.

If we refer to a Λ^+ for which the extremal value in (1.8), or a Λ^- for which the extremal value in (1.10) is reached as being "optimal" (it is not necessarily unique), this means in particular that if both $N_{d_1}^-$ and $N_{d_1}^+$ are redundant, there exist optimal Λ^+ and Λ^- with $\lambda_{d_1} = 0$.

For a Λ^+ we have $\lambda_1 \geq 1$ and for a Λ^- that $\lambda_1 \leq 1$. It is obvious that for an optimal Λ^+ we must always have $\lambda_1 = 1$. Also, if our $M^-(S, P) > 0$, an optimal Λ^- must again have $\lambda_1 = 1$. We may, therefore, in the following always assume $\lambda_1 = 1$ for our Λ without any real loss. With the convention $\lambda_1 = 1$ it may happen that the expression

$$\max_{\Lambda^-(P)} \sum_{d \in (P)} \lambda_d N_d^{\operatorname{sgn} - \lambda_d}$$

becomes negative. This means that the real $M^-(S, P) = 0$. If we relax our original requirement $w_n \geq 0$ for all n in s and require $w_n \geq 0$ only for $(n, P) > 1$, then the negative value of the above expression still would retain its meaning as the best lower bound for $M(S, P)$.

We introduce the notation

$$M^{=}(S, P) = \max_{\Lambda^{-}(P)} \sum_{d \in (P)} \lambda_d N_d^{\operatorname{sgn} - \lambda_d}, \tag{1.10'}$$

where Λ^{-} has $\lambda_1 = 1$, and have

$$M^{-}(S, P) = \max (0, M^{=}(S, P)).$$

2. The determination of $M^{+}(S, P)$ and $M^{-}(S, P)$ from the given data (1.4) is the basic problem of the general sieve. In the following we make our assumptions more definite in order to develop a theory that is general enough to cover the applications we have in mind, but not so general that no significant results can be proved.

We now consider sets S that depend on a parameter x, and P that also will depend on x in some fashion. Usually we will have $P = P(\xi)$ the set of all primes $p \leq \xi$, or $P = P(\xi_1, \xi_2)$, the set of all primes p with $\xi_1 < p \leq \xi_2$; here the ξ are certain functions of x. We wish to investigate what happens to $M^{+}(S, P)$ and $M^{-}(S, P)$ (or $M^{=}(S, P)$) when the parameter x tends to infinity, if we assume that our information (1.4) is now of the form

$$N_d = \frac{x}{f(d)} + O(u(d)),[1] \tag{2.1}$$

where

$$f(d) = d/u(d), \tag{2.2}$$

and $u(d)$ is a multiplicative function with the property that

$$0 \leq u(p) < p,[2] \tag{2.3}$$

for all p, and that (this condition could be relaxed considerably)

$$u(p) < \mathfrak{C}, \tag{2.4}$$

for all p and some large positive constant $\mathfrak{C}$. For convenience, we also require[3] that if $u(p) \neq 0$, we have $u(p) \geq 1$.

---

[1]We shall see later how the results are affected by a relaxation of the requirements on the error term in (2.1). We could actually have handled a somewhat more general main term in (2.1), namely

$$\frac{x}{f(d)} \frac{\partial^r[(x/d)^s v(s, d)]}{\partial s^r} \qquad (s = 0),$$

where r is a fixed integer ≥ 0, $v(s, d)$ a multiplicative function of d, depending analytically on s in such a way that $v(0, d) = 1$. This form of the main term does actually occur in some applications of the sieve method.

[2]If $u(p) = 0$, we put $f(p) = \infty$ and $1/f(p) = 0$.

[3]This requirement could be dropped entirely but some arguments would become rather more complicated. If we modified the remainder term in (2.1) to $O(u^*(d))$, where $u^*(d)$ is the multiplicative function defined by $u^*(p) = \max(u(p), 1)$ much of this complication could be avoided. In nearly all applications $u(p)$ is an integer so that $u(p) \geq 1$ if $u(p) \neq 0$.

From (1.8) and (1.10′) we get, writing now $M_u(x, P)$ for $M(S, P)$ and $M_u^+(x, P)$, $M_u^-(x, P)$ and $M_u^=(x, P)$

$$(2.5)\qquad M_u^+(x, P) = \min_{\Lambda^+(P)} \left\{ x \sum_{d\in(P)} \frac{\lambda_d}{f(d)} + O\left(\sum_{d\in(P)} u(d)|\lambda_d| \right) \right\},$$

and

$$(2.5')\qquad M_u^=(x, P) = \max_{\Lambda^-(P)} \left\{ x \sum_{d\in(P)} \frac{\lambda_d}{f(d)} - O\left(\sum_{d\in(P)} u(d)|\lambda_d| \right) \right\}.$$

Introducing the notation

$$(2.6)\qquad \theta_d = \sum_{d'|d} \lambda_{d'},$$

for $d \in (P)$, we have $\theta_1 = 1$, $\theta_d \geq 0$ for $d > 1$ in case of a $\Lambda^+(P)$, and $\theta_1 = 1$, $\theta_d \leq 0$ for $d > 1$, in case of a $\Lambda^-(P)$.

It is easily shown that

$$(2.7)\qquad \sum_{d\in(P)} \frac{\lambda_d}{f(d)} = \prod_{p\in P} \left(1 - \frac{u(p)}{p}\right) \sum_{d\in(P)} \frac{\theta_d}{f'(d)},$$

where $f'(d)$ is the multiplicative function defined by $f'(p) = f(p) - 1$.

Since our interest is mainly in what happens when $x \to \infty$, we enlarge the concept of an optimal sieve defined in §1, and consider now any sieve $\Lambda^+(P)$ or $\Lambda^-(P)$ which gives a bound which differs from $M_u^+(x, P)$ or $M_u^=(x, P)$ by a quantity of order less than

$$(2.8)\qquad x \prod_{p\in P} \left(1 - \frac{u(p)}{p}\right),$$

as $x \to \infty$, to be optimal in the wider sense.

If P is such that the largest prime in P is $\leq x^a$ with some constant a, then (2.1) and (2.2) together with our assumptions about the $u(p)$, show that the information given about N_d is redundant (in the sense described in §1) when $d > x^{1+a}$, since then d contains a proper divisor $d_1 > x$, and the inequality $0 \leq N_d \leq N_{d_1}$ holds. This implies that there are optimal sieves $\Lambda^+(P)$ and $\Lambda^-(P)$ for which we have $\lambda_d = 0$ for all $d > x^{1+a}$.

We introduce the notations

$$(2.9)\qquad T_u(\Lambda) = \sum_{d\in(P)} \frac{\theta_d}{f'(d)},$$

and

$$(2.10)\qquad R_u(\Lambda) = \sum_{d\in(P)} u(d)|\lambda_d|.$$

If P is a subset of $P(x^a)$ for some fixed constant a, the specific sieve procedures we shall discuss later in §5, enable us to produce a $\Lambda^+(P)$ for which $T_u(\Lambda^+)$ is bounded as $x \to \infty$, while at the same time $R_u(\Lambda^+)$ is of order less than (2.8).

Similarly, if $a < 1$,[4] the procedures given in §5 enable us to produce a $\Lambda^-(P)$ for which $T_u(\Lambda^-)$ is bounded while again $R_u(\Lambda^-)$ is of order less than (2.8) as $x \to \infty$. We may, therefore, without loss restrict our attention to sieves for which $T_u(\Lambda)$ is bounded as $x \to \infty$ and $R_u(\Lambda)$ not growing faster than (2.8).

We shall in the following denote a sieve $\Lambda(P)$ for which

$$R_u(\Lambda) \leq Z, \tag{2.11}$$

by the symbol $\Lambda_u(P, Z)$, a sieve $\Lambda(P)$ for which we have $\lambda_d = 0$ for $d > Z$, we shall denote by $\Lambda[P, Z]$.

We have from (2.6) if $d \in (P)$

$$\lambda_d = \sum_{d'|d} \mu\left(\frac{d}{d'}\right) \theta_{d'}$$

so that

$$|\lambda_d| \leq \sum_{d'|d} |\theta_{d'}|.$$

From this we can deduce that

$$\sum_{d\in(P)} \frac{|\lambda_d|}{f'(d)} \leq \prod_{p\in P} \frac{1}{1 - u(p)/p} \sum_{d\in(P)} \frac{|\theta_d|}{f'(d)}. \tag{2.12}$$

For a Λ^+ we have

$$\sum_{d\in(P)} \frac{|\theta_d|}{f'(d)} = T_u(\Lambda^+),$$

and for a Λ^-

$$\sum_{d\in(P)} \frac{|\theta_d|}{f'(d)} = 2 - T_u(\Lambda^-).$$

Thus for a $\Lambda^+[P, Z]$ or a $\Lambda^-[P, Z]$ with $T_u(\Lambda)$ bounded, we must have, using (2.4)

$$\begin{aligned} R_u(\Lambda) &\leq Z \sum_{d\in(P)} \frac{|\lambda_d|}{f(d)} \\ &< Z \prod_{p\in P} \frac{1}{1 - u(p)/p} (2 + |T_u(\Lambda)|) = O(Z (\log Z)^{e-1}). \end{aligned} \tag{2.13}$$

Thus such a $\Lambda[P, Z]$ is also a $\Lambda_u(P, Z_1)$ with a Z_1 that is not much larger than Z.

Since the class $\Lambda[P, Z]$ is simpler to work with than $\Lambda_u(P, Z)$, one would have liked to have an argument showing that an optimal sieve which is a $\Lambda_u(P, Z)$ with Z of the order of (2.8) or less, also is a $\Lambda[P, Z_1]$ with a Z_1 that is not much larger

---

[4] Under more restrictive conditions on u, namely

$$\limsup_{x\to\infty} \frac{1}{\log x} \sum_p \frac{u(p) \log p}{p} < 1,$$

this holds also for $a = 1$.

than Z. This unfortunately is not true, but we shall see in §4 that with certain new assumptions about the regularity of the distribution of the $u(p)$, we can prove that there do exist sieves $\Lambda[P, Z]$ where Z is of less order than (2.8) which are optimal for our two problems (2.5) and (2.5′).

3. Before we can proceed in our investigation of $M_u^+(x, P)$ and $M_u^-(x, P)$ as defined by (2.5) and (2.5′), we need to give an exposition of certain general principles used in constructing, modifying and improving sieves Λ. These will be used repeatedly in the arguments in §4, as well as in connection with the special sieve procedures dealt with in §5.

I. Suppose we have two sets of primes P_1 and P_2 whose intersection is empty, and two sieves

$$\Lambda_1^+ = \Lambda_u^+(P_1, Z_1) \quad \text{and} \quad \Lambda_2^+ = \Lambda_u^+(P_2, Z_2).$$

We write P for the union of P_1 and P_2, and $Z = Z_1 Z_2$. We now construct a sieve $\Lambda_u^+(P, Z)$ as follows:

Each $d \in (P)$ can be factorized uniquely $d = d_1 d_2$ where $d_1 \in (P_1)$ and $d_2 \in (P_2)$. Define now

$$\lambda_d = \lambda_{d_1}^{(1)} \cdot \lambda_{d_2}^{(2)}, \tag{3.1}$$

where $\lambda^{(1)}$ and $\lambda^{(2)}$ belong to Λ_1^+ and Λ_2^+ respectively. We get

$$\theta_d = \theta_{d_1}^{(1)} \cdot \theta_{d_2}^{(2)}. \tag{3.2}$$

From this we see that

$$T_u(\Lambda^+) = T_u(\Lambda_1^+) T_u(\Lambda_2^+). \tag{3.3}$$

The result holds unchanged if we use the same construction for two $\Lambda^+[P_1, Z_1]$ and $\Lambda^+[P_2, Z_2]$, we get a $\Lambda^+[P, Z]$ for which (3.3) holds.

If we have a $\Lambda_1^+ = \Lambda_u^+(P_1, Z_1)$, a $\Lambda_1^- = \Lambda_u^-(P_1, Z_1)$, a $\Lambda_2^+ = \Lambda_u^+(P_2, Z_2)$ and a $\Lambda_2^- = \Lambda_u^-(P_2, Z_2)$, we can construct in a similar but slightly more complicated way a $\Lambda_u^-(P, 3Z)$ by defining

$$\lambda_d = \lambda_{d_1}^{(1)+} \lambda_{d_2}^{(2)-} + \lambda_{d_1}^{(1)-} \lambda_{d_2}^{(2)+} - \lambda_{d_1}^{(1)+} \lambda_{d_2}^{(2)+}, \tag{3.4}$$

where $\lambda^{(1)+}$, $\lambda^{(1)-}$, $\lambda^{(2)+}$ and $\lambda^{(2)-}$ are taken from Λ_1^+, Λ_1^-, Λ_2^+ and Λ_2^- respectively. We then get

$$T_u(\Lambda^-) = T_u(\Lambda_1^-) T_u(\Lambda_2^-) - \{T_u(\Lambda_1^+) - T_u(\Lambda_1^-)\} \cdot \{T_u(\Lambda_2^+) - T_u(\Lambda_2^-)\}. \tag{3.5}$$

If we use the same construction for sieves $\Lambda^+[P_1, Z_1]$, $\Lambda^-[P_1, Z_1]$, $\Lambda^+[P_2, Z_2]$ and $\Lambda^-[P_2, Z_2]$ a $\Lambda^-[P, Z]$ is produced, and (3.5) holds unchanged.

Principle I is mostly used to get rid of the small primes in P, in that say $P = P(x^\alpha)$ is considered as the union of $P_1 = P(x^\eta)$ and $P_2 = P(x^\eta, x^\alpha)$ and with $Z_1 = x^{\sqrt{\eta}}$ and $Z_2 = Zx^{-\sqrt{\eta}}$ where η is a small positive number. As we shall see from (5.16) and (5.18) proved in §5 we can then find sieves Λ_1^+ and Λ_1^- such that $T_u(\Lambda_1^+)$ and $T_u(\Lambda_1^-)$ both are very close to 1 for large x if η is chosen small enough. (3.3) and (3.5) then essentially imply that $T_u(\Lambda^+)$ and $T_u(\Lambda^-)$ are very close to $T_u(\Lambda_2^+)$ and

$T_u(\Lambda_2^-)$ respectively. Getting rid of the initial section of $P(x^\alpha)$ will make many steps in our proofs much simpler to justify.

Principle I can be modified or generalized to hold in the case when the two sets P_1 and P_2 have elements in common and we may have two functions $u_1(p)$ and $u_2(p)$ and $u(p) = u_1(p) + u_2(p)$ in the intersection of P_1 and P_2 while otherwise $u(p) = u_1(p)$ for $p \in P_1$ and $u(p) = u_2(p)$ for $p \in P_2$. Instead of (3.3) and (3.5) we then get inequalities giving an upper bound for $T_u(\Lambda^+)$ and a lower bound for $T_u(\Lambda^-)$.

II. We now describe a procedure which under certain conditions allows us to improve on a sieve.

Suppose we have a $\Lambda_1 = \Lambda_u^+(P, Z)$ and that for some $d_0 \in (P)$, $d_0 > 1$, we have $\theta_{d_0}^{(1)} > 0$. Denote by $P - d_0$ the set of primes remaining when we remove the prime factors of d_0 from P. If we can find a sieve $\Lambda_2 = \Lambda_u^-(P - d_0, Z/d_0)$, we may define a new sieve $\Lambda = \Lambda_u^+(P, Z')$, as follows:

$$\lambda_d = \lambda_d^{(1)} - \theta_{d_0}^{(1)}\lambda_{d/d_0}^{(2)}, \tag{3.6}$$

for $d \in (P)$, here $\lambda_{d/d_0}^{(2)}$ is interpreted as zero if d_0 does not divide d, and

$$Z' = Z\left(1 + \frac{u(d_0)}{d_0}\,\theta_{d_0}^{(1)}\right). \tag{3.7}$$

If $T_u(\Lambda_2^-) > 0$, the new sieve represents an improvement in that

$$T_u(\Lambda^+) < T_u(\Lambda_1^+),$$

since we have

$$T_u(\Lambda^+) = T_u(\Lambda_1^+) - \frac{\theta_{d_0}^{(1)}}{f'(d_0)}\,T_u(\Lambda_2^-). \tag{3.8}$$

If we had a $\Lambda_1 = \Lambda^+[P, Z]$ and a $\Lambda_2 = \Lambda^-[P - d_0, Z/d_0]$, (3.6) will define a $\Lambda = \Lambda^+[P, Z]$ for which again (3.8) holds true.

For sieves $\Lambda_1 = \Lambda_u^-(P, Z)$ or $\Lambda_1 = \Lambda^-[P, Z]$ with a $\theta_{d_0}^{(1)} < 0$, a similar construction is possible and all statements hold, only $|\theta_{d_0}^{(1)}|$ replaces $\theta_{d_0}^{(1)}$ in (3.7) and (3.8) takes the form

$$T_u(\Lambda^-) = T_u(\Lambda_1^-) - \frac{\theta_{d_0}^{(1)}}{f'(d_0)}\,T_u(\Lambda_2^-), \tag{3.8$'$}$$

so that $T_u(\Lambda^-) > T_u(\Lambda_1^-)$ if $T_u(\Lambda_2^-) > 0$.

III. There is a related procedure that allows one to produce a sieve Λ^+ or Λ^- out of a Λ_1 for which some of the $\theta_d^{(1)}$ where $d \in (P)$ with not too large d, have the wrong sign.

To illustrate this, let us assume that we have $\Lambda_1 = \Lambda_u(P, Z)$ which is so that for a certain $d_0 \in (P)$ with $1 < d_0 < Z$, we have $\theta_{d_0}^{(1)} > 0$, while for the other $d > 1$ we have $\theta_d^{(1)} \leq 0$. By aid of a $\Lambda_2 = \Lambda_u^+(P - d_0, Z/d_0)$ we can then produce a $\Lambda = \Lambda_u^-(P, Z')$, by writing

$$\lambda_d = \lambda_d^{(1)} - \theta_{d_0}^{(1)}\,\lambda_{d/d_0}^{(2)}, \tag{3.9}$$

(subject to the same interpretation as (3.6)). Here again

$$Z' = Z\left(1 + \frac{u(d_0)}{d_0}\,\theta_{d_0}^{(1)}\right).$$

Again, if we replace Λ_1 and Λ_2 by a $\Lambda[P, Z]$ and a $\Lambda^+[P - d_0, Z/d_0]$, our construction gives a $\Lambda^-[P, Z]$.

In the same way we could produce a Λ^+ from a Λ_1 if all the $\theta_d^{(1)}$ are ≥ 0 for $d \in (P)$, except for $d = d_0$, for which $\theta_{d_0}^{(1)} < 0$.

By repeated application we could also take care of cases where several θ_d have the wrong sign.

The primary use of this procedure is to extend the range P of a sieve.

IV. It is clear that when two expressions of the form $\sum_{d|n} \lambda_d$ with the same n but possibly different Λ are multiplied, the product can again be interpreted as an expression of the same kind. From Λ_1 and Λ_2 we may thus form Λ by defining

$$\lambda_d = \sum_{d_1d_2/(d_1,d_2)=d} \lambda_{d_1}^{(1)}\lambda_{d_2}^{(2)}.$$

If Λ_1 is a $\Lambda^+[P, Z_1]$, Λ_2 a $\Lambda^+[P, Z_2]$, then Λ is a $\Lambda^+[P, Z_1Z_2]$. If Λ_2 is instead a $\Lambda^-[P, Z_2]$, we have that Λ is a $\Lambda^-[P, Z_1Z_2]$.

If $\Lambda_1 = \Lambda_2 = \Lambda_1[P, Z]$, then we have that Λ is a $\Lambda^+[P, Z^2]$.

Similar statements hold if we let $\Lambda_1 = \Lambda_u(P, Z_1)$ and $\Lambda_2 = \Lambda_u(P, Z_2)$ but these have hardly any application.

4. Throughout this paragraph we shall assume that we have in addition to the earlier conditions on $u(p)$ that

$$V_u(x) = \sum_{p \leq x} \frac{u(p)}{p} \log p \sim k \log x, \tag{4.1}$$

as $x \to \infty$.[5] Here k is a positive constant.

We now return to (2.5) and (2.5′) and consider the case that $P = P(x^\alpha)$ where α is a number with $0 \leq \alpha \leq 1$. We shall write $M_u(x, x^\alpha)$ for $M_u(x, P)$, and write

$$W_u(x) = \prod_{p \leq x} \left(1 - \frac{u(p)}{p}\right). \tag{4.2}$$

We shall be concerned with what happens to the ratios

$$\frac{M_u^+(x, x^\alpha)}{xW_u(x^\alpha)}, \tag{4.3}$$

and

$$\frac{M_u^-(x, x^\alpha)}{xW_u(x^\alpha)}, \tag{4.3′}$$

[5]If we wanted to keep better track of remainder terms we would require, say,

$$V_u(x) = k \log x + O(1).$$

as $x \to \infty$. The specific sieve procedures given in §5 show us that these two ratios remain bounded as $x \to \infty$ if $\alpha < 1$ and the first ratio also for $\alpha = 1$. Our principal aim in this section is to show that *the ratios* (4.3) *and* (4.3′) *as* $x \to \infty$ *tend to limits that depend only on* k *and* α, and to study these limits as functions of k and α, and in particular to show that *these functions can be effectively computed.*

We define now

$$\liminf_{x\to\infty} T_u(\Lambda^+) = T_u^+(\alpha) \tag{4.4}$$

where $\Lambda^+ = \Lambda_u^+(P(x^\alpha), x)$, and

$$\limsup_{x\to\infty} T_u(\Lambda^-) = T_u^-(\alpha), \tag{4.4′}$$

where $\Lambda^- = \Lambda_u^-(P(x^\alpha), x)$.

We furthermore define, for $0 < \eta < \alpha \leq 1$, $\Lambda^+ = \Lambda_u^+(P(x^\eta, x^\alpha), x)$,

$$\liminf_{x\to\infty} T_u(\Lambda^+) = T_u^+(\eta, \alpha), \tag{4.4″}$$

and with $\Lambda^- = \Lambda_u^-(P(x^\eta, x^\alpha), x)$

$$\limsup_{x\to\infty} T_u(\Lambda^-) = T_u^-(\eta, \alpha). \tag{4.4‴}$$

From (4.1) it follows easily that for $0 < \eta < \alpha \leq 1$ we have

$$\lim_{x\to\infty} \frac{W_u(x^\alpha)}{W_u(x^\eta)} = \left(\frac{\eta}{\alpha}\right)^k. \tag{4.5}$$

We wish to establish the following lemma:

Lemma 1. *$T_u^+(\alpha)$ is continuous in α for $0 \leq \alpha \leq 1$, and $T_u^+(\eta, \alpha)$ in both η and α for $0 < \eta < \alpha \leq 1$. $T_u^-(\alpha)$ is continuous for $0 \leq \alpha < 1$, and $T_u^-(\eta, \alpha)$ for $0 < \eta < \alpha \leq 1$. For $k < 1$, $T_u^-(\alpha)$ is continuous also at $\alpha = 1$.*

We sketch the proof for $T_u^+(\alpha)$ and $T_u^-(\alpha)$.

Let $0 < \alpha_1 < \alpha_2 \leq 1$. From the definition of $T_u^+(\alpha_2)$ it follows that there is a sequence of x tending to infinity and of $\Lambda_{\alpha_2}^+ = \Lambda_u^+(P(x^{\alpha_2}), x)$ such that $\lim T_u(\Lambda_{\alpha_2}^+) = T_u^+(\alpha_2)$, when x tends to infinity through this sequence. If we denote by $\Lambda_{\alpha_1}^+$ the restriction of $\Lambda_{\alpha_2}^+$ to the smaller range $P(x^{\alpha_1})$, this is clearly a $\Lambda_u^+(P(x^{\alpha_1}), x)$ and we have obviously $T_u(\Lambda_{\alpha_1}^+) \leq T_u(\Lambda_{\alpha_2}^+)$. From this follows

$$T_u^+(\alpha_1) \leq T_u^+(\alpha_2). \tag{4.6}$$

Again from (4.4) follows that there exists a sequence of x tending to infinity and of $\Lambda_{\alpha_1}^+ = \Lambda_u^+(P(x^{\alpha_1}), x)$ such that $\lim T_u(\Lambda_{\alpha_1}^+) = T_u^+(\alpha_1)$, when x tends to infinity through this sequence. Let us denote by $\Lambda_{\alpha_2}^+$ the $\Lambda_u^+(P(x^{\alpha_2}), x)$ which is just $\Lambda_{\alpha_1}^+$ used over the larger range $P(x^{\alpha_2})$ (the inequalities (1.6) would still be satisfied for the $d_1 \in (P(x^{\alpha_2}))$). From (2.9) it is easily seen that we have

$$T_u(\Lambda_{\alpha_1}^+) \prod_{x^{\alpha_1} < p \leq x^{\alpha_2}} \left(1 + \frac{1}{f'(p)}\right) = T_u(\Lambda_{\alpha_2}^+),$$

or

$$T_u(\Lambda_{\alpha_1}^+) \frac{W_u(x^{\alpha_1})}{W_u(x^{\alpha_2})} = T_u(\Lambda_{\alpha_2}^+).$$

Letting $x \to \infty$ through the sequence mentioned above, we get, remembering (4.5), that $T_u^+(\alpha_1)\alpha_2^k/\alpha_1^k \geq T_u^+(\alpha_2)$ or

$$\alpha_1^{-k} T_u^+(\alpha_1) \geq \alpha_2^{-k} T_u^+(\alpha_2). \tag{4.7}$$

From (4.6) and (4.7) the continuity of $T_u^+(\alpha)$ is clear for $\alpha > 0$. From our result (5.16) it is evident that $T_u^+(\alpha)$ tends to 1 as $\alpha \to 0$ (since $1 \leq T_u^+(\alpha) \leq 1 + 2e^{-1/(10\alpha)}$ for $\alpha \leq 1/(7(k+1))$), since by the definition $T_u^+(0) = 1$, we get continuity also at $\alpha = 0$.

For $T_u^+(\eta, \alpha)$ arguments similar to the above give, if $0 < \eta_2 \leq \eta_1 < \alpha_1 \leq \alpha_2 \leq 1$,

$$T_u^+(\eta_1, \alpha_1) \leq T_u^+(\eta_2, \alpha_2),^6 \tag{4.6'}$$

and

$$\left(\frac{\eta_1}{\alpha_1}\right)^k T_u^+(\eta_1, \alpha_1) \geq \left(\frac{\eta_2}{\alpha_2}\right)^k T_u^+(\eta_2, \alpha_2). \tag{4.7'}$$

From these inequalities the statements about $T_u^-(\eta, \alpha)$ follow easily.

Turning to $T_u^-(\alpha)$, we can show in much the same way as we showed (4.6), that for $0 < \alpha_1 < \alpha_2 < 1$, we have

$$T_u^-(\alpha_1) \geq T_u^-(\alpha_2). \tag{4.8}$$

There seems, however, to be no way to get an inequality that would play a similar role to (4.7), so we have now to proceed differently.

We use principle III of §3. By definition there exists a sequence of $x \to \infty$ and of $\Lambda_1 = \Lambda_{\alpha_1}^- = \Lambda_u^-(P(x^{\alpha_1}), x)$, such that $\lim T_u(\Lambda_{\alpha_1}^-) = \overline{T_u^-}(\alpha_1)$.

We now write $\beta = \min\{(1 - \alpha_2)/(7(k+1)), \alpha_1\}$. From (5.16) we see that for x large enough exists a $\Lambda_2^+ = \Lambda_u^+(P(x^\beta), x^{1-\alpha_2})$, such that $T_u(\Lambda_2^+) < 2$.

We define a sieve

$$\Lambda = \Lambda_{\alpha_2}^- = \Lambda_u^-(P((2x)^{\alpha_2}), 2x),$$

in the following way. For $d \in (P((2x)^{\alpha_2}))$, we put

$$\lambda_d = \lambda_d^{(1)} - \sum_{x^{\alpha_1} < p \leq (2x)^{\alpha_2}} \lambda_{d/p}^{(2)},$$

where $\lambda_d^{(1)}$ and $\lambda_{d/p}^{(2)}$ are from Λ_1 and Λ_2 respectively, and are taken as zero when

---

[6] A slightly refined argument would give this under the weaker assumption $\eta_1 < \alpha_1$, $\eta_2 < \alpha_2$, $\alpha_1 \leq \alpha_2$ and $\eta_1\alpha_2 \geq \eta_2\alpha_1$.

not defined (that is, if d does not belong to $(P(x^{\alpha_1}))$, or if d/p is not an integer or does not belong to $(P(x^\beta))$. For x large, we have

$$R_u(\Lambda) \leq R_u(\Lambda_1) + R(\Lambda_2) \sum_{p \leq (2x)^{\alpha_2}} u(p) < x + x^{1-\alpha_2} 2\mathfrak{C} \frac{(2x)^{\alpha_2}}{\alpha_2 \log 2x} < 2x,$$

so that Λ really is of the form stated. Now

$$T_u(\Lambda_{\alpha_2}^-) = T_u(\Lambda_{\alpha_1}^-) \frac{W_u(x^{\alpha_1})}{W_u((2x)^{\alpha_2})} - \frac{W_u(x^\beta)}{W_u((2x)^{\alpha_2})} T_u(\Lambda_2^+) \sum_{x^{\alpha_1} < p \leq (2x)^{\alpha_2}} \frac{u(p)}{p},$$

letting x tend to infinity through the sequence mentioned above, we get

$$T_u^-(\alpha_2) \geq \left(\frac{\alpha_2}{\alpha_1}\right)^k T_u^-(\alpha_1) - 2k\left(\frac{\alpha_2}{\beta}\right)^k \log \frac{\alpha_2}{\alpha_1}. \tag{4.9}$$

Here we have used (4.5) and, what also follows simply from (4.1), that

$$\sum_{x^{\alpha_1} < p < (2x)^{\alpha_2}} \frac{u(p)}{p} \to k \log \frac{\alpha_2}{\alpha_1},$$

as $x \to \infty$.

Combining (4.9) and (4.8) we get the continuity of $T_u^-(\alpha)$ for $0 < \alpha < 1$. Continuity at $\alpha = 0$ (where $T_u^-(0) = 1$) follows from (5.18), which gives $1 \geq T_u^-(\alpha) \geq 1 - 3e^{-1/(10\alpha)}$, for $0 < \alpha < 1/(7k + 8)$.

That $T_u^-(1)$ is finite for $k < 1$, and that $T_u^-(\alpha)$ is then continuous at $\alpha = 1$, can be shown by using a refinement of the construction of Λ used in the proof of (4.9). Instead of using the same Λ_2^+ for all the p in the interval $x^{\alpha_1} < p \leq (2x)^{\alpha_2}$, we use for $p > \max(\sqrt{x}, x^{1-\alpha_1})$ a $\Lambda_u^+(P(x/p), x/p)$. This gives a sharper form of (4.9) from which the statement follows.

For $T_u^-(\eta, \alpha)$ with $0 \leq \eta < \alpha \leq 1$, we use arguments similar to those used above to prove our statement.

It is also easy to see that in the cases covered by Lemma 1 a measure of continuity can be effectively computed.

Lemma 2. *We have for* $0 < \alpha \leq 1$, *and* $0 < \eta < \min(\alpha, 10^{-5}, 1/200k^2)$,

$$T_u^+(\eta, \alpha) \leq T_u^+(\alpha) \leq (1 - \sqrt{\eta})^{-(k+1)} T_u^+(\eta, \alpha). \tag{4.10}$$

Also we have for $0 < \alpha < 1$, *writing* $\beta = \min\{(1-\alpha)/(7(k+1)), \alpha(1-\sqrt{\eta})\}$, *that*

$$\begin{aligned} T_u^-(\eta, \alpha) \geq T_u^-(\alpha) \geq (1 - \sqrt{\eta})^{-k} T_u^-(\eta, \alpha) \\ - (k+1)\eta^{-k-1} e^{-1/(20\sqrt{\eta})} - 3k\sqrt{\eta}(\alpha/\beta)^k. \end{aligned} \tag{4.10'}$$

Before sketching the proof, let us remark that by using the sharper form of (4.9) referred to above, we could obtain for $k < 1$ a version of (4.10′) which would hold also for $\alpha = 1$.

Essentially this lemma shows that if we define $T_u^+(0, \alpha) = T_u^+(\alpha)$ and $T_u^-(0, \alpha) = T_u^-(\alpha)$, then $T_u^+(\eta, \alpha)$ and $T_u^-(\eta, \alpha)$ are continuous also at $\eta = 0$.

The first part of (4.10) is clear. Next, we have, because of (4.7), that

$$(4.11) \qquad T_u^+(\alpha) \leq (1 - \sqrt{\eta})^{-k} T_u^+(\alpha(1 - \sqrt{\eta})).$$

We are going to use principle I of §3, with $P_1 = P(x^\eta)$, $Z_1 = x^{\sqrt{\eta}/2}$ and $P_2 = P(x^\eta, x^{\alpha(1-\sqrt{\eta})})$, $Z_2 = x^{1-\sqrt{\eta}}$.

By definition, there exists a sequence of $x \to \infty$ and of

$$\Lambda_2^+ = \Lambda_u^+(P(x^\eta, x^{\alpha(1-\sqrt{\eta})}), x^{1-\sqrt{\eta}}),$$

such that

$$\lim T_u(\Lambda_2^+) = T_u^+\left(\frac{\eta}{1 - \sqrt{\eta}}, \alpha\right).$$

Also by (5.16) there exists a $\Lambda_1^+ = \Lambda_u^+(P(x^\eta), x^{\sqrt{\eta}/2})$ such that $T_u(\Lambda_1^+) < 1 + 2e^{-1/(20\sqrt{\eta})}$, for x large enough. By the construction (3.1) we now obtain a

$$\Lambda^+ = \Lambda_u^+(P(x^{\alpha(1-\sqrt{\eta})}), x)$$

such that

$$T_u(\Lambda^+) < (1 + 2e^{-1/(20\sqrt{\eta})}) T_u(\Lambda_2^+);$$

letting now x tend to infinity through the sequence mentioned above, we get

$$T_u^+(\alpha(1 - \sqrt{\eta})) \leq (1 + 2e^{-1/(20\sqrt{\eta})}) T_u^+\left(\frac{\eta}{1 - \sqrt{\eta}}, \alpha\right) \leq \frac{1}{1 - \sqrt{\eta}} T_u^+(\eta, \alpha).$$

Combining this with (4.11), we get the second part of (4.10).

For (4.10′) the first part is again clear. Application of (4.9) gives

$$(4.12) \qquad T_u^-(\alpha) \geq (1 - \sqrt{\eta})^{-k} T_u^-(\alpha(1 - \sqrt{\eta})) - 3k\sqrt{\eta}(\alpha/\beta)^k.$$

We now apply principle I, again with $P_1 = P(x^\eta)$, $Z_1 = x^{\sqrt{\eta}/2}$ and $P_2 = P(x^\eta, x^{\alpha(1-\sqrt{\eta})})$, $Z_2 = x^{1-\sqrt{\eta}}$. There exists by the definition a sequence of $x \to \infty$ and of

$$\Lambda_2^- = \Lambda_u^-(P(x^\eta, x^{\alpha(1-\sqrt{\eta})}), x^{1-\sqrt{\eta}})$$

such that when $x \to \infty$ through this sequence we have

$$\lim T_u(\Lambda_2^-) = T_u^-\left(\frac{\eta}{1 - \sqrt{\eta}}, \alpha\right).$$

Furthermore, let $\Lambda_2^+ = \Lambda_u^+(P(x^\eta, x^{\alpha(1-\sqrt{\eta})}), x^{1-\sqrt{\eta}})$ be the sieve where $\lambda_1 = 1$ all other $\lambda = 0$. We have

$$T_u(\Lambda_2^+) = \frac{W_u(x^\eta)}{W_u(x^{\alpha(1-\sqrt{\eta})})}$$

so that

$$\lim_{x\to\infty} T_u(\Lambda_2^+) = \left(\frac{\alpha(1-\sqrt{\eta})}{\eta}\right)^k.$$

From (5.16) we see that there exists a

$$\Lambda_1^+ = \Lambda_u^+(P(x^\eta), x^{\sqrt{\eta}/2})$$

such that $1 \le T_u(\Lambda_1^+) < 1 + 2e^{-1/(20\sqrt{\eta})}$ for x large enough. And by (5.18) there exists a $\Lambda_1^- = \Lambda_u^-(P(x^\eta), x^{\sqrt{\eta}/2})$ such that $T_u(\Lambda_1^-) > 1 - 3e^{-1/(20\sqrt{\eta})}$, for x large enough. The construction (3.4) now produces a

$$\Lambda^- = \Lambda_u^-(P(x^{\alpha(1-\sqrt{\eta})}), 3x^{1-\sqrt{\eta}/2}) = \Lambda_u^-(P(x^{\alpha(1-\sqrt{\eta})}), x)$$

for x large enough, such that

$$T_u(\Lambda^-) \ge (1 + 2\omega e^{-1/(20\sqrt{\eta})})T_u(\Lambda_2^-) - 5e^{-1/(20\sqrt{\eta})}T_u(\Lambda_2^+),$$

where $0 \le \omega \le 1$.

Letting $x \to \infty$ through the special sequence referred to above, we get

$$\text{(4.13)}\qquad \begin{aligned} T_u^-(\alpha(1-\sqrt{\eta})) \ge{}& (1 + 2\omega e^{-1/(20\sqrt{\eta})})T_u^-\left(\frac{\eta}{1-\sqrt{\eta}}, \alpha\right) \\ &- 5e^{-1/(20\sqrt{\eta})}\left(\frac{\alpha(1-\sqrt{\eta})}{\eta}\right)^k. \end{aligned}$$

It can be seen, by considering the sieve $\Lambda_u^-(P(x\eta/(1-\sqrt{\eta}), x^\alpha), x)$ that arises by taking $\lambda_1 = 1$, $\lambda_p = -1$ for all $p \in P(x\eta/(1-\sqrt{\eta}), x^\alpha)$ and all other $\lambda_d = 0$, that

$$T_u^-\left(\frac{\eta}{1-\sqrt{\eta}}, \alpha\right) \ge \left(1 - k\log\frac{\alpha(1-\sqrt{\eta})}{\eta}\right)\left(\frac{\alpha(1-\sqrt{\eta})}{\eta}\right)^k.$$

Using this, we get from (4.13)

$$\begin{aligned} T_u^-(\alpha(1-\sqrt{\eta})) &\ge T_u^-\left(\frac{\eta}{1-\sqrt{\eta}}, \alpha\right) - (k+1)\eta^{-k-1}(1-\sqrt{\eta})^k e^{-1/(20\sqrt{\eta})} \\ &\ge T_u^-(\eta, \alpha) - (k+1)\eta^{-k-1}(1-\sqrt{\eta})^k e^{-1/(20\sqrt{\eta})}. \end{aligned}$$

Combining this with (4.12), we get the second part of (4.10′).

In order to carry out a certain construction which enters as a step in the proof of our Theorems 1 and 2, we have to define some new concepts and derive a few results relating to these.

We introduce the notation $\langle P\rangle$ to denote the set of all positive integers (square free or not) all of whose prime factors belong to P. We shall use $\hat{d}$ to denote elements of $\langle P\rangle$, and we consider now systems $\tilde{\Lambda} = \tilde{\Lambda}(P)$ of $\lambda_{\hat{d}}$ defined for all $\hat{d} \in \langle P\rangle$. We define for $\hat{d} \in \langle P\rangle$ the expression $\theta_{\hat{d}}$ as follows: If $\hat{d} = p_1^{\mu_1}p_2^{\mu_2}\cdots p_r^{\mu_r}$, where p_i, $1 \le i \le r$, are different primes from P, we put

$$\text{(4.14)}\qquad \theta_{\hat{d}} = \sum_{0\le\nu_i\le\mu_i;1\le i\le r}\binom{\mu_1}{\nu_1}\cdots\binom{\mu_r}{\nu_r}\lambda_{p_1^{\nu_1}p_2^{\nu_2}\cdots p_r^{\nu_r}},$$

where $\binom{\mu}{\nu}$ denotes the binomial coefficient. We write (4.14) briefly as

$$\theta_{\hat{d}} = \sum_{\hat{d}_1 \mid \hat{d}} \lambda_{\hat{d}_1}. \tag{4.14'}$$

If $\tilde{\Lambda}$ is such that $\theta_{\hat{d}} \geq 0$ for all $\hat{d}$, we denote $\tilde{\Lambda}$ by $\tilde{\Lambda}^+(P)$, and if we have $\theta_{\hat{d}} \leq 0$ for all $\hat{d} > 1$ (we still keep $\lambda_1 = 1$, so $\theta_1 = 1$), we denote $\tilde{\Lambda}$ by $\tilde{\Lambda}^-(P)$.

We extend the definition of $u(d)$ to $u(\hat{d})$ by making it a totally multiplicative function, so that

$$u(p_1^{\mu_1} \ldots p_r^{\mu_r}) = (u(p_1))^{\mu_1} \ldots (u(p_r))^{\mu_r}.$$

We define

$$R_u(\tilde{\Lambda}) = \sum_{\hat{d} \in \langle P \rangle}^* u(\hat{d})|\lambda_{\hat{d}}|, \tag{4.15}$$

where here and in the following $\sum_{\hat{d} \in \langle P \rangle}^*$ means that the term corresponding to $\hat{d} = p_1^{\mu_1} \ldots p_r^{\mu_r}$, where $p_1, \ldots, p_r$ are different primes from $\langle P \rangle$, is counted with the weight $1/(\mu_1! \ldots \mu_r!)$.

If a $\tilde{\Lambda}(P)$ has $R_u(\tilde{\Lambda}) \leq Z$, we denote it by $\tilde{\Lambda}_u(P, Z)$. If we have $\lambda_{\hat{d}} = 0$ for $\hat{d} > Z$, we use the notation $\tilde{\Lambda}[P, Z]$.

If we define as before $f(\hat{d}) = \hat{d}/u(\hat{d})$, and define $f'(\hat{d})$ now as the totally multiplinative function with $f'(p) = f(p) - 1$ for p in P, it is easily seen that

$$\sum_{\hat{d} \in \langle P \rangle}^* \frac{\theta_{\hat{d}}}{f'(\hat{d})} = \exp\left\{\sum_{p \in P} \frac{1}{f'(p)}\right\} \cdot \sum_{\hat{d} \in \langle P \rangle}^* \frac{\lambda_{\hat{d}}}{f'(\hat{d})}. \tag{4.16}$$

We denote the expression on the left-hand side of (4.16) by $T_u(\tilde{\Lambda})$.

The restriction of a $\tilde{\Lambda}$ to the subscripts $d \in (P)$ gives a Λ. Clearly the restriction to (P) of a $\tilde{\Lambda}^+$ leads to a Λ^+, and that of a $\tilde{\Lambda}^-$ to a Λ^-. Also the restriction to (P) of a $\tilde{\Lambda}_u(P, Z)$ leads to a $\Lambda_u(P, Z)$, and that of a $\tilde{\Lambda}[P, Z]$ to a $\Lambda[P, Z]$.

If Λ^+ is the restriction to (P) of $\tilde{\Lambda}^+$ we have also

$$T_u(\Lambda^+) \leq T_u(\tilde{\Lambda}^+), \tag{4.17}$$

and similarly if Λ^- is the restriction of $\tilde{\Lambda}^-$

$$T_u(\Lambda^-) \geq T_u(\tilde{\Lambda}^-). \tag{4.17'}$$

The restriction of a $\tilde{\Lambda}$ is unique, while there of course are many ways of extending a Λ^+ to a $\tilde{\Lambda}^+$ or a Λ^- to a $\tilde{\Lambda}^-$. One way of constructing such an extension which will be very useful for us later, can be defined as follows:

If $\hat{d} = pd$, where $d \in (P)$ and $p \mid d$, we define $\lambda_{\hat{d}} = |\lambda_d|$ for a Λ^+, and $\lambda_{\hat{d}} = -|\lambda_d|$, in case we have a Λ^-. For the $\hat{d}$ which are not in (P) and not of the form just considered, we define $\lambda_{\hat{d}} = 0$.

That this construction in case of a Λ^+ leads to a $\tilde{\Lambda}^+$ and in the case of a Λ^- to a $\tilde{\Lambda}^-$ can be seen by comparing the expression $\theta_{\hat{d}}$ for $\hat{d} = p_1^{\mu_1} \ldots p_r^{\mu_r}$ where the μ_i, $1 \leq i \leq r$ are positive integers, with θ_d for $d = p_1 \ldots p_r$. One finds, using the inequality

$$\mu_1 \ldots \mu_r \left(\frac{\mu_1 - 1}{2} + \cdots \frac{\mu_r - 1}{2}\right) \geq \mu_1 \ldots \mu_r - 1$$

which holds when the μ_i are nonnegative integers, that in the first case $\theta_{\hat{d}} \geq \theta_d$, and in the second case $\theta_{\hat{d}} \leq \theta_d$, which proves our assertion.

If we assume that P is a subset of $P(\xi_1, \xi_2)$ and that Λ is a $\Lambda[P, Z_1]$ and a $\Lambda_u(P, Z_2)$, we have that the $\tilde{\Lambda}$ constructed is a $\tilde{\Lambda}[P, Z_1\xi_2]$. Also, by (2.4),

$$R_u(\tilde{\Lambda}) = \sum_{d\in(P)} u(d)|\lambda_d| \left\{1 + \tfrac{1}{2}\sum_{p|d} u(p)\right\}$$
$$< \left(1 + \mathfrak{C}\frac{\log Z_1}{\log \xi_1}\right) R_u(\Lambda) \leq \left(1 + \mathfrak{C}\frac{\log Z_1}{\log \xi_1}\right) Z_2.$$

Thus $\tilde{\Lambda}$ is a $\tilde{\Lambda}_u(P, Z_2')$ with

$$Z_2' = \left(1 + \mathfrak{C}\frac{\log Z_1}{\log \xi_1}\right) Z_2. \tag{4.18}$$

Assuming now $\xi_1 > 2\mathfrak{C}$, we have further using (2.12)

$$\sum_{\hat{d}\in\langle P\rangle}^{*} \frac{|\lambda_{\hat{d}}|}{f'(\hat{d})} = \sum_{d\in(P)} \frac{|\lambda_d|}{f'(d)}\left\{1 + \frac{1}{2}\sum_{p|d}\frac{1}{f'(p)}\right\}$$
$$< \left(1 + \frac{\mathfrak{C}}{\xi_1}\frac{\log Z_1}{\log \xi_1}\right)\sum_{d\in(P)}\frac{|\lambda_d|}{f'(d)}$$
$$\leq \left(1 + \frac{\mathfrak{C}}{\xi_1}\frac{\log Z_1}{\log \xi_1}\right)\prod_{p\in(P)}\frac{1}{1 - u(p)/p}\sum_{d\in(P)}\frac{|\theta_d|}{f'(d)}.$$

Thus, if Λ is a Λ^+, we get

$$\sum_{\hat{d}\in\langle P\rangle} \frac{|\lambda_{\hat{d}}|}{f'(\hat{d})} < \left(1 + \frac{\mathfrak{C}}{\xi_1}\frac{\log Z_1}{\log \xi_1}\right)\frac{W_u(\xi_1)}{W_u(\xi_2)} T_u(\Lambda^+), \tag{4.19}$$

and if Λ is a Λ^-

$$\sum_{\hat{d}\in\langle P\rangle}^{*} \frac{|\lambda_{\hat{d}}|}{f'(\hat{d})} < \left(1 + \frac{\mathfrak{C}}{\xi_1}\frac{\log Z_1}{\log \xi_1}\right)\frac{W_u(\xi_1)}{W_u(\xi_2)}(2 + |T_u(\Lambda^-)|). \tag{4.19'}$$

Under the same assumptions about Λ^+, we have by (4.16)

$$T_u(\tilde{\Lambda}) = \exp\left\{\sum_{p\in P}\frac{1}{f'(p)}\right\}\sum_{\hat{d}\in\langle P\rangle}^{*}\frac{\lambda_{\hat{d}}}{f'(\hat{d})}. \tag{4.20}$$

Here

$$\sum_{\hat{d}\in\langle P\rangle}^{*}\frac{\lambda_{\hat{d}}}{f'(\hat{d})} = \sum_{d\in(P)}\frac{\lambda_d}{f'(d)} + \frac{1}{2}\sum_{d\in(P)}\frac{|\lambda_d|}{f'(d)}\sum_{p|d}\frac{1}{f'(p)}. \tag{4.21}$$

For $d \leq Z_1$, $d \in (P)$, we have

$$1 \leq \frac{f(d)}{f'(d)} = \prod_{p|d} \left(1 - \frac{u(p)}{p}\right)^{-1} \leq \left(1 - \frac{\mathfrak{C}}{\xi_1}\right)^{-\log Z_1/\log \xi_1}$$
$$< 1 + 2^{\log Z_1/\log \xi_1 + 1} \cdot \frac{\mathfrak{C}}{\xi_1},$$

and

$$\sum_{p|d} \frac{1}{f'(p)} \leq \frac{2\mathfrak{C}}{\xi_1} \frac{\log Z_1}{\log \xi_1} \leq 2^{\log Z_1/\log \xi_1 + 1} \cdot \frac{\mathfrak{C}}{\xi_1}.$$

(4.21) now gives, using (2.7) and (2.12)

$$\sum_{\hat{d} \in \langle P \rangle}^{*} \frac{\lambda_{\hat{d}}}{f'(\hat{d})} < \sum_{d \in (P)} \frac{\lambda_d}{f(d)} + 2^{\log Z_1/\log \xi_1 + 2} \frac{\mathfrak{C}}{\xi_1} \sum_{d \in (P)} \frac{|\lambda_d|}{f'(d)}$$
$$\leq \prod_{p \in P} \left(1 - \frac{u(p)}{p}\right) T_u(\Lambda^+) + 2^{\log Z_1/\log \xi_1 + 2} \frac{\mathfrak{C}}{\xi_1} \prod_{p \in P} \frac{1}{1 - u(p)/p} T_u(\Lambda^+).$$

Inserting this in (4.20), we get

$$(4.22) \quad T_u(\tilde{\Lambda}^+) < T_u(\Lambda^+) \left\{ \prod_{p \in P} \left(1 - \frac{u(p)}{p}\right) e^{1/f'(p)} + 2^{\log Z_1/\log \xi_1 + 2} \frac{\mathfrak{C}}{\xi_1} \prod_{p \in P} \frac{e^{1/f'(p)}}{1 - u(p)/p} \right\}.$$

Here

$$1 - u(p)/p \leq e^{-1/f(p)},$$

thus

$$\prod_{p \in P} \left(1 - \frac{u(p)}{p}\right) e^{1/f'(p)} \leq \prod_{p \in P} e^{1/(f(p)f'(p))}$$
$$\leq \exp\left\{2\mathfrak{C}^2 \sum_{p \in P} \frac{1}{p^2}\right\} \leq e^{2\mathfrak{C}^2/\xi_1} < 1 + \frac{4\mathfrak{C}^2}{\xi_1},$$

if we now also assume that $\xi_1 > 2\mathfrak{C}^2$. Again

$$\prod_{p \in P} \frac{e^{1/f'(p)}}{1 - u(p)/p} = \prod_{p \in P} \left(1 - \frac{u(p)}{p}\right) e^{1/f'(p)} \prod_{p \in P} \frac{1}{(1 - u(p)/p)^2}$$
$$< \left(1 + \frac{4\mathfrak{C}^2}{\xi_1}\right) \left(\frac{W_u(\xi_1)}{W_u(\xi_2)}\right)^2 < 3 \left(\frac{W_u(\xi_1)}{W_u(\xi_2)}\right)^2.$$

Using these inequalities we get from (4.22)

$$(4.23) \qquad T_u(\tilde{\Lambda}^+) < T_u(\Lambda^+) \left\{ 1 + \frac{2^{\log Z_1/\log \xi_1 + 3}(1 + \mathfrak{C})^2}{\xi_1} \left(\frac{W_u(\xi_1)}{W_u(\xi_2)}\right)^2 \right\}.$$

In a similar way we can prove

$$(4.23') \quad T_u(\tilde{\Lambda}^-) > T_u(\Lambda^-) - \frac{2^{\log Z_1/\log \xi_1 + 3}(1 + \mathfrak{C})^2}{\xi_1} \left(\frac{W_u(\xi_1)}{W_u(\xi_2)}\right)^2 (2 + |T_u(\Lambda^-)|).$$

LEMMA 3. *If $0 < \alpha \leq 1$, there exists a sequence of $x \to \infty$ and of $\Lambda^+ = \Lambda_u^+(P(x^\alpha), x) = \Lambda^+[P(x^\alpha), x^{1+\alpha}]$, such that when $x \to \infty$ through this sequence*

$$\lim T_u(\Lambda^+) = T_u^+(\alpha).$$

If $0 < \alpha < 1$, there exists a sequence of $x \to \infty$ and of $\Lambda^- = \Lambda_u^-(P(x^\alpha), x) = \Lambda^-[P(x^\alpha), x^{1+\alpha}]$, such that when $x \to \infty$ through this sequence

$$\lim T_u(\Lambda^-) = T_u^-(\alpha).$$

For $k < 1$, the last statement holds also for $\alpha = 1$.

To prove this we have to go back to the expressions (4.3) and (4.3′) as well as the equations (2.5) and (2.5′). Let $0 < \alpha \leq 1$, we have by (2.5) if we denote by A the constant implied by the O in (2.1),

$$(4.24) \qquad \frac{M_u^+(x, x^\alpha)}{xW_u(x^\alpha)} = \min_{\Lambda^+(P(x^\alpha))} \left\{T_u(\Lambda^+) + A\frac{R_u(\Lambda^+)}{xW_u(x^\alpha)}\right\}.$$

If we let $\Lambda^+(P(x^\alpha))$ range over those Λ^+ for which $R_u(\Lambda^+) > x$, the expression in the curly brackets is $> A/W_u(x^\alpha)$ and thus would tend to infinity for $x \to \infty$. If we let $\Lambda^+(P(x^\alpha))$ range over those for which $R_u(\Lambda^+) \leq x$, that is over the $\Lambda_u^+(P(x^\alpha), x)$, the expression in the curly brackets is $> T_u(\Lambda^+)$.

Thus we get from (4.4)

$$(4.25) \qquad \liminf_{x\to\infty} \frac{M_u^+(x, x^\alpha)}{xW_u(x^\alpha)} \geq T_u^+(\alpha).$$

Next, let δ be a small positive number. There exists by the definition (4.4) a sequence of $x \to \infty$ and of $\Lambda^+ = \Lambda_u^+(P(x^\alpha), x^{1-\delta})$ such that when $x \to \infty$ through this sequence

$$(4.26) \qquad \lim T_u(\Lambda^+) = T_u^+\left(\frac{\alpha}{1-\delta}\right),$$

(if $\alpha < 1$, we may assume δ small enough to make $\alpha/(1 - \delta) \leq 1$, if $\alpha = 1$ we would seem to have some trouble since $\alpha/(1 - \delta) > 1$, however, the definition (4.4) can be used also for $\alpha > 1$ and the function $T_u^+(\alpha)$ is then essentially trivial since it is easy to see that $T_u^+(\alpha) = \alpha^k T_u^+(1)$ if $\alpha > 1$).

Using now this $\Lambda_u^+(P(x^\alpha), x^{1-\delta})$ in the expression in the curly brackets in (4.24) we have as $x \to \infty$

$$\frac{R_u(\Lambda^+)}{xW_u(x^\alpha)} \leq \frac{x^{-\delta}}{W_u(x^\alpha)} \to 0.$$

Thus, letting $x \to \infty$ through the sequence for which (4.26) holds we get from (4.24)

$$\liminf_{x\to\infty} \frac{M_u^+(x, x^\alpha)}{xW_u(x^\alpha)} \le T_u^+\left(\frac{\alpha}{1-\delta}\right).$$

Letting here $\delta \to 0$, we get because of the continuity of $T_u^+(\alpha)$ and because of (4.25), that

$$\liminf_{x\to\infty} \frac{M_u^+(x, x^\alpha)}{xW_u(x^\alpha)} = T_u^+(\alpha). \tag{4.27}$$

A $\Lambda^+(P(x^\alpha))$ that minimizes the expression in the curly brackets on the right-hand side of (4.24) is an optimal sieve in the sense explained earlier. From the statements in §2 it follows that in this case there exist such optimal sieves of the type $\Lambda^+[P(x^\alpha), x^{1+\alpha}]$. Since for such a sieve

$$\frac{M_u^+(x, x^\alpha)}{xW_u(x^\alpha)} > T_u(\Lambda^+), \tag{4.28}$$

and also as $x \to \infty$ through a sequence for which

$$\lim \frac{M_u^+(x, x^\alpha)}{xW_u(x^\alpha)} = T_u^+(\alpha), \tag{4.29}$$

we must have that $R_u(\Lambda^+)/(xW_u(x^\alpha))$ must remain bounded, thus for x sufficiently large $R_u(\Lambda^+) < x$. Thus this Λ^+ is also a $\Lambda_u^+(P(x^\alpha), x)$. Letting now x tend to infinity through the sequence for which (4.29) holds, we get $\lim T_u(\Lambda^+) = T_u^+(\alpha)$.

Thus this sequence and the optimal

$$\Lambda^+ = \Lambda_u^+(P(x^\alpha), x) = \Lambda^+[P(x^\alpha), x^{1+\alpha}]$$

fills the requirements of the first statement in Lemma 3.

For the second statement, we proceed in a similar way, since we assume $\alpha < 1$, we may assume δ so small that also $\alpha/(1 - \delta) < 1$.

The case when $k < 1$ and $\alpha = 1$, does require some other arguments; one uses the statement already proved for all fixed $\alpha < 1$ and principle III to extend the range of a sieve with range $P(x^\alpha)$ to $P(x)$.

In the course of the proof of the second statement one obtains of course as a byproduct the analogue of (4.27)

$$\limsup_{x\to\infty} \frac{M_u^-(x, x^\alpha)}{xW_u(x^\alpha)} = T_u^-(\alpha). \tag{4.30}$$

There is a similar set of statements that can be made about $T_u^+(\eta, \alpha)$ and $T_u^-(\eta, \alpha)$ where $0 < \eta < \alpha \le 1$. The main modification in the argument is that since now the factor $W_u(x^\alpha)/W_u(x^\eta) \to (\eta/\alpha)^k$ replaces $W_u(x^\alpha)$, we consider the analogue of (4.24) with A such that $A(\alpha/\eta)^k$ is larger than some trivial bound for $T_u^+(\eta, \alpha)$ or $|T_u^-(\eta, \alpha)|$ ($A = 2$ for the T_u^+ case, and $A = 1 + k \log(\alpha/\eta)$ for the T_u^- case, would be sufficient). In case of T_u^-, the case $\alpha = 1$, requires slightly special treatment where principle III is applied in a simple way.

We obtain in this way

Lemma 3′. *For* $0 < \eta < \alpha \leq 1$, *there exists a sequence of* $x \to \infty$ *and of* $\Lambda^+ = \Lambda_u^+(P(x^\eta, x^\alpha), x) = \Lambda^+[P(x^\eta, x^\alpha), x^{1+\alpha}]$, *such that when* $x \to \infty$ *through this sequence*

$$\lim T_u(\Lambda^+) = T_u^+(\eta, \alpha).$$

There also exists a sequence $x \to \infty$ *and of*

$$\Lambda^- = \Lambda_u^-(P(x^\eta, x^\alpha), x) = \Lambda^-[P(x^\eta, x^\alpha), x^{1+\alpha}],$$

such that when $x \to \infty$ *through this sequence*

$$\lim T_u(\Lambda^-) = T_u^-(\eta, \alpha).$$

Lemma 4. *Let* $u_1(p)$ *be a function satisfying all our earlier assumptions about* $u(p)$, *let* η *and* α *be numbers with* $0 < \eta < \alpha \leq 1$, ϵ *a small positive number. There then exists a sieve* $\Lambda^+ = \Lambda^+[P(x^\eta, x^\alpha), x^{1-\delta}]$, *where* $\delta > 0$ *is a number depending only on* ϵ, η, α *and* k, *and such that the* λ_d *for* $d \in (P(x^\eta, x^\alpha))$, *and of the form* $d = p_1 \ldots p_r$ *can be expressed as*

$$\lambda_{p_1 \ldots p_r} = l_r\left(\frac{\log p_1}{\log x}, \ldots, \frac{\log p_r}{\log x}\right)$$

where the l_r *are functions of* r *variables defined when these are in the interval* (η, α), *and not otherwise dependent on* d *or* x. *The* $|\lambda|$ *are bounded by quantities depending on* ϵ, η, α *and* k *only. Furthermore, for any function* $u(p)$ *with* $0 \leq u(p) < p$ *which satisfies* (4.1), *we have that the limit* $\lim_{x\to\infty} T_u(\Lambda^+)$ *exists, and that*

$$\lim_{x\to\infty} T_u(\Lambda^+) < T_{u_1}(\eta, \alpha) + \epsilon.$$

Similarly there exists under the added assumption $\alpha < 1$, *a* $\Lambda^- = \Lambda^-[P(x^\eta, x^\alpha), x^{1-\delta}]$ *for which the same statements are true about the* λ_d, *and such that the limit* $\lim_{x\to\infty} T_u(\Lambda^-)$ *exists, and that*

$$\lim_{x\to\infty} T_u(\Lambda^-) > T_{u_1}^-(\eta, \alpha) - \epsilon.$$

We shall prove the existence of a Λ^+ with these properties, the other half of the lemma can be proved in a quite similar way.

We write for some large positive integer N:

$$t = (\alpha/\eta)^{1/N} \quad \text{and} \quad t' = (\alpha/\eta)^{1/(N-1)},$$

also we put $\eta' = \eta t'^2$ and $\alpha' = \alpha t'^3$. By (4.6′) and (4.7′), we have

$$T_{u_1}^+(\eta', \alpha') \leq T_{u_1}^+(\eta, \alpha') \leq T_{u_1}^+(\eta, \alpha)t'^{3k} = T_{u_1}^+(\eta, \alpha)(\alpha/\eta)^{3k/(N-1)}$$
$$< T_{u_1}^+(\eta, \alpha) + 3k(\alpha/\eta)^{k+1}/(N-1) < T_{u_1}^+(\eta, \alpha) + \epsilon/2,$$

if N is chosen large enough.

From Lemma 3′ we know that there exist a sequence of $x \to \infty$, and of $\Lambda_1^+ =$

$\Lambda_{u_1}^+(P(x^{\eta'}, x^{\alpha'}), x) = \Lambda^+[P(x^{\eta'}, x^{\alpha'}), x^{1+\alpha'}] = \Lambda^+[P(x^{\eta'}, x^{\alpha'}), x^3]$ such that

$$\lim T_{u_1}(\Lambda_1^+) = T_{u_1}^+(\eta', \alpha'),$$

when $x \to \infty$ through this special sequence. We shall in the following use the symbol x^* to denote numbers of this sequence. Thus we have

$$T_{u_1}(\Lambda_1^+) < T_{u_1}^+(\eta, \alpha) + \epsilon/2, \tag{4.31}$$

for x^* large enough.

We shall write for simplicity $P^* = P(x^{*\eta'}, x^{*\alpha'})$, and look at the extension of Λ_1^+ to a $\tilde{\Lambda}_1^+$ by the procedure described earlier in this paragraph. (4.18) gives

$$R_{u_1}(\tilde{\Lambda}_1^+) \leq (1 + 3\mathcal{C}/\eta)x^*. \tag{4.32}$$

Also, from (4.19)

$$\begin{aligned} \sum_{\hat{d}\in\langle P^*\rangle}^{*} \frac{|\lambda_{\hat{d}}^{(1)}|}{f_1'(\hat{d})} &\leq \left(1 + \frac{3\mathcal{C}}{\eta x^{*\eta'}}\right) \frac{W_{u_1}(x^{*\eta'})}{W_{u_1}(x^{*\alpha'})} T_{u_1}(\Lambda_1^+) \\ &\leq 2\left(\frac{\alpha'}{\eta'}\right)^k T_{u_1}^+(\eta', \alpha') < \frac{3}{\eta^{2k}}, \end{aligned} \tag{4.33}$$

for x^* large enough. $f_1'(d)$ and later $f_1(d)$ denotes here the functions related in the same way to $u_1(d)$, as $f'(d)$ and $f(d)$ are related to $u(d)$.

From (4.23) and (4.7) we get

$$\begin{aligned} T_{u_1}(\Lambda_1^+) &\leq T_{u_1}(\tilde{\Lambda}_1^+) \\ &\leq T_{u_1}(\Lambda_1^+)\left\{1 + \frac{2^{6/\eta}(1+\mathcal{C})^2}{x^{*\eta'}}\left(\frac{W_u(x^{*\eta'})}{W_u(x^{*\alpha'})}\right)^2\right\} \\ &\leq T_{u_1}(\Lambda_1^+)\left\{1 + \frac{1}{x^{*\eta}}\right\}, \end{aligned}$$

for x^* large enough. Thus for x^* sufficiently large, we have by (4.31),

$$T_u(\tilde{\Lambda}_1^+) \leq (T_{u_1}(\eta, \alpha) + \epsilon/2)(1 + 1/x^{*\eta}) < T_{u_1}(\eta, \alpha) + \epsilon. \tag{4.34}$$

We consider now besides x^* a variable x, and shall for brevity denote by P the expression $P(x^\eta, x^\alpha)$.

We divide the interval (η, α) in N parts by the points ηt^i, we say that a $p \in P$ belongs to the interval Δ_i if $\eta t^{i-1} < \log p/\log x \leq \eta t^i$ for some i with $1 \leq i \leq N$. Similarly we divide the interval (η', α') in N parts by the points $\eta' t'^i$, and say that a prime $q \in P^*$ (for once we will use another letter than p to designate prime numbers) belongs to the interval Δ_i^* if we have

$$\eta' t'^{i-1} < \log q/\log x^* \leq \eta' t'^i.$$

We write

$$\sigma_i^* = \sum_{q\in\Delta_i^*} \frac{1}{f_1'(q)}, \tag{4.35}$$

and have for $x^* \to \infty$,

$$\sigma_i^* \to k \log t' = \frac{k}{N-1} \log \frac{\alpha}{\eta}. \tag{4.36}$$

For $i_1, \ldots, i_r$ positive integers $\leq N$ we define

$$l^*(i_1, \ldots, i_r) = \frac{1}{\sigma_{i_1}^* \ldots \sigma_{i_r}^*} \sum_{q_\nu \in \Delta_{i_\nu}^*; 1 \leq \nu \leq r} \frac{\lambda_{q_1 \ldots q_r}^{(1)}}{f_1'(q_1) \ldots f_1'(q_r)} \tag{4.37}$$

where the $\lambda^{(1)}$ are taken from $\tilde{\Lambda}_1^+$. Also we define a $\tilde{\Lambda}_*(P)$ (the * to remind us that it depends on x^*), where the $\lambda_{\hat{d}}^*$ for $\hat{d} \in \langle P \rangle$, $P = P(x^\eta, x^\alpha)$, are defined as follows: for $\hat{d} = p_1 p_2 \ldots p_r$ we put

$$\lambda_{\hat{d}}^* = l^*(i_1, \ldots, i_r) \tag{4.38}$$

where for $1 \leq \nu \leq r$, i_ν is the interval to which p_ν belongs.
From (4.38) and (4.37) we get, using (4.35)

$$\theta_{\hat{d}}^* = \theta_{p_1 \ldots p_r}^* = \frac{1}{\sigma_{i_1}^* \ldots \sigma_{i_r}^*} \sum_{q_\nu \in \Delta_{i_\nu}^*; 1 \leq \nu \leq r} \frac{\theta_{q_1 \ldots q_r}^{(1)}}{f_1'(q_1) \ldots f_1'(q_r)}, \tag{4.39}$$

thus $\theta_{\hat{d}}^* \geq 0$ for $\hat{d} \in \langle P \rangle$, so that $\tilde{\Lambda}_*(P)$ is a $\tilde{\Lambda}_*^+$.
From (4.37) and (4.33), we get

$$|l^*(i_1, \ldots, i_r)| \leq \frac{r!}{\sigma_{i_1}^* \ldots \sigma_{i_r}^*} \sum_{\hat{d} \in \langle P^* \rangle}^* \frac{|\lambda_{\hat{d}}^{(1)}|}{f_1'(\hat{d})} \leq \frac{3r!\,\eta^{-2k}}{\sigma_{i_1}^* \ldots \sigma_{i_r}^*}$$
$$\leq \frac{3r!\,\eta^{-2k}}{(k \log t)^r} = \frac{3r!\,N^r}{\eta^{2k}(k \log \alpha/\eta)^r}, \tag{4.40}$$

for x^* large enough.
Since Λ_1^+ is a $\Lambda_1^+[P^*, x^{*3}]$ we get that the right-hand side of (4.37) vanishes for $r \geq 1 + 3/\eta$, so that then $l^*(i_1, \ldots, i_r) = 0$.
Let us assume that $p \in \Delta_i$ and $q \in \Delta_i^*$, we have then

$$\log q/\log x^* > \eta t'^{i+1} > \eta t^{i+1} \geq t \log p/\log x,$$

thus if $p_\nu \in \Delta_{i_\nu}$ and $q_\nu \in \Delta_{i_\nu}^*$ for $1 \leq \nu \leq r$, we have

$$\log q_1 \ldots q_r/\log x^* > t \log p_1 \ldots p_r/\log x.$$

We put now

$$\delta = \frac{1}{3N} \log \frac{\alpha}{\eta},$$

then if $p_1 p_2 \ldots p_r > x^{1-\delta}$, we see that $q_1 q_2 \ldots q_r > x^{*\,1+\delta}$. For large x^* we have for $q > x^{*\eta'}$

$$\frac{1}{f_1'(q)} = \frac{u_1(q)}{q - u_1(q)} < \frac{u_1(q)}{q^{1-\delta/4}},$$

thus if $q_1 \dots q_r > x^{*\,1+\delta}$, we get

$$\frac{1}{f'_1(q_1)\dots f'_1(q_r)} < \frac{u_1(q_1\dots q_r)}{x^{*\,1+\delta/2}}.$$

From these inequalities we get that if $i_1, \dots, i_r$ are such that there exist $p_1, \dots, p_r$ with $p_\nu \in \Delta_{i_\nu}$ for $1 \le \nu \le r$ for which

$$p_1 \dots p_r > x^{1-\delta}, \tag{4.41}$$

then by (4.37) and (4.32)

$$\begin{aligned} |l^*(i_1, \dots, i_r)| &\le \frac{r!x^{*-1-\delta/2}}{\sigma^*_{i_1}\dots\sigma^*_{i_r}} \sum\nolimits^*_{\hat d \in \langle P^*\rangle} u_1(\hat d)|\lambda^{(1)}_{\hat d}| \\ &< \frac{r!(1+3\mathcal{C}/\eta)x^{*-\delta/2}}{\sigma^*_{i_1}\dots\sigma^*_{i_r}} \\ &\le \frac{r!(1+3\mathcal{C}/\eta)x^{*-\delta/2}}{(k\log t)^r} \end{aligned} \tag{4.42}$$

for x^* large enough.

Since the l^* vanish for $r > 1 + 3/\eta$ and each i can only take N values, there are only a bounded number of the l^* to consider as x^* tends to infinity. By (4.40) the l are uniformly bounded as $x^* \to \infty$, thus we can take out a subsequence of the x^* such that each $l^*(i_1, \dots, i_r)$ tends to a limit $l(i_1, \dots, i_r)$ as $x^* \to \infty$ through this subsequence. (4.42) shows that if the $i_1 \dots i_r$ are such that the condition (4.41) can be fulfilled, then $l(i_1, \dots, i_r) = 0$ since the right hand side of (4.42) tends to zero for $x^* \to \infty$. From l we get a $\tilde\Lambda(P)$ which is the limit of $\tilde\Lambda_*(P)$, from what we have already proved it is clear that it is a $\tilde\Lambda^+[P(x^\eta, x^\alpha), x^{1-\delta}]$.

Now let $u(p)$ be a function of which we shall only assume that $0 \le u(p) < p$ for all p and that (4.1) holds. We wish to obtain estimates for $T_u(\tilde\Lambda^+)$. First we need such estimates for $T_u(\tilde\Lambda^+_*)$.

We write

$$\sigma_i = \sum_{p\in\Delta_i} \frac{1}{f'(p)}, \tag{4.43}$$

and have then as $x \to \infty$

$$\sigma_i \to k\log t = \frac{k}{N}\log\frac{\alpha}{\eta}. \tag{4.44}$$

Comparing this with (4.36) we see that when both x^* and x are large enough, we have $\sigma_i < \sigma^*_i$. If we define, for $q \in \Delta^*_i$, $\rho(q) = \sigma_i/\sigma^*_i$ we have $0 < \rho(q) < 1$. Writing for brevity $\rho(q_1 \dots q_r)$ for $\rho(q_1) \dots \rho(q_r)$, we now get from (4.39)

$$\theta^*_{p_1\dots p_r} = \frac{1}{\sigma_{i_1}\dots\sigma_{i_r}} \sum_{q_\nu\in\Delta^*_{i_\nu};1\le\nu\le r} \frac{\rho(q_1\dots q_r)\theta^{(1)}_{q_1\dots q_r}}{f'_1(q_1\dots q_r)},$$

or remembering the definition of the σ_i and that $\theta^*_{p_1\dots p_r}$ depends only on $i_1, \dots, i_r$, we get

$$\sum_{p_\nu \in \Delta_{i_\nu}; 1 \le \nu \le r} \frac{\theta^*_{p_1 \ldots p_r}}{f'(p_1 \ldots p_r)} = \sum_{q_\nu \in \Delta^*_{i_\nu}; 1 \le \nu \le r} \frac{\rho(q_1 \ldots q_r)\theta^{(1)}_{q_1 \ldots q_r}}{f'_1(q_1 \ldots q_r)},$$

dividing by $r!$ and adding for all combinations $(i_1, \ldots, i_r)$ and all $r > 0$ we get

$$\sum^*_{\hat{d} \in \langle P \rangle} \frac{\overset{*}{\theta_{\hat{d}}}}{f'(\hat{d})} = \sum^*_{\hat{d} \in \langle P^* \rangle} \frac{\rho(\hat{d})\theta^{(1)}_{\hat{d}}}{f'_1(\hat{d})},$$

or since $0 < \rho(\hat{d}) < 1$ for $\hat{d} > 1$, $\rho(1) = 1$, we get, using (4.34),

$$T_u(\tilde{\Lambda}^+_*) \le T_u(\tilde{\Lambda}^+_1) \le T_{u_1}(\eta, \alpha) + \epsilon.$$

Letting $x^* \to \infty$ in such a way that $\tilde{\Lambda}^+_*$ tends to $\tilde{\Lambda}^+$, we get $T_u(\tilde{\Lambda}^+) \le T_{u_1}(\eta, \alpha) + \epsilon$, since $T_u(\Lambda^+) \le T_u(\tilde{\Lambda}^+)$, this proves that $T_u(\Lambda^+) \le T_{u_1}(\eta, \alpha) + \epsilon$ for x sufficiently large. It remains to show that $\lim_{x \to \infty} T_u(\Lambda^+)$ exists.

We have

$$T_u(\Lambda^+) = \frac{W_u(x^\eta)}{W_u(x^\alpha)} \sum_{d \in \langle P \rangle} \frac{\lambda_d}{f(d)}. \tag{4.45}$$

We compare the two expressions

$$\sum_{d \in \langle P \rangle} \frac{\lambda_d}{f(d)} \quad \text{and} \quad \sum^*_{\hat{d} \in \langle P \rangle} \frac{\lambda_{\hat{d}}}{f(\hat{d})}.$$

From (4.1) follows that $1/f(p) \to 0$. Thus if we denote by $\tau(x)$ the maximum of $1/f(p)$ for $p \in P(x^\eta, x^\alpha)$, we have $\tau(x) \to 0$ as $x \to \infty$. Also let B denote the largest of the numbers

$$\frac{3r!\,N^r}{\eta^{2k}(k \log \alpha/\eta)^r}$$

for $0 \le r \le 1/\eta$. Since the $|\lambda_{\hat{d}}| \le B$ for all $\hat{d} \in \langle P \rangle$, we get easily

$$\begin{aligned} \left| \sum^*_{\hat{d} \in \langle P \rangle} \frac{\lambda_{\hat{d}}}{f(\hat{d})} - \sum_{d \in \langle P \rangle} \frac{\lambda_d}{f(p)} \right| &\le B\tau(x) \prod_{p \in P} \frac{1}{1 - 1/f(p)} \sum_{p \in P} \frac{1}{f(p)} \\ &\le B\tau(x) \frac{W_u(x^\eta)}{W_u(x^\alpha)} \frac{V_u(x^\alpha)}{\eta \log x} < 2kB\left(\frac{\alpha}{\eta}\right)^{k+1} \tau(x) \end{aligned} \tag{4.46}$$

for x sufficiently large.

Also

$$\sum^*_{\hat{d} \in \langle P \rangle} \frac{\lambda_{\hat{d}}}{f(\hat{d})} = \sum_{0 \le r \le 1/\eta} \frac{1}{r!} \sum_{1 \le i_\nu \le N; 1 \le \nu \le r} l(i_1, i_2, \ldots, i_r)\sigma'_{i_1} \ldots \sigma'_{i_r}, \tag{4.47}$$

where we have written

$$\sigma_i' = \sum_{p \in \Delta_i} \frac{1}{f(p)} \cdot$$

We have as $x \to \infty$

$$\sigma_i' \to k \log t = \frac{k}{N} \log \frac{\alpha}{\eta} \cdot$$

Letting $x \to \infty$, we have that the right-hand side of (4.46) tends to zero, and the right-hand side of (4.47) to

$$\sum_{0 \le r < 1/\eta} \frac{k^r}{r!} \left(\frac{1}{N} \log \frac{\alpha}{\eta}\right)^r \sum_{1 \le i_\nu \le N; 1 \le \nu \le r} l(i_1, i_2, \ldots, i_r).$$

(4.45) now gives

$$\text{(4.48)} \quad \lim_{x \to \infty} T_u(\Lambda^+) = \left(\frac{\eta}{\alpha}\right)^k \sum_{0 \le r < 1/\eta} \frac{k^r}{r!} \left(\frac{1}{N} \log \frac{\alpha}{N}\right)^r \sum_{1 \le i_\nu \le N; 1 \le \nu \le r} l(i_1, i_2, \ldots, i_r),$$

which proves that the limit of $T_u(\Lambda^+)$ exists, and so completes the first part of Lemma 4.

The statements about Λ^- are proved in a quite similar manner, the only new point is that N has to be taken so large that $\alpha' = \alpha t'^3 < 1$.

It follows from this lemma that the functions $T_u^+(\eta, \alpha)$ and $T_u^-(\eta, \alpha)$ are independent of the particular u, and only depend on η, α and k. This follows by considering besides $u_1(p)$ another function $u_2(p)$ satisfying the same conditions, we have then

$$T_{u_2}^+(\eta, \alpha) < T_{u_1}^+(\eta, \alpha) + \epsilon$$

or since ϵ can be taken arbitrarily small, we get $T_{u_2}^+(\eta, \alpha) \le T_{u_1}^+(\eta, \alpha)$. Since the roles of u_1 and u_2 can be interchanged it follows that $T_{u_2}^+(\eta, \alpha) = T_{u_1}^+(\eta, \alpha)$. Because of Lemma 2 we get by making $\eta \to 0$, that also $T_{u_2}^+(\alpha) = T_{u_1}^+(\alpha)$. The corresponding result holds for $T_u^-(\eta, \alpha)$ and $T_u^-(\alpha)$.

We shall therefore in the following write $T_k^+(\alpha)$, $T_k^-(\alpha)$, $T_k^+(\eta, \alpha)$ and $T_k^-(\eta, \alpha)$ for these functions.

THEOREM 1. *For $k > 0$, $0 \le \alpha \le 1$, there exists a function $T_k^+(\alpha)$, such that if u satisfies our previous conditions*[7] *we have*

$$\text{(4.49)} \qquad \lim_{x \to \infty} \frac{M_u^+(x, x^\alpha)}{x W_u(x^\alpha)} = T_k^+(\alpha).$$

[7]We could show that (4.49), (4.49′), (4.49″), and (4.49‴) hold under the weaker conditions on $u(p)$ that only (2.3) and (4.1) hold; the above gives only that

$$\limsup_{x \to \infty} \frac{M_u^+(x, x^\alpha)}{x W_u(x^\alpha)} \le T_k^+(\alpha),$$

instead of (4.49). Similar statements hold about the others.

Similarly there exists a function $T_k^-(\alpha)$ *defined for* $k > 0$, $0 \le k < 1$ *(and for* $0 \le \alpha \le 1$ *if* $0 < k < 1$*), such that*

$$\lim_{x\to\infty} \frac{M_u^-(x, x^\alpha)}{xW_u(x^\alpha)} = T_k^-(\alpha). \tag{4.49'}$$

For $k > 0$, $0 < \eta < \alpha \le 1$, *there exists a function* $T_k^+(\eta, \alpha)$ *such that*

$$\lim_{x\to\infty} \frac{M_u^+(x; x^\eta, x^\alpha)}{x(\eta/\alpha)^k} = T_k^+(\eta, \alpha),^{8} \tag{4.49''}$$

under the same assumptions there exists a function $T_k^-(\eta, \alpha)$ *such that*

$$\lim_{x\to\infty} \frac{M_u^-(x; x^\eta, x^\alpha)}{x(\eta/\alpha)^k} = T_k^-(\eta, \alpha). \tag{4.49'''}$$

The function $T_k^+(\alpha)$ *is continuous for* $0 \le \alpha \le 1$, *and if we define* $T_k^+(0, \alpha) = T_k^+(\alpha)$, *then* $T_k^+(\eta, \alpha)$ *is continuous for* $0 \le \eta < \alpha \le 1$. *The function* $T_k^-(\alpha)$ *is continuous for* $0 \le \alpha < 1$ *(for* $k < 1$ *also at* $\alpha = 1$*), and if we define* $T_k^-(0, \alpha) = T_k^-(\alpha)$, *then* $T_k^-(\eta, \alpha)$ *is continuous for* $0 \le \eta < \alpha < 1$, *and for* $0 < \eta < \alpha \le 1$. *Measures of continuity can be explicitly given in all cases.*

The functions $T_k^+(\alpha)$ *and* $T_k^+(\eta, \alpha)$ *are continuous in k (and obviously nondecreasing) for* $0 \le \alpha \le 1$ *and* $0 \le \eta < \alpha \le 1$ *respectively. Similarly* $T_k^-(\alpha)$ *is continuous in k for* $0 \le \alpha < 1$ *(and for* $k < 1$ *for* $0 \le \alpha \le 1$*), and for* $0 < \eta < \alpha \le 1$, $T_k^-(\eta, \alpha)$ *is continuous in k, both of these functions are nonincreasing as functions of k.*

The functions $T_k^+(\alpha)$, $T_k^+(\eta, \alpha)$, $T_k^-(\alpha)$ *and* $T_k^-(\eta, \alpha)$ *can be effectively computed.*

Of the statements (4.49), (4.49′), (4.49″), and (4.49‴), the two last are immediate consequences of Lemma 4, as they follow at once from application of the sieves Λ^+ and Λ^- of Lemma 4.

The first two need more effort. We use Lemma 4 with $\eta/(1-\sqrt{\eta})$, $\alpha/(1-\sqrt{\eta})$ and $x^{1-\sqrt{\eta}}$ in place of η, α and x, and apply principle I in a simple way as in the proof of Lemma 2, to produce from a sieve

$$\Lambda^+[P(x^\eta), x^{\sqrt{\eta}/2}] \quad \text{and a} \quad \Lambda^+[P(x^\eta, x^\alpha), x^{1-\sqrt{\eta}}]$$

a $\Lambda^+[P(x^\alpha), x^{1-\sqrt{\eta}/2}]$ which gives that

$$\frac{M_u^+(x, x^\alpha)}{xW_u(x^\alpha)} \le T_k^+(\alpha) + \epsilon$$

for all x sufficiently large, where ϵ is an arbitrary fixed positive number. (4.49′) is handled similarly.

The statements about the continuity follow from Lemmas 1 and 2, with the exception of the statement about continuity in k. This can be seen as follows: if

[8] We write here $M_u^+(x; x^\eta, x^\alpha)$ instead of $M_u^+(x, P(x^\eta, x^\alpha))$ and similarly for M_u^-.

we apply the sieve Λ^+ of Lemma 4 for a function $u(p)$ which belongs to a different k, say k_1, from that of $u_1(p)$ we see easily that the formula (4.48) is still true with k_1 in place of k on the right-hand side. Denoting the right-hand side by $(\eta/\alpha)^{k_1}Q(k_1)$ where Q is a polynomial of degree $< 1/\eta$, we see that for $0 \leq k_1 - k < \delta$ where δ is sufficiently small we have

$$|(\eta/\alpha)^{k_1}Q(k_1) - (\eta/\alpha)^k Q(k)| < \epsilon,$$

thus $T_{k_1}(\eta, \alpha) \leq T_k(\eta, \alpha) + 2\epsilon$, that in addition $T_k(\eta, \alpha) \leq T_{k_1}(\eta, \alpha)$ is trivial. This proves continuity of $T_k^+(\eta, \alpha)$ on the right in k, to prove continuity on the left, we consider a $k_1 < k$ but sufficiently close and reverse the roles of k and k_1. For $T_k^+(\alpha)$ continuity in k follows from the fact that $T_k^+(\alpha) = \lim_{\eta\to 0} T_k^+(\eta, \alpha)$ uniformly for k in a closed interval, something that is evident from (4.10). For $T_k^-(\eta, \alpha)$ and $T_k^-(\alpha)$ similar arguments work.

To finally show that these functions can be effectively computed, it is enough to consider $T_k^+(\eta, \alpha)$ and $T_k^-(\eta, \alpha)$ with $\eta > 0$, since the inequalities (4.10) and (4.10′) show how we can then also effectively compute $T_k^+(\alpha)$ and $T_k^-(\alpha)$.

We consider $T_k^+(\eta, \alpha)$ and go back to the formula (4.48).

We consider, with $t = (\alpha/\eta)^{1/N}$, quantities $l'(i_1, \ldots, i_r)$, $1 \leq i_\nu \leq r$ which satisfy the following conditions

$$(4.50) \quad l' = 1, \qquad l'(i_1, \ldots, i_r) = 0 \quad \text{for } \eta(t^{i_1} + t^{i_2} + \cdots + t^{i_r}) > 1 - \delta.$$

$$(4.51) \qquad |l'(i_1, \ldots, i_r)| \leq \frac{3r!N^r}{\eta^{2k}(k \log \alpha/\eta)^r}.$$

We define the polynomial in N variables $y_1 \ldots y_N$

$$(4.52) \qquad \theta(y_1, \ldots, y_N) = \sum_{0 \leq \nu_1, \ldots, \nu_N} \binom{y_1}{\nu_1}\binom{y_2}{\nu_2} \cdots \binom{y_N}{\nu_N} l'(i_1, \ldots, i_r)$$

where among the indices $i_1 \ldots i_r$, i occurs ν_i times for $1 \leq i \leq N$, and we require

$$(4.53) \qquad \theta(y_1, \ldots, y_N) \geq 0$$

for all nonnegative integral values of the N variables $y_1, \ldots, y_N$. We have then

$$(4.54) \quad T_k^+(\eta, \alpha) \leq \min_{l'} \left(\frac{\alpha}{\eta}\right)^k \sum \frac{1}{r!} l'(i_1, \ldots, i_r) \left(\frac{k}{N} \log \frac{\alpha}{\eta}\right)^r \leq T_k^+(\eta, \alpha) + \epsilon,$$

where ϵ can be made arbitrarily small if N is chosen large enough.

Since there is a finite range (4.51) for each l', and there are only a finite number, say R, of these l' that are $\neq 0$, we can divide each range into L equal parts and test all possible combinations of l' corresponding to our grid. The number of these is $(L + 1)^R$.

For each such combination we would like to test whether (4.53) is satisfied. There is of course no effective way of doing this. However, if we denote the right hand side of (4.51) by B_r, it is possible to decide effectively that one of the following two things takes place: a. We have that (4.53) does not hold for our set of $l'(i_1 \ldots i_r)$; b. We have that (4.53) does hold if we replace the $l'(i_1, \ldots, i_r)$

by

$$l'(i_1, \ldots, i_r) + 2B_r/L.$$

(Of course both of these things may be true.)

For the cases where we find b to hold, we compute the expression

$$\left(\frac{\alpha}{\eta}\right)^k \sum \frac{1}{r!}\left(l'(i_1, \ldots, i_r) + \frac{2B_r}{L}\right)\left(\frac{k}{N}\log\frac{\alpha}{\eta}\right)^r. \tag{4.55}$$

The smallest of these values must differ from

$$\min_{l'} \left(\frac{\alpha}{\eta}\right)^k \sum \frac{1}{r!}\, l'(i_1, \ldots, i_r)\left(\frac{k}{N}\log\frac{\alpha}{\eta}\right)^r,$$

by less than

$$\frac{4}{L}\left(\frac{\alpha}{\eta}\right)^k R\,\frac{3N^{1/\eta}}{\eta^{2k}} \leq \frac{12}{L}\,\frac{N^{1/\eta}}{\eta^{3k}}\,R.$$

Choosing L large enough we can make this $\leq \epsilon$. Thus the smallest of the values of (4.55) that we compute is $\leq T_k^+(\eta, \alpha) + 2\epsilon$ and also $\geq T_k^+(\eta, \alpha)$, because of (4.54). Since ϵ can be chosen arbitrarily small this shows that $T_k^+(\eta, \alpha)$ can be computed to any given degree of accuracy.

The case of $T_k^-(\eta, \alpha)$ is treated in a similar way.

The nature of the sieves Λ^+ and Λ^- in Lemma 4 makes it natural to look at sieves where the λ_d for $d = p_1 \ldots p_r$ are given as functions of $\log p_1/\log x, \ldots, \log p_r/\log x$. We let $F(v_1, \ldots, v_r)$ denote functions of r arguments, symmetric in these and such that $F(v_1, \ldots, v_r) = 0$ for $v_1 + v_2 + \cdots + v_r > 1$, and defined for the variables v in a range $(0, \alpha)$ or (η, α) (which we may indicate by a subscript of F; F_α or $F_{\eta,\alpha}$), and we define

$$\begin{aligned}\theta(v_1, \ldots, v_r) = 1 - \sum_{1\leq i\leq r} F(v_i) + \sum_{1\leq i\leq j\leq r} F(v_i, v_j)\\ - \sum_{1\leq i\leq j\leq k\leq r} F(v_i, v_j, v_k) + \cdots + (-1)^r F(v_1, v_2, \ldots, v_r)\end{aligned} \tag{4.56}$$

and consider the class of such functions F for which

$$\theta(v_1, v_2, \ldots, v_r) \geq 0, \tag{4.57}$$

for all r and all choices of $v_1, v_2, \ldots, v_r$ within the range of definition. We denote these F by the symbol F^+.

Similarly we may consider the class of such functions for which

$$\theta(v_1, v_2, \ldots, v_r) \leq 0 \tag{4.58}$$

for $r \geq 1$, and all choices of $v_1, \ldots, v_r$ within the range of v. We denote these F by the symbol F^-.

We can then prove

THEOREM 2. *We have*

$$T_k^+(\alpha) = \text{l.u.b.}\left\{1 + k\int_0^\alpha \frac{\theta(v)}{v}\,dv + \frac{k^2}{2!}\int_0^\alpha\int_0^\alpha \frac{\theta(v_1, v_2)}{v_1 v_2}\,dv_1\,dv_2 + \cdots + \frac{k^n}{n!}\int_0^\alpha \cdots \int_0^\alpha \frac{\theta(v_1, \ldots, v_n)}{v_1 \ldots v_n}\,dv_1 \ldots dv_n + \cdots\right\}, \tag{4.59}$$

where F varies over the class F_α^+.

$$T_k^-(\alpha) = \text{g.l.b.}\left\{1 + k\int_0^\alpha \frac{\theta(v)}{v}\,dv + \cdots + \frac{k^n}{n!}\int_0^\alpha \cdots \int_0^\alpha \frac{\theta(v_1, \ldots, v_n)}{v_1 \ldots v_n}\,dv_1 \ldots dv_n + \cdots\right\}, \tag{4.59$'$}$$

where F varies over the class F_α^-.

$$\begin{aligned} T_k^+(\eta, \alpha) &= \text{l.u.b.}\left\{1 + k\int_\eta^\alpha \frac{\theta(v)}{v}\,dv + \cdots + \frac{k^n}{n!}\int_\eta^\alpha \cdots \int_\eta^\alpha \frac{\theta(v_1, \ldots, v_n)}{v_1 \ldots v_n}\,dv_1 \ldots dv_n \ldots\right\} \\ &= \text{l.u.b.}\left(\frac{\alpha}{\eta}\right)^k \left\{1 - k\int_\eta^\alpha \frac{F^+(v)}{v}\,dv + \cdots + (-1)^n \frac{k^n}{n!}\int_\eta^\alpha \cdots \int_\eta^\alpha \frac{F^+(v_1, \ldots, v_n)}{v_1 \ldots v_n}\,dv_1 \ldots dv_n \ldots\right\}, \end{aligned} \tag{4.59$''$}$$

where F^+ varies over the class $F_{\eta,\alpha}^+$.

$$\begin{aligned} T_k^-(\eta, \alpha) &= \text{g.l.b.}\left\{1 + k\int_\eta^\alpha \frac{\theta(v)}{v}\,dv + \cdots + \frac{k^n}{n!}\int_\eta^\alpha \cdots \int_\eta^\alpha \frac{\theta(v_1, \ldots, v_n)}{v_1 \ldots v_n}\,dv_1 \ldots dv_n + \cdots\right\} \\ &= \text{g.l.b.}\left(\frac{\alpha}{\eta}\right)^k \left\{1 - k\int_\eta^\alpha \frac{F^-(v)}{v}\,dv + \cdots + (-1)^n \frac{k^n}{n!}\int_\eta^\alpha \cdots \int_\eta^\alpha \frac{F^-(v_1, \ldots, v_n)}{v_1 \ldots v_n}\,dv_1 \ldots dv_n + \cdots\right\}, \end{aligned} \tag{4.59$'''$}$$

where F^- varies over the class $F^-_{\eta,\alpha}$. It should be noted that series in the last curly brackets of (4.59″) *and of* (4.59‴) *are finite, as they break off when $n\eta \geq 1$, the terms being from then on identically zero.*

The statements (4.59″) and (4.59‴) are fairly obvious from the preceding. For (4.59″) say, the sieve Λ^+ constructed in Lemma 4, gives if we put $F(v_1, \ldots, v_r) = (-1)^r l(i_1, \ldots, i_r)$ if $\eta t^{i_\nu - 1} < v_\nu \leq \eta t^{i_\nu}$, for $1 \leq \nu \leq r$, that the right-hand side of (4.59″) is $\leq T_k^+(\eta, \alpha) + \epsilon$. Also by constructing from a F^+ a sieve $\Lambda_F^+[P(x^{\eta(1-\epsilon)}, x^{\alpha(1-\epsilon)}), x^{1-\epsilon}]$ by putting λ_d for $d = p_1 \ldots p_r$ with $p_\nu \in P(x^{\eta(1-\epsilon)}, x^{\alpha(1-\epsilon)})$, $1 \leq \nu \leq r$, to be

$$\lambda_d = (-1)^r F^+\left(\frac{\log p_1}{(1-\epsilon)\log x}, \ldots, \frac{\log p_r}{(1-\epsilon)\log x}\right),$$

where F^+ belongs to the class $F^+_{\eta,\alpha}$, we get easily that the right-hand side of (4.59″) is $\geq T_k^+(\eta(1-\epsilon), \alpha(1-\epsilon))$. Letting $\epsilon \to 0$ we establish the equality of the two sides. (4.59‴) can be shown in a similar manner.

To establish (4.59) and (4.59′) is a little bit more involved, and uses beside the above ideas principle I in a somewhat similar way as in the proof of Lemma 2, it is complicated by the fact that we have to show that the $\Lambda^+[P(x^\eta), x^{1/2\sqrt{\eta}}]$ and $\Lambda^-[P(x^\eta), x^{1/2\sqrt{\eta}}]$ used can be chosen both such that $T_u(\Lambda^+)$ and $T_u(\Lambda^-)$ are close to 1, while the λ_d, for these sieves when $d = p_1 \ldots p_r$ with the $p \in P(x^\eta)$, can be chosen as "nice" functions (functions constant in intervals will do) of the ratios $\log p_1/\log x, \ldots, \log p_r/\log x$. This requires going over the arguments that lead up to (5.16) and (5.18) from the beginning, with this restriction imposed on the λ. This is not really difficult, but rather tedious.

There seems to be no real use one can make of the formulas given in Theorem 2. One may, however, feel that it is nice to have some kind of analytic expressions for these four functions, which do not in any way involve the prime numbers or other concepts from number theory.

Of particular interest in connection with the functions $T_k^+(\alpha)$ and $T_k^-(\alpha)$ is for a given k, whether $T_k^-(\alpha)$ vanishes for any $0 \leq \alpha \leq 1$. As will appear from results given in §5, we have for $0 \leq k < \frac{1}{2}$ that $T_k^-(\alpha) > 0$ for $0 \leq \alpha \leq 1$, for $k = \frac{1}{2}$, we have $T_{1/2}^-(1) = 0$, and $T_{1/2}^-(\alpha) > 0$ for $0 \leq \alpha < 1$. For $k > \frac{1}{2}$ there is a value $\alpha(k) < 1$ for which $T_k^-(\alpha(k)) = 0$, we refer to this as the "sieving limit" for k or for the k-residue sieve. For $k \leq \frac{1}{2}$ we put $\alpha(k) = 1$; although for $k < \frac{1}{2}$ we have that $T_k^-(\alpha(k)) > 0$, $\alpha(k)$ is still a "sieving limit", since there is no way of carrying the sieving process for the lower bounds beyond $\alpha = 1$, even if $T_k^-(1) > 0$. We shall give a more complete account of what is known about $\alpha(k)$ in §5, in connection with the specific sieve procedures discussed there.

Going back to (4.59″) and (4.59‴), it is possible to show that there exist "reasonable" functions for which the extremal value is attained. They are not necessarily unique. We can show that in (4.59) and (4.59′) we may impose the restriction $F(v_1, \ldots, v_r) = 1$, on F^+ and F^-, if $\alpha < \alpha(k)$ and $v_1 + \cdots + v_k < 1 - \alpha/\alpha(k)$. In fact, it can be shown that an F^+ or F^- for which this restriction would be violated outside of a set of measure zero, could not be one for which the extremal value was attained. For the sieve corresponding to F the condition means that

$$\lambda_d = \mu(d) \quad \text{for} \quad d < x^{1-\alpha/\alpha(k)}.$$

For applications of the general sieve it is of course not really the function $T_k^-(\alpha)$ but max $(0, T_k^-(\alpha))$ which is of interest, since it is not really $M_u^{=}(x, x^\alpha)$ but rather

$$M_u^-(x, x^\alpha) = \max(0, M_u^{=}(x, x^\alpha)),$$

which enters in these applications.

Before closing this section about the general sieve for the k-residue case, we want to make some remarks about the sensitivity of the general sieve to changes in the remainder terms in the formulas (2.1). We have seen already that one could give up some of the restrictions imposed on $u(p)$ in our earlier arguments.

It is easily seen, by using the sieve constructed in the proof of Lemma 4, that if we replace (2.1) by

$$(4.60) \qquad N_d = \frac{x}{f(d)} + O\left(\frac{x}{f(d)(\log x/d)^h}\right),$$

for $d \leq x^{1-\delta}$ and every $\delta > 0$, with some constant $h > 2k$, and require no information about the N_d for $d > x$, the results remain unaltered with the exception of those relating to M^- and $M^=$ for $\alpha = 1$.

Again, it may be sufficient to have only an estimate of some average of the remainder terms in (2.1). If we write

$$(4.61) \qquad N_d = x/f(d) + R_d,$$

and for some constant $0 < \mu \leq 1$ and every $\epsilon > 0$, and some $h > k$, have that

$$(4.62) \qquad \sum_{d \leq x^{\mu-\epsilon}} |R_d| = O\left(\frac{x}{(\log x)^h}\right),$$

then we can show

$$(4.63) \qquad \limsup_{x\to\infty} \frac{M(x, x^\alpha)}{xW_u(x^\alpha)} \leq T_k^+\left(\frac{\alpha}{\mu}\right)$$

for $0 < \alpha \leq \mu$, and

$$(4.64) \qquad \liminf_{x\to\infty} \frac{M(x, x^\alpha)}{xW_u(x^\alpha)} \geq T_k^-\left(\frac{\alpha}{\mu}\right)$$

for $0 < \alpha < \mu$.

This is what lies behind the great importance of Bombieri's results about primes in arithmetic progressions[9] for the applications of the sieve method to certain problems.

5. The theory of the general sieve does not at present give us any simple expressions for $T_k^+(\alpha)$ or $T_k^-(\alpha)$, nor does it even give us any practical algorithm or procedure for computing these functions. The "effective method of computation" sketched in the proof of Theorem 1, has of course only theoretical interest and is unusable for practical purposes.

---

[9]Bombieri [3].

What we could call "practical" sieve methods or procedures are therefore still needed, and we shall in this section give a brief account of such methods. The methods enable us in general to obtain useful upper bounds for $T_k^+(\alpha)$ (or $T_k^+(\eta, \alpha)$) or lower bounds for $T_k^-(\alpha)$ (or $T_k^-(\eta, \alpha)$). Also we have to substantiate certain estimates for which we earlier in these pages have referred to §5.

The first really effective sieve procedures were introduced by Viggo Brun around 1920. He depended mainly on ideas of a combinatorial nature, like the fact that

$$\sum_{0\leq s\leq t} (-1)^s \binom{r}{s}$$

is always nonnegative if r and t are positive integers and t even, and nonpositive if r and t are positive integers and t odd.

Brun's first sieve was constructed from this fact and utilization of principle I of §3. This was not yet a very effective instrument, but Brun improved his method by introducing refinements. Later further improvements were made notably by H. Rademacher also in the 1920's. Even for larger values of k the method is very good if α is small enough.

An important step was made by Buchstab in 1937. Buchstab used a special case of the principles II and III in §3, to improve successively the bounds obtained by Brun's method.

In the early 1950's Barkley Rosser devised a sieve procedure that essentially represents the limit of the Buchstab procedure. Rosser was able by his method to determine the functions $T_k^+(\alpha)$ and $T_k^-(\alpha)$ explicitly for $k = 1$, to be precise, his method (combined with certain examples given in [7] which show that his results are best possible) gives $T_1^+(\alpha)$ for $0 \leq \alpha \leq 1$ and $T_1^-(\alpha)$ for $0 \leq \alpha \leq \frac{1}{2}$. It is not clear whether $T_1^-(\alpha)$ has been determined yet for $\frac{1}{2} < \alpha < 1$ where it is negative. Rosser, like his predecessors, was anyway only concerned with the quantity $M_u^-(x, x^\alpha)$ and so with max $(0, T_k^-(\alpha))$ rather than $T_k^-(\alpha)$.

I shall refer to this sieve as the Buchstab-Rosser sieve. It is still like the original Brun method in that the λ_d have to be either $\mu(d)$ or zero, no other value can enter. This seems to be almost unavoidable for sieves that are basically combinatorial, and when k is large and α not small it is the reason for the weakness of these methods.

Buchstab's idea can briefly be explained as follows: He uses principle II to extend the range of a $\Lambda^+[P, Z]$ by adding a new prime p^* to the range P. The original sieve used over the longer range has $\theta_{p*} = 1$, taking p^* as our d_0 we can then use II to construct a new sieve over the extended range which has $\theta_{p*} = 0$. To extend the range of a $\Lambda^-[P, Z]$ by again adding a prime p^* to P, he uses principle III since the original sieve used over the larger range has $\theta_{p*} = 1$, which has the wrong sign. Using III we can construct a Λ^- over the larger range with $\theta_{p*} = 0$.

Using these two procedures one arrives at the following:[10] Let $u \geq 1$ and suppose we have found bounds for $T_k^+(1/u)$ and $T_k^*(1/u)$ as follows: $f_k^+(u) \geq T_k^+(1/u)$, and $f_k^-(u) \leq T_k^*(1/u)$.

[10]We introduce the notation $T_k^*(\alpha) = \max(0, T_k^-(\alpha))$ which will be used throughout this section.

If $1 \leq u < v$ we may then replace $u^k f_k^+(u)$ by

$$v^k f_k^+(v) - k \int_u^v t^{k-1} f_k^-(t-1)\, dt,$$

if the latter expression is smaller.

Similarly $u^k f_k^-(u)$ can be replaced by

$$v^k f_k^-(v) - k \int_u^v t^{k-1} f_k^+(t-1)\, dt,$$

if the latter expression is larger.

One starts with the bounds obtained by Brun's method, these are very good for large u, and by successive improvements one is in the limit led to bounds $f_k^+(u)$ and $f_k^-(u)$ which now satisfy the difference-differential equations

$$(5.1) \qquad (u^k f_k^+(u))' = k u^{k-1} f_k^-(u-1)$$

(where the right hand side is interpreted as zero for $u - 1 \leq 1$), and

$$(5.2) \qquad (u^k f_k^-(u))' = k u^{k-1} f^+(u-1),$$

(where the right-hand side is interpreted as zero if $f_k^-(u) = 0$).

Rosser constructed a sieve that by a suitable choice of a parameter $\beta \geq 1$ as a function of k, gives directly the limit case of the Buchstab method. Briefly described, his method consists in determining the λ_d for $\Lambda^+[P, Z]$ in the following way:

If $d = p_1 p_2 \ldots p_r$ where $p_1 < p_2 < \cdots < p_r$, then if all the inequalities

$$(5.3) \qquad p_{r-2t}^{\beta+1} p_{r-2t+1} \ldots p_r \leq Z,$$

are satisfied, for $0 \leq t \leq [r/2]$, we put $\lambda_d = \mu(d)$, otherwise we put $\lambda_d = 0$.

For the construction of $\Lambda^-[P, Z]$ it is required that P be a subset of $P(Z^{1/\beta})$. Again, for $d = p_1 p_2 \ldots p_r$ where $p_1 < p_2 < \cdots < p_r$, then if all the inequalities

$$(5.4) \qquad p_{r-2t-1}^{\beta+1} p_{r-2t} \ldots p_r \leq Z,$$

hold for $0 \leq t \leq [(r-1)/2]$, we put $\lambda_d = \mu(d)$, otherwise we put $\lambda_d = 0$.

The estimation of the expressions involved and the determination of the optimal choice of β were problems that Rosser solved completely for $k = 1$. If we denote the optimal choice of β by $\beta(k)$, Rosser showed that $\beta(1) = 2$.

Rosser's work was never published, but a paper about 10 years later by Jurkat and Richert [4] obtained essentially the same results for $k = 1$, by methods which combine Buchstab's ideas with estimations obtained by Brun's method and by a method mentioned in [6] and [7] for the upper bound. If one carefully analyzed their method, it would probably turn out that this is really the Buchstab-Rosser sieve again.

Going back to Buchstab's formulation and the two functions satisfying (5.1) and (5.2), the optimal choice of β is determined by the conditions $f_k^-(\beta) = 0$ and

$f_k^-(u) > 0$ for $u > \beta$; if $f_k^-(u) > 0$ for all $u > 1$, we take $\beta = 1$. $f_k^+(u)$ and $f_k^-(u)$ both tend very strongly to 1 as $u \to \infty$. From (5.16) and (5.18) which we prove later, it follows that they both are equal to $1 + O(e^{-u/20})$, and much sharper estimates are easily given. It can also be shown by Brun's method. The functions $f_k^+(u)$ and $f_k^-(u)$ can then be determined from (5.1) and (5.2) and the "boundary behaviour" at $u = \infty$ and at $u = \beta$. The equations can be solved most conveniently by passing to the Laplace transforms of the functions which satisfy ordinary differential equations that can be integrated by standard means. The $\beta(k)$ in general has no simple expression in terms of k, being determined by rather complicated transcendental relations.

However, if k is the half of an integer, it turns out that $\beta(k)$ satisfies a simpler relation. We shall sketch a procedure for determining $\beta(k)$ if $2k = n$, where n is a positive integer:

We consider the function $\Delta_k(u) = f_k^+(u) - f_k^-(u)$, and its Laplace transform

$$K(Z) = \int_\beta^\infty e^{-uZ}\,\Delta_k(u)\,du. \tag{5.5}$$

It can be shown that $K(Z)$ satisfies a differential equation

$$(Z(K(Z))' = k(1 + e^{-Z})K(Z) + c_k e^{-Z}\,U_k(Z), \tag{5.6}$$

where c_k is a certain positive constant, and

$$U_k(Z) = k\int_\beta^{\beta+1} \frac{e^{-(t-1)Z}}{(t-1)^k}\,dt - \beta^{1-k}e^{(1-\beta)Z}. \tag{5.7}$$

Since $\Delta_k(u) = O(e^{-u/20})$ as $u \to \infty$, $K(Z)$ must be regular at $Z = 0$. If we differentiate (5.6) ν times for $0 \le \nu \le n = 2k$, and put $Z = 0$, the $(n + 1)$ equations we get contain the expressions $K^{(\nu)}(0)$ for $0 \le \nu \le n - 1$, but no $K^{(n)}(0)$. Eliminating these n quantities, we end up with a linear relation between the $U_k^{(\nu)}(0)$ for $0 \le \nu \le n$. This is a relation that contains only β as unknown quantity, since the factor c_k can be divided out.

One finds in this way:

$$\beta(\tfrac{1}{2}) = 1, \qquad \beta(1) = 2, \qquad \beta(\tfrac{3}{2}) = \frac{5 + \sqrt{3}}{2},$$

$\beta(2) \approx 4.79\ldots$, and is the largest root of the equation:

$$\beta^4 - 9\beta^3 + 24\beta^2 - \tfrac{55}{3}\beta - 1 = 0.$$

$\beta(\tfrac{5}{2}) \approx 6.2\ldots$, and is the largest root of the equation:

$$(\beta - 1)^4 - 10(\beta - 1)^3 + 30(\beta - 1)^2 - \tfrac{85}{3}(\beta - 1) + \tfrac{55}{12} = 0.$$

In general $\beta(k)$ is an algebraic number if k is half of an odd integer, since the $U_k^{(\nu)}(0)$ then all are algebraic expressions in β. If k is a positive integer the $U_k^{(\nu)}(0)$ are algebraic expressions in β if $\nu \ne k - 1$, $U_k^{(k-1)}(0)$ however contains a term $k \log (\beta/(\beta - 1))$. Only if this derivative drops out from the relation between the $U_k^{(\nu)}(0)$ is $\beta(k)$ algebraic, otherwise it satisfies an equation $\log (\beta/(\beta - 1)) = R(\beta)$, where R is a rational function with rational coefficients, thus in this case $\beta(k)$ will be a transcendental number. For $k = 1$ and $k = 2$ the derivative $U_k^{(k-1)}(0)$ does drop out. I do not know if this can happen for any other positive integer k.

Once $\beta(k)$ is determined, it remains to determine the constant c_k. This can be done by looking at

$$f_k^+(u) + f_k^-(u)$$

which satisfies a simple difference-differential equation also, and its Laplace transform which satisfies again a comparatively simple ordinary differential equation. For $Z \to 0$ this Laplace transform must behave like $2/Z$. This actually determines c_k. The functions $f_k^+(u)$ and $f_k^-(u)$ are then fully determined. For $k = \frac{1}{2}$ and $k = 1$, we can find simple closed expressions for c_k. This does not seem possible in general when k is the half of an integer. Only when $\beta(k)$ also is an integer do such expressions seem to be obtainable, and it is quite possible that $2k$ and $\beta(k)$ are integers at the same time only for $k = \frac{1}{2}$ and $k = 1$.

For $k = 1$, one can show that the Buchstab-Rosser sieve gives the best possible result, so that $\alpha(1) = 1/\beta(1) = \frac{1}{2}$, $T_1^+(\alpha) = f_1^+(1/\alpha)$ for $0 \leq \alpha \leq 1$, and $T_1^*(\alpha) = f_1^-(1/\alpha)$ for $0 \leq \alpha < 1$ (or $T_1^-(\alpha) = f_1^-(1/\alpha)$ for $0 \leq \alpha \leq \frac{1}{2}$).

This can be seen by looking at the following two examples of S_x: (a) $0 < n \leq x$, number of prime factors of n (counted with their multiplicities) $\nu(n)$ odd. And (b) $0 < n \leq x$, $\nu(n)$ even, unfortunately this second example, which shows us that $f_1^-(u)$ is best possible, can tell us nothing about the values of $T_1^-(\alpha)$ for $\frac{1}{2} < \alpha < 1$.

For $k = \frac{1}{2}$, the Buchstab-Rosser sieve again gives the best possible result, so that

$$\alpha(\tfrac{1}{2}) = \frac{1}{\beta(\frac{1}{2})} = 1, \qquad T_{1/2}^+(\alpha) = f_{1/2}^+(\alpha)$$

for $0 \leq \alpha \leq 1$, and $T_{1/2}^-(\alpha) = T_{1/2}^*(\alpha) = f_{1/2}^-(\alpha)$, for $0 \leq \alpha \leq 1$.

This can be seen by looking at the following two examples of S_x: (a) The sequence we get by dropping from the numbers of the form $4m + 1 < x$ all their prime factors of the form $4m + 1$, and (b) the sequence we get by dropping from the numbers of the form $4m + 3 < x$ all their prime factors of the form $4m + 1$. Here $u(p) = 1$ for $p \equiv 3$ (4) and $u(p) = 0$ otherwise.

It is certainly possible that the Buchstab-Rosser sieve is optimal for other values of k in the interval $0 < k < 1$, but this would seem impossible to show in a similar way. I do not wish to venture any guesses about this.

For the larger values of k the Buchstab-Rosser sieve is not very effective unless α is small. For these k the methods which I have indicated in [5], [6] and [7] are both more convenient and efficient.

For the upper bound the first step of this method depends on the remark at the end of IV in §3 on how to produce a $\Lambda^+[P, Z^2]$ from a $\Lambda[P, Z]$.

We get then as upper bound for $M(x, P)$,

$$(5.8) \qquad x \sum_{d,d' \leq Z} \frac{f((d, d'))}{f(d)f(d')} \lambda_d \lambda_{d'} + O\left(\sum_{d,d' < Z} \frac{u(d)u(d')}{u((d, d'))} |\lambda_d| \cdot |\lambda_{d'}| \right).^{11}$$

[11] d and d' in these and following expressions are members of (P).

Minimizing the quadratic form in the leading term of (5.8) under the condition $\lambda_1 = 1$, we get a quite simple expression

$$\frac{x}{\sum_{d \le Z} \frac{1}{f'(d)}} \tag{5.9}$$

for the minimum, and the expressions

$$\lambda_d = \mu(d) f(d) \frac{\sum_{d|d' \le Z} \frac{1}{f'(d')}}{\sum_{d' \le Z} \frac{1}{f'(d')}}, \tag{5.10}$$

for the λ's. These make it easy to obtain estimates for the expression in the Θ-term of (5.8). If we assume that the $u(p)$ satisfy our conditions of the earlier paragraphs we can show that

$$\sum_{d,d' \le Z} \frac{u(d)u(d')}{u((d, d'))} |\lambda_d| \cdot |\lambda_{d'}| = O\left(\frac{Z^2}{\log^2 Z}\right), \tag{5.11}$$

where the constant implied by the O in (5.11) depends on the particular u. For most purposes somewhat cruder estimates are as useful, and easier to obtain. It is of interest that we can obtain estimates with absolute constants which are valid if we assume nothing about $u(p)$ beyond $0 \le u(p) < p$ for all p, and which are not much worse than (5.11). Thus for instance

$$4Z^2(1 + \log Z)^3, \tag{5.11$'$}$$

or

$$16Z^{2+\epsilon}/\epsilon^3, \tag{5.11$''$}$$

where $0 < \epsilon < 1$, are both upper bounds for the left hand side of (5.11).

For the k residue sieve this leads to upper bounds for $T_k^+(\alpha)$ which for the larger k are much better than those given by the $f_k^+(1/\alpha)$ obtained by the Buchstab-Rosser method. We may denote this new bound for $T_k^+(1/u)$ by $g_k^+(u)$.

For the lower bound one can use for instance first a method mentioned in [6] and [7], which essentially consists of combining the upper bound method with principle III from §3. This gives a lower bound for $T_k^-(1/u)$ which can be expressed as

$$g_k^-(u) = 1 - ku^{-k} \int_u^\infty t^{k-1}(g_k^+(t-1) - 1)\, dt. \tag{5.12}$$

The two functions g_k^+ and g_k^- are, however, only so to say the first approximation. Namely, when u is not too small (more precisely for $u > 1/\alpha(k)$) it is certain that $g_k^+(u)$ can be improved upon. There are then enough small d_0 with $\theta_{d_0} > 0$ and large enough that principle II leads to improvement on $g_k^+(u)$. In addition, one may use II also in the way it is done in the Buchstab procedure. The improved

g_k^+ in turn leads to an improved g_k^-, which again (since any use of II requires the use of lower bounds) allows us further to improve the g_k^+ and so forth.

It seems difficult to follow up this process to the limit, partly because of the flexibility. If one uses principles II and III only the way it is done in the Buchstab procedure, one gets something that is simpler to follow explicitly, but the limit may then for general k (for $k = \frac{1}{2}$ and $k = 1$ we know that it would lead to the best possible results) possibly not be quite as good as what one could get from the more general way of using II.

The "first approximation", the $g_k^+(u)$ and $g_u^-(u)$ obtained from (5.8) and (5.12) have been investigated for large k by Ankeny and Onishi in [2] where also some numerical tables are given.

As approximations to $T_k^+(1/u)$ and $T_k^-(1/u)$, these $g_k^+(u)$ and $g_k^-(u)$ have the advantage of being tolerably good over a much larger range in k and u than $f_k^+(u)$ and $f_k^-(u)$.

In particular, the bound one gets for $\alpha(k)$ from $g_k^-(u)$ is substantially better than the bound $1/\beta(k)$ already for $k = 3/2$ for instance. For $0 < k < 1$ or k sufficiently close to 1, the g_k^+ and g_k^- would not give as good bounds as f_k^+ and f_k^-. We would in these cases have to go through the procedure of successive improvements, and in case $0 < k \leq 1$, go to the limit in order to equal the f_k^+ and f_k^- bounds.

The bounds we get for $\alpha(k)$ from $1/\beta(k)$ or from $g_k^-(u)$ are lower bounds. To get upper bounds one may apply the general sieve to certain sequences specially constructed for the purpose. It was in this way it could be established that for $k = 1$ or $k = \frac{1}{2}$ we had determined the correct value of $\alpha(k)$. For other k this method leads to upper bounds for $\alpha(k)$, which in most cases probably are not very good.

For $\alpha(2)$ we get from $g_2^-(u)$ that $\alpha(2) > 1/4.43$, in [7] it was stated that $\alpha(2) < 1/(1 + \exp \frac{3}{4})$. This was obtained by applying the sieve to the sequence $x < n < 2x$, with each n counted with the weight $w_n = \mu^2(n)\{t(n)(1 - \mu(n)) + 16\mu(n)\}$, where $\mu(n)$ is Möbius function and $t(n)$ the number of divisors of n. The resulting N_d is not quite of the form (2.1), both main term and remainder term being more complicated, but one can show that any sieve Λ^- which gives a nonnegative lower bound for the 2-residue sieve also would give a nonnegative lower bound when applied to the above sequence.

Since for this sequence, if $P = P(x^\alpha)$, $\frac{1}{4} < \alpha < \frac{1}{2}$, the lower bound could not be larger than

$$-12(\pi(2x) - \pi(x)) + 16 \sum_{x^\alpha < p < \sqrt{2x}} \left(\pi\left(\frac{2x}{p}\right) - \pi\left(\frac{x}{p}\right)\right),$$

for the above sequence, one gets easily $\alpha(2) \leq 1/(1 + \exp \frac{3}{4})$ since when α is larger the expression above is negative for large x. A closer analysis leads to $\alpha < 1/(1 + \exp \frac{3}{4})$ and even to the replacement of this upper bound by a smaller one. My guess is that $\alpha(2) < \frac{1}{4}$.

In order to justify certain steps in §4, we need to obtain certain bounds, an upper bound for $M_u^+(x, P)$ and a lower bound for $M_u^-(x, P)$, where $P = P(\xi)$ where ξ is comparatively small (that is, less than some x^t where t is a small positive constant). We assume that $u(p)$ satisfies our condition (4.1) and otherwise the conditions imposed on $u(p)$ in §2.

From (5.8) and (5.11″) we have for $0 < \epsilon < \frac{1}{2}$

$$M_u^+(x, P(\xi)) \le \frac{x}{\sum_{d\le Z} \frac{1}{f'(d)}} + O\left(\frac{Z^{2+\epsilon}}{\epsilon^3}\right), \tag{5.13}$$

where the constant in the O-term depends only on the constant involved in the O-term of (2.1). We choose $Z = x^{1/2-\epsilon}$, and have then $Z^{2+\epsilon} < x^{1-3\epsilon/2}$. Furthermore, we have

$$\begin{aligned}\sum_{d\le Z}\frac{1}{f'(d)} &\ge \sum_{d\in(P)}\frac{1}{f'(d)} - Z^{-\vartheta}\sum_{d\in(P)}\frac{d^{\vartheta}}{f'(d)} \\ &= \frac{1}{W_u(\xi)}\left\{1 - Z^{-\vartheta}\prod_{p\in P}\left(1+\frac{p^{\vartheta}-1}{f(p)}\right)\right\},\end{aligned} \tag{5.14}$$

where $\vartheta > 0$ is to be determined later. For $p \le \xi$,

$$1 + \frac{p^{\vartheta}-1}{f(p)} \le e^{(p^{\vartheta}-1)/f(p)} \le \exp\left(\frac{\xi^{\vartheta}}{\log \xi}\,\frac{u(p)\log p}{p}\right).$$

Thus

$$\prod_{p\in P(\xi)}\left(1+\frac{p^{\vartheta}-1}{f(p)}\right) \le \exp\left\{\frac{\xi^{\vartheta}}{\log\xi}V_u(\xi)\right\},$$

this gives us

$$Z^{-\vartheta}\prod_{p\in P}\left(1+\frac{p^{\vartheta}-1}{f(p)}\right) \le \exp\left\{-\vartheta\log Z + \frac{\xi^{\vartheta}}{\log\xi}V_u(\xi)\right\}. \tag{5.15}$$

We choose

$$\vartheta = \frac{\log\dfrac{\log Z}{V_u(\xi)}}{\log\xi}$$

which is positive for $V_u(\xi) < \log Z$, and get for the right-hand side of (5.15)

$$\exp\left\{-\frac{\log Z}{\log\xi}\log\frac{\log Z}{eV_u(\xi)}\right\}. \tag{5.15′}$$

If we take $\xi \le x^{1/(7(k+1))}$, we get from (4.1) that for $x \ge x_0$,

$$\log\frac{\log Z}{eV_u(\xi)} > \frac{1}{5},$$

if our fixed ϵ is chosen small enough.

We get from (5.13) using (5.14), (5.15) and (5.15′) that

$$M_u^+(x, P(\xi)) \le xW_u(\xi)\{1 + 2e^{-\log x/(10\log\xi)} + O(x^{-\epsilon})\}. \tag{5.16}$$

We can also easily get the estimate valid for $u > 2ek$

$$T_k^+\left(\frac{1}{u}\right) \le g_k^+(u) \le \frac{1}{1 - e^{-(u/2)\log(u/2ek)}}. \tag{5.17}$$

To get an estimate for $M_u^-(x, P(\xi))$, or actually $M_u^=(x, P(\xi))$, with similar assumptions about ξ as before, we can use the method indicated in [6] or [7]. This gives

$$M_u^=(x, P(\xi)) \ge N_1 - \sum_{p \le \xi} u(p) M_u^+\left(\frac{x}{p}, P(p-1)\right).$$

Using now (5.16), we find if we now assume $\xi \le x^{1/(7k+8)}$, that

$$M_u^=(x, P(\xi)) \ge xW_u(\xi)\{1 - 3e^{-\log x/(10 \log \xi)} - O(x^{-\epsilon})\} \tag{5.18}$$

where again the constant in the O depends on u and on ϵ only.

Inserting (5.17) in (5.12) we could get a lower bound for $g_k^-(u)$ and $T_k^-(1/u)$ valid for $u > 2ke + 1$.

The inequalities (5.16) and (5.18) are of course not very sharp, but they are sufficient for the needs of the previous section.

6. The applications of the sieve method to the actual problems of number theory have been directed more towards showing that integers n of some particular form can be found (and also, are not too thinly distributed) whose number of prime factors (counted with multiplicities) lie below certain bounds; rather than to show that they not too rarely have no prime factor $\le n^\alpha$ for some positive fixed α. This means that even a precise knowledge of the functions $T_k^+(\alpha)$ and $T_k^-(\alpha)$, would not necessarily give us best possible results (obtainable by sieve methods) for these applications.

For such applications it will usually be better to apply sieves which count also numbers which have a certain number of prime factors from P, but count them with weights that depend on these prime factors. For the lower bound sieve of this type, we would thus require $\theta_d \le 0$ only for $d \in (P)$, with more than a certain number of prime factors, and for the d with less prime factors, the θ_d for $d > 1$ might be > 0.

Since the possibilities for different choice of weights are so many, it seems difficult to analyze this type of sieve in its full generality along the lines of §4, although if we make sufficiently restrictive assumptions about the weights, methods similar to those of the previous sections will work to some extent. Anyway, the application of the sieve to problems of this kind remains an art where one depends on lucky guesses and on trial and error to some extent.

Since in most actual problems considered, the sets S_x consist of numbers that are bounded by some fixed power of x, say x^a, where $a \ge 1$, it would seem that one might obtain better results by applying what I in [7] have referred to as a "local sieve". That is: a sieve where the inequality (1.6) or (1.9) would be required to hold only for numbers $d_1 \le x^a$. Although it does not seem easy to prove this in general, it is unlikely that the local sieve would present any advantage. For the k-residue sieve with $k = \frac{1}{2}$ or $k = 1$, this can be seen to be so by considering the examples of S_x used in connection with these k in the previous section. Local sieves have of course been useful in proving the asymptotic formulae that form the basis for the elementary proof of the prime number theorem.

The sieve, while it is in some sense a crude tool and of little value where finer analytical techniques will work, has the advantage that almost any application of it kills several birds with one stone so to say, since several quite different number-theoretic problems may be quite equivalent with respect to the sieve. Thus the problem of proving that an irreducible integral-valued polynomial of degree r with no fixed prime divisors, represents (and quite often) numbers which have at most s prime factors, is equivalent by these methods, to showing that between x and $x + x^{1/r}$, there are for large x always integers with at most s prime factors.

For the classical problems, like Goldbach's problem or the twin-prime problem, the major new fact since I gave my lecture at Harvard is that Bombieri's remarkable theorem on primes in arithmetic progressions became available (see [3]). This makes it possible to treat these problems with the 1-residue sieve instead of the 2-residue sieve (while sacrificing by a factor $\frac{1}{2}$ on the exponent α one can use, but still coming out ahead). By Bombieri's result the problem of proving that $p + 2$ infinitely often has not more than s prime factors, becomes equivalent to proving the same for numbers of the form $n^2 + 1$ (this one can do for $s = 3$, see footnote on p. 17 in [6], a very nice version is given in [2]).

Since Bombieri's result is not elementary, however, the result stated in [7] remains the best that has been proved by elementary means. My first proof of this was along similar lines but not so streamlined as the elegant version given in [2].

Let me finally, to indicate the variety of ways in which one may adapt the ideas of the sieve method to a particular problem, by using weights depending on the actual numbers and not only on their small prime factors, sketch a second proof (dating from 1951) for the result that there are infinitely many integers n such that of the pair n and $n + 2$, one has at most two, the other at most three prime factors.

We consider the two expressions

$$Q_1 = \sum_{x<n\leq 2x} (t(n) + t(n+2)) \left\{ \sum_{d/n(n+2)} \lambda_d \right\}^2,$$

and

$$Q_2 = \sum_{x<n\leq 2x} \left\{ \sum_{d/n(n+2)} \lambda_d \right\}^2,$$

where $t(n)$ is the number of divisors of n, and we assume $\lambda_d = 0$ for $d > Z$, where Z is to be chosen as a function of x later.

We try to choose our Z and our λ's in such a way that we can estimate Q_1 and Q_2 well, and such that the ratio Q_1/Q_2 becomes as small as possible.

It is not hard to show that one can make $Q_1/Q_2 < 14$ for all sufficiently large x.

Since the numbers for which either n or $n + 2$ are not squarefree can be shown to contribute too little to influence Q_1 or Q_2 with our choice of λ's, this means that there are pairs n and $n + 2$ with $x < n \leq 2x$, such that if n has s and $n + 2$ has t prime factors we have $2^s + 2^t < 14$, or of the numbers s and t, one is at most 2, the other at most 3.

By a slight modification we could also add the requirement that the smallest prime factor of n and $n + 2$ be $> x^\delta$ with some small positive constant δ, and still show that there are more than $cx/\log^2 x$ such pairs n and $n + 2$, where c is a positive constant depending on δ.

The ratio 14 is, as I said, rather simple to reach, it can be pressed further down, especially if one replaces the expression

$$\sum_{d/n(n+2)} \lambda_d,$$

in Q_1 and Q_2 by

$$\sum_{d/n\,;d'/n+2} \lambda_{d,d'},$$

and let $\lambda_{d,d'}$ be symmetric in d and d', but not necessarily a function of the product dd' only.

If one could bring the ratio below 12, one could improve the result qualitatively since it would exclude the possibility that one number of the pair has 2 and the other 3 prime factors. G. Hofmeister has been able to bring the ratio quite far down, without yet exhausting the existing possibilities for further improvements. However, the calculations get extremely complicated, and it remains uncertain whether 12 can be reached without introducing some new device, that would allow one to extend further the summation range of the d, d' without interfering with the remainder terms in the estimation of Q_1 (which is the more sensitive of the two because of the presence of $t(n)$ and $t(n + 2)$).

So far this seems the most promising approach to these problems (the Goldbach problem can be treated in a similar way with identical results) if one does not use Bombieri's theorem on primes in arithmetical progressions.

REFERENCES

(We have listed here only the relevant papers since 1946, and not the earlier papers by Brun, Rademacher and Buchstab.)

1. N. C. Ankeny, *Applications of the sieve*, Proc. Sympos. Pure Math., vol. VIII, Amer. Math. Soc., Providence, R.I., 1965, pp. 113–118.
2. N. C. Ankeny and H. Onishi, *The general sieve*, Acta Arith. **10** (1964/65), 31–62.
3. E. Bombieri, *On the large sieve*, Mathematika **12** (1965), 201–225.
4. W. B. Jurkat and H.-E. Richert, *An improvement of Selberg's sieve method.* I, Acta Arith. **11** (1965), 217–240.
5. A. Selberg, *On a new method*, Norske Vid. Selsk. Forh. Kgl. (1946).
6. ———, *On elementary methods in primenumber-theory and their limitations*, Den 11te Skand. Mat. Kongr. (Trondheim, 1949), Johan Grundt Tanums Forlag, Oslo, 1952, pp. 13–22.
7. ———, *The general sieve-method and its place in prime number theory*, Proc. Internat. Congress Math. (Cambridge, Mass., 1950) vol. 1, Amer. Math. Soc., Providence, R.I., 1952, pp. 286–292.

THE INSTITUTE FOR ADVANCED STUDY
PRINCETON, NEW JERSEY

DENSITY THEOREMS FOR THE ZETA FUNCTION

ENRICO BOMBIERI

I.

Let $N(\alpha, T)$ denote the number of zeros of the Riemann Zeta function $\zeta(s)$ in the rectangle

$$\alpha \leq \sigma \leq 1, \qquad 0 \leq t \leq T.$$

Let

$$\mu_1(\alpha) = \limsup_{T\to\infty} \frac{\log^+ N(\alpha, T)}{\log T}$$

and let

$$\mu(\alpha) = \limsup_{|t|\to\infty} \frac{\log^+ |\zeta(\alpha + it)|}{\log |t|}.$$

Consider the following condition

$$(\mathrm{T}_\alpha) \qquad \zeta(s) \neq 0 \quad \text{for } \sigma > \alpha \text{ and } |t - T| < (\log T)^2.$$

In this case something more can be said on the order of magnitude of $\zeta(s)$ inside the critical strip. Let

$$\mu(\sigma; \alpha) = \limsup_{|t|\to+\infty}{}' \frac{\log^+ |\zeta(\sigma + it)|}{\log |t|}$$

where the accent means that we restrict the values of t to intervals $|t - T| < \frac{1}{2}(\log T)^2$, where T satisfies condition (T_α).

It is easily seen using the functional equation of $\zeta(s)$ and the fact that $\zeta(s)$ is real on the real axis that

(i) $\mu(\sigma; \alpha) = \frac{1}{2} - \sigma + \mu(1 - \sigma; \alpha)$,
(ii) $\mu(\sigma; \alpha)$ is a convex function of σ,
(iii) $\mu(\sigma; \alpha) = 0$ for $\sigma \geq \alpha$,

property (ii) being an easy consequence of Hadamard's three circle theorem and property (iii) being essentially due to Littlewood.

By (ii) the function $\mu(\sigma;\alpha)$ is continuous and has right and left derivatives everywhere. We define $-\gamma(\alpha)$ to be the left derivative of $\mu(\sigma;\alpha)$ at the point $\sigma = \alpha$:

$$\gamma(\alpha) = -\mu'_{-}(\alpha;\alpha).$$

It follows at once from (i), (ii), (iii) that

$$\gamma(\alpha) \leq \tfrac{1}{2}.$$

THEOREM. *Let $\alpha > (\theta + 1)/2$, $\frac{1}{2} < \alpha \leq 1$, where θ is any number $\theta < 1$. Then we have*

$$\mu_1(\alpha) \leq \left[\frac{\mu(\theta)}{2\alpha - 1 - \theta} + \gamma(\alpha)\right] 2(1-\alpha).$$

In particular, letting $\theta \to -\infty$ we have

$$\mu_1(\alpha) \leq [2 + 2\gamma(\alpha)](1-\alpha).$$

This theorem contains as special cases some recent results of Turán and Halász on the function $\mu_1(\alpha)$.

COROLLARY 1. *We have*

$$\mu_1(\alpha) = 0 \quad \textit{for } \alpha > \tfrac{3}{4}$$

if the Lindelöf hypothesis $\mu(\frac{1}{2}) = 0$ is true.

COROLLARY 2. *If $\gamma(\alpha) = 0$ then the density hypothesis $\mu_1(\alpha) \leq 2(1-\alpha)$ is true.*

COROLLARY 3. *If $\alpha > (\theta + 1)/2$ and $\alpha > \psi$ we have*

$$\mu_1(\alpha) \leq \left[\frac{\mu(\theta)}{2\alpha - 1 - \theta} + \frac{\mu(\psi)}{\alpha - \psi}\right] 2(1-\alpha);$$

taking $\theta = \psi = 1 - k(1-\alpha)$ with $k > 2$ one has

$$\mu_1(\alpha) \leq \left(\frac{2}{k-1} + \frac{2}{k-2}\right)\mu(\alpha - (k-1)(1-\alpha))$$

for each $k > 2$.

PROOF. We have $\gamma(\alpha) \leq \mu(\psi)/(\alpha - \psi)$ by (i), (ii), (iii) and the obvious inequality $\mu(\sigma;\alpha) \leq \mu(\sigma)$.

It follows from Corollary 3 and the known results for the Lindelöf μ function that $\mu_1(\alpha) = o(1-\alpha)$ as $\alpha \to 1-$, and in fact rather more than this.

The results of Corollaries 1 and 2 and the consequence of Corollary 3 are already in the paper [1] of Halász and Turán.

Our proof is different but makes use of the ideas developed in [1] as well as of a method of Montgomery [2]. The author also wishes to express his indebtedness to Mr. Montgomery for some helpful discussions on this subject, and to P. X. Gallagher for suggesting the use of Bellman's inequality in the proof of (13).

II. Proof of the theorem

Let

$$M_{X,k}(s) = \sum_{n=1}^{\infty} \frac{\mu(n)}{n^s} e^{-(n/X)^k} \tag{1}$$

and note that for $c > \max(0, 1 - \sigma)$ we have

$$M_{X,k}(s) = \frac{1}{2\pi i k} \int_{(c)} \frac{1}{\zeta(s+w)} \Gamma\left(\frac{w}{k}\right) X^w \, dw \tag{2}$$

where $\int_{(c)}$ denotes an integral along the vertical line $\operatorname{Re}(w) = c$. We have

$$\zeta(s) M_{X,k}(s) = \sum_{n=1}^{\infty} \frac{b_n(X; k)}{n^s}$$

where

$$b_n(X; k) = \sum_{d \mid n} \mu(d) e^{-(d/X)^k} \tag{3}$$

and it follows that

$$\begin{aligned} \sum_{n=1}^{\infty} \frac{b_n(X;k)}{n^s} e^{-(n/Y)^k} &= \frac{1}{2\pi i k} \int_{(c)} \zeta(s+w) M_{X,k}(s+w) \Gamma\left(\frac{w}{k}\right) Y^w \, dw \\ &= \zeta(s) M_{X,k}(s) + \frac{1}{k} M_{X,k}(1) \Gamma\left(\frac{1-s}{k}\right) Y^{1-s} \\ &\quad + \frac{1}{2\pi i k} \int_{(-\delta)} \zeta(s+w) M_{X,k}(s+w) \Gamma\left(\frac{w}{k}\right) Y^w \, dw \end{aligned} \tag{4}$$

for δ such that $\max(\sigma - 1, 0) < \delta < k$.

Let $\rho = \beta + i\gamma$ be a zero of $\zeta(s)$ with the properties

(i) $\alpha \leq \beta$,

(ii) $T < \gamma \leq 2T$,

(iii) condition $\gamma_{\alpha+\eta}$ is satisfied, hence $\zeta(s) \neq 0$ if $\sigma \geq \alpha + \eta$, $|t - \gamma| < (\log \gamma)^2$. On applying first the Borel-Carathéodory theorem to $\log \zeta(s)$ and then the three circle theorem one finds that

$$\zeta(s) = O(T^\epsilon) \tag{5}$$

$$1/\zeta(s) = O(T^\epsilon) \tag{6}$$

in the range

$$\sigma \geq \alpha + 2\eta, \qquad |t - \gamma| \leq \tfrac{1}{2}(\log \gamma)^2.$$

In the integral in equation (2) we move the contour of integration to the left, in the broken line (here $w = u + iv$)

$$\begin{aligned} u &= c, & v &< -\tfrac{1}{4}(\log\gamma)^2, \\ \alpha + 2\eta - \sigma \le u &\le c, & v &= -\tfrac{1}{4}(\log\gamma)^2, \\ u &= \alpha + 2\eta - \sigma, & -\tfrac{1}{4}(\log\gamma)^2 &< v < \tfrac{1}{4}(\log\gamma)^2, \\ \alpha + 2\eta - \sigma \le u &\le c, & v &= \tfrac{1}{4}(\log\gamma)^2, \\ u &= c, & v &> \tfrac{1}{4}(\log\gamma)^2 \end{aligned}$$

consisting of five line segments $L_1, L_2, \ldots, L_5$. If $\sigma \le \alpha + \eta$, we get

$$M_{X,k}(s) = \frac{1}{2\pi ik} \sum_{j=1}^{5} \int_{L_j} \frac{1}{\zeta(s+w)} \Gamma\left(\frac{w}{k}\right) X^w \, dw$$

provided $t = \operatorname{Im}(s)$ is such that $|t| > (\log\gamma)^2$. The integrals over L_1 and L_5 are easily estimated because of the Γ factor in the integrand, which is exponentially decreasing in its argument. The integrals over L_2, L_3, L_4 can also be estimated using (6). It follows easily from this that

$$(7) \qquad M_{X,k}(s) = O(X^{\alpha+2\eta-\sigma}T^{\epsilon})$$

provided $X = O(T^N)$ for some N,

$$\sigma \le \alpha + \eta, \qquad |t - \gamma| \le \tfrac{1}{4}(\log\gamma)^2$$

and $\gamma \ge \gamma_0$.

By the definition of $\gamma(\alpha)$ there is a function $\lambda = \lambda(\delta) \to 0$ as $\delta \to +0$ such that

$$(8) \qquad \zeta(s) = O(T^{[\gamma(\alpha+\eta)+\lambda](\alpha+\eta-\sigma)+\epsilon})$$

for

$$\alpha + \eta - \delta < \sigma \le \alpha + \eta, \qquad |t - \gamma| \le \tfrac{1}{2}(\log\gamma)^2$$

(the function $\lambda(\delta)$ depends also on $\alpha + \eta$ but $\lambda(\delta) \to 0$ as $\delta \to 0$ uniformly with respect to the second variable $\alpha + \eta$). Now the same argument used in the proof of inequality (7), along with (4), (7), (8) shows that

$$(9) \quad \sum_{n=1}^{\infty} \frac{b_n(X;k)}{n^s} e^{-(n/Y)^k} = \zeta(s)M_{X,k}(s) + O(Y^{\eta-\delta}X^{\eta+\delta}T^{[\gamma(\alpha+\eta)+\lambda(\delta)]\delta+\epsilon})$$

if

$$X, Y = O(T^N) \quad \text{for some } N$$

and

$$\alpha + \eta - \delta < \sigma \le \alpha + \eta, \qquad |t - \gamma| \le \tfrac{1}{4}(\log\gamma)^2, \qquad \gamma \ge \gamma_1.$$

We take $\delta = \sqrt{\eta}$, $Y = T^{\gamma(\alpha)+\epsilon_0}X$ where ϵ_0 is any fixed small positive number, and let $\eta \to 0$. If η is sufficiently small the error term in (9) will be $O(T^{-\epsilon_1})$ for some $\epsilon_1 = \epsilon_1(\alpha, \eta, \epsilon_0)$. It follows that for some $\eta_0 > 0$ one has

$$\sum_{n=1}^{\infty} \frac{b_n(X; k)}{n^s} e^{-(n/Y)^k} = \zeta(s)M_{X,k}(s) + O(T^{-\epsilon_1}) \tag{10}$$

if

$$Y = T^{\gamma(\alpha)+\epsilon_0}X, \qquad X = O(T^N),$$
$$\alpha - \eta_0 \le \sigma \le \alpha + \eta_0, \qquad |t - \gamma| \le \tfrac{1}{4}(\log \gamma)^2, \qquad T < \gamma \le 2T.$$

By (3) we have

$$b_1(X; k) = 1 + O(X^{-k}),$$
$$b_n(X; k) = \sum_{d|n} \mu(d)[e^{-(d/X)^k} - 1] \quad \text{if } n > 1,$$

whence

$$b_n(X; k) = O((n/X)^k) \quad \text{if } 1 < n \le X,$$
$$b_n(X; k) = O(d(n)) \quad \text{if } n > X.$$

Hence

$$\sum_{n=1}^{X^{1-\lambda}} \frac{b_n(X; k)}{n^s} e^{-(n/Y)^k} = 1 + O\left(\sum_{n=1}^{X^{1-\lambda}} \frac{1}{n^\sigma}\left(\frac{n}{X}\right)^k\right)$$
$$= 1 + O(X^{1-\lambda k})$$

if $\sigma \ge 0$, and we conclude that

$$\sum_{n=X^{1-\lambda}}^{\infty} \frac{b_n(X; k)}{n^s} e^{-(n/Y)^k} = -1 + \zeta(s)M_{X,k}(s) + O(T^{-\epsilon_1}) + O(X^{1-\lambda k}) \tag{11}$$

if

$$Y = T^{\gamma(\alpha)+\epsilon_0}X, \qquad X = O(T^N),$$

and

$$\alpha - \eta_0 \le \sigma \le \alpha + \eta_0, \qquad |t - \gamma| \le \tfrac{1}{4}(\log \gamma)^2, \qquad T < \gamma \le 2T.$$

We apply (11) with $s = \beta + i\gamma = \rho$ and get

$$\left| \sum_{n=X^{1-}}^{\infty} \frac{b_n(X; k)}{n^\rho} e^{-(n/Y)^k} \right| \ge \frac{1}{2} \tag{12}$$

if

$$Y = T^{\gamma(\alpha)+\epsilon_0}X, \qquad X = O(T^N),$$
$$\alpha \le \beta \le \alpha + \eta_0, \qquad T_0 \le T < \gamma \le 2T, \qquad k \ge 2/\lambda$$

and $X \ge X_0$, the constants η_0, X_0, T_0 depending on ϵ_0 and k; this provided that condition $(\gamma_{\alpha+\eta_0})$ is satisfied.

Let $\rho_1, \rho_2, \ldots, \rho_M$ be M zeros of $\zeta(s)$ such that

$$\alpha \le \operatorname{Re}(\rho_j) \le \alpha + \eta_0, \tag{A$_1$}$$
$$\zeta(s) \ne 0 \quad \text{if } \sigma \ge \alpha + \eta_0 \tag{A$_2$}$$

and

$$|t - \gamma_j| \le (\log \gamma_j)^2,$$

(A$_3$) $$\gamma_{j+1} - \gamma_j > (\log \gamma_j)^2,$$

(A$_4$) $$T < \gamma_j \le 2T.$$

We start with the inequality (Bellman's inequality)[1]

$$\sum_i \Big|\sum_j a_{ij}\xi_j\Big|^2 \le \Big(\sum_j |\xi_j|^2\Big) \max_i \sum_j \Big|\sum_h a_{ih}\bar{a}_{jh}\Big|$$

which is quickly proved as follows. In the Hilbert space of sequences $\xi = (\xi_1, \xi_2, \ldots)$ and scalar product $\langle \xi; \eta\rangle = \sum_i \xi_i \bar{\eta}_i$, consider the linear operator A such that

$$(A\xi)_i = \sum_j a_{ij}\xi_j.$$

Our inequality will follow if we show that

$$\|A\|^2 \le \max_i \sum_j \Big|\sum_h a_{ih}\bar{a}_{jh}\Big| ;$$

however if A' is the adjoint operator we have $\|A\| = \|A'\|$ and noting that $\|A'\xi\|^2 = \langle AA'\xi; \xi\rangle$, we obtain Bellman's inequality as a consequence of the fact that the largest eigenvalue of a matrix (A_{hk}) is bounded by $\max_h \sum_k |A_{hk}|$, applied to the matrix corresponding to the linear operator AA'.

We apply Bellman's inequality with

$$\begin{aligned}\xi_n &= \frac{b_n(X;k)}{n^{1/2}}[e^{-(n/Y)^k} - e^{-(n/X^{1-\lambda})^k}]^{-1/2} e^{-(n/Y)^k} \quad \text{if } n \ge X^{1-\lambda}\\ &= 0 \quad \text{if } n < X^{1-\lambda}\end{aligned}$$

and

$$\begin{aligned}a_{hn} &= [e^{-(n/Y)^k} - e^{-(n/X^{1-\lambda})^k}]^{1/2} n^{1/2-\rho_h} \quad \text{if } h \le M,\ n = 1, 2, 3, \ldots,\\ &= 0 \quad \text{if } h > M,\ n = 1, 2, 3, \ldots.\end{aligned}$$

The left hand side of Bellman's inequality is at least $\frac{1}{4}M$ by inequality (12). Also

$$\sum |\xi_n|^2 = O((\log Y)^4)$$

because $b_n(X; k) = O(d(n))$, and we conclude that

$$M = O((\log Y)^4) \max_j \sum_{h=1}^{M} \Big|\sum_{n=1}^{\infty} [e^{-(n/Y)^k} - e^{-(n/X^{1-\lambda})^k}] n^{1-\rho_j-\bar{\rho}_h}\Big| . \tag{13}$$

The same argument used in the proof of (10) shows that

$$\sum_{n=1}^{\infty} e^{-(n/Z)^k} \frac{1}{n^s} = \zeta(s) + O(|t|^{\mu(\theta)+\epsilon} Z^{\theta-\sigma}) \tag{14}$$

[1] *Note.* The fact that Bellman's inequality includes both the large sieve inequality and Halász' idea has been discovered by P. X. Gallagher. I am indebted to him for useful discussions on this subject, which led to this simple proof of Bellman's inequality.

provided $Z = O(T^N)$ and $|t| > (\log T)^2$ and $-k < \theta - \sigma < 0$. In any case we have, if $\sigma < 1$,

$$\sum_{n=1}^{\infty} e^{-(n/Z)^k} \frac{1}{n^s} = O(Z^{1-\sigma})$$

whence combining this inequality with (13) and (14) we get, using conditions (A_3), (A_4)

$$M = O((\log Y)^4) \cdot \{Y^{2-2\alpha} + MT^{\mu(\theta)+\epsilon}(X^{1-\lambda})^{\theta+1-2\alpha}\}$$

and from this we obtain $M = O(Y^{2-2\alpha}(\log Y)^4)$ provided

$$X > T^{\frac{\mu(\theta) + \epsilon_0}{(1-\lambda)(2\alpha - \theta - 1)}}, \qquad 2\alpha - 1 - k < \theta < 2\alpha - 1,$$
$$Y = T^{\gamma(\alpha)+\epsilon_0}X, \qquad X = O(T^N), \qquad \lambda \geq 2/k,$$

and $X \geq X_0, T \geq T_0$.

Let us count the number of zeros of $\zeta(s)$ such that $\mathrm{Re}(\rho) \geq \alpha, T < \mathrm{Im}(\rho) \leq 2T$. We divide the interval $(T, 2T)$ in horizontal strips of width $4(\log T)^2$, and let us call a strip "good" if there are no zeros in the strip for $\sigma \geq \alpha + \eta_0$, and "bad" otherwise. Every "bad" strip contributes at least one to $N(\alpha + \eta_0, 2T) - N(\alpha + \eta_0, T)$ and so there are at most $O(T^{\mu_1(\alpha+\eta_0)+\epsilon})$ such strips.

Now each strip cannot contain more than $O((\log T)^3)$ zeros. Hence

$$N(\alpha, 2T) - N(\alpha, T) = O(M(\log T)^3) + O(T^{\mu_1(\alpha+\eta_0)+\epsilon}).$$

It follows, using the estimate we have obtained for M, that

$$\mu_1(\alpha) \leq \max\left\{\mu_1(\alpha + \eta_0); \left[\frac{\mu(\theta) + \epsilon_0}{(1-\lambda)(2\alpha - \theta - 1)} + \gamma(\alpha) + \epsilon_0\right] 2(1-\alpha)\right\}$$

where η_0 depends on ϵ_0 and k and where λ is restricted by $\lambda \geq 2/k$, and θ by $2\alpha - 1 - k < \theta < 2\alpha - 1$.

On the other hand, $\mu_1(1) = 0$ and $\mu_1(\alpha)$ is nonincreasing. It follows easily from the previous inequality that[2]

$$\mu_1(\alpha) \leq \left[\frac{\mu(\theta) + \epsilon_0}{(1-\lambda)(2\alpha - \theta - 1)} + \gamma(\alpha) + \epsilon_0\right] 2(1-\alpha)$$

and the result of our theorem is easily obtained by letting $\epsilon_0 \to 0$, $k \to +\infty$, $\lambda \to 0$.

REFERENCES

1. G. Halász and P. Turán, *On the distribution of roots of Riemann zeta and allied functions*. I, J. Number Theory **1** (1969), 121–137. MR **38** #4422.

2. H. L. Montgomery, *Zeros of L-functions*, Invent. Math. (to appear).

UNIVERSITY OF PISA
PISA, ITALY

[2] We use here the fact that $\gamma(\alpha) = \lim_{\sigma\to\alpha-} \gamma(\sigma)$, which allows one to replace $\mu_1(\alpha)$ by

$$\tilde{\mu}_1(\alpha) = \lim_{\sigma\to\alpha-} \mu_1(\sigma) \geq \mu_1(\alpha).$$

ON SOME RECENT RESULTS IN THE ANALYTICAL THEORY OF NUMBERS

PAUL TURAN

The *arithmetical* significance of the *complex* behaviour of Riemann ζ-function was discovered by him more than 100 years ago. As he realised this comes out via the so called nontrivial zeros in the strip $0 < \sigma < 1$, $s = \sigma + it$; if $N(T)$ stands for the number of zeros in this strip satisfying $0 < t \leqq T$ then Riemann-Mangoldt theorem states that for $T \to \infty$

$$(1) \qquad N(T) \sim \frac{T}{2\pi} \log \frac{T}{2\pi} \cdot$$

He realised the arithmetical significance of the function-theoretic supposition that $\zeta(s) \neq 0$ for

$$(2) \qquad \sigma > \tfrac{1}{2}.$$

70 years passed in the belief that without proving it, at least in a somewhat weakened form, no progress could be made in the arithmetical applications of Riemann's method; this belief was definitely strengthened by the half-success of Hardy-Littlewood's magnificent attack to prove the ternary Goldbach problem. It was G. Hoheisel who made the breakthrough in 1930 with the problem of whether or not the asymptotic relation

$$(3) \qquad \sum_{x<p<x+h} \log p \sim h, \qquad h = x^{\theta}$$

can be proved with a *numerical* $\theta < 1$. His success in proving it with

$$(4) \qquad \theta = 1 - 1/33000$$

sounds funny and does not reflect at all the significance of his work. In an age when pathbreakers are mostly forgotten and only world records and recorders are known, it is worth saying something about this. His success was due to the discovery that two types of function-theoretical results, which were found previously for the purpose of approaching the conjecture (2) only, can be used for his aim. The first of these refers to the order of magnitude of $\zeta(s)$ in the critical strip. Defining the Lindelöf μ-function belonging to $\zeta(s)$ by

$$\mu(\alpha) = \overline{\lim_{t\to\infty}} \frac{\log|\zeta(\alpha + it)|}{\log|t|}$$

we have evidently $\mu(\alpha) = 0$ for $\alpha > 1$ and, from the functional equation, $\mu(\alpha) = \frac{1}{2} - \alpha$ for $\alpha < 0$. As shown by Littlewood, Riemann's conjecture (2) implies

$$\mu(\alpha) = 0 \quad \text{for } \alpha \geqq \tfrac{1}{2} \tag{5}$$

as conjectured by Lindelöf in 1908. Now it was an important discovery of Hardy and Littlewood that the graph of $\mu(\alpha)$ *touches* the α-axis at $\alpha = 1$; their result stood even in the strongest form

$$|\zeta(\sigma + it)| \ll |t|^{4(1-\sigma)/\log(1/(1-\sigma))} \frac{\log|t|}{\log\log|t|}, \quad \frac{63}{64} \leqq \sigma \leqq 1, \qquad |t| \geqq 10 \tag{6}$$

at their disposal. (For later aims I remark that since then, the powerful method of I. M. Vinogradov with its improvements by Korobov and Richert led to the inequality

$$|\zeta(\sigma + it)| \ll |t|^{100(1-\sigma)^{3/2}} \log^{2/3}|t|, \quad \tfrac{1}{2} \leqq \sigma \leqq 1, \qquad |t| \geqq 10 \tag{7}$$

and according to a communication of Professor L. Schoenfeld the important constant 100 can be replaced by 39.) The second of these referred to the so called density theorems. Denoting by $N(\alpha, T)$ the number of zeros of $\zeta(s)$ in the parallelogram

$$\sigma \geqq \alpha, \qquad 0 < t \leqq T \tag{8}$$

these theorems refer to upper bounds for them in the case $\alpha > \frac{1}{2}$ which are "small" compared to $N(T)$ in (1). The first such theorem was discovered by Bohr and Landau in 1914; Hoheisel had at his disposal the Carlson form which—with his slight improvement—asserted uniformly for $\frac{1}{2} \leqq \alpha \leqq 1$ the inequality

$$N(\alpha, T) \ll T^{4\alpha(1-\alpha)} \log^6 T. \tag{9}$$

(6) and (9) were enough for Hoheisel to prove his theorem in (4); it was clear that improvements of (6) and (9) will improve Hoheisel's exponent. It turned out soon that the inequality (7) is ample enough to replace (6); an improvement of (9)—the best published up to now—waited 15 years after Carlson. Ingham proved in 1937 resp. 1940 the inequalities[1]

$$\begin{aligned} N(\alpha, T) &\ll T^{2(1+\mu(1/2))(1-\alpha)} \log^5 T, \\ N(\alpha, T) &\ll T^{3/(2-\alpha)(1-\alpha)} \log^5 T \end{aligned} \tag{10}$$

[1]During this conference Montgomery proved for $\frac{1}{2} \leqq \alpha \leqq 1$ the inequality $N(\alpha, T) \ll T^{(2/\alpha)(1-\alpha)} \log^{10} T$ which improved Hoheisel's exponent after an interval of 30 years to $\theta = 3/5$.

both valid uniformly for $\frac{1}{2} \leqq \alpha \leqq 1$. These were enough to push down Hoheisel's exponent—using the Hardy-Littlewood estimation

$$\mu(\tfrac{1}{2}) \leqq \tfrac{1}{6} \tag{11}$$

to $\theta = 5/8 + \epsilon$ and—as an important consequence—that for $n > n_0$ there are primes between n^3 and $(n + 1)^3$. Taking in account among others also the fact that using density theorems for the Dirichlet L-functions Linnik proved the ternary Goldbach conjecture eliminating the difficulties in the Hardy-Littlewood attack, it is clear that the density-method of Hoheisel-Linnik operating with density theorems often can be substituted for the existence of a nontrivial zero-free half-plane and is of great significance for arithmetical applications.

The idea of Bohr-Landau which was basic for all attempts to prove density theorems for more than 30 years can be described shortly as follows. Let A be a closed domain bounded by the rectifiable curve $l, f(s)$ be regular in A and $\neq 0$ on l. Denoting the number of its zeros in A by R we have

$$R = \frac{1}{2\pi i} \int_{(l)} \frac{f'}{f}(s)\, ds.$$

But if $\varphi(s)$ is regular in A and k is an arbitrary positive integer we have also

$$R = \frac{1}{2\pi i} \int_{(l)} \frac{f'}{f}(s)(1 - f(s)\varphi(s))^k\, ds.$$

So if l^* is another rectifiable closed curve outside A and such that on it

$$\left|\frac{f'}{f}(s)\right| \leqq M \tag{12}$$

then—choosing $k = 2$—we get

$$R \leqq \frac{M}{2\pi} \int_{(l^*)} |1 - f(s)\varphi(s)|^2 |ds|. \tag{13}$$

In the case of $f(s) = \zeta(s)$ and A being the parallelogram

$$\alpha \leqq \sigma \leqq 1, \qquad T \leqq t \leqq 2T$$

standard theorems give the possibility of choosing l^* in the form

$$\begin{aligned} &\alpha \leqq \sigma \leqq 2, \quad t = \tau_1, \quad 2T \leqq \tau_1 \leqq 2T + 1/\log T \\ &\alpha \leqq \sigma \leqq 2, \quad t = \tau_2, \quad T \leqq \tau_2 \leqq T + 1/\log T \\ &\sigma = 2, \quad \tau_2 \leqq t \leqq \tau_1 \end{aligned}$$

and completed in the vertical strip

$$\alpha - 1/\log T \leqq \sigma \leqq \alpha$$

by a broken line l_1^* consisting alternately of horizontal and vertical segments each of the vertical ones having a length $\geqq \frac{1}{2}$ so that $M = \log^3 T$. Choosing $\varphi(s)$ as a function which approximates in some sense $1/\zeta(s)$ it turns out quite trivially that only the integral along l_1^* matters. Applying the artifice of Bohr according to which the absolute value of a function at a point s_0 cannot exceed its *areal* mean value on a disc with center at $s = s_0$ and radius r_0 lying in its regularity domain and choosing $r_0 = \log^{-1} T$, the integral along l_1^* can be estimated from above by

$$(14)\qquad \int_{\alpha-\log^{-1}T}^{\alpha+\log^{-1}T} d\sigma \int_{T/2}^{3T} |1 - \zeta(s)\varphi(s)|^2\, dt.$$

In Bohr-Landau's original proof they choose $\varphi(s) = \prod_{p\leqq\omega}(1 - p^{-s})$, ω large; Carlson has chosen $\varphi(s) = \sum_{n\leqq T/100} \mu(n)n^{-s}$ and estimated the inner integral in (14) directly. Ingham's new device was the use of convexity theorems for the estimation of the inner integral in (14).

The lower limit of θ in (4) obtainable by Hoheisel's method is $\frac{1}{2}$ as pointed out by Ingham. This could be attained by a density theorem

$$(15)\qquad N(\alpha, T) \ll T^{2(1-\alpha)} \log^c T$$

or slightly weaker by

$$(16)\qquad N(\alpha, T) < c(\epsilon)T^{(2+\epsilon)(1-\alpha)}$$

valid for $\frac{1}{2} \leqq \alpha \leqq 1$ uniformly. This is called the density hypothesis. Near to $\alpha = \frac{1}{2}$ this is trivial owing to (1). Near to $\alpha = 1$, where it is in a sense the deepest, I proved it by a completely different method in 1949. More exactly I proved for certain positive c the inequality

$$(17)\qquad N(\alpha, T) < T^{2(1-\alpha)+(1-\alpha)^{101/100}}$$

is valid for $1 - c \leqq \alpha \leqq 1$. Introducing the $\mu_1(\alpha)$ function defined by

$$(18)\qquad \mu_1(\alpha) = \overline{\lim_{T\to+\infty}} \frac{\log(1 + N(\alpha, T))}{\log T}$$

we have $\mu_1(\alpha) = 0$ for $\alpha \geqq 1$; Riemann's conjecture (2) would imply $\mu_1(\alpha) = 0$ for $\alpha > \frac{1}{2}$. The density hypothesis is equivalent to

$$\mu_1(\alpha) \leqq 2(1 - \alpha), \qquad \tfrac{1}{2} \leqq \alpha \leqq 1;$$

my inequality (17) gave at least that the graph of $\mu_1(\alpha)$ touches the line $y = 2(1 - \alpha)$ in an (α, y)-system from above. So with $\mu_1(\alpha)$ the situation was until 1968 much as it was with $\mu(\alpha)$ before Hardy-Littlewood's discovery (6). Then we found in collaboration with G. Halász that the graph of $y = \mu_1(\alpha)$ also touches the α-axis at $\alpha = 1$. More exactly we proved the

THEOREM I. *For suitable explicitly calculable positive constants, c_1, c_2 and c_3 the inequality*

$$(19)\qquad N(\alpha, T) < c_1 T^{c_2(1-\alpha)^{3/2}\log^3(1/(1-\alpha))}$$

for[2] $1 - c_3 \leqq \alpha \leqq 1$.

[2]From this it follows at once the existence of an (explicitly calculable) positive constant $c_4 < 1$ so that for $c_4 \leqq \alpha \leqq 1$ the inequality $N(\alpha, T) < T^{2(1-\alpha)}$ holds. This c_4 falls most probably short of $c_4 = 9/10$ as indicated by Montgomery in his talk.

Our proof holds mutatis mutandis for Dirichlet L-functions as well as for all Dedekind zetafunctions; only in this case c_1, c_2 and c_3 may depend upon the modulus of the character resp. upon the constants of the field.

Generally speaking Riemann's method transfers the arithmetical difficulties to the investigation of sums depending on zeros of certain "zeta like" functions. In some problems zeros "near" the line $\sigma = 1$ are essential, for some others the ones "near" to $\sigma = \frac{1}{2}$. Theorem I seems to be particularly significant when dealing with problems of the first type. E.g. a short reflection on a recent paper of Hooley indicates the Artin's well-known conjecture on primitive roots mod p is a problem of the first type. The first phase of this work—refinement of the zerofree domain of Dedekind zetafunctions given by A. V. Sokolovskij with respect to the constants of the field—is in work by Dr. W. Stas in Poznań.

The Lindelöf conjecture in (5) was thought to be of little influence on the vertical distribution of the nontrivial zeros. It was a surprise when Ingham proved in 1937 that the density hypothesis (16) can be deduced from it. When proving Ingham's result (in a slightly stronger form) in 1954, I expressed the view that much more than this is true. As a partial fulfilment of this conjecture we proved with G. Halász:

THEOREM II. *The Lindelöf conjecture* (5) *implies for arbitrary small positive* ϵ *and* δ *and* $T > T_0(\epsilon, \delta)$ *the inequality* $N(3/4 + \delta, T) < T^{\epsilon}$.

The full conjecture which asserts the same for $N(\frac{1}{2} + \delta, T)$ is still open. If it is true it would mean that Lindelöf's conjecture is "almost" as strong as Riemann's. Theorem I is more perfect in the sense that this is the best one can expect, at the present state, of the zerofree domain; when this can be improved to

(20) $$\sigma > 1 - c_5 \log^{-\beta} (2 + |t|)$$

with a $0 < \beta < \frac{2}{3}$ then our method would give immediately for the right side of (19)

(21) $$T^{c_2(1-\alpha)^{1/\beta}}.$$

Before passing to other results concerning density theorems, we begin first with a simple remark. Let $0 < \eta < 1/10$, $0 < \delta \leqq \eta$

(22) $$\tfrac{1}{2} + 2\eta \leqq \alpha \leqq 1$$

and we suppose that $\zeta(s) \neq 0$ in a parallelogram

(23) $$\sigma \geqq \alpha, \qquad |t - T| \leqq \log T, \qquad T \geqq 3.$$

Then it is easy to show that the number of zeros in the half-disc

(24) $$|s - (\alpha + iT)| \leqq \delta, \qquad \sigma \leqq \alpha$$

cannot exceed

(25) $$\delta \log T$$

provided $T > T_1(\delta, \eta)$.

Now suppose that (22)–(23)–(24)–(25) is true instead of (25) with

$$A\ \delta \log T \tag{26}$$

with a positive numerical constant A; owing to what was said above it is certainly $A \leqq 1$.

I proved more than ten years ago the inequality

$$N(\alpha, T) \ll T^{2(1+15\pi A)(1-\alpha)} \log^{10} T \tag{27}$$

uniformly for $\frac{1}{2} + 3\eta \leqq \alpha \leqq 1$ (unpublished). If with an $\frac{1}{2} < \theta < 1$ (22) holds only for $\theta + 2\eta \leqq \alpha \leqq 1$ then (27) holds uniformly for $\theta + 3\eta \leqq \alpha \leqq 1$. It is more interesting to observe that when A in (26) can be chosen arbitrarily small then the density hypothesis follows. This result I published in 1958. Since—as proved by Littlewood—Lindelöf's conjecture is equivalent to the inequality

$$N(\alpha, T+1) - N(\alpha, T) = o(\log T), \qquad \alpha > \tfrac{1}{2} \text{ fixed}, \quad T \to \infty \tag{28}$$

it is clear that our assumption

$$A \quad \text{in (26) arbitrarily small} \tag{29}$$

is much weaker than Lindelöf conjecture and might be called "weak Lindelöf conjecture". Shorter—and somewhat unprecisely expressed—whereas Ingham deduced the density hypothesis from the assumption $\mu(\alpha) = 0$ $(\alpha \geqq \frac{1}{2})$ we did it from the assumption that $\mu(\alpha)$ is *continuously* derivable for $\alpha > \frac{1}{2}$. To give an unconditional proof of the density hypothesis along these lines seems to be within possibility; first one would need to have a *function theoretical* proof for the fact that $\mu(\alpha)$ touches the α-axis at $\alpha = 1$ (like Lindelöf's *general* convexity theorem improved at once the previous estimations of $|\zeta(s)|$ based on some analytical representation).

After the strong conclusion of Theorem II it was plausible to ask what stronger conclusions can be drawn for $N(\alpha, T)$ from the weak Lindelöf conjecture? We found with G. Halász in 1968 that supposing the weak Lindelöf hypothesis for $\frac{3}{4} \leqq \alpha \leqq 1$ only[3] the inequality

$$N(\alpha, T) \ll T^{(2/3)(1-\alpha)/(4\alpha-3)+\epsilon} \tag{30}$$

can be deduced. Especially for $\alpha \geqq 11/12$ this means the inequality

$$N(\alpha, T) \ll T^{1-\alpha}. \tag{31}$$

I lectured on Theorems I and II in May, 1968, in Cambridge, Nottingham and Edmonton with detailed proofs; since several of the participants of that meeting participated in one of the above talks and moreover complete proofs have appeared, I shall not go into details of the proofs. Instead I shall try to make the ideas of the proofs in nontechnical terms more plausible; this will be useful in understanding the sketches of proofs of certain results on L-functions which are the main aim of these lectures.

---

[3]This could be weakened considerably.

Let $\sigma_0 \geqq 2$ and fixed, with t_0 so large that on putting $s_0 = \sigma_0 + it_0$ we have

(32) $$|s_0 - 1| > \sigma_0 - \tfrac{1}{2}.$$

The radius R of regularity of $\zeta'/\zeta(s)$ around s_0 is given by

$$\frac{1}{R} = \overline{\lim_{\nu\to\infty}} \left(\frac{1}{\nu!} \left| \frac{\zeta'}{\zeta}(s)^{(\nu)} \right|_{s=s_0} \right)^{1/\nu}.$$

Replacing heuristically $\zeta'/\zeta(s)^{(\nu)}_{s=s_0}$ by the quadratic mean-value of $\zeta'/\zeta(s)^{(\nu)}$ along $\sigma = \sigma_0$ the right side is

$$\overline{\lim_{\nu\to\infty}} \left(\frac{1}{\nu!^2} \sum_n \frac{\Lambda(n)^2 \log^{2\nu} n}{n^{2\sigma_0}} \right)^{1/2\nu}.$$

As is easy to see that the quantity behind lim sup actually has a limit and

(33) $$R = \sigma_0 - \tfrac{1}{2}.$$

Choosing t_0 in succession as $t_1, t_2, \ldots$ so that $t_{\nu+1} - t_\nu \leqq \sqrt{\sigma_0}/\log \sigma_0$, say, the corresponding discs will cover for arbitrarily small $\epsilon > 0$ the halfplane $\sigma \geqq \frac{1}{2} + \epsilon$ —apart from a finite part of it—provided σ_0 is sufficiently large.

This heuristic reasoning suggests several ways out. Immediately it suggests trying to find a "dense" subset on $\sigma = \sigma_0$ for which all derivatives of sufficiently large index are "comparable nontrivially" with their quadratic mean-value along this line. Certainly this gives the idea of investigating high order derivatives of $\zeta'/\zeta(s)$ and of trying to compare them with their quadratic mean-value, if not on a dense subset, at least everywhere with exception of a "small" subset on the segment

(34) $$I\colon \sigma = \sigma_0, \qquad T \leqq t \leqq 2T;$$

it is obvious ab ovo that this can lead to estimations of $N(\alpha, T)$. To work with all sufficiently large ν's is hopeless but is it enough to work with finitely many ν's only? If we succeed in comparing a *fixed* "large" ν $|\zeta'/\zeta(s)^{(\nu)}|$ with its quadratic mean-value on all but a "small" set H_ν, then the partial fraction representation gives easily that

(35) $$\left| \sum_\rho \frac{1}{(s-\rho)^{\nu+1}} \right|$$

is "small" on the segment (34), with exception of H_ν; here ρ runs over all nontrivial zeros of $\zeta(s)$. Let, for a fixed s in $I - H_\nu$, ρ^* be one of the nearest nontrivial zeros; since ν is large, the contribution to the sum in (35) of ρ's "far" from ρ^* is "small". If we could conclude that our remaining finite sum

(36) $$\left| {\sum}' \frac{1}{(s-\rho)^{\nu+1}} \right|$$

is greater than $|s - \rho^*|^{-\nu-1}$ or a "not too small percentage" of it then we could conclude that for $s \in I - H_\nu$, $|s - \rho^*|^{-\nu-1}$ is small, i.e. $|s - \rho^*|$ "large" which is only another form of an upper bound for $N(\alpha, T)$.

But such an inference does not exist in the complex field. However there is a way out of this difficulty. The critical sum in (36) is of the form

$$\sum_{j=1}^{n} z_j^{\nu}$$

with fixed complex numbers z_j. If ν runs over n consecutive integers it is clear that not all of them can be "too small". It is easy to get a lower bound depending on the configuration of the z_j vectors but this is useless since the role of z_j's are played by the $1/(s - \rho)$ numbers and we know very little on their configuration. But it was possible to get rid of this difficulty by proving the following theorem.

If $z_1, z_2, \ldots, z_n$ are complex numbers such that

$$\max_{j=1,\ldots,n} |z_j| = 1$$

and m is a positive integer then there is an integer ν_0 with $m + 1 \leqq \nu_0 \leqq m + n$ so that[4]

$$\left|\sum_{j=1}^{n} z_j^{\nu_0}\right| \geqq \left(\frac{n}{8e(m + n)}\right)^n. \tag{37}$$

A further difficulty might be that allowing ν to change the H_ν-sets are different and the set of "bad" s-values is their union; will this not be "too large" even if the single H_ν's are "small"? Since ν is "large" the terms in (35) fall down rather quickly and hence the number of terms in (36) is "small"; hence n is "small" and even the sum of the measures of the H_ν sets is "small".

It remains to analyse how one could estimate $|\zeta'/\zeta(s)^{(\nu)}|$ for a fixed ν by its quadratic mean-value apart from a "small" set H_ν in I. I used previously in various forms the remark that if $f(t)$ is continuous for $[T, 2T]$ and $0 < \beta < 1$ is fixed then in this interval the inequality

$$|f(t)| \leqq T^{-\beta/2}\left(\int_T^{2T} |f(x)|^2\, dx\right)^{1/2} \tag{38}$$

holds apart from a subset with measure $\leqq T^{\beta}$. Applying it to

$$\sum_n \frac{\Lambda(n)\log^{\nu} n}{n^{\sigma_0}} e^{-it\log n} \tag{39}$$

only $\Lambda(n) \leqq \log n$ can be used; an ingenious new idea of Halász, found originally by him in his investigation of mean-values of multiplicative functions, made it possible to go much further, reducing the investigation of (39) to the investigation of

$$\sum_n \frac{\log^{\nu+1} n}{n^s},$$

which is much easier. How this can be done will be more clear when we give proofs

[4]My original constant instead of $8e$ was larger; this one is contained in a paper written in collaboration with Vera T. Sós. It would be of significance to improve it further.

of results concerning L-functions (inequality (62) and its proof). Its improving effect can be visualized shortly as follows. Let ν be fixed, $0 < \lambda \leqq \frac{1}{4}$ say, we consider the $H_\nu(\lambda)$-set on the segment $[2 + iT, 2 + 2iT]$ satisfying the inequality

$$(40) \qquad \frac{1}{\nu!}\left|\sum_n \frac{\Lambda(n)\log^\nu n}{n^{2+it}}\right| \geqq T^{-\lambda}.$$

Then applying (38) we get for the measure of $H_\nu(\lambda)$ the upper bound $T^{2\lambda}$ if only

$$(41) \qquad \nu > \log T/\log (9/4)$$

whereas Halász's idea already gives the same for

$$(42) \qquad \nu > \frac{\lambda}{\log (5/4)} \log T$$

which is essentially smaller and for our aims much better.

The idea expressed in the form of (37) is rather direct and leads often to *first* proofs. It is also rather flexible and fit to extend the applicability of Riemann's method to arithmetic. A suitable extension led the late Knapowski and myself to the following theorem.

For an explicitly calculable positive constant c_6 and $T > c_6$ there are U_1 and U_2 so that

$$(43) \qquad \log\log\log T \leqq U_1 < U_2 \leqq U_1 e^{\log^{9/10} U_1} \leqq T$$

and[5]

$$(44) \qquad \sum_{p>2;\ p \text{ primes};\ U_1 \leqq p \leqq U_2} (-1)^{(p-1)/2} \log p > \sqrt{U_2}.$$

This theorem was announced in a recently published paper of ours.[6] It means that there are again and again intervals containing "much more" primes $\equiv 1 \bmod 4$ than $\equiv 3 \bmod 4$. Besides the independent interest of this theorem this is perhaps enhanced by the fact that it raises some doubts as to the truth of Riemann-Piltz conjecture concerning $L(s, 4, \chi)$, χ denoting the nonprincipal character belonging to mod 4. As we have shown namely as a development of ideas of Chebyshev, Hardy-Littlewood and Landau, the truth of this conjecture would imply the inequality[7]

$$(45) \qquad \sum_{p>2;\ p \text{ primes}} (-1)^{(p-1)/2} \log p e^{-c_7 \log^2 p/x} < -c_8\sqrt{x}$$

valid for $x > c_9$; here again the c_ν's stand for explicitly calculable positive numerical constants. These constants can be chosen in several ways; for not too large c_9 the constant c_7 might be relatively big, about 30. Let us choose $x > c_9$ in the "middle" of the interval (U_1, U_2); then (44) gives a fair chance that the

[5]The right side could be replaced if necessary by a bigger one.

[6]"Abhandlungen aus Zahlentheorie und Analysis, zur Erinnerung an Edmund Landau (1877–1938)", VEB Deutscher Verlag der Wissenschaften, Berlin 1968.

[7]Actually it is equivalent to it.

contribution of the primes in this interval will *not* give a "large negative" contribution as required by (45). But what is the contribution of the primes outside (U_1, U_2)? First we have to get an idea on the size of (U_1, U_2). Since Riemann-Piltz conjecture implies, as is well-known,

$$\sum_{n \text{ odd};\, n \leqq U_2} (-1)^{(n-1)/2} \Lambda(n) < c_{10} \sqrt{U_2} \log^2 U_2 \tag{46}$$

one is inclined to think that the interval in (43) is of the form (cU_2, U_2) $c < 1$. But then for the outside primes the corresponding terms in (45) are $\leqq \log p e^{-c_7 \log^2 c}$ i.e. "small"; about equally many are positive, resp. negative, and they are quite small outside $(xe^{-\sqrt{\log x}}, xe^{\sqrt{\log x}})$. As D. Shanks[8] found we have, e.g.

$$\sum_{2 < p \leqq 359{,}327} (-1)^{(p-1)/2} = +105;$$

thus $x \sim 35 \cdot 10^4$ offers a possibility for a first orientation of a machine attack.

Next we turn to Dirichlet L-functions, which we shall denote by $L(s, q, \chi)$ (instead of the generally used $L(s, \chi)$). As told before it would be easy to prove the analogue of Theorem I and II for fixed q and χ and $T \to \infty$ (but c_1, c_2, c_3 depending on the modulus q). As it turned out in the last twenty years, for many important problems of number theory, it is not so much the number of zeros of a fixed L-function for $T \to \infty$, but the total number of zeros of all L-functions with moduli $q \leqq Q$ for $Q \to \infty$ that are of importance. (Here the domain is fixed or increases slowly with Q.) That is, one wants the quantity

$$M_T(Q, \alpha) \overset{\text{def}}{=} \sum_{q \leqq Q} \sum_{\chi \bmod q}{}^{*} N(\alpha, q, \chi, T) \tag{47}$$

where $\frac{1}{2} \leqq \alpha < 1$, $N(\alpha, q, \chi, T)$ stands for the number of zeros of $L(s, q, \chi)$ in the parallelogram

$$\sigma \geqq \alpha, \qquad |t| \leqq T \tag{48}$$

and the star means that the summation is to be extended only to primitive characters. Perhaps the latest such example is my "exact" formula[9] for the number $\nu(n)$ of the binary Goldbach decompositions of even n in the interval $M/2 \leqq n \leqq M$ (uniformly in n):

$$\begin{aligned} \nu(n) = A_0 \frac{n}{\log^2 n} \prod_{p>2;\, p/n} \frac{p-1}{p-2} \left\{1 + O\left(\frac{1}{\log\log M}\right)\right\} \\ - \left\{1 + O\left(\frac{1}{\sqrt{\log M}}\right)\right\} \sum_{q \leqq M} \frac{\mu(q) \log M/q}{\varphi(q)} \\ \sum_{\chi \bmod q} \bar{\chi}(n, q) \sum_{|\gamma| \leqq \log^2 M;\, 1/3 \leqq \beta \leqq 5/6} \frac{n^\rho - n^{\rho/100}}{\rho \left(1 + \dfrac{\rho}{\log M}\right)^{1+[2\log M/\log\log M]}} \end{aligned} \tag{49}$$

---

[8]"Quadratic residues and the distribution of primes", Mathematical Tables and other Aids to Computation, Vol. XIII (1959) p. 272–294.

[9]This formula holds unconditionally; the main term is the one conjectured by Hardy and Littlewood. It shows the new feature of the problem that it depends only upon the "small" zeros of L-functions, off the line $\sigma = 1$. Even in this form—which is by far not the last one—it fits by averaging with respect to n to obtain "almost all"-type results concerning the binary Goldbach problem.

where

$$A_0 = 2 \sum_{p>2} \left(1 - \frac{1}{(p-1)^2}\right)$$

is the Hardy-Littlewood constant and $\rho = \rho(q, \chi) = \beta + i\gamma$ stand for the nontrivial zeros of $L(s, q, \chi)$. The first results on $M_T(Q, \alpha)$ appeared in the papers of Linnik and Rényi; the strongest published result is due to Bombieri who proved—with a slight but important improvement of Davenport-Halberstam—that

$$M_T(Q, \alpha) \ll T(Q^2 + QT)^{4(1-\alpha)/(3-2\alpha)} \log^{10}(X + T) \tag{50}$$

uniformly for $\frac{1}{2} \leqq \alpha \leqq 1$. In his paper he made the conjecture that uniformly for $\frac{1}{2} \leqq \alpha \leqq 1$ the inequality

$$M_T(Q, \alpha) \ll T^{1+\epsilon} Q^{4(1-\alpha)+\epsilon} \tag{51}$$

holds, which is, as is easy to see, evident for α's near[10] to $\frac{1}{2}$.

Bombieri remarked, with full right, that (51) is a new type of density theorem. This makes plausible, for the sake of orientation, the question how the conjecture (51) is connected with other known conjectures, notably with the natural analogue of Lindelöf's conjecture, which we shall state for the sake of simplicity in the form

$$\begin{gathered} |L(s, q, \chi)| < c(\epsilon, \Omega) q^{\epsilon} \\ \sigma \geqq \tfrac{1}{2}, \qquad -\Omega \leqq t \leqq \Omega, \qquad \chi \neq \chi_0. \end{gathered} \tag{52}$$

We proved with G. Halász two theorems in November, 1968, in this direction which were announced in December in Rome without proofs. They run as follows.

THEOREM III. *The Lindelöf conjecture in* (52) *implies Bombieri's conjecture* (51)

Again for $\alpha > \frac{3}{4}$ much more can be said. For the sake of simplicity we state it as

THEOREM IV. *The Lindelöf conjecture in* (52) *implies for arbitrary small positive* ϵ *and* δ *the inequality*

$$M_1(Q, \tfrac{3}{4} + \delta) < c(\epsilon, \delta) Q^{\epsilon}.$$

Barban expressed the further conjecture

$$\sum_{q \leqq Q} \max_{\chi \bmod q} N(\alpha, q, \chi, T) < c(\epsilon, T) Q^{2(1-\alpha)} \tag{53}$$

uniformly for $\frac{1}{2} \leqq \alpha \leqq 1$. Theorem IV gives of course that this follows amply from the conjecture (52) at least for $\alpha > \frac{3}{4}$.

---

[10]During this conference Montgomery proved the inequalities

$$M_T(Q, \alpha) \ll (Q^2 T)^{(2/\alpha)(1-\alpha)} \log^{10} QT,$$
$$M_T(Q, \alpha) \ll (Q^2 T)^{(3/2-\alpha)(1-\alpha)} \log^{10} QT$$

which are stronger than (50). The first of these shows the truth of (51) for the first time at least for α's near to 1.

Before going into the sketches of proofs of Theorem III and IV we shall note a consequence of Theorem IV on the smallest prime $P(q, l)$ which is $\equiv l \bmod q$ where $(q, l) = 1$. As S. Chowla remarked the Riemann-Piltz conjecture implies for arbitrarily small $\epsilon > 0$ the inequality

$$P(q, l) < c(\epsilon)q^{2+\epsilon}; \tag{54}$$

I showed that the same conjecture implies

$$P(q, l) < c(\epsilon)\varphi(q) \log^{2+\epsilon} q \tag{55}$$

for every q and almost all l's.[11] As Elliott proved, without conjectures, the inequality

$$P(q, l) < c\varphi(q) \log q$$

holds for almost all q's and almost all l's.[12] Now it is easy to deduce from Theorem IV and from the Lindelöf conjecture (52) a new type of result that, for an arbitrarily small $\epsilon > 0$, the inequality

$$P(q, l) < c(\epsilon)q^{4+\epsilon} \tag{57}$$

holds for *all* l's dropping a "few" q-moduli (the number of which is $< Q^\epsilon$ if $q \leqq Q$).[13]

Finally we shall sketch the proofs of Theorem III and IV; we shall start with the latter. We shall use (52) in a slightly weaker form

$$|L(s, q, \chi)| < c(\lambda)q^{\lambda^2}, \quad \lambda > 0 \text{ arbitrarily small}, \quad \sigma \geqq \tfrac{1}{2}, \quad |t| \leqq 6. \tag{58}$$

Let further

$$s_j = 2 + i\frac{j}{[\log^3 Q]} \stackrel{\text{def}}{=} 2 + it_j, \qquad j = 0, \pm 1, \pm 2, \ldots, \pm[\log^3 Q] \tag{59}$$

and suppose the integer ν restricted at present only by

$$\frac{\lambda + 6\lambda^2}{\log (5/4)} \log Q \leqq \nu \leqq \left(\frac{\lambda}{\log (5/4)} + 40\lambda^2\right) \log Q. \tag{60}$$

Let $\chi(n, q)$ be a primitive character belonging to the modulus q and we ask for any fixed ν and j-value and $q \leqq Q$ how many times the inequality

$$\frac{1}{\nu!}\left|\frac{L'}{L}(s, q, \chi)^{(\nu)}\right|_{s=s_j} \geqq Q^{-\lambda} \tag{61}$$

---

[11] I.e. denoting for fixed q the number of l's for which (55) is false, by $F(q)$, the relation

$$\lim_{q\to\infty} \varphi(q)^{-1}F(q) = 0 \tag{56}$$

holds.

[12] I.e. the number of exceptional q's with $q \leqq Q$ is $o(Q)$ if $Q \to \infty$ and for the remaining ones (56) holds.

[13] Linnik proved unconditionally the inequality $P(q, l) < cq^{c_{11}}$. M. Jutila proved in his recent thesis that $c_{11} < 630$. Erdös proved unconditionally that for all positive D's and integer q's the inequality $P(q, l) < D\varphi(q) \log q$ holds for at least $c(D) > 0$ percentage of all residue classes mod q.

holds. If it holds M-times then we assert first that

$$M < c(\lambda)Q^{2\lambda+10\lambda^2}. \tag{62}$$

If namely in some order the corresponding L-functions are $L(s, q_h, \chi_h)$ $(h = 1, 2, \ldots, M)$ then summation with respect to h gives

$$\begin{aligned} \nu! Q^{-\lambda} M &\leqq \sum_{h=1}^{M} \left| \sum_n \frac{\Lambda(n) \log^\nu n \chi_h(n, q_h)}{n^{s_j}} \right| \\ &= \left| \sum_{h=1}^{M} \eta_h \sum_n \frac{\Lambda(n) \log^\nu n}{n^{s_j}} \chi_h(n, q_h) \right| \end{aligned}$$

with suitable $|\eta_h| = 1$. This gives

$$\begin{aligned} \nu! Q^{-\lambda} M &\leqq \left| \sum_n \frac{\Lambda(n) \log^\nu n}{n^{s_j}} \sum_{h=1}^{M} \eta_h \chi_h(n, q_h) \right| \\ &\leqq \sum_n \frac{\log^{\nu+1} n}{n^2} \left| \sum_{h=1}^{M} \eta_h \chi_h(n, q_h) \right|. \end{aligned}$$

Applying Schwarz's inequality we get

$$\nu!^2 Q^{-2\lambda} M^2 \leqq c(\lambda) \sum_n \frac{\log^{2\nu} n}{n^{3-2\lambda}} \left| \sum_{h=1}^{M} \eta_h \chi_h(n, q_h) \right|^2.$$

Performing the squaring, changing the order of summation and using $|\eta_h| \leqq 1$ we get

$$\nu!^2 Q^{-2\lambda} M^2 \leqq c(\lambda) \{ M |\zeta^{(2\nu)}(3 - 2\lambda)| + M^2 \max |L^{(2\nu)}(3 - 2\lambda, q, \chi)| \},$$

where the max refers to all primitive characters χ belonging to a modulus $q \leqq Q$. Using (58) this takes the form

$$Q^{-2\lambda} M^2 \leqq c(\lambda) \left\{ \frac{M}{(1-\lambda)^{2\nu}} + M^2 \frac{Q^{2\lambda^2}}{(5/4 - \lambda)^{2\nu}} \right\}. \tag{63}$$

The restriction (60) gives for $Q > c(\lambda)$

$$c(\lambda) \frac{Q^{2\lambda^2}}{(5/4 - \lambda)^{2\nu}} \leqq \tfrac{1}{2} Q^{-2\lambda}$$

and hence from (63)

$$M \leqq c(\lambda) Q^{2\lambda} (1 - \lambda)^{-2\nu}$$

which gives—using again (60)—the estimation in (62).[14]

Since ν, resp. j, can have at most $O(\log Q)$, resp. $O(\log^3 Q)$, values (from (60), resp. (59)) we get from (62) that throwing away at most

$$c(\lambda) Q^{2\lambda+10\lambda^2} \log^4 Q \tag{64}$$

[14] I am indebted to Mr. Montgomery for calling my attention to a paper of Heilbronn in Mathematika 1958 where—for completely different aims—a somewhat similar reasoning was applied.

"bad" L-functions belonging to primitive moduli $q \leqq Q$ we have for the remaining "good" ones the inequality

$$\frac{1}{\nu!}\left|\frac{L'}{L}(s, q, \chi)^{(\nu)}\right|_{s=s_j} \leqq Q^{-\lambda} \tag{65}$$

for all ν's in (60) and all j's in (59).

The contribution of the "bad" L-functions to $M_1(Q, \frac{3}{4} + \delta)$ is evidently at most

$$c(\lambda)Q^{2\lambda+10\lambda^2}\log^5 Q < Q^{2\lambda+11\lambda^2} \tag{66}$$

for $Q > c(\lambda)$. If we succeed in proving that no "good" function vanishes for

$$\sigma \geqq \tfrac{3}{4} + 2\sqrt{\lambda}; \qquad |t| \leqq 1 \tag{67}$$

then choosing λ sufficiently small Theorem IV will follow from (66).

Using the partial fraction representation of L'/L (65) gives easily

$$\left|\sum_{\rho} \frac{1}{(s_j - \rho)^{\nu+1}}\right| < 2Q^{-\lambda}$$

and also—simultaneously for all of our j's and ν's—

$$\left|\sum_{|s_j-\rho|\leqq 7/5}^{(\rho)} (s_j - \rho)^{-\nu-1}\right| \leqq 3Q^{-\lambda} \leqq Q^{-\lambda+\lambda^2}. \tag{68}$$

Let us form the horizontal strips

$$l_j\colon |t - t_j| \leqq 1/2[\log^3 Q] \tag{69}$$

$j = 0, \pm 1, \ldots, \pm[\log^3 Q]$. Fixing j we shall determine ν by using the theorem in (37) in the form that for real m and complex $w_1, \ldots, w_n$ there is an integer ν_0 with

$$m \leqq \nu_0 \leqq m + n \tag{70}$$

so that

$$\left|\sum_{r=1}^{n} w_r^{\nu_0}\right| \geqq \left(\frac{N^*}{8e(m + N^*)}\right)^{N^*} |w_d|^{\nu_0} \tag{71}$$

where N^* is any majorant of n and w_d is any one of the w's (of course ν_0 does not depend on d). The role of the w_r's is played by the quantities $1/(s_j - \rho)$, j fixed. We choose

$$m = \frac{\lambda + 7\lambda^2}{\log \frac{5}{4}} \log Q. \tag{72}$$

The use of the Lindelöf conjecture in (58) gives easily that

$$N^* = 16\lambda^2 \log Q \tag{73}$$

can be chosen. As w_d we choose $(s_j - \rho^*)^{-1}$ where $\rho^* = \sigma^* + it^*$ is one of the zeros in the strip l_j with the maximal real parts (if there is a zero at all). Determining $\nu + 1$ as ν_0 in the theorem (70)–(71) the restriction (60) is evidently satisfied. This gives from (68), after easy estimations,

$$Q^{-\lambda+\lambda^2} \geqq \left(\frac{1}{2-\sigma^*}\right)^{\lambda \log Q(1/\log(5/4)+41\lambda)} \cdot Q^{-32\lambda^{3/2}}$$

or

$$-1+\lambda \geqq \log \frac{1}{2-\sigma^*}\left(\frac{1}{\log(5/4)}+41\lambda\right) - 32\sqrt{\lambda},$$

from which (67) follows at once. This completes the sketch of proof of Theorem IV.

For Theorem III we shall sketch two proofs. One is based on the large sieve method and does not use Halász's idea, the other one does not use the large sieve but uses Halász's idea and also the functional equation of L-functions. The comparison of the two proofs will offer a possibility to see *equivalent* arithmetical and analytical ideas.

We shall use the Lindelöf conjecture in the form (52) and the large sieve in the Bombieri-Gallagher form

$$\sum_{q \leqq Q} \sum_{\chi \bmod q}^{*} \left| \sum_{r=M+1}^{M+R} a_r \chi(r,q) \right|^2 \leqq (Q^2+R) \sum_{r=M+1}^{M+R} |a_n|^2, \tag{74}$$

where the a_n's stand for arbitrary complex constants.

Let $\omega = \omega(\epsilon) > 4$ be a (large) constant,

$$s_j = \omega + \frac{j}{[\log^3 Q]} i \qquad (j = 0, \pm 1, \ldots, \pm[\log^3 Q]), \tag{75}$$

and let the integer ν be restricted, for the time being, by

$$2\omega \log Q \leqq \nu \leqq (2\omega + 10/\sqrt{\omega}) \log Q. \tag{76}$$

For fixed j and ν we investigate the quantity

$$Z_j(\nu) = \sum_{q \leqq Q} \sum_{\chi \bmod q}^{*} \left| \frac{L'}{L}(s, q, \chi)^{(\nu)} \right|_{s=s_j}^2 . \tag{77}$$

Using the Dirichlet series representation we get

$$Z_j(\nu) = \sum_{q \leqq Q} \sum_{\chi \bmod q}^{*} \left| \sum_{d=0}^{\infty} \sum_{r=dQ^2+1}^{(d+1)Q^2} \frac{\Lambda(r) \log^\nu r}{r^{s_j}} \chi(r,q) \right|^2,$$

which is, by Schwarz's inequality,

$$\leqq c(\epsilon) \sum_{q \leqq Q} \sum_{\chi \bmod q}^{*} \sum_{d=0}^{\infty} (d+1)^{1+2\epsilon} \left| \sum_{r=dQ^2+1}^{(d+1)Q^2} \frac{\Lambda(r) \log^\nu r}{r^{s_j}} \chi(r,q) \right|^2 .$$

Using the large sieve inequality (74) with $R = Q^2$ this gives

$$Z_j(\nu) \leqq c(\epsilon) Q^2 \sum_{d=0}^{\infty} (d+1)^{1+2\epsilon} \sum_{r=dQ^2+1}^{(d+1)Q^2} \frac{\log^{2\nu+2} r}{r^{2\omega}}$$

from which routine estimations give

$$Z_j(\nu) \leqq c(\epsilon) \frac{\nu!^2 \log^2 Q}{(\omega - 1 - 2\epsilon)^{2\nu}}. \tag{78}$$

This gives, as before, the inequality

$$\frac{1}{\nu!}\left|\frac{L'}{L}(s, q, \chi)^{(\nu)}\right|_{s=s_j} \leqq \frac{Q^{-2(1-\alpha)}}{(\omega - 1 - \epsilon)^\nu} \tag{79}$$

for fixed j and ν with exception of at most

$$Q^{4(1-\alpha)} \log^2 Q \tag{80}$$

$L(s, q, \chi)$ functions with $q \leqq Q$ and primitive χ. Having the inequality (79)–(80) the concluding part of the proof follows the line of the previous sketch with some technical adjustments, choosing $\omega = \omega(\epsilon)$ sufficiently large.

As to the second proof of Theorem III we have to prove the inequality (79)–(80) without the large sieve. Following Halász's reasoning in the proof of (62) (on the line $\sigma = \omega$ instead of $\sigma = 2$ however) we get again an inequality of type

$$AM^2 \leqq BM + CM^2 \tag{81}$$

and everything depends on whether C can be made $< A/2$ say. This in turn depends on the estimation of $|L^{(2\nu)}(2\omega - 1 - \epsilon, q, \chi)|$, $\chi \neq \chi_0$, $q \leqq Q^2$ (instead of $|L^{(2\nu)}(3 - 2\lambda, q, \chi)|$). It would be plausible to use Cauchy's inequality for the disc $|s - (2\omega - 1 - \epsilon)| \leqq 2\omega - \frac{3}{2} - \epsilon$ and use Lindelöf's hypothesis. This does not work however. Now using the disc

$$|s - (2\omega - 1 - \epsilon)| \leqq 2\omega + \sqrt{\omega}$$

we cannot use Lindelöf's conjecture any more, but using the functional equation instead we can squeeze out the relation $C < A/2$ in (81). So instead of a conditional step an unconditional one helps; so in the proof of Theorem III Lindelöf's conjecture could be used *only once* (whereas in the proof of Theorem IV it could be used *twice*). Probably this is the basic reason for the great success of Halász's idea. The only slight change necessary in the further course of this second proof is that instead of (76) we have to require

$$2\omega \log Q \leqq \nu \leqq (2\omega + \sqrt{\omega}) \log Q.$$

MATHEMATICAL INSTITUTE OF THE HUNGARIAN ACADEMY OF SCIENCES
BUDAPEST, HUNGARY

CHARACTERISATION OF THE LOGARITHM AS AN ADDITIVE FUNCTION

EDUARD WIRSING

1. Summary

Recently I. Kátai [2] succeeded in proving the following hypothesis of P. Erdös [1]:

If a function f defined on the natural numbers is additive (*i.e.* $f(mn) = f(m) + f(n)$ *if* $(m, n) = 1$) *and if*

$$\sum_{n \leq x} |f(n+1) - f(n)| = o(x) \tag{1}$$

holds, then $f(n) = c \log n$ *with a constant c.*

We shall give another proof of the theorem (as an outline in 2.1.) and sharpen the theorem by replacing (1) by the following condition:

(2) *There is a constant* $\gamma > 1$ *and a sequence of numbers* $x_i \to \infty$ *such that, as* $i \to \infty$,

$$\sum_{x_i \leq n \leq \gamma x_i} |f(n+1) - f(n)| = o(x_i).$$

2. Outlines

2.1. Our proof with the assumption (1) is based on a consideration of the sum

$$S(x) := \sum_{n \leq x} f(n),$$

for which we show the relation

$$S(x) = qS(x/q) + xf(q) + o(x) \tag{3}$$

(q an arbitrary natural number, $x \to \infty$). To show the idea let f be completely additive first. For any residue classes r, s mod q we get from (1) (adding suitable intermediate terms and using that $f(1) = 0$)

$$\left| \sum_{n \leq x;\, n \equiv r \bmod q} f(n) - \sum_{n \leq x;\, n \equiv s \bmod q} f(n) \right| \leq \sum_{n \leq x} |f(n+1) - f(n)| = o(x).$$

Taking $s = 0$ and summing over r mod q gives

$$\begin{aligned} S(x) &= q \sum_{n \leq x;\, n \equiv 0 \bmod q} f(n) + o(x) \\ &= q \sum_{m \leq x/q} (f(m) + f(q)) + o(x) \\ &= qS(x/q) + qf(q)[x/q] + o(x) \\ &= qS(x/q) + xf(q) + o(x). \end{aligned}$$

Replacing x in (3) by $x/q, \ldots, x/q^{k-1}$ ($k = [\log x/\log q]$) and combining these inequalities we find

$$\begin{aligned} S(x) &= kxf(q) + o(kx) \\ &= \frac{f(q)}{\log q} x \log x + o(x \log x). \end{aligned}$$

This same iteration, done with a different number p, must produce the same constant. Hence

$$\frac{f(p)}{\log p} = \frac{f(q)}{\log q} \qquad (p, q \text{ arbitrary}),$$

which means $f(n) = c \log n$.

2.2. Complete additivity, which was used here for $f(n) = f(m) + f(q)$, can be avoided by not using the residue class $0 \bmod q$ but $q \bmod q^2$ instead. If $n \equiv q \bmod q^2$ then $n = mq$ with $m \equiv 1 \bmod q$. Here $(m, q) = 1$ and $f(n) = f(m) + f(q)$ under the assumption of restricted additivity.

2.3. It is easy to use $\sum_{n \leq x} (1/n) f(n)$ instead of S. One obtains

$$\sum_{n \leq x} \frac{1}{n} f(n) = \frac{f(q)}{2 \log q} \log^2 x + o(\log^2 x) \tag{4}$$

with only the condition

$$\sum_{n \leq x} \left| \frac{f(n+1)}{n+1} - \frac{f(n)}{n} \right| = o(\log x). \tag{5}$$

It is clear that (4) again implies $f(n) = c \log n$. Another condition that can be used instead of (5) is

$$\sum_{n \leq x} \frac{1}{n} |f(n+1) - f(n)| = o(\log x). \tag{6}$$

This formula is obtained from (1) by partial summation and on the other hand implies (5): from

$$\left|\frac{f(n+1)}{n+1}-\frac{f(n)}{n}\right| \le \frac{1}{n+1}|f(n+1)-f(n)| + \frac{1}{n(n+1)}|f(n)|$$
$$\le \frac{1}{n+1}|f(n+1)-f(n)|$$
$$+\frac{1}{n(n+1)}\sum_{m<n}|f(m+1)-f(m)|$$

we see

$$\sum_{n\le x}\left|\frac{f(n+1)}{n+1}-\frac{f(n)}{n}\right| \le \sum_{n\le x}\frac{1}{n+1}|f(n+1)-f(n)|.$$

It is worth emphasizing that the sketched proof still works, and without additional complications, if (5) or (6) are given for a sequence of numbers $x_i \to \infty$ only (of course we get (4) for these x_i only). There is, however, no immediate generalization of the proof that would employ the still weaker

(7) $$\sum_{n\le x_i}|f(n+1)-f(n)| = o(x_i), \qquad x_i \to \infty,$$

because the later steps in the iteration (x_i/q^m much smaller than x_i) cannot be controlled sufficiently.

2.4. C. Ryavec **[3]**, who learned about the proof sketched in 2.1. found it sufficient to do a large but fixed number of steps with p and q in (3) (down to x/p^a and x/q^b, say) where p^a and q^b are taken close to each other provided one can estimate

(8) $$\left|p^a S\left(\frac{x}{p^a}\right) - q^b S\left(\frac{x}{q^b}\right)\right| \quad \text{suitably.}$$

Thus he proved that (7) combined with $f(n) = O(\log n)$ again implies $f(n) = c \log n$. He conjectured that the condition $f(n) = O(\log n)$ might be unnecessary.

2.5. In the present paper we use (3) with p and q alternating in such a way that $S(x)$ is linked to $S(xp^a q^{-b})$ while every step is done in the neighborhood of x_i. This way we can do arbitrarily many steps without loss of accuracy, and with $a, b \to \infty$ the influence of the term corresponding to (8) vanishes.

Actually, replacing (7) by (2), we go beyond Ryavec's hypothesis. To be able to work in the narrow range from x_i to γx_i we use rational numbers p, q that are little larger than 1 instead of natural ones. This can be done without essentially changing the structure of the proof.

3. Detailed proof

We assume now that f is an additive function and that (2) holds. We shall prove $f(n) = c \log n$.

3.1. *Notations.* Let

$$T_i := \frac{1}{x_i}\sum_{x_i\le n\le \gamma x_i}|f(n+1)-f(n)|,$$

$$F_i := \text{mean value of } f \text{ on } [x_i, \gamma x_i], \text{ hence}$$

$$\sum_{x_i \le n \le \gamma x_i} (f(n) - F_i) = 0.$$

$$\gamma_1 := \gamma^{1/3}, \qquad \delta := \gamma_1 - 1,$$

$$U(y) := \frac{1}{\delta y} \sum_{y \le n \le \gamma_1 y} (f(n) - F_i)$$

$$U(y; q, r) := \frac{1}{\delta y} \sum_{\substack{y \le n \le \gamma_1 y; \\ n \equiv r \bmod q}} (f(n) - F_i) \quad \text{for } y \in [x_i, \gamma_1^2 x_i],$$

$$f(p/q) := f(p) - f(q) \quad \text{if } (p, q) = 1.$$

3.2. We shall need the following analogue of (3): *If p, q are natural numbers,*

$$\gamma_1^{-1} \le p/q \le \gamma_1,$$

then to each $\epsilon > 0$ there is an $i_0 = i_0(\epsilon, p, q)$ such that

$$i \ge i_0, \qquad y, y' \in [x_i, \gamma_1^2 x_i], \qquad y'/y = p/q$$

imply

(9) $$|U(y') - U(y) - f(p/q)| \le \epsilon.$$

Without loss of generality we assume $(p, q) = 1$.

The function $f(m) - F_i$ cannot be positive or negative throughout the interval $[x_i, \gamma x_i]$. For each m we have a suitable m', therefore, to conclude

(10) $$\begin{aligned} |f(m) - F_i| &\le |f(m) - f(m')| \\ &\le \sum_{x_i \le n \le \gamma x_i} |f(n+1) - f(n)| \\ &= T_i x_i, \end{aligned}$$

$$|f(m) - F_i| \le T_i x_i \quad \text{for } m \in [x_i, \gamma x_i].$$

For $y \in [x_i, \gamma_1^2 x_i]$ the summation in $U(y; q, r)$ goes over part of the interval $[x_i, \gamma x_i]$. The numbers of terms in $U(y; q, r)$ and $U(y; q, s)$ differ by no more than one. We estimate the possible odd term by (10). The others can be paired in the natural order to estimate the difference with terms $|f(n+1) - f(n)|$. Thus

$$\begin{aligned} \delta y |U(y; q, r) - U(y; q, s)| &\le \sum_{y \le n \le \gamma_1 y} |f(n+1) - f(n)| + T_i x_i \\ &\le 2T_i x_i \le 2T_i y. \end{aligned}$$

After summing over $s \bmod q$ we have

$$\delta |U(y) - qU(y; q, r)| \le 2qT_i \quad \text{for } y \in [x_i, \gamma_1^2 x_i].$$

Specializing we get for y, y'

(11)
$$\delta |U(y) - pq^2 U(y; pq^2, q)| \le 2pq^2 T_i, \quad \delta |U(y') - p^2 q U(y'; p^2 q, p)| \le 2p^2 q T_i.$$

The numbers n:

$$y \leq n \leq \gamma_1 y, \qquad n \equiv q \bmod pq^2$$

are the products $n = mq$ with

$$y/q \leq m \leq \gamma_1 y/q, \qquad m \equiv 1 \bmod pq.$$

Since $(m, q) = 1$ it is $f(n) = f(m) + f(q)$. We get

$$\delta pq^2 U(y; pq^2, q) = \frac{pq^2}{y} \sum_{\substack{y/q \leq m \leq \gamma_1 y/q; \\ m \equiv 1 \bmod pq}} (f(m) + f(q) - F_i)$$

and similarly

$$\delta p^2 q U(y'; p^2 q, p) = \frac{p^2 q}{y'} \sum_{\substack{y'/p \leq m \leq \gamma_1 y'/p; \\ m \equiv 1 \bmod pq}} (f(m) + f(p) - F_i).$$

By assumption $y'/p = y/q$, $pq^2/y = p^2q/y'$. The two estimates therefore combine to

$$\begin{aligned} \delta p^2 q U(y'; p^2 q, p) - \delta pq^2 U(y; pq^2, q) & \\ &= \frac{pq^2}{y} \sum_{\substack{y/q \leq m \leq \gamma_1 y/q; \\ m \equiv 1 \bmod pq}} (f(p) - f(q)) \\ &= \frac{pq^2}{y} f\left(\frac{p}{q}\right)\left((\gamma_1 - 1)\frac{y}{pq^2} + \vartheta\right) \qquad (|\vartheta| \leq 1) \\ &= \delta f\left(\frac{p}{q}\right) + \vartheta \frac{pq^2}{y} f\left(\frac{p}{q}\right). \end{aligned} \tag{12}$$

According to (2) $T_i \to 0$ as $i \to \infty$. So we infer from (11) and (12)

$$\begin{aligned} \left| U(y') - U(y) - f\left(\frac{p}{q}\right) \right| &\leq \frac{2}{\delta}(pq^2 + p^2 q) T_i + \frac{pq^2}{\delta x_i} \left| f\left(\frac{p}{q}\right) \right| \\ &\leq \epsilon \quad \text{if } i \geq i_0, \end{aligned}$$

that is (9).

3.3. *The iteration.* Now we consider two pairs of numbers p, q and r, s, with

$$1 < p/q \leq \gamma_1, \qquad 1 < r/s \leq \gamma_1.$$

We take $i \geq i_0(\epsilon, p, q)$, $i_0(\epsilon, r, s)$ and define a sequence of numbers y_k by

$$\begin{aligned} y_0 &= x_i \\ y_k &= \frac{p}{q} y_{k-1} \quad \text{if} \quad x_i \leq y_{k-1} < \gamma_1 x_i \\ &= \frac{s}{r} y_{k-1} \quad \text{if } \gamma_1 x_i \leq y_{k-1} < \gamma_1^2 x_i. \end{aligned}$$

Then we have $y_k \in [x_i, \gamma_1^2 x_i]$ for all k. In the first case (9) yields

$$U(y_k) - U(y_{k-1}) = f(p/q) + r_k, \qquad |r_k| \leq \epsilon,$$

in the other case similarly

$$U(y_k) - U(y_{k-1}) = f(s/r) + r_k, \qquad |r_k| \leq \epsilon.$$

In the formation of y_k let the first and the second case occur a_k and b_k times resp., then our iteration gives

$$U(y_k) - U(y_0) = a_k f\left(\frac{p}{q}\right) + b_k f\left(\frac{s}{r}\right) + \sum_{j=1}^{k} r_j.$$

If we let $k \to \infty$ the left-hand side stays bounded. Therefore

$$(13) \qquad |a_k f(p/q) - b_k f(r/s)| \leq \epsilon k + O(1).$$

On the other hand from

$$x_i \leq y_k = x_i \left(\frac{p}{q}\right)^{a_k} \left(\frac{s}{r}\right)^{b_k} \leq \gamma x_i$$

we infer

$$0 \leq a_k \log \frac{p}{q} - b_k \log \frac{r}{s} \leq \log \gamma.$$

This and the trivial $a_k + b_k = k$ give

$$\frac{a_k}{k} \to \log \frac{r}{s} \left(\log \frac{pr}{qs}\right)^{-1}, \qquad \frac{b_k}{k} \to \log \frac{p}{q} \left(\log \frac{pr}{qs}\right)^{-1}$$

as $k \to \infty$. Inserting this into (13) we find

$$\left| f\left(\frac{p}{q}\right) \log \frac{r}{s} - f\left(\frac{r}{s}\right) \log \frac{p}{q} \right| \leq \epsilon \log \frac{pr}{qs}.$$

Since ϵ was arbitrary,

$$\frac{f\left(\frac{p}{q}\right)}{\log \frac{p}{q}} = \frac{f\left(\frac{r}{s}\right)}{\log \frac{r}{s}}.$$

3.4. *Finish.* So far we have proved that there is a constant c such that

$$f(p) - f(q) = c\,(\log p - \log q) \quad \text{if } 1 < p/q \leq \gamma_1.$$

In particular

$$f(m+1) - f(m) = c\,(\log (m+1) - \log m) \quad \text{if } m \geq 1/\delta.$$

This makes

$$f(m) = c \log m + c' \quad \text{for } m \geq 1/\delta$$

with another constant c'.

Now let n be arbitrary. We choose $m \geq 1/\delta$, $(m, n) = 1$ and see

$$\begin{aligned} f(n) &= f(mn) - f(m) \\ &= c(\log mn - \log m) \\ &= c \log n. \end{aligned}$$

REFERENCES

1. Erdös, P., *On the distribution function of additive functions*, Ann. of Math. (2) **47** (1946), 1–20. MR **7**, 416.
2. Kátai, I., (to appear in Journal of Number Theory).
3. Ryavec, C., Some results concerning additive functions. (Publication in preparation.)

MATHEMATISCHES INSTITUT DER UNIVERSITÄT MARBURG
MARBURG DER LAHN, FEDERAL REPUBLIC OF GERMANY

SOME RESULTS CONCERNING RECIPROCITY LAW AND REAL ANALYTIC AUTOMORPHIC FUNCTIONS

TOMIO KUBOTA

1. Two kinds of special functions related to the reciprocity law

In this and in the following two articles, the author will emphasize that there are various possibilities of investigations in a territory of the number theory where functional analysis on a topological group and the reciprocity law in a number field are combined. To see this, we shall observe several analytical results which are deduced as applications of an arithmetical theorem concerning both the reciprocity law and the congruence subgroup problem of some matric groups.

Throughout this and following articles, F will stand for a totally imaginary number field of degree $2r$ containing the n-th roots of unity for a fixed $n \geqq 2$. On the other hand, we shall denote by $\mathfrak{o}$ the ring of integers of F, and by (a/b) the n-th power residue symbol in F. The theorem mentioned above means now:

THEOREM 1-1. *Denote by* Γ *the congruence subgroup* mod N *of* $SL(2, \mathfrak{o})$ *for a natural number* N, *and define the function* $\chi(\sigma)$ *of*

$$\sigma = \begin{pmatrix} a & b \\ c & d \end{pmatrix} \in \Gamma$$

by $\chi(\sigma) = (c/d)$ *or* 1 *according to* $c \neq 0$ *or* $= 0$. *Then* χ *is a homomorphism of* Γ *onto the group of n-th roots of unity, whenever* $N \equiv 0 \pmod{n^2}$. *Furthermore the kernel of* χ *contains no congruence subgroup of* $SL(2, \mathfrak{o})$ [1], [4], [6] .

In many cases, the condition $N \equiv 0 \pmod{n^2}$ can be far weakened.

While Theorem 1-1 concerns all other articles, the purpose of §1 is to construct by means of the theorem two kinds of special functions with some number theoretical meaning. One is a real analytic automorphic form, and the other is a Dirichlet series with functional equation.

First of all, we give a brief illustration about the real three dimensional upper half space as an analogy of the complex upper half plane. The upper half space means the space H of all matrices

$$u = \begin{pmatrix} z & -v \\ v & \bar{z} \end{pmatrix} \quad \text{with } z \in \mathbf{C}, v > 0.$$

If no confusion is possible, u will also be denoted by (z, v). For a $\sigma \in SL(2, \mathbf{C})$, we define the operation of σ on H by the linear transformation

$$(1) \qquad \sigma u = (\tilde{a}u + \tilde{b})(\tilde{c}u + \tilde{d})^{-1}, \qquad \sigma = \begin{pmatrix} a & b \\ c & d \end{pmatrix}, u \in H$$

where $\tilde{\omega} = \begin{pmatrix} \omega & 0 \\ 0 & \bar{\omega} \end{pmatrix}$ for any $\omega \in \mathbf{C}$, and the operations in the right-hand side of (1) are all matric ones. The stabilizer of $(0, 1) \in H$ is $SU(2)$, and therefore H is a realization of the homogeneous space $SL(2, \mathbf{C})/SU(2)$. The operation (1) has a well-defined meaning even if $v \leqq 0$, so $SL(2, \mathbf{C})$ operates on the whole space $\mathbf{R}^3$, or more precisely on $\mathbf{R}^3 \cup \{\infty\}$, where ∞ is a symbolical point at infinity, and in a natural way we get three orbits of the operation; they are the upper half space, the lower half space and the complex projective line $\mathbf{C} \cup \{\infty\}$.

The group $SL(2, F)$ is considered as a subgroup of $SL(2, \mathbf{C})^r$ by the imbedding $\sigma \to (\sigma^{(1)}, \ldots, \sigma^{(r)})$, where $\sigma^{(i)}$ is the i-th conjugate of $\sigma \in SL(2, F)$. Thus σ defines an operation of Hilbert type on H^r, and in particular $SL(2, \mathfrak{o})$ becomes a discontinuous group of Hilbert type operating on H^r whose fundamental domain is of finite invariant volume. If we identify a number $\gamma \in F$ with the point $(\gamma^{(1)}, \ldots, \gamma^{(r)}) \in \mathbf{C}^r$, then γ is a boundary point of $H^r \subset (\mathbf{R}^3)^r$. The set F is exactly the set of all cusps of $SL(2, \mathfrak{o})$, or more generally of an arbitrary subgroup Γ' of finite index of $SL(2, \mathfrak{o})$. Here, a cusp of Γ' means a point $\gamma \in \mathbf{C}^r \subset (\mathbf{R}^3)^r$ such that the stabilizer of γ in Γ' contains a group of unipotent elements isomorphic to $\mathbf{Z}^{2r}$. The fundamental domain $\Gamma' \backslash H^r$ has only a finite number of cusps.

Now we introduce some Eisenstein series attached to the group Γ in Theorem 1-1. We denote by $\mathfrak{D} = \Gamma \backslash H^r$ a fundamental domain of Γ in H^r such that $\mathfrak{D}$ is a convex polyhedron in the hyperbolic space H^r. Furthermore, for each cusp κ of $\mathfrak{D}$, we denote by Γ_κ the stabilizer of κ in Γ. If the character χ in Theorem 1-1 is trivial on Γ_κ, then κ will be called an essential cusp. In each of the following observations, we fix a suitable N in Theorem 1-1 so that Theorem 1-1 holds, and denote by $\kappa_1, \ldots \kappa_h$ all the essential cusps of $\mathfrak{D}$. For each κ_i, take a $\sigma_i \in SL(2, \mathbf{C})^r$ with $\sigma_i \infty = \kappa_i$, where ∞ means $(\infty, \ldots, \infty)$. Then, Γ_i being the stabilizer of κ_i in Γ, the group $\sigma_i^{-1}\Gamma_i\sigma_i$ is the largest triangular subgroup of $\sigma_i^{-1}\Gamma\sigma_i$, and the group of diagonal elements in $\sigma_i^{-1}\Gamma_i\sigma_i$ is naturally isomorphic to a subgroup $\mathbf{e}_i$ of the unit group of F. The group $\mathbf{e}_i$ does not depend on the choice of σ_i, and there exists a common subgroup $\mathbf{e}$ of $\mathbf{e}_i$ such that the index $(\mathbf{e}_i : \mathbf{e}) = k_i$ is finite. On the other hand, the group of all unipotent matrices in $\sigma_i^{-1}\Gamma_i\sigma_i$ is canonically isomorphic to a submodule $\mathfrak{m}_i$ of $\mathbf{C}^r$. More precisely, $\mathfrak{m}_i$ is a lattice group in $\mathbf{C}^r$ such that $\mathfrak{m}_i \backslash \mathbf{C}^r$ is compact. The euclidean volume V_i of $\mathfrak{m}_i \backslash \mathbf{C}^r$ depends on σ_i, but we choose σ_i such that $V_1 : V_2 : \ldots : V_n = k_1 : k_2 : \ldots k_n$. If some κ_i, say κ_1, is ∞, then we take $\sigma_1 = 1$.

We define then the Eisenstein series E_i for the essential cusp κ_i by

$$(2) \qquad E_i(u, s) = \sum_{\sigma} \bar{\chi}(\sigma) v(\sigma_i^{-1}\sigma u)^s, \qquad (\sigma \in \Gamma_i \backslash \Gamma),$$

where $u = (u_1, \ldots, u_r) \in H^r$, $u_i = (z_i, v_i) \in H$, $v = \prod_i v_i$, and s is a complex variable. If $\operatorname{Re} s > 2$, then (2) converges absolutely, and defines an automorphic function on H^r with respect to Γ and the automorphic factor $\chi(\sigma)$, i.e. $E_i(\sigma u, s) = \chi(\sigma)E_i(u, s)$. On the other hand, E_i is an eigenfunction of all Laplacians of H^r, and therefore E_i is a real analytic automorphic form.

To formulate further important properties of E_i, we have to consider the Fourier expansion of E_i. For this purpose, we put first $e(z) = \exp(2\pi\sqrt{(-1)}(z + \bar{z}))$ for $z \in \mathbf{C}$. Next, for $u = (z, v) \in H$, we put $e(u) = e(z)$, and if $u = (u_1, \ldots, u_r) \in H^r$, then we put $e(u) = \prod_i e(u_i)$. Moreover, we define mu for $m \in \mathbf{C}$, $u = (z, v) \in H$ by $mu = (mz, |m|v)$; this is equal to the image of u by the transformation

$$\begin{pmatrix} m^{1/2} & 0 \\ 0 & m^{-1/2} \end{pmatrix}.$$

If $m \in \mathbf{C}^r$, and $u \in H^r$, then mu is defined componentwisely.

Since $E_i(\sigma_j u, s)$ is a periodic function with respect to $\mathfrak{m}_j$, it allows the Fourier expansion

$$E_i(\sigma_j u, s) = \sum_m a_{ij,m}(v_1, \ldots v_r, s)e(mu) \qquad (m \in \mathfrak{m}_j^*), \tag{3}$$

with

$$a_{ij,m}(v_1, \ldots, v_r, s) = V_j^{-1} \int_{P_j} E_i(\sigma_j u, s)e(-mu)\, dV(z), \tag{4}$$

where $P_j = \mathfrak{m}_j \backslash \mathbf{C}^r$, $dV(z)$ is the euclidean measure, and $\mathfrak{m}_j^*$ the dual of $\mathfrak{m}_j$, i.e. the module of all $m \in \mathbf{C}^r$ with $e(mb) = 1$ for all $b \in \mathfrak{m}_j$. The series (3) will be called the Fourier expansion of E_i at κ_j. In particular, the constant term $a_{ij,0}$ of the expansion has the form

$$a_{ij,0}(v_1, \ldots v_r, s) = \delta_{ij} v(u)^s + \varphi_{ij}(s) v(u)^{2-s} \tag{5}$$

with Kronecker's δ. By the theory of Selberg [9], [10], (see also [2]), the function E_i has a one valued, meromorphic continuation on the whole s-plane, and the function vector $\mathcal{E}(u, s) = {}^t(E_1, \ldots, E_h)$ satisfies the functional equation

$$\mathcal{E}(u, s) = \Phi(s)\mathcal{E}(u, 2 - s), \qquad \Phi(s) = (\varphi_{ij}(s)), \tag{6}$$

which is a consequence of the relation $\Phi(s)\Phi(2 - s) = I$, the identity matrix. This is proved by using the fact that $\Phi(s)$ is unitary on $\operatorname{Re} s = 1$, and is hermitian on the real axis. While the former assertion consists of the most important part of Selberg's theory, the latter can be shown rather elementarily. For, $\mathbf{C}^\times$ being the multiplicative group of nonzero numbers of $\mathbf{C}$, and $\mathfrak{C}$ being a set of representatives of $\mathbf{C}^{\times r}/\mathbf{e}$, we have first

$$E_i(\sigma_j u, s, \chi) = \delta_{ij} v(u)^s + k_i^{-1} \sum_{(c,d)} \bar{\chi}_{ij}(c, d) \frac{v(u)^s}{j(c, d; u)^s}, \qquad ((c; d) \in M_{ij}), \tag{7}$$

where $j(c, d; u) = \prod_i |c^{(i)}z_i + d^{(i)}|^2 + |c^{(i)}|^2 v_i^2$ for $u = (u_1, \ldots, u_r)$, M_{ij} is the set of pairs (c, d) of complex r-dimensional vectors $c = (c^{(1)}, \ldots, c^{(r)})$, $d =$

$(d^{(1)}, \ldots, d^{(r)})$ such that there exists an element in $\sigma_i^{-1}\Gamma\sigma_j$ which is of the form

$$\sigma = \begin{pmatrix} * & * \\ c & d \end{pmatrix},$$

and $\chi_{ij}(c, d) = \chi(\sigma_i \sigma \sigma_j^{-1})$. Hence we have an explicit formula

$$\varphi_{ij}(s) = k_i^{-1} V_j^{-1} \sum_{c \in \mathfrak{C}} \frac{1}{\|c\|^s} \left(\sum \chi_{ij}(c, d)\right) \cdot \frac{\pi^r}{(s-1)^r}, \tag{8}$$

$$((c, d) \in M_{ij},\ d \bmod cm_j),$$

with $\|c\| = \prod_i |c^{(i)}|^2$ of $\varphi_{ij}(s)$, and from $\chi(\sigma_i \sigma \sigma_j^{-1}) = \bar{\chi}(\sigma_j \sigma^{-1} \sigma_i^{-1})$ follows the hermitness of $\Phi(s)$ for $s \in \mathbf{R}$.

To obtain a special kind of automorphic form on H^r, we denote by $\mathcal{E}_s$ the space of all linear combinations $\sum_{i=1}^h w_i(s) E_i(u, s)$ by entire functions $w_i(s)$ of s.[1] $\mathcal{E}_s$ does not depend on the choice of σ_i, and we can prove

THEOREM 1-2. *In general, $\mathcal{E}_s \cap L^2(\mathfrak{D})$ is 0, but for the only one exceptional value $s = (n+1)/n$, the space $\Theta = \mathcal{E}_{n+1/n} \cap L^2(\mathfrak{D})$ is not trivial and finite dimensional over* $\mathbf{C}$ [6, Theorem 6].

This theorem is proved by using the fact that the Dirichlet series in (8) coincides essentially with the quotient $\zeta(ns - n)/\zeta(ns - (n-1))$ of Dedekind's zeta functions of F. This quotient has a pole at $s = (n+1)/n$, and the residues of Eisenstein series produce there, as is mentioned in p. 183 of [10], square integrable automorphic forms which are not cusp forms. It should also be noted that a nontrivial space as Θ is obtained only when a character χ of Γ as in Theorem 1-1 is applied. A function $\theta \in \Theta$ satisfies the transformation formula $\theta(\sigma u) = \chi(\sigma)\theta(u)$ for each $\sigma \in \Gamma$. Therefore, if one can construct such a θ by an analytical method without using the reciprocity law, then the above transformation implies $\chi(\sigma\sigma') = \chi(\sigma)\chi(\sigma')$, which is the assertion of Theorem 1-1. Since, however, the theorem is essentially equivalent to the reciprocity law, we get in this way an analytical proof of the reciprocity law. Although the direct construction of Θ is still unknown for general cases, the space Θ reduces, as will be mentioned again in §2, to a space of classical theta-functions if $n = 2$, and the Hecke's proof of the quadratic reciprocity law in a general number field by means of theta-functions is based upon exactly the same arguments. Further meanings of Θ will be described in §2.

To obtain a special kind of Dirichlet series satisfying a functional equation, we observe the nonconstant terms of (3); since the functional equation (6) holds termwisely for (3), a conconstant term of (3) can contain a new functional equation. In fact, we have

$$\begin{aligned} &a_{ij,m}(v_1, \ldots, v_r, s) \\ &\quad = \varphi_{ij}(s, m)\|m\|^{(s-1)/2} \cdot \prod_{i=1}^r K_{s-1}(4\pi|m^{(i)}|v_i) \cdot (2\pi)^{rs}\Gamma(s)^{-r}v(u), \end{aligned} \tag{9}$$

[1] A pole of $E_i(s)$ may be cancelled by a zero of $w_i(s)$.

where K_s is the modified Bessel function, and

$$(10) \qquad \varphi_{ij}(s, m) = k_i^{-1} V_j^{-1} \sum_{c \in \mathfrak{C}} \frac{1}{\|c\|^s} \left(\sum_{(c,d)} \bar{\chi}_{ij}(c, d) e\left(\frac{md}{c}\right)\right),$$

$$((c, d) \in M_{ij},\ d \bmod cm_j).$$

This formula, as well as (8), can be shown by direct computations (see [6]).

Let now d_F be the absolute value of the discriminant of F, and define Φ_1 by

$$(11) \qquad \Phi(s) = \Phi_1(s) \cdot n^{-r} d_F^{-1/2} \frac{(2\pi)^r \zeta(ns - n)}{(s-1)^r \zeta(ns - (n-1))}.$$

Then, it follows from $\Phi(s)\Phi(2 - s) = I$ and from the functional equation of ζ, i.e. the invariance under $s \to 1 - s$ of $(2\pi)^{-rs} d_F^{s/2} \Gamma(s)^r \zeta(s)$, that $\Phi_1(s)\Phi_1(2 - s) = I$, and hence $(I + \Phi_1(2 - s))\Phi_1(s) = \Phi_1(s) + I$. Because of this fact, (3), (6), and (11) imply

$$(12) \qquad \begin{aligned} &(I + \Phi_1(2 - s))\,{}^t(a_{i1,m}, \ldots, a_{ih,m})(s) \\ &= (I + \Phi_1(s)) n^{-r} d_F^{-1/2} \frac{(2\pi)^r \zeta(ns - n)}{(s-1)^r \zeta(ns - (n-1))}\,{}^t(a_{il,m}, \ldots, a_{ih,m})(2 - s). \end{aligned}$$

Therefore, if we put

$$(13) \qquad \begin{aligned} (I + \Phi_1(2 - s))\,{}^t(\varphi_{i1}(s, m), \ldots, \varphi_{ih}(s, m))\zeta(ns - (n-1)) \\ = {}^t(\zeta_{i1}(s, m), \ldots, \zeta_{ih}(s, m)), \end{aligned}$$

then a calculation using in particular (9), (12), (13) and the symmetry $K_s = K_{-s}$ gives rise to a functional equation

$$(14) \qquad \xi(s, m) = \xi(2 - s, m)$$

with

$$\xi(s, m) = \|m\|^{s/2} (2\pi)^{-(n-1)rs} d_F^{ns/2} \frac{\Gamma(n(s-1))^r}{\Gamma(s-1)^r} \zeta(s, m)$$

for any $\zeta(s, m) = \zeta_{ij}(s, m)$.

By using (8) and (10), it is not hard to see that the function $\zeta(s, m)$ in (14) is actually a Dirichlet series, and that the coefficients (numerators in usual notation) are Gauss sums. An explicit explanation of this situation by a special example is in [5]. In §3, we shall show that Dirichlet series obtained by generalizing in a certain sense our $\zeta(s, m)$ here have an effective application in a problem concerning Gauss sums.

2. A unitary representation of a metaplectic group

The functions belonging to the space Θ in Theorem 2 of §1 are in a certain sense generalizations of usual theta-functions. In fact, it is not hard to show that Θ actually consists of theta-functions when $n = 2$. Here, a theta-function means a function on H^r induced by an ordinary theta-function on a Siegel's space through a suitable imbedding of H^r into the Siegel's space (see [6, §7, 3]). To show more clearly that the functions in Θ are, at least in a number theoretical sense, legitimate generalizations of ordinary theta-functions, we shall prove in this article that the space Θ gives a special kind of unitary representation of a generalized metaplectic group, i.e. an n-fold topological covering group of the adèle group of $SL(2, F)$, in such a way that the representation for $n = 2$ coincides essentially with the representation constructed in [11] by means of theta-functions. The method in [11] is purely functional analytic, so that one can obtain an analytic proof of the quadratic reciprocity at the same time. Our method is applicable for general n, but requires the reciprocity law of degree n.

To begin with, we have to state a construction of a metaplectic group. In the sequel, we denote by G_F the group $SL(2, F)$ over the same number field F as in §1, and by G_A the adèle group of G_F. The finite resp. infinite part of G_A will be denoted by G_0 resp. G_∞, and $G_\mathfrak{p}$ will denote the $\mathfrak{p}$-component of G_A for a prime divisor of F. Of course, $G_\mathfrak{p} = SL(2, F_\mathfrak{p})$, $F_\mathfrak{p}$ being the $\mathfrak{p}$-adic field, and $G_\infty = SL(2, \mathbf{C})^r$.

Now we have the following

THEOREM 2-1. *Let $\mathfrak{p}$ be a prime divisor of F, and let $(\alpha, \beta/\mathfrak{p})$ be the norm residue symbol of degree n of F in the sense of Hilbert-Hasse. For an element*

$$\sigma = \begin{pmatrix} a & b \\ c & d \end{pmatrix} \in G_\mathfrak{p} = SL(2, F_\mathfrak{p}),$$

put $x(\sigma) = c$ or d, according to $c \neq 0$ or $c = 0$, and put

$$a_\mathfrak{p}(\sigma, \tau) = (x(\sigma), x(\tau)/\mathfrak{p})(-x(\sigma)^{-1}x(\tau), x(\sigma\tau)/\mathfrak{p}), \tag{15}$$

$\sigma, \tau \in G_\mathfrak{p}$. Then, $a_\mathfrak{p}(\sigma, \tau)$ is a factor set which determines a central, topological covering group $\tilde{G}_\mathfrak{p}$ of $G_\mathfrak{p}$. The kernel $\mathfrak{z}$ of the covering $\tilde{G}_\mathfrak{p} \to G_\mathfrak{p}$ is the group of n-th roots of unity, and the covering has no splitting part whenever $\mathfrak{p}$ is finite [6], [7], [8].

As usual, an element of $\tilde{G}_\mathfrak{p}$ will be denoted by the symbol (σ, ζ), $(\sigma \in G_\mathfrak{p}, \zeta \in \mathfrak{z})$, with the multiplication $(\sigma, \zeta)(\sigma', \zeta') = (\sigma\sigma', \zeta\zeta' a_\mathfrak{p}(\sigma, \sigma'))$.

Denote now by $\mathfrak{o}_\mathfrak{p}$ the valuation ring of $F_\mathfrak{p}$ for finite $\mathfrak{p}$ and denote by $K_\mathfrak{p}$ the principal congruence subgroup mod N of $SL(2, \mathfrak{o}_\mathfrak{p})$, N being a natural number. For infinite $\mathfrak{p}$, put $K_\mathfrak{p} = SU(2)$. Then, G_A is the group of elements in the direct product of all the $G_\mathfrak{p}$ whose components are in $K_\mathfrak{p}$ for almost all $\mathfrak{p}$. Furthermore we have

PROPOSITION 2-1. *If $s_\mathfrak{p}(\sigma)$ is a function of*

$$\sigma = \begin{pmatrix} a & b \\ c & d \end{pmatrix} \in K_\mathfrak{p}$$

defined by

$$\begin{aligned} s_\mathfrak{p}(\sigma) &= (c, d/\mathfrak{p})^{-1}, \quad \text{if } c \text{ is not } 0, \text{ and is not a unit}, \\ &= 1, \quad \text{otherwise}. \end{aligned} \tag{16}$$

Then, $a_\mathfrak{p}(\sigma, \tau) = s_\mathfrak{p}(\sigma)s_\mathfrak{p}(\tau)s_\mathfrak{p}(\sigma\tau)^{-1}$, provided that $N \equiv 0 \pmod{n^2}$ [6, Theorem 2].

This proposition shows that the nontrivial covering $\tilde{G}_{\mathfrak{p}} \to G_{\mathfrak{p}}$ splits on a compact subgroup of $G_{\mathfrak{p}}$, contrary to the situation of ordinary Lie groups.

The definition (2) of $s_{\mathfrak{p}}$ can be extended over the whole group $G_{\mathfrak{p}}$ by setting

$$(17) \qquad \begin{aligned} s_{\mathfrak{p}}(\sigma) &= (c, d/\mathfrak{p})^{-1}, \quad \text{if } cd \neq 0, \text{ and } \operatorname{ord}_{\mathfrak{p}} c \not\equiv 0 \pmod{n}, \\ &= 1, \quad \text{otherwise.} \end{aligned}$$

If we put $s_{\mathfrak{p}}(\sigma) = 1$ for all infinite prime divisors $\mathfrak{p}$, and put $b_{\mathfrak{p}}(\sigma, \tau) = a_{\mathfrak{p}}(\sigma, \tau) s_{\mathfrak{p}}(\sigma)^{-1} s_{\mathfrak{p}}(\tau)^{-1} s_{\mathfrak{p}}(\sigma, \tau)$, then Proposition 2-1 implies that $b_{\mathfrak{p}}(\sigma, \tau) = 1$ for almost all $\mathfrak{p}$ when σ, τ are $\mathfrak{p}$-components of two fixed adèles. Therefore, a factor set $b_A(g, g')$ of G_A can be defined by

$$(18) \qquad b_A(g, g') = \prod_{\mathfrak{p}} b_{\mathfrak{p}}(g, g'), \qquad (g, g' \in G_A),$$

where $b_{\mathfrak{p}}(g, g') = b_{\mathfrak{p}}(g_{\mathfrak{p}}, g'_{\mathfrak{p}})$, and $g_{\mathfrak{p}}$ means the $\mathfrak{p}$-component of an adèle g. In this way, we get an n-fold, topological covering $\tilde{G}_A$ such that $1 \to \mathfrak{z} \to \tilde{G}_A \to G_A \to 1$ is exact, which will be called a (generalized) metaplectic group. As in the case of $G_{\mathfrak{p}}$, an element of $\tilde{G}_A$ will be denoted by (g, ζ), $(g \in G_A, \zeta \in \mathfrak{z})$.

The compact group $K = \prod_{\mathfrak{p}} K_{\mathfrak{p}}$ has a natural injection into $\tilde{G}_A$ given by $k \to (k, 1)$, $(k \in K)$. The image of the injection will be identified with K, and will be denoted by the same notation K. Similarly, the group $G_\infty \subset G_A$, elements and subgroups of G_∞, as well as of K, will be identified with corresponding objects in $\tilde{G}_A$.

If $\alpha \in G_F$ is a principal adèle, then $s_{\mathfrak{p}}(\alpha)$ in (17) is 1 for almost all $\mathfrak{p}$ so that $s_A(\alpha) = \prod_{\mathfrak{p}} s_{\mathfrak{p}}(\alpha)$ has a well-defined meaning. It is proved by using the product formula of the norm residue symbol that $\alpha \to \hat{\alpha} = (\alpha, s_A(\alpha)) \in \tilde{G}_A$ gives an isomorphism. The group $\hat{G}_F$ of all $\hat{\alpha}$ is discrete in $\tilde{G}_A$ and the existence of such G_F signifies that $\tilde{G}_A$ is a global covering.

Since $\tilde{G}_A$ contains reasonable discrete and compact subgroups $\hat{G}_F$ and K, and since G_∞/K_∞, $(K_\infty = K \cap G_\infty = SU(2)^r)$, is the direct product H^r of the upper half space H, the general theory of automorphic functions on a unimodular, locally compact topological group is applicable to $\tilde{G}_A$, and in general an "adelized" automorphic function on $\tilde{G}_A$ must correspond to an automorphic function of Hilbert's type on H^r. As is well known and basic in the general theory of automorphic functions, functions f_A on $G_F \backslash G_A / K$ are in one-to-one correspondence with functions f on $\Gamma \backslash H^r$, Γ being as in Theorem 1-1, and the correspondence is given by

$$(19) \qquad f_A(\xi k g_\infty) = f(g_\infty), \qquad (\xi \in G_F, k \in K, g_\infty \in G_\infty).$$

It should be noted here that an element of G_A can always be expressed in the form $\xi k g_\infty$ by the approximation theorem. For $\tilde{G}_A$, it is evident that $\tilde{G}_A = \hat{G}_F K G_\infty \mathfrak{z}$, but as a matter of fact we have $\tilde{G}_A = \hat{G}_F K G_\infty$, and quite analogously to (19) one obtains

PROPOSITION 2-2. *The functions f_A on $\hat{G}_F \backslash \tilde{G}_A / K$ satisfying $f_A(g\zeta) = f_A(g)\zeta^{-1}$, $(\zeta \in \mathfrak{z})$, are in one-to-one correspondence with functions f on H^r satisfying $f(\sigma u) =$*

$\chi(\sigma)f(u)$, $(\sigma \in \Gamma)$, *through the relation*

$$f_A(\hat{\xi}kg_\infty) = f(g_\infty), \qquad (\hat{\xi} \in \hat{G}_F,\ k \in K,\ g_\infty \in G_\infty), \tag{20}$$

where Γ *and* χ *are as in Theorem* 1-1 [6, Proposition 5].

It is remarkable that the functions on H^r corresponding to automorphic functions on the metaplectic group $\tilde{G}_A$ are not Γ-invariant functions but functions which are left invariant by a noncongruence subgroup of Γ. This may be regarded as an explanation of the general situation that the existence of a noncongruence subgroup of an arithmetic matric group is equivalent with the existence of a nontrivial covering of the adèle group.

By Proposition 2, the functions θ in the space Θ in Theorem 1-2 are regarded as functions on $\hat{G}_F \backslash \tilde{G}_A / K$ satisfying $\theta(g\zeta) = \theta(g)\zeta^{-1}$. To construct a unitary representation of $\tilde{G}_A$ by using Θ, there remains one thing to introduce; it is Hecke operators of $\tilde{G}_A$. The Hecke ring $\mathcal{H}_\chi(\tilde{G}_A, K\mathfrak{z})$ means the algebra over $\mathbf{C}$ of all continuous, complex valued functions ψ on $\tilde{G}_A$ with a compact support such that $\psi(kgk'\zeta) = \psi(g)\zeta^{-1}$ for $k, k' \in K$, $g \in \tilde{G}_A$, $\zeta \in \mathfrak{z}$; the multiplication in $\mathcal{H}_\chi$ is the convolution $*$ on $\tilde{G}_A$. If f is a function on $\tilde{G}_A$, and if $\psi \in \mathcal{H}_\chi(\tilde{G}_A, K\mathfrak{z})$, then $f * \psi$ is denoted by $f^{T(\psi)}$, and $T(\psi)$ is called the Hecke operator determined by ψ. Let $\mathcal{E}_s$ be, as in §1, the space of linear combinations of Eisenstein series by entire functions; then a basic property of Eisenstein series under the operation of Hecke ring is

THEOREM 2-2. $\mathcal{E}_s^{T(\psi)} \subset \mathcal{E}_s$ *for any* $\psi \in \mathcal{H}_\chi(G_A, K\mathfrak{z})$ [6, Theorem 7].

This theorem follows quite mechanically from the definition of Eisenstein series.

Let $\mathfrak{h}$ be the Hilbert space spanned by all the functions on $\tilde{G}_A$ of the form $\theta(xg_0)$ with $\theta \in \Theta$ and with a $g_0 \in \tilde{G}_A$. Furthermore, denote by U_g the operator which maps a function $f(x)$ on $\tilde{G}_A$ on $f(xg)$. Then, $\mathfrak{h}$ is a closed subspace of $L^2(\hat{G}_F \backslash \tilde{G}_A)$, and U_g is a unitary operator of $\mathfrak{h}$. Since furthermore Θ is finite dimensional over $\mathbf{C}$, and since Theorem 2-2 implies $\Theta^{T(\psi)} \subset \Theta$ for $\psi \in \mathcal{H}_\chi(\tilde{G}_A, K\mathfrak{z})$, general arguments in the theory of unitary representations yield now

THEOREM 2-3. *The set of operators* $\{U_g\}$ *of* $\mathfrak{h}$ *gives a unitary representation of* $\tilde{G}_A$ *which is faithful on* $\mathfrak{z}$, *and is decomposed into the sum of a finite number of irreducible representations.*

3. On cubic Gauss sums

In §1, we observed Eisenstein series containing a special type of character χ given by Theorem 1-1. Here we propose to investigate Eisenstein series which contain not only the character χ of the discrete subgroup Γ of $SL(2, \mathbf{C})$ but also a representation of the compact group $SU(2)$, and show that in this way we get some reasonable Dirichlet series whose coefficients are composed of both Gauss sums and Grössencharacters. For the sake of simplicity, we treat here only the case in which the basic field F in the sense of §1 is $\mathbf{Q}(\sqrt{-3})$. It is a difficult problem to determine the exact distribution of values of Gauss sums; for example there is a conjecture of Kummer on cubic Gauss sums (see [3]). But, as an application of our results in this article, we can obtain some information on the value distribution of Gauss sums.

We put $\rho = \exp(2\pi\sqrt{(-1)}/3)$, and $F = \mathbf{Q}(\rho)$, $\mathfrak{o} = \mathbf{Z}[\rho]$. Furthermore, we denote by Γ resp. χ the same group resp. character as in Theorem 1-1, but with $n = N = 3$. It is easy to verify that Theorem 1-1 is valid for $N = 3$. Let

$$u = \begin{pmatrix} z & -v \\ v & \bar{z} \end{pmatrix}$$

be as usual a point of the upper half space H, and let

$$\sigma = \begin{pmatrix} a & b \\ c & d \end{pmatrix}$$

be an element of $SL(2, \mathbf{C})$. Then, it is easily shown that

$$j(\sigma, u) = (\tilde{c}u + \tilde{d})/|\tilde{c}u + \tilde{d}|$$

is an automorphic factor, i.e.

$$(21) \qquad j(\sigma, \sigma' u) j(\sigma', u) = j(\sigma\sigma', u), \qquad (\sigma, \sigma' \in SL(2, \mathbf{C})),$$

holds, where the meaning of $\sim$ is as in (1), and we put $|\tilde{c}u + \tilde{d}| = (|cz + d|^2 + |c|^2 v^2)^{1/2}$. If σ is a complex matrix, we denote by σ_l the tensor product $\sigma \otimes \cdots \otimes \sigma$ (l-times), and by σ_* the matrix ${}^t\bar{\sigma}$. We have $(\sigma_l)_* = (\sigma_*)_l$ and this will be denoted simply by σ_{l*}. If σ is a 2×2 matrix, then σ is a $2^l \times 2^l$ matrix, and if σ is of the form $\begin{pmatrix} a & -b \\ \bar{b} & \bar{a} \end{pmatrix}$, i.e. a quaternion, then $\sigma \to \sigma_*$ coincides with the standard involution, and we have $\sigma\sigma_* = |\sigma|^2 I$. The absolute value means here that of a quaternion, although $|\tilde{c}u + \tilde{d}|$ defined above was also nothing else than the absolute value of a quaternion because $\tilde{c}u + \tilde{d}$ is in fact a quaternion. The tensor product σ_l is the ordinary one, and not the symmetric; it is not necessary for us to have irreducible representations of $SU(2)$.

We now put $j_l(\sigma, u) = j(\sigma, u)_l$, denote by Γ_∞ the group of triangular matrices in Γ, and define the (matric) Eisenstein series $\mathbf{E}(u, s, \chi)$ by

$$(22) \qquad \mathbf{E}(u, s, \chi) = \sum_{\sigma} \bar{\chi}(\sigma) v(\sigma u)^s j_l(\sigma, u) \qquad (\sigma \in \Gamma_\infty \backslash \Gamma),$$

where v is as in (2). Since $j_l(\sigma, u)$ is a unitary matrix so that $j_l(\sigma, u)_* = j_l(\sigma, u)^{-1}$ holds, we have

$$(23) \qquad \mathbf{E}(\sigma u, s, \chi) = \chi(\sigma) \mathbf{E}(u, s, \chi) j_l(\sigma, u)_* \qquad (\sigma \in \Gamma).$$

Similarly, for the series

$$(24) \qquad \mathbf{E}_*(u, s, \chi) = \sum_{\sigma} \bar{\chi}(\sigma) v(\sigma u)^s j_l(\sigma, u)_* \qquad (\sigma \in \Gamma_\infty \backslash \Gamma),$$

we have

$$\mathbf{E}_*(\sigma u, s, \chi) = \chi(\sigma) j_l(\sigma, u) \mathbf{E}_*(u, s, \chi) \qquad (\sigma \in \Gamma). \tag{25}$$

The series (22), (24) converge absolutely for $\operatorname{Re} s > 2$, and admit a Fourier expansion. Precisely speaking, since Γ_∞ is naturally isomorphic to $\mathfrak{m} = 3\mathfrak{o}$, we have

$$\mathbf{E}(u, s, \chi) = \sum_m a_m(v, s, \chi) e(mu) \qquad (m \in \mathfrak{m}^* = (-3)^{-3/2}\mathfrak{o}), \tag{26}$$

with

$$\begin{aligned} a_m(v, s, \chi) &= V(\mathfrak{m})^{-1} \int_P \mathbf{E}(u, s, \chi) e(-mu)\, dV(z) \\ &= \delta_{0m} \cdot v^s I + V(\mathfrak{m})^{-1} \sum_c \frac{(\bar{c}/|c|)_l}{\|c\|^s} \left(\sum_d (c/d) e\left(\frac{md}{c}\right)\right) \\ &\quad \cdot \int_{\mathbf{C}} u_l \frac{v^s}{|u|^{2s+l}} e(-mu)\, dV(z) \qquad (c \equiv 0 \pmod 3,\ d \bmod c\mathfrak{m}), \end{aligned}$$

where (c/d) is the cubic power residue symbol, $P = \mathfrak{m}\backslash\mathbf{C}$, and $V(\mathfrak{m})$ $(= 9)$ is the volume of P with respect to the euclidean measure $dV(z)$. For $\mathbf{E}_*$, the corresponding formula to (26) is obtained only by replacing u_l and $(c/|c|)_l$ with u_{l*} and $(c/|c|)_{l*}$. In particular, the constant term $a_0(v, s, \chi)$ of $\mathbf{E}$ is given by

$$a_0(v, s, \chi) = v^s I + v^{2-s} \varphi(s, \chi) M(s), \tag{27}$$

where $M(s)$ is determined by the formula

$$\int_C u_l \frac{v^s}{|u|^{2s+l}}\, dV(z) = v^{2-s} M(s), \tag{28}$$

and $\varphi(s, \chi)$ is a diagonal, matric Dirichlet series. The constant term $a_{0*}(v, s, \chi)$ of $\mathbf{E}_*$ is similarly given by

$$a_{0*}(v, s, \chi) = v^s I + v^{2-s} M_*(s) \varphi_*(s, \chi), \tag{29}$$

where M_* is determined by (28) after replacing u_l with u_{l*}. It is not difficult to show by means of (26) that $\varphi(s, \chi)$ consists of L-functions of F with Grössencharacters. It is also not difficult to calculate the matrices $M(s)$, $M_*(s)$ explicitly. What we need later is, however, that the 1–1 entry, i.e. the entry in the first line and first column, of $M(s)$ as well as $M_*(s)$ is 0.

Let now $\mathfrak{D}$ be a fundamental domain of Γ which is a convex polyhedron in H, and in which the column $\mathfrak{D}'_Y$ consisting of all points $u = (z, v) \in H$ with $z \in P$, $v > Y$ is contained for a suitable positive number Y. Here P means as above a parallelogram $\mathfrak{m}\backslash\mathbf{C}$. Put $\mathfrak{D}_Y = \mathfrak{D} - \mathfrak{D}'_Y$, and define $\mathbf{E}^Y(u, s, \chi)$ by

$$\begin{aligned} \mathbf{E}^Y(u, s, \chi) &= \mathbf{E}(u, s, \chi), \qquad u \in \mathfrak{D}_Y, \\ &= \mathbf{E}(u, s, \chi) - v^s I, \qquad u \in \mathfrak{D}'_Y. \end{aligned} \tag{30}$$

Furthermore define $\mathbf{E}^Y_*(u, s, \chi)$ by replacing $\mathbf{E}$ in (30) with $\mathbf{E}_*$. Then, as an important step in our investigation, we have

$$\begin{aligned}
&\int_{\mathfrak{D}} \mathbf{E}^Y(u, s, \chi)\mathbf{E}^Y_*(u, \bar{s}, \bar{\chi})\, du \\
&= \int_{\mathfrak{D}_Y} \mathbf{E}(u, s, \chi)\mathbf{E}_*(u, \bar{s}, \bar{\chi})\, du + \int_{\mathfrak{D}'_Y} (\mathbf{E}(u, s, \chi) - v^s I)(\mathbf{E}_*(u, \bar{s}, \bar{\chi}) - v^{\bar{s}} I)\, du \\
&= \int_{\bigcup_\sigma \sigma\mathfrak{D}_Y} v^s \mathbf{E}_*(u, \bar{s}, \bar{\chi})\, du + \int_{\bigcup_{\sigma \neq 1} \sigma\mathfrak{D}'_Y} v^s \mathbf{E}_*(u, \bar{s}, \bar{\chi})\, du - \int_{\mathfrak{D}'_Y} (\mathbf{E}(u, s, \chi) - v^s I) v^{\bar{s}}\, du \\
&= V(\mathfrak{m})\int_0^Y v^s(v^{\bar{s}} I + v^{2-\bar{s}} M_*(\bar{s})\varphi_*(\bar{s}, \bar{\chi})) \frac{dv}{v^3} - V(\mathfrak{m})\int_Y^\infty v^{2-s}\varphi(s, \chi)M(s) \cdot v^{\bar{s}} \frac{dv}{v^3} \\
&\qquad = V(\mathfrak{m})\left[\frac{Y^{2S-2}}{2S-2} + M_*(\bar{s})\varphi_*(\bar{s}, \bar{\chi})\frac{Y^{2it}}{2it} - \varphi(s, \chi)M(s)\frac{Y^{-2it}}{2it}\right]
\end{aligned}$$ [2]

using (27), (29), where $s = S + it$, $(S, t \in \mathbf{R})$, and $du = dz\, d\bar{z}\, dv/v^3$ is the invariant measure of H. The matrix $\mathbf{E}_*(u, \bar{s}, \bar{\chi})$ is the conjugate of the transposition of $\mathbf{E}(u, s, \chi)$. Therefore, if we denote by $E(u, s, \chi)$ resp. $E^Y(u, s, \chi)$ the first line of $\mathbf{E}(u, s, \chi)$ resp. $\mathbf{E}^Y(u, s, \chi)$, then the above results imply the evaluation

$$\int_{\mathfrak{D}} \|E^Y(u, s, \chi)\|^2\, du = V(\mathfrak{m})\frac{Y^{2S-2}}{2S-2} \tag{31}$$

containing the vector norm. By (31), we can develop the whole theory described in [10] for our Eisenstein series, and in particular we see that $E(u, s, \chi)$ has a holomorphic continuation for $\operatorname{Re} s > 1$. This shows that the 1–1 entry of the Fourier coefficient a_m is a holomorphic function of s for $\operatorname{Re} s > 1$. On the other hand, we have

$$\begin{aligned}
&(w/|w|)^l \int_C z^l \frac{v^s}{|u|^{2s+l}} e(-wz)\, dV(z) \\
&\qquad = \int_C z^l \frac{v^s}{|u|^{2s+l}} e(-|w|z)\, dV(z) \\
&\qquad = (-1)^{-l/2} v^{2-s} \int_0^{2\pi}\int_0^\infty \frac{r^{l+1}}{(r^2+1)^{s+l/2}} \cos(-4\pi|w|vr\sin\theta + l\theta)\, d\theta\, dr \\
&\qquad = 2\pi(-1)^{-l/2} v^{2-s} \int_0^\infty \frac{r^{l+1}}{(r^2+1)^{s+l/2}} J_l(4\pi|w|vr)\, dr \\
&\qquad = 2\pi(-1)^{-l/2} v^{2-s} \frac{(4\pi|w|v)^{s-1+l/2} K_{s-1-l/2}(4\pi|w|v)}{\Gamma(s+l/2)2^{s-1+l/2}} \\
&\qquad = (2\pi)^{s+l/2} v^{1+l/2} |w|^{s-1+l/2} (-1)^{-l/2} K_{s-1-l/2}(4\pi|w|v)\Gamma(s+l/2)^{-1}
\end{aligned}$$

with $W \in C$, $u = (z, v) \in H$. (See Erdélyi's tables of integral transformations,

[2]Some integrals in this calculation do not converge, but the difficulty will disappear if we regard s and $\bar{s}$ as independent variables.

p. 24, (20)). The last formula is of course not identically 0 for any s with $\operatorname{Re} s > 1$. Thus, if we put

$$(32)\qquad \tau_m(c) = \sum_d (c/d)e\left(\frac{md}{c}\right) \qquad (c, d \in \mathfrak{o},\ c \equiv 0 \pmod 3,\ d \bmod 3c),$$

and $\lambda'(c) = (c/|c|)^l$, then, recalling (6), we have

PROPOSITION 3-1. *For any* $m \neq 0$ *in* $\mathfrak{m}^* = (-3)^{-3/2}\mathfrak{o}$, *the function determined by the Dirichlet series* $\varphi_l(s, m) = \sum_c \lambda'(c)\tau_m(c)/\|c\|^s$, $(c \in 3\mathfrak{o},\ c \neq 0)$, *is holomorphic in the region* $\operatorname{Re} s > 1$ *for* $l = 1, 2, \ldots$.

This proposition is proved by observing the 1–1 entry of the matrix **E**. If one observes the entry at the opposite end of the diameter, it is seen at once that the proposition is true also for negative integers l.

For an integer c_0 of $F = \mathbf{Q}(\sqrt{-3})$ with $c_0 \equiv 1 \pmod 3$, we define now the Gauss sum $g(c_0)$ by

$$(33)\qquad g(c_0) = \sum_d (d/c_0)e(d/c_0) \qquad (d \bmod c_0),$$

and we propose to deduce from Proposition 1 some direct results on $g(c_0)$.

Any integer $c \neq 0$ in $3\mathfrak{o}$, ($\mathfrak{o} = \mathbf{Z}[\rho]$, $\rho = \exp(2\pi\sqrt{-1}/3)$), is expressed in the form $c = 3(\sqrt{-3})^e\zeta c_0$, where $e \geqq 0$, and ζ is a power of $-\rho$. Moreover, if d_1 resp. d_2 runs over the residue classes mod c_0 resp. mod $9(\sqrt{-3})^e$, then $d = 9(\sqrt{-3})^e d_1 + c_0 d_2$ runs over the residue classes mod $3c$. Therefore, we have on one hand

$$\frac{md}{c} = m\,\frac{9(\sqrt{-3})^e\, d_1 + c_0\, d_2}{3(\sqrt{-3})^e\zeta c_0} = m\,\frac{d_2}{3(\sqrt{-3})^e\zeta} + m\,\frac{3d_1}{\zeta c_0},$$

and on the other hand, under the additional restriction $d \equiv 1 \pmod 3$ which enables us to make use of the cubic reciprocity $(c_0/d) = (d/c_0)$, we obtain

$$\begin{aligned}(c/d) &= ((\sqrt{-3})^{e+2}\zeta c_0/d) = ((\sqrt{-3})^{e+2}\zeta/c_0\, d_2)(d/c_0)\\ &= ((\sqrt{-3})^{e+2}\zeta/c_0)((\sqrt{-3})^{e+2}\zeta/d_2)((\sqrt{-3})^{e+1}\, d_1/c_0)\\ &= ((\sqrt{-3})^{2e}\zeta/c_0)((\sqrt{-3})^{e+2}\zeta/d_2)(d_1/c_0).\end{aligned}$$

Hence, $\tau_m(c)$ in (32) is transformed as

$$\begin{aligned}\tau_m(c) &= ((\sqrt{-3})^{2e}\zeta/c_0)\sum_{d_2}((\sqrt{-3})^{e+2}\zeta/d_2)e\left(\frac{m\zeta^{-1}\, d_2}{3(\sqrt{-3})^e}\right)\cdot\sum_{d_1}(d_1/c_0)e\\ &\quad\times\left(\frac{3m\zeta^{-1}\, d_1}{c_0}\right) \qquad (d_2 \bmod 9(\sqrt{-3})^e,\ d_2 \equiv 1 \pmod 3,\ d_1 \bmod c_0)\\ &= ((\sqrt{-3})^{e+2}m\zeta/c_0)^{-1}\sum_{d_2}((\sqrt{-3})^{e+2}\zeta/d_2)e\left(\frac{m\zeta^{-1}\, d_2}{3(\sqrt{-3})^e}\right)\cdot g(c_0).\end{aligned}$$

Put here $m = \sqrt{-3}$. Then the conductor of

$$e(m\zeta^{-1}d_2/3(\sqrt{-3})^e) = e(-\zeta^{-1}d_2/(\sqrt{-3})^{e+1})$$

is $(\sqrt{-3})^e$, i.e., $e(-\zeta^{-1}d_2/(\sqrt{-3})^{e+1})$ is constant for all $d_2 \equiv 1 \pmod{(\sqrt{-3})^{e'}}$, if and only if $e \leqq e'$. Since, however, the conductor of $((\sqrt{-3})^{e+2}\zeta/c_0)$ is 9 for $e \not\equiv 1 \pmod 3$, and is 1 or $3\sqrt{-3}$ for $e \equiv 1 \pmod 3$, the sum for d_2 in the above formula is 0 unless $e = 1$. If $e = 1$, the sum is equal to

$$\begin{aligned} \sum_{d_2} (\zeta/d_2) e\left(\frac{\zeta^{-1} d_2}{3}\right) &= 0, \qquad \zeta \neq \pm 1 \qquad (d_2 \equiv 1 \pmod 3),\ d_2 \bmod 9\sqrt{-3}), \\ &= 27\rho^2, \qquad \zeta = 1, \\ &= 27\rho, \qquad \zeta = -1. \end{aligned}$$

So, for $c = 3\sqrt{(-3)}c_0$, the sum $\sum_\zeta \lambda'(c\zeta)^l \tau_m(c\zeta)$ over the 6th roots of unity is $(3/c_0) \cdot (-1)^{-l/2} \lambda'(c_0)^l \cdot 27(\rho^2 + (-1)^l \rho) g(c_0)$ for $m = \sqrt{-3}$. Since $c_0 \equiv 1 \pmod 3$, we can define by $\lambda(c_0) = \lambda'(c_0)^l$ a Grössencharacter λ of the ideal group of F, and of course $\rho^2 + (-1)^l \rho$ is not 0. Thus, Proposition 3-1 yields

THEOREM 3-1. *The Dirichlet series* $\sum \lambda(c_0)(3/c_0)g(c_0)/\|c_0\|^s$, *the sum being extended over all integral ideals* (c_0), $c_0 \equiv 1 \pmod 3$, *of F prime to* 3, *determines a holomorphic function in the region* $\operatorname{Re} s > 1$ *for every Grössencharacter λ of the form* $\lambda(c_0) = (c_0/|c_0|)^l$, $(l \neq 0)$.

For $l = 0$, the method explained in [5] is applicable, and it can be proved that the series $\sum (3/c_0) g(c_0)/\|c_0\|^s$ gives a holomorphic function in the region $\operatorname{Re} s > 4/3$. It is also easy to see that $g(c_0) = 0$ unless $c_0 = \pi_1 \ldots \pi_k$ is a product of different prime numbers with $\pi_i \equiv 1 \pmod 3$, and if this is the case we have $|g(c_0)|^2 = \|c_0\|$. This means in particular that the series in Theorem 1 is absolutely convergent for $\operatorname{Re} s > \frac{3}{2}$. Thus, by virtue of the usual Tauberian theorem, we obtain the following theorem concerning the distribution of the values $g(c_0)$:

THEOREM 3-2. *Let λ be a Grössencharacter as in Theorem* 3-1, *including also* $l = 0$. *Then,*

$$\sum_{c_0} \lambda(c_0)(3/c_0)g(c_0)/|g(c_0)| = o(Y),$$

where the sum is extended over all integers c_0 of F such that $\|c_0\| < Y$, and such that $c_0 = \pi_1 \ldots \pi_k$ is a product of different prime numbers of F with $\pi_i \equiv 1$ (mod 3).

It is known in the general theory of Gauss sums that $g(c_0)$ in Theorem 3-2 satisfies $g(\bar{c}_0) = \bar{g}(c_0)$ and $g(c_0)^2 = \mu(c_0) \cdot c_0/|c_0| \cdot g(\bar{c}_0)$, where $\mu(c_0) = (-1)^k$ is Möbius' function. (See e.g. [3].) Combining these facts with Theorem 3-2, it is not difficult to prove

THEOREM 3-3. *Let c_0 be the product $\pi_1 \ldots \pi_k$ of different prime numbers of F with $\pi_i \equiv 1$* (mod 3), *and let $g(c_0)$ be as in* (33). *Then, we have*

$$\sum_{c_0} ((3/c_0)g(c_0)/|g(c_0)|)^l = o(Y),$$

or

$$\sum_{c_0} \mu(c_0)((3/c_0)g(c_0)/|g(c_0)|)^l = o(Y)$$

according as $l \neq 0$ *is* $\equiv -1, 0, 1 \pmod 6$, *or not, the sums being extended over all* c_0 *with* $\|c_0\| < Y$.

The factor $(3/c_0)$ in the above theorems will be removed, if we make longer computations observing several m in Proposition 3-1. In this way, one can also replace $(3/c_0)$ by an arbitrary congruence character of c_0.

REFERENCES

1. H. Bass, J. Milnor and J. P. Serre, *Solution of the congruence subgroup problem for* $SL_n(n \geq 3)$ *and* $S_{p_{2n}}(n \geq 2)$, Inst. Hautes Études Sci. Publ. Math. **33** (1967), 421–499.

2. Harish-Chandra, *Automorphic forms on semisimple Lie groups*, Lecture Notes in Math., no. 62, Springer-Verlag, Berlin and New York, 1968. MR **38** #1216.

3. H. Hasse, *Vorlesungen über Zahlentheorie*, Die Grundlehren der math. Wissenschaften, Band 59, Springer-Verlag, Berlin and New York, 1964. MR **32** #5569.

4. T. Kubota, *Ein arithmetischer Satz über eine Matrizengruppe*, J. Reine Angew. Math. **222** (1966), 55–57. MR **32** #5633.

5. ———, *On a special kind of Dirichlet series*, J. Math. Soc. Japan **20** (1968), 193–207. MR **37** #4035.

6. ———, *On automorphic functions and the reciprocity law in a number field*, Lectures in Math., no. 2, Kyoto University, 1969.

7. H. Matsumoto, *Sur les sous-groupes arithmetiques des groupes semi-simples déployés*, Thèse, Paris, 1969.

8. C. Moore, *Group extensions of p-adic linear groups*, Inst. Hautes Études Sci. Publ. Math. **35** (1969), 5–74.

9. A. Selberg, *Harmonic analysis and discontinuous groups in weakly symmetric Riemannian spaces with applications to Dirichlet series*, J. Indian Math. Soc. **20** (1956), 47–87. MR **19**, 531.

10. ———, *Discontinuous groups and harmonic analysis*, Proc. Internat. Congress Math. (Stockholm, 1962), Inst. Mittag-Leffler, Djursholm, 1963, pp. 177–189. MR **31** #372.

11. A. Weil, *Sur certains groupes d'opérateurs unitaires*, Acta Math. **111** (1964), 143–211. MR **29** #2324.

NAGOYA UNIVERSITY
NAGOYA, JAPAN

ELLIPTIC CURVES OVER Q: A PROGRESS REPORT

B. J. BIRCH

Let E be an elliptic curve defined over the rationals Q; we may take E in Weierstrass form $y^2 = x^3 + Ax + B$, with A, B integers. Its rational points form a group, which we denote by E_Q; the theorem of Mordell tells us that E_Q is finitely generated. A principal problem of the theory is to determine the group E_Q, and in particular to determine its rank g; well-known conjectures, described in [14], connect E_Q with the zeta function of the curve. Let us recall the appropriate definitions.

If p does not divide $6(27B^2 + 4A^3)$ then the reduction E_p, defined over the finite field k_p, is an elliptic curve; we call such primes *good*. The local zeta function of E_p over k_p is

$$\zeta_p(E_p, s) = \frac{(1 - \alpha_p p^{-s})(1 - \bar{\alpha}_p p^{-s})}{(1 - p^{-s})(1 - p^{1-s})}$$

where α_p, $\bar{\alpha}_p$ are the eigenvalues of the Frobenius transformation; so $\alpha_p \bar{\alpha}_p = p$ and $\alpha_p + \bar{\alpha}_p = 1 + p - N_p$, where N_p is the number of points of E_p with coordinates in k_p. We may define an L-function by

$$L_E^\dagger(s) = \prod (1 - \alpha_p p^{-s})^{-1}(1 - \bar{\alpha}_p p^{-s})^{-1}$$

where the product is taken over good primes. We know that $|\alpha_p| = p^{1/2}$, so $L_E^\dagger(s)$ converges for $\mathrm{Re}\,(s) > 3/2$.

Now let ω be a differential on E; we may take $\omega = dx/2y$. Then ω gives a Haar measure on the various completions of E, and we may form

$$M_p(E) = \int_{E(Q_p)} |\omega|_p, \qquad M_\infty(E) = \int_{E(\mathbf{R})} \omega$$

where the integral $M_p(E)$ is over the p-adic points of E, and $M_\infty(E)$ is an integral over the real points of E. Note that if p is good, then $M_p(E) = N_p/p = (1 - \alpha_p p^{-1})(1 - \bar{\alpha}_p p^{-1})$. For any set S of primes including all primes dividing $6(27B^2 + 4A^3)$ and the infinite prime, define

$$L^*_{E,S}(s) = \prod_{p\in S} M_p(E)^{-1} \prod_{p\notin S} (1 - \alpha_p p^{-s})^{-1}(1 - \bar{\alpha}_p p^{-s})^{-1}.$$

We will be interested in whether $L^*_{E,S}(s)$ may be continued past $s = 1$, and, if it may, in its behavior near $s = 1$; none of this depends on the particular choice of the set S, so in assertions for which S is irrelevant we will write simply L^*_E instead of $L^*_{E,S}$.

Now we state the standard conjectures; in order for the others to make sense, we need (not at full strength)

(I) $L_E(s)$ may be continued as a meromorphic function over the whole plane, and has a functional equation.

Given this, we may state the others.

(II) $L_E(s)$ has a zero of order g at $s = 1$.

(III) If $g = 0$, then $L^*_E(1) = |\text{Ш}|/|E_Q|^2$, where $|\text{Ш}|$ is the order of the Tate-Safarevic group, and $|E_Q|$ is the order of E_Q.

(IV) $L^*_E(s) \sim |\text{Ш}|R|\text{Tors}\,(E_Q)|^{-2}(s - 1)^g$ as $s \to 1$, where now $|\text{Tors}\,(E_Q)|$ is the number of points of E_Q of finite order, and R is an analogue of the regulator of an algebraic number field, and measures the size of the generators of E_Q.

For curves with complex multiplication, (I) is a theorem of Deuring [7]; accordingly, it was natural that such curves should be examined first. Curves of the shape $y^2 = x^3 - Dx$, which have complex multiplication by i, were treated in [2]; we proved

(V) $L^*_E(1)$ is rational, with bounded denominator,

and for a very large number of values of D, we verified (II) and (III) in a weakened form. To be precise, we found that $L^*(E, 1) = 0$ whenever $g > 0$, and $|E_Q|^2L^*_E(1)$ was a positive square when $g = 0$; there is no known way of calculating Ш, so at present we cannot verify (III) as stated, but $|\text{Ш}|$ is a square if it is finite.

Subsequently, Stephens [13] has treated curves of the shape $x^3 + y^3 + Az^3 = 0$, with complex multiplication by $\sqrt[3]{1}$; he proved (V), and verified (II), (III), and (IV) (with the same gloss about Ш, and (IV) only approximately) in several thousand cases. Rajwade [11] and Damerell [6] have completed the theory for all elliptic curves over Q with complex multiplication; they have proved (V), and Damerell has made a few computations.

In all this published work, the methods have followed [2] fairly closely; $L^*_E(1)$ is evaluated in terms of a closed formula involving $\wp$-functions. It seems unlikely that this is really the right way to do it. Curves with complex multiplication are a much smaller class than the class of elliptic curves that may be parametrized by modular functions. For modular function curves, (I) is usually provable and methods of evaluating $L^*_E(1)$ are available which seem much simpler and more effective than those involving $\wp$-functions (the ideas go back essentially to Shimura). Though the methods applicable to curves parametrized by modular functions were in fact sketched in [14], no one seems to have noticed, so it seems worthwhile to publicize them further.

We call E: $y^2 = x^3 + Ax + B$ a 'good' elliptic curve if, for some N, E is parametrized by functions on $H/\Gamma_0(N)$, and 'corresponds to' a differential

$f(z)\,dz = \sum a_n \exp(2\pi inz)\,dz$ on $H/\Gamma_0(N)$. It corresponds in two senses: the curve E is obtained by integrating f along paths on $H/\Gamma_0(N)$, and also the zeta function of E is

$$L_E(s) = \sum a_n n^{-s} = \frac{(2\pi)^s}{\Gamma(s)} \int_0^\infty f(iz) z^{s-1}\,dz.$$

In taking this formula for $L_E(s)$, we have implicitly made a canonical choice for factors of the Euler product corresponding to the bad primes. The analytic conductor of E is the minimal N for which all this is possible.

The above properties are more than enough to characterize 'good' elliptic curves. It is generally believed that if E has analytic conductor N then it also has algebraic conductor N, in the sense of [10] and [16]; in view of results of Igusa [9], a good curve with analytic conductor N has good reduction at primes not dividing N. If E is a good curve, its L-function satisfies a functional equation, for the involution W_N: $z \leftrightarrow -1/Nz$ of $H/\Gamma_0(N)$ takes $f(z)\,dz$ to $\epsilon f(z)\,dz$, with $\epsilon = \pm 1$, and then $\Lambda_E(s) = -\epsilon N^{1-s}\Lambda_E(2-s)$ where $\Lambda_E(s) = (2\pi)^{-s}\Gamma(s)L_E(s)$, and $L_E(s)$ may be continued over the whole plane; in a similar way $L_E(s, x) = \sum a_n \chi(n) n^{-s}$ has a functional equation for any character χ with conductor D prime to N.

Virtue in the sense we have just described appears to be a great deal to ask of a curve. However, no one has yet found an elliptic curve over Q which is not isogenous to a 'good' curve; following Weil [16] we may conjecture that every elliptic curve over Q is isogenous to a good curve. A necessary corollary of such a conjecture would be that two elliptic curves with the same zeta function should be isogenous—this seems likely, in fact Serre [12] goes a long way toward proving it (see also Tate [15]). For fixed N, it is not too difficult to list all isogeny classes of elliptic curves with analytic conductor N—it comes down to a question of factoring the Jacobian of $H/\Gamma_0(N)$ (up to isogeny) as a product of simple abelian varieties, and this may be accomplished by studying the operation of the Hecke algebra on the one-dimensional homology of $H/\Gamma_0(N)$. A computational procedure is described in some detail in [14, pp. 146–148]. (At the time, the procedure was not guaranteed to work, but necessary information, that certain Hecke operators have distinct eigenvalues, has since been supplied by Atkin and Lehner [1].) It is desirable to make a comparison with lists of all curves with algebraic conductor N; using Baker's theorem, it is now possible to make such lists, and we hope to do so.

There are obvious advantages in looking at good curves—they are born equipped with an almost excessively rich structure. One has simply to pull out the information one needs. Let us give a few examples.

First, let us check that $L_E^*(1)$ is rational, and show how to compute it. $L_E^*(1)$ is a rational multiple of $L_E(1)/M_\infty(E)$, and $M_\infty(E)$ is essentially the real period of E. On the other hand, $L_E(1) = -2\pi i \int_0^{i\infty} f(z)\,dz$; there is a function on $H/\Gamma_0(N)$ with all its zeros at 0 and all its poles at $i\infty$, so it is clear that $L_E(1)$ is a rational multiple of a real period of the Jacobian of $H/\Gamma_0(N)$, and reasonable to suppose that this period is actually the real period of E. So $L_E^*(1)$ is computable, using no more than qualitative information about the differential on $H/\Gamma_0(N)$ corresponding to E.

We have control not only over the L-function of E over Q, but also of E over K whenever K is an abelian extension of Q. If χ is a character with conductor D

prime to N, then

$$L_E(s, \chi) = \sum a_n \chi(n) n^{-s} = -2\pi i \int_0^{i\infty} \sum a_n \chi(n) \exp(2\pi i n z)\, dz$$

$$= \sum_{0 \le b < D} \lambda_b I(b/D), \quad \text{say,}$$

where

$$\lambda_b = D^{-1} \sum_{(a,D)=1} \exp(-2\pi i a b/D) \chi(a)$$

is a Gauss sum, and

$$I(b/D) = -2\pi i \int_0^{i\infty} f\left(z + \frac{b}{D}\right) dz = -2\pi i \int_{b/D}^{i\infty} f(z)\, dz$$

is an integral of a differential along a path. To compute $L_E(1, \chi)$ in terms of the periods of E is a matter of one-dimensional homology, expressing the paths $[b/D, i\infty]$ (or, directly, the 1-chain $\sum \lambda_b [b/D, i\infty]$) in terms of the homology cycles on $H/\Gamma_0(N)$. The homology computation is a simple one, reminiscent of the continued fraction expansion of b/D; it is easy to do by hand and not hard to program.

In particular, if χ happens to be the quadratic character corresponding to $Q(\Delta^{1/2})$, then the L-function of the curve $E(\Delta)$: $\Delta y^2 = x^3 + Ax + B$ is essentially $L_E(s, \chi)$; so we may compute all of these. The functional equation of $L_E(s, \chi)$ contains a factor $\epsilon\chi(N)$, so (II) predicts that the parity of g should depend on the quadratic character of N modulo D—this seems consistent with the Selmer conjecture, that g has the same parity as the number of first descents.

One would like very much to evaluate $L'_E(1)$ exactly. Unhappily, even if $L_E(1) = 0$, $L'_E(1) = 2\pi \int_0^{i\infty} f(iz) \log z\, dz$ is a thoroughly unmanageable function, about which we have as yet nothing good to say.

A most enticing hope is that, besides being able to evaluate the L-functions involved in the conjectures more easily, we should actually be able to prove the conjecture sometimes. The idea originates with Heegner [8]; its point is that $H/\Gamma_0(N)$, and hence any 'good' curve, is born equipped with points on it whose coordinates are in predictable class fields. In favorable circumstances, it is possible to pull down these points, so that we may explicitly construct generators for E_Q, just as the conjectures predict. I have given details elsewhere [3], [4], [5] but so far I have only been able to make the 'pull down' argument work for curves $E(p)$ of the pencil $py^2 = x^3 + Ax + B$ for which $\pm p$ is prime.

REFERENCES

1. A. O. L. Atkin and J. Lehner, *Hecke operators on* $\Gamma_0(m)$, Math. Ann.

2. B. J. Birch and H. P. F. Swinnerton-Dyer, *Notes on elliptic curves.* II, J. Reine Angew. Math. **218** (1965), 79–108. MR **31** #3419.

3. B. J. Birch, *Diophantine analysis and modular functions*, Proc. Conf. Algebraic Geometry (Bombay, 1968), pp. 35–42.

4. ———, *Elliptic curves and modular functions*, Proc. Conf. Number Theory (Rome, 1968).

5. ———, *Weber's class invariants*, Mathematika **16** (1969), 283–294.

6. M. Damerell, Ph. D. Thesis, Cambridge, 1969.

7. M. Deuring, *Die Zetafunktion einer algebraischen Kurve von Geschlechte Eins.* I, II, III, IV, Nachr. Akad. Wiss. Göttingen Math.-Phys. Kl. IIa **1953,** 85–94; **1955,** 13–42; **1956,** 37–76; **1957,** 55–80. MR **15,** 779; MR **17,** 17; MR **18,** 113; MR **19,** 637.

8. K. Heegner, *Diophantische Analysis und Modulfunktionen*, Math. Z. **56** (1952), 227–253. MR **14,** 725.

9. J.-I. Igusa, *Kroneckerian model of fields of elliptic modular functions*, Amer. J. Math. **81** (1959), 561–577. MR **21** #7214.

10. A. P. Ogg, *Abelian curves of small conductor*, J. Reine Angew. Math. **226** (1967), 204–215. MR **35** #1592.

11. A. R. Rajwade, *Arithmetic on curves with complex multiplication by* $\sqrt{-2}$, Proc. Cambridge Philos. Soc. **64** (1968), 659–672. MR **37** #4079.

12. J.-P. Serre, *Abelian l-adic representations and elliptic curves*, Benjamin, New York, 1968.

13. N. M. Stephens, *The diophantine equation* $X^3 + Y^3 = DZ^3$ *and the conjectures of Birch and Swinnerton-Dyer*, J. Reine Angew. Math. **231** (1968), 121–162. MR **37** #5225.

14. H. P. F. Swinnerton-Dyer, *The conjectures of Birch and Swinnerton-Dyer, and of Tate*, Proc. Conf. Local Fields (Driebergen, 1966), Springer, Berlin, 1967, pp. 132–157. MR **37** #6287.

15. J. Tate, *Endomorphisms of abelian varieties over finite fields*, Invent. Math. **2** (1966), 134–144. MR **34** #5829.

16. A. Weil, *Über die Bestimmung Dirichletscher Reihen durch Funktionalgleichungen*, Math. Ann. **168** (1967), 149–156. MR **34** #7473.

MATHEMATICS INSTITUTE
OXFORD, ENGLAND

RECENT ADVANCES IN DETERMINING ALL COMPLEX QUADRATIC FIELDS OF A GIVEN CLASS-NUMBER

H. M. STARK

1. Introduction

Let me hasten to say that the problem in the title has been completely solved only for class-number one. Let d be the discriminant of the quadratic field $\mathbf{Q}(\sqrt{d})$, and $h(d)$ be its class-number. There are now several proofs of the fact that if $d < 0$ then $h(d) = 1$ only for $d = -3, -4, -7, -8, -11, -19, -43, -67, -163$. What I wish to do in these lectures is illustrate some of these proofs and the obstacles that lie in the way of extending them even to $h(d) = 2$.

One of the most interesting things about this subject is the wide number of seemingly unrelated items that can be connected by the theory of quadratic fields. Let me give three numerical examples all related to the fact that $h(-163) = 1$.

(i) $x^2 - x + 41$ is a prime for $x = 1, 2, \ldots, 40$. The discriminant of $x^2 - x + 41$ is -163.

(ii) $e^{\pi\sqrt{163}} = 262\ 537\ 412\ 640\ 768\ 743.999\ 999\ 999\ 999\ 2\ldots$ is remarkably close to an integer.

(iii) The continued fraction expansion of the real root of $x^3 - 8x - 10 = 0$ begins

$$[3, 3, 7, 4, 2, 30, 1, 8, 3, 1, 1, 1, 9, 2, 2, 1, 3, 22986, \ldots].$$

The discriminant of $x^3 - 8x - 10$ is $-4 \cdot 163$.

The first of these was shown to be equivalent to $h(-163) = 1$ by Rabinowitsch (the late G. Y. Rainich) in 1913 and is fairly elementary. The second and third of these are explainable by the theory of modular functions. We will not have time to discuss the third of these here but it will be covered in [10]. For a brief survey of some of the earlier results and a wider list of references, see [7].

2. The method of Heegner

Set

$$j(z) = \frac{\{1 + 240\sum_{n=1}^{\infty}(\sum_{d|n} d^3)q^n\}^3}{q\prod_{n=1}^{\infty}(1-q^n)^{24}} \qquad (\operatorname{Im} z > 0)$$
$$= 1/q + 744 + 196884q + 21493760q^2 + \cdots$$

where here and later $q = e^{2\pi i z}$ (and q^α means $e^{2\pi i \alpha z}$).

It is well known that $j(z)$ is invariant under the full modular group:

$$j\left(\frac{\alpha z + \beta}{\gamma z + \delta}\right) = j(z); \quad \alpha, \beta, \gamma, \delta \text{ integers}, \alpha\delta - \beta\gamma = 1.$$

Further, we can evaluate $j(z)$ when z is in a complex quadratic field. Let

(1) $\quad Q(x, y) = ax^2 + bxy + cy^2; \quad a > 0, \quad (a, b, c) = 1, \quad d = b^2 - 4ac < 0$

and let $h(d)$ be the number of such forms which are inequivalent under unimodular substitutions of determinant 1. If d is the discriminant of a quadratic field, then this value of $h(d)$ is the same as the class-number of the field. We know that if $z = (-b + \sqrt{d})/(2a)$ then $j(z)$ is an algebraic integer of degree (exactly) $h(d)$. When $z = (1 + \sqrt{-163})/2$, we get the explanation of (ii) above.

In fact, some rather amazing things are true of j. For example, set

$$\gamma_2(z) = [j(z)]^{1/3},$$

the cube root being chosen which is real on the imaginary axis. If $3 \mid b$, $3 \nmid d$ then $\gamma_2((-b + \sqrt{d})/2a)$ is also an algebraic integer of degree $h(d)$ (rather than the expected $3h(d)$). We will illustrate at the end of these lectures the classical method of proving that such reductions take place for another function.

For notational convenience, we will suppose from now on that d is a negative field discriminant, $|d| = \Delta \equiv 3 \pmod 8$, and we put

$$\tau = (1 + \sqrt{d})/2.$$

If $d < -8$ and $h(d) = 1$ then Δ is a prime and $\Delta \equiv 3 \pmod 8$. A special case of the result about γ_2 is now that if $3 \nmid d$ then $\gamma_2((-3 + \sqrt{d})/2)$ is an algebraic integer of degree $h(d)$. For our purposes, it will be convenient to translate this: since $j(z + 1) = j(z)$, we see that

$$\gamma_2(z + 1) = e^{-2\pi i/3}\gamma_2(z)$$

(the cube root being chosen which matches at $i\infty$), and hence $e^{-2\pi i/3}\gamma_2(\tau)$ is an algebraic integer of degree $h(d)$ when $3 \nmid d$.

Another important function is

$$f(z) = q^{-1/48}\prod_{n=1}^{\infty}(1 + q^{n-1/2}).$$

One way that it is connected to the previous functions is that $f(2z - 1)$ is a root of

(2) $$x^{24} + e^{-2\pi i/3}\gamma_2(z)x^{16} - 256 = 0.$$

Thus when $3 \nmid d$, $f(\sqrt{d})$ is a root of a 24th degree equation with coefficients in $\mathbf{Q}(j(\tau))$. But in many cases, $f(\sqrt{d})$ is a root of an equation of lower degree. This was well known even in the last century.

Set

$$J = j(\sqrt{d}), \qquad F = f(\sqrt{d}), \qquad j = j(\tau), \qquad \gamma = e^{-2\pi i/3}\gamma_2(\tau).$$

If $3 \nmid d$, then Weber [11] proved that F^2 is in $\mathbf{Q}(J)$; he conjectured that F is in $\mathbf{Q}(J)$ also but was unable to prove this (Birch [2] has just recently done so). Since $J = ((F^{24} - 16)/F^8)^3$, it follows that $\mathbf{Q}(F^2) = \mathbf{Q}(J)$. Since $\Delta \equiv 3 \pmod 8$, $h(4d) = 3h(d)$ and hence

$$3h(d) = [\mathbf{Q}(J)\colon \mathbf{Q}] \leq [\mathbf{Q}(J, j)\colon \mathbf{Q}] = [\mathbf{Q}(J, j)\colon \mathbf{Q}(j)]h(d) \tag{3}$$

so that

$$[\mathbf{Q}(J, j)\colon \mathbf{Q}(j)] \geq 3. \tag{4}$$

But we will now prove that there is a cubic equation relating J and j. The functions $j(2z)$, $j(z/2)$, $j((z + 1)/2)$ are permuted by elements of the full modular group. Hence $j(2z - 1) = j(2z)$ is a root of $\phi(x) = 0$ where

$$\phi(x) = [x - j(2z)][x - j(z/2)][x - j((z + 1)/2)].$$

But the coefficients in $\phi(x)$ are invariant under the full modular group (being symmetric functions of the roots) and, having poles only at $i\infty$ in the fundamental domain of j, are in fact polynomials in $j(z)$. These polynomials in $j(z)$ have rational coefficients since the coefficients in the expansion of j in powers of q are rational. We now set $z = \tau$ with the result that J is the root of a cubic equation with coefficients in $\mathbf{Q}(j)$. Thus there is equality in (4) and hence by (3),

$$\mathbf{Q}(F^2) = \mathbf{Q}(J) = \mathbf{Q}(j, J)$$

is a cubic extension of $\mathbf{Q}(j)$. This justifies the reducibility of (2) when $z = \tau$.

It was left to Heegner [4] more than a half century later to make use of this reducibility. We have $\mathbf{Q}(F^2) = \mathbf{Q}(F^4) = \mathbf{Q}(F^8)$ is a cubic extension of $\mathbf{Q}(j)$ and hence there is a unique cubic equation for $F^k (k = 2, 4, 8)$ with coefficients in $\mathbf{Q}(j)$. Thus F is the root of an equation of the form

$$x^{3k} + B_k x^{2k} + A_k x^k + C_k = 0 \tag{5}$$

where A_k, B_k, C_k are in $\mathbf{Q}(j)$. But these equations are related. If we square

$$x^{3k} + A_k x^k = -(B_k x^{2k} + C_k),$$

we get

$$x^{6k} + (2A_k - B_k^2)x^{4k} + (A_k^2 - 2B_kC_k)x^{2k} + (-C_k^2) = 0.$$

Since (5) is unique for $k = 4, 8$, we see that

$$\begin{aligned} &2A_2 - B_2^2 = B_4, \qquad A_2^2 - 2B_2C_2 = A_4, \qquad -C_2^2 = C_4;\\ &2A_4 - B_4^2 = B_8 = \gamma, \qquad A_4^2 - 2B_4C_4 = A_8 = 0, \qquad -C_4^2 = C_8 = -256. \end{aligned} \tag{6}$$

Hence $C_4 = \pm 16$ and since C_2 is real ($\mathbf{Q}(j)$ is a real field), $C_4 = -16$ and $C_2 = \pm 4$ (in actual fact, $C_2 = -4$). Therefore,

$$[A_2^2 - 2B_2(\pm 4)]^2 + 32(2A_2 - B_2^2) = 0$$

which simplifies to

$$A_2^4 - 64A_2 = 2(A_2^2 \mp 4B_2)^2.$$

In case $d < -8$ and $h(d) = 1$, A_2 and B_2 are rational integers. Thus $2 \mid A_2$ and now $4 \mid A_2$ and $2 \mid B_2$. Let us set $A_2 = -4\alpha$, $B_2 = \pm 2\beta$, our last equation now reads,

(7) $$2\alpha(\alpha^3 + 1) = (2\alpha^2 - \beta)^2.$$

To solve (7) we must consider either $\alpha = \pm x^2$ or $\alpha = \pm 2x^2$ and hence we are led to solve

(8) $$x^6 \pm 1 = 2y^2 \quad \text{and} \quad 8x^6 \pm 1 = y^2.$$

This is quite simple and the result is that the only solutions to (7) in rational integers are

$$(\alpha, \beta) = (0, 0), (1, 0), (-1, 2), (2, 2), (1, 4), (2, 14).$$

We then find γ from (6) and get,

(9) $$\gamma = 0, -32, -96, -960, -5280, -640320.$$

These correspond to $d = -3, -11, -19, -43, -67, -163$ respectively.

This settles the problem for $h(d) = 1$. We also have $\Delta \equiv 3 \pmod 8$, $3 \nmid d$ in the difficult cases of $h(d) = 2$. However here we must solve (7) with α and β being integers in an unknown real quadratic field. Further, the very nature of (7) guarantees infinitely many quadratic fields with solutions.

The α and β that we have used are the same as Heegner's but we have derived them from a cubic equation that is more easily related to results of Weber. Lastly, the reader interested in Heegner's paper may also wish to refer to Birch [2], Deuring [3], and Stark [8].

3. The methods of Baker and Stark

Heegner's work was considered to be either false or incomplete and thus it was that 15 years later, Stark and Baker gave undisputed proofs of the class-number one problem. Their methods are different but based on the same equations. Let k be the discriminant of a quadratic field and let χ_k be the real primitive character (mod k) defined by the Kronecker symbol,

$$\chi_k(n) = (k/n).$$

We will use the following functions,

$$L_k(s) = \sum_{n=1}^{\infty} \chi_k(n)n^{-s},$$

$$L(s, \chi_k, Q, z) = \frac{(z - \bar{z})^s}{2} \sum_{m,n \neq 0,0} \frac{\chi_k(Q(m, n))}{[(m + zn)(m + \bar{z}n)]^s} \qquad (\operatorname{Im} z > 0),$$

$$L(s, \chi_k, Q) = d^{-s/2} L\left(s, \chi_k, Q, \frac{b + \sqrt{d}}{2a}\right) = \frac{1}{2} \sum_{m,n \neq 0,0} \frac{\chi_k(Q(m, n))}{[Q(m, n)]^s}.$$

Here Q is the quadratic form given in (1) and which we will now often write as $Q = (a, b, c)$. The three series converge absolutely for $\operatorname{Re} s > 1$ and possess a limit from the right at $s = 1$. The basic relationship is

$$L_k(s)L_{kd}(s) = \sum_{Q} L(s, \chi_k, Q) \tag{10}$$

where the summation is over a complete set of inequivalent forms of discriminant d.

For convenience, we will deal here with $k > 0$ only. We let ϵ_k be the fundamental unit of $\mathbf{Q}(\sqrt{k})$ and lastly, we assume throughout that $(k, d) = 1$. At $s = 1$, we may evaluate the left side of (10) by Dirichlet's formula,

$$\frac{2\pi\omega h(k)h(kd) \log \epsilon_k}{k\sqrt{\Delta}} = \sum_{Q} L(1, \chi_k, Q) \qquad \begin{pmatrix} \omega = 3, & d = -3 \\ = 2, & d = -4 \\ = 1, & d < -4 \end{pmatrix}. \tag{11}$$

The functions $L(s, \chi_k, Q, z)$ possess a rapidly convergent expansion at $s = 1$:

$$\frac{6L(1, \chi_k, Q, z)}{\pi i k \prod_{p|k} (1 - p^{-2})} = \frac{1}{2}\{f_Q(z) + f_{Q'}(-\bar{z}) - 2\pi i \chi_k(a)/k\} \tag{12}$$

where $Q' = (a, -b, c)$ and

$$\begin{aligned} f_Q(z) = f_{Q,k}(z) = {} & -\chi_k(a)\frac{2\pi i(z - \frac{1}{2})}{k} - \frac{12\chi_k(a)}{k^2\prod_{p|k}(1 - p^{-2})} \cdot \begin{cases} \log p & \text{if } k = p^r \\ 0 & \text{otherwise} \end{cases} \\ & + \frac{24}{k^2\prod_{p|k}(1 - p^{-2})} \sum_{n=1}^{\infty} q^{n/k} \sum_{y|n} y^{-1} \sum_{j=1}^{k} \chi_k(Q(j, y))e^{2\pi i n j/ky} \end{aligned} \tag{13}$$

and p denotes primes only. Further, if $\operatorname{Re} z = \frac{1}{2}$ and $b = a$,

$$f'_Q(z) = f_{Q'}(-\bar{z}) - 2\pi i\chi_k(a)/k. \tag{14}$$

If we use $Q = (1, 1, (\Delta + 1)/4)$ and $h(d) = 1$, then we may assemble (11), (12), (13) and (14) to get

$$\frac{12\omega h(k)h(kd) \log \epsilon_k}{k^2\prod_{p|k}(1 - p^{-2})} = f_{(1,1,(\Delta+1)/4),k}(\tau). \tag{15}$$

Note that for this Q, the coefficients in the series on the right depend on $(\Delta + 1)/4$ (mod k).

We can now show how Baker's work applies. Gelfond and Linnik had in 1949 a connection between our problem and linear forms in the logarithms of algebraic numbers. At that time the expansion in (13) was known only for prime k. They therefore chose to put $k = 5$ and $k = 13$ in (15); if $h(d) = 1$ and $d < -8$ (so that Δ is a prime and $(d, k) = 1$) then

$$\begin{aligned} &\frac{1}{2} h(5d) \log\left(\frac{1+\sqrt{5}}{2}\right) = \frac{\pi\sqrt{\Delta}}{5} - \frac{1}{2}\log 5 + O(e^{-\pi\sqrt{\Delta}/5}), \\ (16)\qquad & \\ &\frac{1}{14} h(13d) \log\left(\frac{3+\sqrt{13}}{2}\right) = \frac{\pi\sqrt{\Delta}}{13} - \frac{1}{14}\log 13 + O(e^{-\pi\sqrt{\Delta}/13}). \end{aligned}$$

We see further from (16) that

$$(17)\qquad h(5d) = O(\sqrt{\Delta}), \qquad h(13d) = O(\sqrt{\Delta}).$$

We may also eliminate the $\sqrt{\Delta}$ term from (16),

$$\begin{aligned} (18)\qquad & 35h(5d)\log\left(\frac{1+\sqrt{5}}{2}\right) \\ & - 13h(13d)\log\left(\frac{3+\sqrt{13}}{2}\right) + \log\left(\frac{3^{35}}{13^{13}}\right) = O(e^{-\pi\sqrt{\Delta}/13}). \end{aligned}$$

While Gelfond had an effective theorem on linear forms in two logarithms, he had only an ineffective theorem saying that (18) with the restriction (17) is impossible for large Δ. Thus it was that Baker [1] completed the Gelfond-Linnik attack by giving an effective theorem for linear forms in three logarithms. In this case, effective means $h(d) > 1$ if $\Delta > 10^{500}$. (This really is effective: it was known that if $\Delta > 163$ and $h(d) = 1$ then $\Delta > 10^{9000000}$.)

This brings us to another historical curiosity of the type that so often occurs in mathematics. We note in the expansion (13) that the log term disappears if k has two distinct prime divisors. Thus if the expansion of (13) had been known for composite k in 1949 and had Gelfond and Linnik used $k = 12$ and 24 in (15), they would have arrived at

$$h(24d)\log(5 + 2\sqrt{6}) - 2h(12d)\log(2 + \sqrt{3}) = O(e^{-\pi\sqrt{\Delta}/24})$$

for which Gelfond had an effective method of solution! In fact, Baker has noted that $k = 12$ would have sufficed by itself since $\pi = -i\log(-1)$.

When it comes to $h(d) = 2$, however, there is a second quadratic form (a, b, c) that enters the picture. The rate of convergence of the series in (13) at $z = (b + \sqrt{d})/2a$ is determined by how small $\exp(-\pi\sqrt{\Delta}/ka)$ turns out to be. It is possible that the unknown value of a is almost $\sqrt{\Delta/3}$ in size and then the series in (13) converges far too slowly to apply Baker's method. However, if a is specified, then Baker's method is applicable. For example, Baker has recently shown that if $h(d) = 2$ and $\Delta \not\equiv 3 \pmod 8$ ($a = 2$ except for small Δ) then $\Delta < 10^{500}$.

Thus we come to my solution [6] of the class-number one problem. It was motivated by the idea of Heegner to find a Diophantine equation and also by the desire to completely avoid modular functions so that the doubts cast upon Heegner would not be turned in my direction.

The basic idea is to construct integers by exponentiating (15). Let

$$F_Q(z) = F_{Q,k}(z) = \exp[f_{Q,k}(z)].$$

When $k = 8$ and $h(d) = 1$, we showed that $\omega h(8d) = 4N + 2$ where N is an integer. It follows from (15) that if $h(d) = 1$,

$$\epsilon_8^{-1/2} F_{(1,1,(\Delta+1)/4),8}(\tau) = \epsilon_8^N, \tag{19}$$

where $\epsilon_8 = 1 + \sqrt{2}$. Set

$$Y_n = \frac{1}{2\sqrt{2}}[\epsilon_8^n - (-\epsilon_8^{-1})^n], \qquad Z_n = \tfrac{1}{2}[\epsilon_8^n + (-\epsilon_8^{-1})^n]$$

so that Y_n and Z_n are rational integers.

The combination $Z_{2N+1} - 4Y_N$ came up when $(\Delta + 1)/4 \equiv 1 \pmod 8$. In order to translate this into terms of F_Q, it is necessary to know whether N is odd or even. For example, it developed later that if $(\Delta + 1)/4 \equiv 1 \pmod 8$ then N is odd (at least when $h(d) = 1$). In this way, we are led to define for $Q = (1, 1, (\Delta + 1)/4)$ and $k = 8$,

$$\begin{aligned} \text{(20)}\quad H_Q(z) &= \tfrac{1}{2}[F_Q(z)^2 - F_Q(z)^{-2}] - \sqrt{2}[\epsilon_8^{-1/2}F_Q(z) + \epsilon_8^{1/2}F_Q(z)^{-1}], \\ &\qquad\qquad (\Delta + 1)/4 \equiv 1 \pmod 8 \\ &= \tfrac{1}{2}[F_Q(z)^2 - F_Q(z)^{-2}] + \sqrt{2}[\epsilon_8^{1/2}F_Q(z) + \epsilon_8^{-1/2}F_Q(z)^{-1}], \\ &\qquad\qquad (\Delta + 1)/4 \equiv 3 \pmod 8 \\ &= \tfrac{1}{2}[F_Q(z)^2 - F_Q(z)^{-2}] + \sqrt{2}[\epsilon_8^{-1/2}F_Q(z) + \epsilon_8^{1/2}F_Q(z)^{-1}], \\ &\qquad\qquad (\Delta + 1)/4 \equiv 5 \pmod 8 \\ &= \tfrac{1}{2}[F_Q(z)^2 - F_Q(z)^{-2}] - \sqrt{2}[\epsilon_8^{1/2}F_Q(z) + \epsilon_8^{-1/2}F_Q(z)^{-1}], \\ &\qquad\qquad (\Delta + 1)/4 \equiv 7 \pmod 8. \end{aligned}$$

This definition of $H_Q(z)$ is such that if $h(d) = 1$

$$\begin{aligned} \text{(21)}\quad H_Q(\tau) &= Z_{2N+1} - 4Y_N \qquad (\Delta + 1)/4 \equiv 1 \pmod 8, \\ &= Z_{2N+1} + 4Y_{N+1} \qquad (\Delta + 1)/4 \equiv 3 \pmod 8, \\ &= Z_{2N+1} + 4Y_N \qquad (\Delta + 1)/4 \equiv 5 \pmod 8, \\ &= Z_{2N+1} - 4Y_{N+1} \qquad (\Delta + 1)/4 \equiv 7 \pmod 8. \end{aligned}$$

We should emphasize that whenever we see H_Q, we are dealing with $Q = (1, 1, (\Delta + 1)/4)$ and $k = 8$. We may also find the first few terms of the series expansion of $H_Q(z)$. In fact, even though we dealt only with $z = \tau$ in [6], we may read off this information from there (this is true elsewhere also; the series expansions we need are available in [6] when interpreted correctly). We find (in all four cases) that

$$H_Q(z) = \frac{1}{2\sqrt{2}} e^{\pi i/4} q^{-1/4} + O(q^{1/8}) \tag{22}$$

where O refers to $\operatorname{Im} z \to \infty$, or what is the same thing, $q \to 0$.

But we may also get to $q^{-1/4}$ from $k = 12$. In this case $\epsilon_{12} = 2 + \sqrt{3}$ and if $d < -11$, $h(d) = 1$, then $(\Delta + 1)/4 \equiv 5 \pmod 6$ and $h(12d) = 8M + 4$, so that

$$F_{(1,1,(\Delta+1)/4),12}(\tau) = \epsilon_{12}^{M+1/2} = \sqrt{2}((1 + \sqrt{3})/2)\epsilon_{12}^{M}.$$

Define the rational integer

$$W_m = \frac{1 + \sqrt{3}}{2}\epsilon_{12}^{m} + \frac{1 - \sqrt{3}}{2}\epsilon_{12}^{-m},$$

and set $a = W_M \pm 1$; here and in the next few lines the top sign refers to $(\Delta + 1)/4 \equiv 1 \pmod 4$ and the bottom sign refers to $(\Delta + 1)/4 \equiv 3 \pmod 4$. Define for $Q = (1, 1, (\Delta + 1)/4)$, $(\Delta + 1)/4 \equiv 5 \pmod 6$, and $k = 12$,

$$a_Q(z) = \frac{1}{\sqrt{2}}[F_Q(z) - F_Q(z)^{-1}] \pm 1.$$

Then for $d < -11$ and $h(d) = 1$, we have

$$a_Q(\tau) = a. \tag{23}$$

Further

$$a_Q(z) = \frac{1}{\sqrt{2}} e^{\pi i/12} q^{-1/12}[1 \mp 2\sqrt{2}\, e^{-\pi i/4} q^{1/4} + O(q^{1/3})],$$

so that

$$a_Q(z)^3 \pm 3 = \frac{1}{2\sqrt{2}} e^{\pi i/4} q^{-1/4} + O(q^{1/12}). \tag{24}$$

We see from (22) and (24) that if Δ is sufficiently large (in [6], $\Delta > 200$ is sufficiently large) and $h(d) = 1$ then

$$\begin{aligned} H_Q(z) &= a_Q(z)^3 + 3, \qquad (\Delta + 1)/4 \equiv 1 \pmod 4, \\ &= a_Q(z)^3 - 3, \qquad (\Delta + 1)/4 \equiv 3 \pmod 4) \qquad (z = \tau) \end{aligned} \tag{25}$$

since both sides are rational integers differing by less than one. As an example, if $(\Delta + 1)/4 \equiv 1 \pmod 8$, then (21), (23) and (25) give

$$Z_{2N+1} - 4Y_N = a^3 + 3. \tag{26}$$

It turns out that we may find all solutions to (26). It is at this point that N turns out to be odd (in this case). If we set $N = 2n + 1$ then (26) ultimately reduces to

$$8C^6 + 1 = Z_n^2 \quad (n \text{ even}), \qquad D^6 + 1 = 2Y_n^2 \quad (n \text{ odd}).$$

When the other cases are included, we must solve the two Diophantine equations

$$8x^6 \pm 1 = y^2, \qquad x^6 \pm 1 = 2y^2. \tag{27}$$

There are no solutions that can correspond to a value of $\Delta > 200$.

The same thing goes wrong for $h(d) = 2$ here that goes wrong in Baker's method. We now have series involving combinations of $\exp(\sqrt{\Delta}/k)$ and $\exp(\sqrt{\Delta}/ak)$ where a is unknown and possibly large. Thus we no longer know how many terms of our series are necessary when we try to construct a Diophantine equation and hence this approach is hopeless. If a is fixed then our method is possible. For example, all the fields with $h(d) = 2$ and d even ($a = 2$) have been determined by Peter Weinberger and Monsur Kenku independently.

Note that equations (27) are precisely the equations (8) that we must solve by Heegner's method! Thus we may well ask the following questions:

(I) Are the functions of z here the same as in Heegner's method?

(II) Is (25) an identity valid for all z?

(III) Can we determine $F_{Q,8}(\tau)$ if $h(d) > 1$?

The answers to I, II, and III are in fact, "yes". As an example of III, let $Q = (1, 1, 107)$ ($d = -427$, $h(d) = 2$). Then

$$L(1, \chi_8, Q) = \frac{\pi}{2\sqrt{427}} \log\,[(1 + \sqrt{2})^{11}(11 + \sqrt{122})^2].$$

4. Kronecker's limit formula

Our first problem is to introduce modular functions into my method and then we can attempt to relate them to Heegner's functions. Modular functions in connection with the class-number one problem were first used by Siegel [5]. Using $k = 5$, he obtained a Diophantine equation similar to (8) without having to restrict himself to sufficiently large Δ. While $k = 5$ is simpler, $k = 8$ and $k = 12$ lead to the connections between Stark and Heegner.

As far as getting modular functions into the act, let us recall Kronecker's limit formula,

$$\lim_{s\to 1;(\sigma>1)} \left\{\frac{1}{2} \sum_{m,n\neq 0,0} \left(\frac{-i(z-\bar{z})}{(m+nz)(m+n\bar{z})}\right)^s - \frac{\pi}{s-1}\right\} = C - 2\pi \log\,[(-i(z-\bar{z}))^{1/2}\eta(z)\eta(-\bar{z})]$$

where C is a constant and

$$\eta(z) = q^{1/24} \prod_{n=1}^{\infty} (1 - q^n).$$

This leads us to hope that $L(1, \chi_k, Q, z)$ may be expressed in terms of η-functions also. This is true; for $k = 8$ we find from (13) that

$$F_{Q,8}(z) = \zeta_{Q,8} \cdot 2^{-\chi_8(a)/2} \cdot \left[\frac{\eta(8z)}{\eta((4z+3)/2)}\right]^{-\chi_8(a)/2} \cdot \prod_{j=0}^{3} \left[\eta\left(\frac{2z+3j}{4}\right)\right]^{-\chi_8(Q(3j,2))/2} \cdot \prod_{j=0}^{7} \left[\eta\left(\frac{z+3j}{8}\right)\right]^{-\chi_8(Q(3j,1))/2}$$

where $\zeta_{Q,8}$ is a 16th root of unity (in fact, if $b = a$, it is the 16th root of unity which makes $F_{Q,8}(z)$ real and positive on $\operatorname{Re} z = \frac{1}{2}$).

We can now determine how $F_{Q,8}(z)$ transforms under the full modular group.

Set

$$(28)\quad G_{Q,8}(z) = \chi_8(a)\chi_8\left(\frac{a^2-b^2}{4}+ac\right)\exp\left[\frac{2\pi i}{8}\left(\frac{b-a}{2}\right)a\chi_8(a)\right]F_{Q,8}(z).$$

Then (remember, $\Delta \equiv 3 \pmod 8$),

$$(29)\qquad G_{Q,8}\left(\frac{\alpha z+\beta}{\gamma z+\delta}\right) = G_{Q_1,8}(z)$$

where α, β, γ, δ are integers, $\alpha\delta - \beta\gamma = 1$ and

$$Q(\alpha x - \beta y, -\gamma x + \delta y) \equiv Q_1(x, y) \pmod k.$$

I fear that this was originally proved by the very painful way of checking it for $z \to z + 1$ and $z \to -1/z$. In particular, we see that if

$$\begin{pmatrix}\alpha & \beta\\ \gamma & \delta\end{pmatrix} \equiv \begin{pmatrix}1 & 0\\ 0 & 1\end{pmatrix} \pmod 8$$

then

$$F_{Q,8}\left(\frac{\alpha z+\beta}{\gamma z+\delta}\right) = F_{Q,8}(z).$$

Now we can investigate how $H_Q(z)$ transforms. The first thing that we prove is

THEOREM 1. $H_{1,1,1}(z) = H_{1,1,5}(z)$.

This is proved in the standard manner of showing that the difference has no poles and is 0 at $i\infty$. By transforming z in this theorem in various ways, we get the

COROLLARY 1. $H_{1,1,1}(z + 4) = H_{1,1,1}(z)$,

$$H_{1,1,3}(z) = H_{1,1,7}(z) = -H_{1,1,1}(z + 2).$$

It is this theorem and its corollary that motivated the combination in (28) and (29). The expressions in (28) and (29) are necessary if we are to successfully derive the theorem. And the truth of the theorem is suspected because it is a necessary condition for (25) to be an identity. To see this, if (25) is an identity then

$$H_{1,1,1}(z) = H_{1,1,17}(z) = a_{1,1,17}(z)^3 + 3 = a_{1,1,5}(z)^3 + 3 = H_{1,1,5}(z).$$

In fact the techniques outlined above now suffice to prove that (25) is an identity. Thus we discovered what we had to do by proceeding from (25) to the theorem to (28) and (29) and then we proved everything by going in the opposite direction.

We now have enough information to further investigate the function H_Q.

THEOREM 2. *The functions* $H_{1,1,1}(z + l)$ $(l = 0, 1, 2, 3)$ *are permuted among themselves by the full modular group.*

SKETCH OF PROOF. Since $z \to z + 1$ and $z \to -1/z$ generate the full modular group, we need only check the theorem for these transformations. But since $H_{1,1,1}(z + 4) = H_{1,1,1}(z)$, the result for $z \to z + 1$ is obvious. Finally, we may calculate that

$$H_{1,1,1}(-1/z) = H_{1,1,1}(z + 1), \qquad H_{1,1,1}(-1/z + 1) = H_{1,1,1}(z),$$
$$H_{1,1,1}(-1/z + 2) = -H_{1,1,3}(z) = H_{1,1,1}(z + 2),$$
$$H_{1,1,1}(-1/z + 3) = H_{1,1,5}(z - 1) = H_{1,1,1}(z + 3).$$

Here we used (29) and the definition of $H_Q(z)$ in (20) and, in the last two lines, Theorem 1 and its corollary.

THEOREM 3. *The four functions* $H_{1,1,1}(z + l)$ $(l = 0, 1, 2, 3)$ *are the four solutions in* x *to*

$$(x - 3)(x + 1)^3 = -\tfrac{1}{64}j(z).$$

PROOF. Set

$$P(x) = \prod_{l=0}^{3} [x - H_{1,1,1}(z + l)]$$
$$= x^4 + A(z)x^3 + B(z)x^2 + C(z)x + D(z).$$

Then, by Theorem 2, $A(z)$, $B(z)$, $C(z)$, and $D(z)$ are invariant under the full modular group. Further, since their only possible poles in the fundamental domain of $j(z)$ are at $i\infty$, these functions are polynomials in $j(z)$. We see from (22) that

$$A(z) = O(q^{1/8}), \qquad B(z) = O(q^{-1/2}), \qquad C(z) = O(q^{-3/4}),$$
$$D(z) = \tfrac{1}{64}q^{-1} + O(q^{-3/4}),$$

and hence $P(x)$ takes the form,

$$P(x) = x^4 + Bx^2 + Cx + \tfrac{1}{64}j(z) + D$$

where B, C, D are constants.

If we knew the expansion in (22) for several more terms, we could actually evaluate B, C, D directly, but this is not desirable. Instead, we recall that $H_{1,1,1}(z + l)$ is a root of $P(x) = 0$ and that we know the values of $H_{1,1,1}(z + l)$ at several locations from [6]. In particular, we found in [6] that

$$H_{1,1,1}(\tau) = H_{1,1,5}(\tau) = 11 \quad \text{if } d = -19,$$
$$H_{1,1,1}(\tau + 2) = -H_{1,1,3}(\tau) = -61 \quad \text{if } d = -43,$$
$$H_{1,1,1}(\tau) = 219 \quad \text{if } d = -67,$$
$$H_{1,1,1}(\tau) = 8003 \quad \text{if } d = -163.$$

The first three of these together with the known values of $j(\tau)$ (their cube roots are given in (9)) give the equations

$$\begin{aligned} 121B + 11C + D &= -817, \\ 3721B - 61C + D &= -21841, \\ 47961B + 219C + D &= -289521, \end{aligned}$$

which have the unique solution $B = -6$, $C = -8$, $D = -3$. The equation $P(x) = 0$ now factors as in the theorem.

The value of $H_{1,1,1}(\tau)$ for $d = -163$ [illegible] check; we may also verify the theorem for $d = -3$ and $d = -11$ where we did not calculate the values of $H_Q(\tau)$ in [6] since the use of $k = 12$ restricted us to $\Delta \equiv 19 \pmod{24}$. We now note that if $3 \nmid d$, then $j(\tau)$ is a perfect cube in $\mathbf{Q}(j(\tau))$ and hence we may do without $k = 12$ completely in [6]. We find directly from Theorem 3 and Theorem 1 and its corollary that

$$\begin{aligned} H_Q(\tau) &= \text{cube} + 3 \quad \text{if } (\Delta + 1)/4 \equiv 1 \pmod 4, \\ &= \text{cube} - 3 \quad \text{if } (\Delta + 1)/4 \equiv 3 \pmod 4. \end{aligned}$$

Siegel used these same methods for $k = 5$ to get a Diophantine equation. If we were to stop here, $k = 5$ has the distinct advantage of being decidedly simpler than $k = 8$. The usefullness of $k = 8$ will become apparent in the next section.

5. An isomorphism between Stark and Heegner

We now have a fair knowledge of how modular functions occur in the Stark proof of the class-number one problem and we can return to the question of relating Stark and Heegner. Our first problem is to connect the modular function $H_Q(z)$ with Heegner's method and this turns out to be not too difficult.

The cubic equation for $f(2z - 1)^{24}$ follows from (2); $f(2z - 1)$ is a root of

$$x^{72} - (3\cdot 2^8 - j(z))x^{48} + 3\cdot 2^{16}x^{24} - 2^{24} = 0. \tag{30}$$

This equation may be reduced to one of degree 36,

$$x^{36} + B_{12}(z)x^{24} + A_{12}(z)x^{12} - 2^{12} = 0 \tag{31}$$

(a cubic equation in x^{12} satisfied by either $f(2z - 1)^{12}$ or $-f(2z - 1)^{12}$) with the relations

$$2A_{12}(z) - B_{12}(z)^2 = j(z) - 3\cdot 2^8, \qquad A_{12}(z)^2 + 2^{13}B_{12}(z) = 3\cdot 2^{16}.$$

In order to rid ourselves of some of the factors of 2, let

$$A_{12}(z) = 2^8C(z), \qquad B_{12}(z) = 2^4D(z).$$

Then our relations are

$$2C(z) - D(z)^2 = 2^{-8}j(z) - 3, \qquad C(z)^2 + 2D(z) = 3. \tag{32}$$

We may eliminate $D(z)$ in (32) and get the quartic equation for $C(z)$,

$$[C(z)^2 - 3]^2 = 4[2C(z) - 2^{-8}j(z) + 3]$$

which simplifies to

$$(C(z) - 3)(C(z) + 1)^3 = -\tfrac{1}{64}j(z).$$

This is the same equation satisfied by $H_{1,1,1}(z + l)$. There are four possible reductions from (30) to (31); according to which reduction is made, we get the coefficient of x^{12} in (31) is $A_{12}(z) = 2^8 H_{1,1,1}(z + l)$ with $l =$ either 0, 1, 2, or 3. Thus we have identified $H_Q(z)$ with the coefficients of the cubic equation for $f(2z - 1)^{12}$ and it is now a matter of algebraic manipulation to show that equations (8) and (27) really involve the same modular functions.

But are they really the same? We illustrate the meaning of this question with $H_Q(z)$. We know that exactly one of the four values of $H_Q(\tau + l)$ $(0 \le l \le 3)$ is in $\mathbf{Q}(j(\tau))$ (even if $3 \mid d$) but which one is it? If $h(d) = 1$, we know the answer thanks to Dirichlet's formula (11), but if $h(d) > 1$, we can no longer use (11) to give us the answer. However, we are certainly entitled to conjecture that $H_Q(\tau)$ is in $\mathbf{Q}(j(\tau))$ even when $h(d) > 1$. After Theorem 1 and its corollary, our conjecture becomes

THEOREM 4. *$H_{1,1,1}(\tau)$ is in $\mathbf{Q}(j(\tau))$ if $(\Delta + 1)/4 \equiv 1 \pmod 4$ and $H_{1,1,1}(\tau + 2)$ is in $\mathbf{Q}(j(\tau))$ if $(\Delta + 1)/4 \equiv 3 \pmod 4$.*

Sketch of proof in the case $(\Delta + 1)/4 \equiv 1 \pmod 4$. Define for $n > 0$,

$$\phi_n(x, H_{1,1,1}(z)) = \prod_{\substack{\alpha,\beta,\delta;\alpha\delta=n,\alpha>0\\ 0\le\beta<4\delta;\beta\equiv 0 \pmod 4}} \left[x - H_{1,1,1}\left(\frac{\alpha z + \beta}{\delta}\right)\right]. \tag{33}$$

If $n \equiv 1 \pmod 4$, then ϕ_n is a polynomial in x and $H_{1,1,1}(z)$ with rational coefficients. This is shown by proving that any unimodular transformation which preserves $H_{1,1,1}(z)$ permutes the factors on the right side of (33) and hence the coefficients of the powers of x in ϕ_n are invariant under transformations which preserve $H_{1,1,1}(z)$. But the fundamental domain of $H_{1,1,1}(z)$ has genus zero and as we might hope, $H_{1,1,1}(z)$ has exactly one pole (of first order) in this domain. With the proper choice of the domain, this pole is at $i\infty$ and now the coefficients of the powers of x in (33) have poles only at $i\infty$ in the fundamental domain of $H_{1,1,1}(z)$ and hence are polynomials in $H_{1,1,1}(z)$. A close examination of the algebraic nature of the coefficients in the series expansion of $H_{1,1,1}(z)$ show that the coefficients in ϕ_n are rational.

Now let us examine the roots of

$$\phi_n(x, x) = 0. \tag{34}$$

Here we use the following idea. If x is a root of (34) then there is an ω, $\operatorname{Im}\omega > 0$, such that $x = H_{1,1,1}(\omega)$. For these values of ω and x, $\phi_n(x, H_{1,1,1}(\omega)) = 0$ and hence by (33), for some α, β, δ $(\alpha\delta = n, \alpha > 0, \beta \equiv 0 \pmod 4, 0 \le \beta < 4\delta)$,

$$H_{1,1,1}(\omega) - H_{1,1,1}\left(\frac{\alpha\omega + \beta}{\delta}\right) = x - H_{1,1,1}\left(\frac{\alpha\omega + \beta}{\delta}\right) = 0. \tag{35}$$

But since $H_{1,1,1}(z)$ has only one (first order) pole in its fundamental domain,

$H_{1,1,1}(z)$ takes on every value exactly once in its fundamental domain. It follows from (35) that there is a transformation

$$\begin{pmatrix} A & B \\ C & D \end{pmatrix} \text{ preserving } H_{1,1,1}(z) \text{ sending } \frac{\alpha\omega + \beta}{\delta} \text{ to } \omega.$$

This gives a quadratic equation for ω which gives, if we take $C > 0$,

$$\omega = \frac{A\alpha - C\beta - D\delta + ((A\alpha + C\beta + D\delta)^2 - 4\alpha\delta)^{1/2}}{2C\alpha}.$$

It is clear that we can now establish the following fact:

If $(\Delta + 1)/4 \equiv 1 \pmod 4$, $x = H_{1,1,1}(\tau + l)$ is a root of $\phi_{(\Delta+1)/4}(x, x) = 0$ if and only if $4 \mid l$.

But now we are done. When $(\Delta + 1)/4 \equiv 1 \pmod 4$, the two equations

$$(x - 3)(x + 1)^3 = -\tfrac{1}{64} j(\tau) \quad \text{and} \quad \phi_{(\Delta+1)/4}(x, x) = 0$$

have only the root $x = H_{1,1,1}(\tau)$ in common. Since the coefficients of both equations are in $\mathbf{Q}(j(\tau))$ (and in fact in $\mathbf{Q}$ except for $j(\tau)$), the Euclidean algorithm guarantees that $H_{1,1,1}(\tau)$ is in $\mathbf{Q}(j(\tau))$.

Thanks to the definition of $H_Q(z)$ in (20), it is clear that we are close to eliminating the need of using Dirichlet's evaluation of $L(1, \chi_k)$ in discussing the algebraic nature of $F_{Q,8}(z)$. In fact we need no longer restrict ourselves to $h(d) = 1$:

THEOREM 5. $\epsilon_8^{-1/2} F_{(1,1,(\Delta+1))/4,8}(\tau)$ *is a unit in* $\mathbf{Q}(j(\tau), \sqrt{2})$.

The proof is a more complicated version of the methods used in Theorem 4. For more details about this section in some directions (and less in others), the reader is referred to [9].

REFERENCES

1. A. Baker, *Linear forms in the logarithms of algebraic numbers*, Mathematika **13** (1966), 204–216. MR **36** #3732.

2. B. J. Birch, *Diophantine analysis and modular functions*, Proc. Conference on Algebraic Geometry (Tata Institute, Bombay, 1968) pp. 35–42.

3. Max Deuring, *Imaginäre quadratische Zahlkörper mit der Klassenzahl Eins*, Invent. Math. **5** (1968), 169–179. MR **37** #4044.

4. Kurt Heegner, *Diophantische Analysis und Modulfunktionen*, Math. Z. **56** (1952), 227–253. MR **14,** 725.

5. C. L. Siegel, *Zum Beweise des Starkschen Satzes*, Invent. Math. **5** (1968), 180–191. MR **37** #4045.

6. H. M. Stark, *A complete determination of the complex quadratic fields of class-number one*, Michigan Math. J. **14** (1967), 1–27. MR **36** #5102.

7. ———, *On the problem of unique factorization in complex quadratic fields*, Amer. Math. Soc. Proceedings of the Symposium in Pure Mathematics held in Houston, Jan. 1967. Vol. 12.

8. ———, *On the "gap" in a theorem of Heegner*, J. Number Theory **1** (1969), 16–27.

9. ———, *The role of modular functions in a class-number problem*, J. Number Theory **1** (1969), 252–260.

10. ———, *An explanation of some exotic continued fractions found by Brillhart*, Proc. Sympos. Computers in Number Theory (Oxford, 1969) (to appear).

11. H. Weber, *Lehrbuch der Algebra*. Vol. 3, 3rd ed., Chelsea, New York, 1961.

MASSACHUSETTS INSTITUTE OF TECHNOLOGY
CAMBRIDGE, MASSACHUSETTS

CLASS NUMBER, A THEORY OF FACTORIZATION, AND GENERA

DANIEL SHANKS

1. Introduction

Consider the real Dirichlet series $L(1, \chi_\Delta)$ having a character χ_Δ given by the Kronecker symbol:

$$\chi_\Delta = (-\Delta/n). \tag{1}$$

One has

$$L(1, \chi_\Delta) = \sum_{n=1}^{\infty} \chi_\Delta \frac{1}{n} = \prod_{p=2}^{\infty} \frac{p}{p - (-\Delta/p)} = \frac{\pi h(-\Delta)}{\sqrt{\Delta}} \tag{2}$$

where the class number $h(-\Delta)$ is the number of equivalence classes of primitive binary quadratic forms

$$F = (A, B, C) = Au^2 + Buv + Cv^2 \tag{3}$$

of negative discriminant

$$B^2 - 4AC = -\Delta. \tag{4}$$

If $-\Delta$ is a fundamental discriminant, cf. [1], $h(-\Delta)$ is also the number of ideal classes in the imaginary quadratic field $Q(\sqrt{-\Delta})$, but since we will also want other Δ, and since the forms (3) occur directly in the algorithms we develop, we prefer to use this earlier (Gaussian) formulation.

About a year ago, D. H. and Emma Lehmer and I were interested in this problem: Given Δ of a given size, say, $\Delta \leqq N$, how small (or how large) can $L(1, \chi_\Delta)$ be? What Δ have these extreme values, and how can one compute these values? Such extreme $L(1, \chi_\Delta)$ relate to several questions of interest, in particular [2] to the question whether the $L(s, \chi)$ satisfy the generalized Riemann Hypothesis.

Lehmer's sieve DLS 127 enables one [3] to readily find numbers $-\Delta$ having a prescribed quadratic character, that is, the $\epsilon(p)$ in

$$(-\Delta/p) = \epsilon(p)$$

can be assigned values $\epsilon(p) = \pm 1$ for any, or all, primes $p = 2, 3, \ldots, 127$. Choosing all $\epsilon(p) = -1$ (or $+1$) minimizes (or maximizes) the corresponding factors in the product in (2), and therefore gives values $L(1, \chi_\Delta)$ that tend to be especially small (or large). For example, the smallest Δ having $\epsilon(p) = -1$ for all $p = 2, \ldots, 127$ is [4]

$$\Delta = 71837718283$$

and we can evaluate the small value $L(1, \chi_\Delta) = 0.17821$ from (2) if we can compute $h(-\Delta) = 15204$.

If we try to prescribe $\epsilon(p)$ for still more p, $(p > 127)$, Δ will grow rapidly and the computation of $h(-\Delta)$ threatens to become unfeasible. The classical methods of calculating $h(-\Delta)$ require $O(\Delta)$ operations. Whether one counts all reduced triples (A, B, C) in (4), or evaluates Dirichlet's famous sums of Jacobi symbols, $O(\Delta)$ operations are needed, and the $h(-\Delta) = 15204$ above already requires a lengthy computation. Suppose Δ is substantially larger. For the 22-digit prime

$$S_{36} = (2^{36} + 3)^2 - 8$$

and field $Q(\sqrt{-S_{36}})$, one can say, unequivocally, that if either of these classical methods were used on the world's fastest computer, and if the program ran an entire year, its completion would be nowhere in sight. To compute

$$h(-4S_{36}) = 50866650112 \tag{5}$$

and $L(1, \chi_\Delta) = 1.16271\ldots$ therefore requires an entirely new method.

We will first describe a new method that uses composition of the forms (3) and requires only $O(\Delta^{1/4})$ operations. This much greater efficiency enables one to compute (5) in some minutes on a computer. The $h(-\Delta) = 15204$ case above takes almost no time on an electronic computer, and could even be done on a desk machine—with some effort. The method works best for imaginary fields, but, with some adaptation, at least some real fields can be computed also, and we will discuss these modifications briefly.

The h equivalence classes form an Abelian group under composition; the method is based upon this, and with little or no extra computation one also obtains the structure of this group. For example, the class group of $Q(\sqrt{-S_{36}})$ is cyclic, having the quadratic form

$$F = (3, 2, 1574122161093987358038) \tag{5a}$$

as a generator.

If p^α is the largest power of p dividing the class number, the corresponding subgroup of the class group is its p-Sylow subgroup. Our second main topic is a new theory of factorization. To factor an integer N, we compute $h(-N)$, or $h(-4N)$ if $N \not\equiv 0, 3 \pmod 4$, and thereby obtain its 2-Sylow subgroup. This yields all ambiguous forms having the appropriate discriminant, and therefore a complete factorization of N. We discuss this theory in some detail.

Finally, the "and genera" in our title refers to the use of these class number-class group techniques in an algorithm that computes numbers $-\Delta$ having a prescribed quadratic character. This returns us to the original problem.

2. The new method

From (2) we may approximate $h(-\Delta)$ by the formula

$$\frac{\sqrt{\Delta}}{\pi} \prod_{p=2}^{P} \frac{p}{p - (-\Delta/p)} \approx h(-\Delta). \tag{6}$$

Originally, this was done by hand, and if one takes $P = 400$, say, the approximations will usually be correct to 5%. This would suffice for Δ that are not too large, but for large Δ a better estimate, while not mandatory, would reduce the number of subsequent operations. The Lehmers have programmed the left side of (6) using the Reciprocity Law for $(-\Delta/p)$, and with this program an IBM 7094 can compute the partial product for $P = 132{,}000$ in only a few seconds. These estimates are usually correct to 1 part in 1000.

As an important by-product one simultaneously determines all small primes p having $(-\Delta/p) = +1$. For each such p we have

$$-\Delta \equiv B_p^2 \pmod{p},$$

or $\Delta + B_p^2 = 4pC_p$ provided we choose B_p even or odd according as $\Delta \equiv 0$ or 3 (mod 4). For each such p we thus obtain a form

$$F_p = (p, B_p, C_p) \tag{7}$$

of discriminant $-\Delta$, and we save a small stock of these forms. (A handy algorithm for computing B_p was recently devised [5], but this is hardly needed for small p such as 2, 3, ...)

Suppose we know a priori that

$$h \equiv k \pmod{b}. \tag{8}$$

Then let us choose as our estimate for h that integer H which is closest to (6) and also satisfies

$$H \equiv k \equiv h \pmod{b}. \tag{8a}$$

There are many such relations (8) known, and the larger b is, the faster will be our subsequent calculation, as we shall see. But if nothing is known about h we can simply set $b = 1$, $k = 0$.

A few examples of (8) are these: If $\Delta = 2^p - 1$, it will be shown that

$$h(-\Delta) \equiv 0 \pmod{p - 2}. \tag{9}$$

If $\Delta = 4p$ with $p \equiv 1, 5 \pmod{8}$, then $h(-\Delta) \equiv 0, 2 \pmod{4}$. Thus, for the S_{36} above, we would know a priori that $4 \mid h$. Further analysis would also show that $8 \mid h$ in this case, and had we been able to determine that $2^{10} \mid h$ in this case, but $2^{11} \nmid h$, as in fact can be done by the technique developed in [6], the computation of (5) would have been substantially accelerated by using $k = 2^{10}$ and $b = 2^{11}$.

If $h < 1000$, our (Lehmer) estimate H is now probably exact, although some verification is desirable. In any case, if $h > 1000$ further computations are needed. Since the forms (3) constitute a group under composition, any form

$$F = (A, B, C)$$

satisfies

$$F^h = I \tag{10}$$

where F^h is the h'th power of F by composition, and I is the principal form, the identity of the group. This form I represents 1 and is given by

$$I = (1, 1, (\Delta + 1)/4) \quad \text{or } (1, 0, \Delta/4) \tag{11}$$

according as Δ is odd or even. Every form F is of some *order* e

$$F^e = I \qquad (e \text{ minimal}) \tag{12}$$

and there is some *multiple* m such that

$$em = h. \tag{13}$$

If the group is cyclic and F is primitive, then $m = 1$.

Now, from our stock of forms (7) we choose $F = F_p$ for the smallest p where we have no a priori knowledge that its multiple m is large.[1] Usually, such knowledge does not exist and we simply take the smallest p (first). If it does exist, as in (9) where

$$F_2 = (2, 1, 2^{p-3})$$

may be seen to have $e = p - 2$ (thus explaining (9)), we would avoid F_2 and choose instead the next F_p. Up to this point, we have computed H, b, and F.

Now compute F^H—see Appendix 1 for an efficient formulation of composition—by expressing H in binary:

$$H = \sum_0^N a_n 2^n \qquad (a_n = 0 \text{ or } 1),$$

and computing the forms

$$G_N = F, G_{N-1}, G_{N-2}, \ldots, G_0 = F^H$$

recursively by

$$G_{i-1} = G_i^2 \quad \text{or} \quad G_{i-1} = F \cdot G_i^2 \tag{14}$$

according as $a_{i-1} = 0$ or 1. After each composition (14), the form G_{i-1} is reduced, if necessary. (See Appendix 1.) Approximately $\frac{3}{2}N = O(\log h)$ compositions, and about an equal number of reductions, are thus needed to compute F^H in reduced form.

[1]See footnote, p. 420 for an exception.

If $F^H \neq I$ some correction

$$h = H + C \tag{15}$$

is needed to obtain h from our estimate. To minimize the number of compositions now needed to compute C we proceed as follows. For an integer s, to be determined soon, we compute and store the coefficients A_n, B_n in the first s powers of F^b:

$$F^{bn} = (A_n, B_n, C_n) \qquad (n = 1, 2, \ldots, s). \tag{16}$$

One has, absolutely free, A_n, B_n for $n = 0, -1, -2, \ldots, -s$, since $F^0 = I$, and therefore $A_0, B_0 = 1, 0$ or $1, 1$ according as $\Delta =$ even or odd, and

$$F^{-bn} = (A_n, -B_n, C_n), \tag{17}$$

since these inverses are, very luckily, obtained merely by changing the sign of the middle coefficient.

If $|H - h| \leqq sb$, we now find a match:

$$F^H = (A_H, B_H, C_H) = F^{bn} = (A_n, B_n, C_n) \tag{18}$$

for some $n = -s, \ldots, +s$. That is,

$$A_H = A_n \quad \text{and} \quad B_H = \pm B_n. \tag{18a}$$

(The C's are redundant since all the forms have the same discriminant.) It is important to note that we are using the fact that each equivalence class has a *unique reduced form* for these negative discriminants; this enables the identification (18) to be readily made. From (18) we then have $F^{H-bn} = I$.

If $|H - h| > sb$, we set

$$g = 2sb \tag{19}$$

and compute, successively, F^{H+rg} for $r = +1, -1, +2, -2, \ldots$ until, for some r, we obtain a match

$$F^{H+rg} = F^{bn}. \tag{20}$$

Then

$$F^{H+rg-bn} = I. \tag{21}$$

Graphically, what we have done here is to compute F^x for s "baby steps" b. This lays out a patch of width $g = 2bs$ symmetrically centered on the identity $F^0 = I$. Then we take $|2r|$ or $|2r| - 1$ "giant steps" g, equal to the width of this patch, symmetrically around our estimated F^H, until we step into the patch and find the match (20). To minimize the number of compositions we would like

$$|2r| \approx s: \tag{22}$$

that is, an equal number of baby steps and giant steps. Since the $rg - bn$ of (21) equals the correction C of (15)—subject to some verification discussed soon—and since we expect $|C| \leqq H/1000$, we estimate

$$s^2 b \approx \tfrac{1}{1000} H,$$

and therefore select

$$s = (H/1000b)^{1/2} \tag{23}$$

as the optimum number of steps.

Since $h = O(\Delta^{1/2})$, except possibly for a low order logarithmic factor, we deduce that the

$$\mathcal{C} = \tfrac{3}{2}N + s + |2r| \tag{24}$$

compositions needed to locate the match (20) satisfies the relation $\mathcal{C} = O(\Delta^{1/4})$, as claimed. Some other (verification) operations—as discussed below—are $O(\log h)$ in number, and do not change this conclusion. Note, in (23), that Lehmer's factor 1000, and the baby step b both diminish the implied coefficient in the expression $O(\Delta^{1/4})$.

But is

$$H + C = h = h' = H + rg - bn? \tag{25}$$

It is clear that

$$em' = h' \tag{26}$$

for some multiple m', but does $m' = m$, the correct multiple? This will almost always be true, but if e is very small relative to h, specifically, if $m > 1000$,[2] there is a possibility that $m' \neq m$ and, therefore, that $h' = H + rg - bn \neq h$. To guard against this (very rare) possibility, one wishes to compute the order e of our form F, and, more generally, to compute the whole group structure. This is discussed in the next section.

For now, we note that the immense cyclic group (5) has billions of equivalence classes, and billions of operations would be needed to compute each form (3) by factoring the $B^2 + \Delta$ of (4). The present efficiency stems entirely from the use of the group structure here: only several hundred compositions (24) are needed to obtain a closure (20). Only the several hundred corresponding forms F^x are examined explicitly, and, at that, are computed relatively easily, by composition. The vast number of classes (5) are only counted implicitly—via the group structure.

3. Briefly, the group

We may wish to determine the class group for (a), the verification of $h = h'$ mentioned above, (b), its own sake, or (c), the use of the 2-Sylow subgroup for factorization as described below. In any case, the presumed class number $h = h'$

---

[2]If Δ is highly composite, m must be large. See paragraph 3 of §9 for the replacement then of m by one of its factors M.

may be factored $h = \prod_i p_i^{\alpha_i}$, and the true order of F:

$$e = \prod_i p_i^{\beta_i} \qquad (\beta_i \leqq \alpha_i) \tag{27}$$

found as follows. For each p_i dividing h we compute $G = F^n$ with $n = h/p_i^{\alpha_i}$. We now raise G to the p_i power, then the p_i^2 power, etc., until we find the correct exponent β_i by

$$G^{p_i^{\beta_i}} = I \qquad (\beta_i \text{ minimal}). \tag{28}$$

Note that there is no trial-and-error here of the type needed to find the match (20), and the number of compositions required to find β_i is only $O(\log h)$. In this way we determine e.

If $e = h$, the group is cyclic and F is a generator. If not, to complete the group structure we dip into our stock of forms (7) and compute the order e for each form F_p until every p-Sylow subgroup $\mathcal{G}_{p_i}$ has been determined as a product of its cyclic factors. That is,

$$\mathcal{G}_{p_i} = C(p_i^t) \times C(p_i^u) \times \cdots \tag{29}$$

with $t + u + \cdots = \alpha_i$. For brevity, we cannot discuss techniques in detail here. Suffice it to say (A) that usually only a few forms F_p are needed to compute the complete structure, and (B) the whole process has been encoded into the same computer program (called CLASNO) that finds the match (20).

For the verification question ($h = h'$), if it would happen that the original e satisfied $e < h'/1000$, and that will almost never happen in practice,[3] there would be a doubt that $h = h'$. In that case we would have to go through another search (20) with some other form F_p. (We could take advantage of the now known factor $e \mid h$ to increase the baby step b, and thereby diminish the computations for the new search (20).)

There is another minor point concerning $h = h'$ that is discussed in Appendix 3. Appendix 2 deals briefly with the modifications needed for real fields and Appendix 4 deals briefly with questions of technique.

4. Several examples

We show a few examples, mostly taken from [4], that are of interest both for their extreme values of $L(1, \chi_\Delta)$, as was discussed in the introduction, and as examples of factorization to be discussed in our next section. Consider these discriminants $-\Delta$:

	Δ	$h(-\Delta)$	$L(1, \chi_\Delta)$
(30)	328878692999	1499699	8.21554
(31)	85702502803	16259	0.17448
(32)	84148631888752647283	496652272	0.17009
(33)	928185925902146563	52739552	0.17198

[3]See footnote, p. 420 for an exception.

The Δ in (30) is the smallest satisfying $(-\Delta/p) = +1$ for all $p = 2, 3, \ldots, 127$. It, of course, has a large value of $L(1, \chi_\Delta)$; in fact, it is the largest presently known to us. The Δ in (31) is a prime and its $L(1, \chi_\Delta)$ is the smallest presently known to us for a negative prime discriminant. The smallest $L(1, \chi_\Delta)$ presently known to us is listed as (32) and has a much larger, composite discriminant. The Δ in (33) happens to be the first really large Δ examined by this method. It also is composite, and we will show presently how it was factored.

5. A theory of factorization

The *ambiguous* forms $\mathfrak{A}$ are those special forms (3) that satisfy one of the three conditions:

$$B = 0, \qquad B = A \quad \text{or} \quad C = A$$

when reduced. That is,

(34) $$\mathfrak{A} = (A, 0, C), \qquad (A, A, C) \quad \text{or} \quad (A, B, A).$$

Each $\mathfrak{A}$ yields a factorization of Δ. Specifically, the three types in (34) give the factorizations

(35) $$\Delta = 4AC, \qquad \Delta = A(4C - A), \qquad \Delta = (2A - B)(2A + B).$$

Conversely, every factorization $\Delta = fg$, with f prime to g, corresponds to a single ambiguous form. For example, if $\Delta \equiv 3 \pmod 4$ a factorization having

$$3f < g, \qquad (f, g) = 1$$

corresponds to the reduced form

(34a) $$\mathfrak{A} = (f, f, \tfrac{1}{4}(f + g)),$$

while one having

$$f < g < 3f, \qquad (f, g) = 1$$

corresponds to the reduced form

(34b) $$\mathfrak{A} = (\tfrac{1}{4}(f + g), \tfrac{1}{2}(g - f), \tfrac{1}{4}(f + g)).$$

The first type in (34) cannot occur for Δ odd since then B is odd also. Since our forms are all primitive there are no forms corresponding to any factorization having $(f, g) > 1$ if Δ is odd. (See (42) below for an exception with Δ even.) If Δ is square-free, the set of all $\mathfrak{A}$ yields all factorizations of Δ; otherwise it yields all factorizations having $(f, g) = 1$, in particular, all factorizations into prime-powers.

Each ambiguous form $\mathfrak{A}$ satisfies

(36) $$\mathfrak{A}^2 = I$$

and is, therefore, a square-root of the identity. The number of $\mathfrak{A}$ equals the number of genera. To factor Δ, therefore, in effect is to determine the square-roots of the identity, and this may be accomplished if we compute the class number and the 2-Sylow subgroup.

Our method of factorization is this: we compute $h(-\Delta)$ as in §2 and the 2-Sylow subgroup as in §3 with $p_i = 2$. This yields the complete set of $\mathcal{A}$, and therefore the complete factorization of Δ (into prime-powers).

If

$$h(-\Delta) = 2^S \cdot T \tag{37}$$

with T odd, the mapping

$$G = F^T \tag{38}$$

maps the h classes F onto the 2^S classes that constitute the 2-Sylow subgroup $\mathcal{S} = \mathcal{G}_2$. Suppose $\mathcal{S}$ is a product of r cyclic groups $C(2^{s_i})$ having generators F_i of order 2^{s_i}. One has

$$S = \sum_{i=1}^{r} s_i. \tag{39}$$

There are 2^r genera and 2^r forms $\mathcal{A}$. While $\mathcal{S}$ is given by

$$\mathcal{S} = \prod_i F_i^{\alpha_i} \tag{40}$$

for arbitrary α_i, the 2^r ambiguous forms are given by

$$\mathcal{A} = \prod_i F_i^{\alpha_i} \tag{41}$$

for $\alpha_i = 2^{s_i}$ or $2^{s_i - 1}$. But it is generally not necessary to determine all the F_i explicitly for our purpose. Given any F, we compute its image G by (38) and then square repeatedly: $G, G^2, G^4, \ldots$ until we obtain an $\mathcal{A}$. We call this form the $\mathcal{A}$ *appertaining* to F.

Consider (30) above. We have at once that $S = 0$; there is only one genus and one ambiguous form:

$$I = (1, 1, \tfrac{1}{4}(1 + \Delta)).$$

This yields only the degenerate factorization: $\Delta = 1 \cdot \Delta$. Therefore, and also by known theory, Δ must be an odd power of a prime p^{2k+1}. But if $k > 0$ it is also known that $p \mid h$. By the easy check $(\Delta, h(-\Delta)) = 1$ in this case, we therefore find that Δ is prime. This technique is therefore a primality test which requires only $O(\Delta^{1/4})$ operations to perform. Further, if Δ had *not* been prime we could have continued and factored it. This is an important feature not included in many primality tests. Similarly, the Δ of (31) is prime, as was already asserted, while those in (32) and (33) must be composite. Before we factor them let us examine the even discriminant $-4S_{36}$ in (5).

Here we have $h = 2^{10} \cdot 49674463$ and therefore $S = 10$. But the F of (5a) is a generator, as was stated. Therefore, $\mathcal{S} = C(2^{10})$, and the only $\mathcal{A}$ besides

$$I = (1, 0, S_{36})$$

is the "parasitic" ambiguous form:

$$F^{h/2} = (2, 2, \tfrac{1}{2}(S_{36} + 1)) \tag{42}$$

which gives a factorization of $4S_{36}$ but not of S_{36}. Since there is no other $\mathcal{A}$, S_{36} is a prime-power, and, as before, a prime.

Now consider (33). One has $S = 5$, $T = 1648111$, and a priori all one knows is that $\mathcal{S}$ is *one of the seven* Abelian groups of order 32. We apply the mapping (38) to the first F_p in (7) which is

$$F = F_{73} = (73, 71, 3178718924322437).$$

As was indicated, we compute $G = F^T$, G^2 and then G^4 which is ambiguous. It is the $\mathcal{A}$ appertaining to F:

$$\mathcal{A} = F^{4 \cdot 1648111} = (1189633, 1189633, 195057496961).$$

This gives us a factorization

$$\Delta = 1189633 \cdot 780228798211 \tag{43}$$

together with the knowledge that G is of order 8. This eliminates several of the aforementioned seven possibilities, and we now must have

$$\mathcal{S} = C(32) \quad \text{or} \quad C(16) \times C(2) \quad \text{or} \quad C(8) \times C(4) \quad \text{or} \quad C(8) \times C(2) \times C(2). \tag{44}$$

Until we determine the correct structure we do not know if the factorization (43) is complete. We could now select another F_p to find its $\mathcal{A}$, but additional theory saves time. The Jacobi symbols

$$\left(\frac{73}{1189633}\right) = \left(\frac{73}{780228798211}\right) = +1$$

suggest that F_{73} is in the principal genus, and therefore is a square. If so, its cycle of order 8 could be enlarged by choosing $F' = \sqrt{F}$ instead. In that case, the last two possibilities in (44) also vanish. We reach into our stock (7) and select the form

$$F_{199} = (199, 101, 1166062720982609)$$

since

$$\left(\frac{199}{1189633}\right) = -1,$$

and F_{199} is *not* in the principal genus. We now find that F_{199}^T is of order 32 and appertains to the same $\mathcal{A}$. Therefore, $\mathcal{S} = C(32)$ and no other $\mathcal{A}$ (except I) exists. The factors in (43) are clearly not squares, and checking $(\Delta, h(-\Delta)) = 1$, as before, we conclude that (43) is the complete factorization of Δ.

The factorization of (32) is more elaborate. One has $S = 4$ and finds that $\mathcal{S} = C(2) \times C(2) \times C(4)$ with 8 genera and $8\mathcal{A}$. But any three independent $\mathcal{A} \neq I$, that is, any three satisfying

$$\mathcal{A}_1 \neq \mathcal{A}_2 \cdot \mathcal{A}_3,$$

would suffice to find the complete factorization:

$$\Delta = 6079 \cdot 30469 \cdot 132137 \cdot 3438209.$$

We forego the details. All these techniques have been mechanized and are incorporated into the program CLASNO mentioned above. In §7 we will return to one feature of considerable theoretical interest.

6. The method characterized

For factoring very small Δ we would not want to use such a sophisticated method, but for large Δ it must be taken seriously. We may characterize the new method as follows.

A. The method is $O(\Delta^{1/4+\epsilon})$, that is, it is $O(\Delta^{1/4})$ except perhaps for a low-order logarithmic factor. A factorization of Δ based on trial-and-error divisors, or one based on finding a representation $\Delta = x^2 - y^2$ by trial-and-error substitutions y requires $O(\Delta^{1/2})$ operations. It is true that much technique has been developed to improve these two methods through the use of exclusion moduli, sieves, etc., but that merely reduces the coefficient in $O(\Delta^{1/2})$, the order remains unchanged. In the new method, the coefficient for $O(\Delta^{1/4})$ is relatively large since composition (Appendix 1) is rather involved. Nonetheless, beyond some "crossover" Δ the present method must become faster. One cannot say flatly what that Δ is since it depends entirely on the efficiency of the routines and equipment used. But, even now, a 20-digit number can be class-numbered in a few minutes, and we must expect that with faster routines and machines the lower order $O(\Delta^{1/4})$ will prevail.

B. The new method gives a *complete* factorization: the factors found require no primality test—if they were composite there would be other ambiguous forms. The method is *both* a primality test and a factorization technique. The complication mentioned above concerning prime-power factors is not a serious one.

C. The trial divisor method on a $\Delta = fg$ with $f < g$ works fastest when the ratio f/g is small, while the method based on $\Delta = x^2 - y^2$ works fastest if $f/g \approx 1$. The new method is indifferent to this ratio and treats both cases simultaneously; in the first case, the ambiguous form, when reduced, is of type (34a), and in the second case, of type (34b).

D. The classical factorization methods are fundamentally trial-and-error. One works in the dark and has no idea where the factors are, or, sometimes, even if they exist. That is clearly unsatisfactory. Scientifically, and also aesthetically, the new method has certain merits. By the use of the group structure, and $h = 2^S T$, we can determine an "address" F^T where we then go to find the desired factorization. True, some trial-and-error remains in searching for the match (20), but there is a lesser amount both because of our very good estimate of h, and because h is already $O(\Delta^{1/2+\epsilon})$. That the individual operations now appear more meaningful, relates, of course, to the lower order $O(\Delta^{1/4})$ already discussed.

All these features A through D stem from our use of the group. The individual factorizations of, say, $\Delta = fgk$, namely, $f \cdot (gk)$, $g \cdot (fk)$, etc., are "lying around loosely", so to speak. By imbedding the corresponding ambiguous forms (34) into the structure of *all* the forms (3), we create order where it was previously lacking.

7. The dominant factorization

Consider the example $h(-71837718283) = 15204$ mentioned in §1. One finds that $\mathcal{S} = C(2) \times C(2)$. There are four ambiguous forms, $I = \mathcal{A}_0$,

$$\mathcal{A}_1 = (281, 281, 63912631),$$
$$\mathcal{A}_2 = (3709, 3709, 4843049),$$
$$\mathcal{A}_3 = (68927, 68927, 277789).$$

One finds, for this group $C(2) \times C(2)$, that the mapping $G = F^{3801}$ plays no favorites. Exactly 1/4 of the h forms F appertain to each $\mathcal{A}$. An F chosen at random will, one time out of four, appertain to I and yield only the useless $\Delta = 1 \cdot \Delta$. The remaining choices would yield, equally, the remaining $\mathcal{A}$ and the three proper factorizations: $281 \mid \Delta$, etc.

But for some Δ there is a *dominant* factorization, a concept that does not occur in other theories. Suppose, for simplicity, that $\mathcal{S} = C(2) \times C(4)$. There is a $G = G_1$ of order 4 and a $G = A_1$ of order 2. The group $\mathcal{S}$ is the product of these two cycles and has the cycle graph (see [7]) shown in Figure 1.

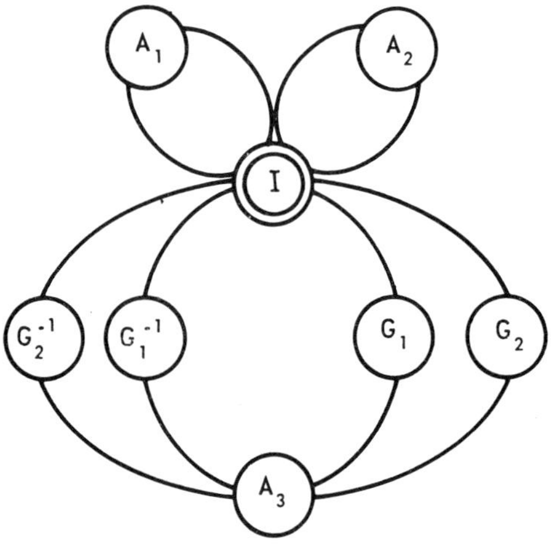

FIGURE 1

In this circumstance, 5/8 of the forms F appertain to A_3, namely, those having $G = G_1^{\pm 1}$, $G_2^{\pm 1}$, and $G = A_3$ itself, the latter being the common square of the other four forms mentioned. Only 1/8 of the forms F appertain to each of the other three $\mathcal{A}$. The factorization corresponding to A_3 is therefore dominant. Such dominance would be much more extreme if $\mathcal{S} = C(2) \times C(2^n)$ for a large n, or for any 2-Sylow subgroup $\mathcal{S}$ having a highly dominant cyclic factor.

Is this a serious defect of our method? If, probabilistically, successive forms F_p from (7) repeatedly appertain to the same $\mathcal{A}$, how will we locate the others? An especially gruesome example is

$$h(-4 \cdot 10721) = 128$$

which has $\mathcal{S} = C(2) \times C(64)$. The dominant $\mathcal{A}$, which appertains to 125 of the 128 forms, is

$$\mathcal{A}_1 = (2, 2, 5361),$$

a parasitic form similar to (42) that only factors $4 \cdot 10721$ and not 10721 itself. Only 2 times out of 128, only by hitting directly upon

$$\mathcal{A}_2 = (71, 0, 151) \quad \text{or} \quad \mathcal{A}_3 = (111, 80, 111)$$

would we really factor $10721 = 71 \cdot 151$. What can be done about such highly recessive forms $\mathcal{A}$?

The solution is very simple. The very fact that one $\mathcal{A}$ is dominant, such as A_3 in Figure 1, means that a number, or even very many, $G = F^T$ will lead to this $\mathcal{A}$ through different paths. We thereby encounter distinct square-roots such as

$$G_1^2 = G_2^2 = A_3 \tag{45}$$

in Figure 1. Now, consider the composite

$$P = G_1 \cdot G_2^{-1}. \tag{46}$$

Since $P^2 = I$ because the group is Abelian, P must be ambiguous. But $P \neq I$ since $G_1 \cdot G_1^{-1} = I$, and $P \neq A_3$ since $G_1 \cdot G_1 = A_3$. Therefore, P is one of the desired recessive forms A_1 or A_2.

Further, in general the right side of (45) need not be itself ambiguous. Any relation

$$F_1^2 = F_2^2 = F_3$$

suffices since $F_1 \cdot F_2^{-1}$ is ambiguous. In this way we may easily uncover any recessive $\mathcal{A}$ that occur, and this compositing technique (46) has also been incorporated into the program CLASNO.

The generalization of this technique to other p-Sylow subgroups is also clear. If, for example, we have $C(3) \times C(81)$, there will be a similar dominance and a similar remedy for computing the recessive solutions of $X^3 = I$. From $F_1^3 = F_2^3 = F_3$ we compute $X = F_1 \cdot F_2^{-1}$.

8. Briefly, other p-Sylow subgroups

Just as the 2-Sylow subgroup relates to factorization and genera, the 3-Sylow subgroup, for certain discriminants, relates to the set of quadratic forms representing all primes that have a prescribed composite integer a as a cubic residue. For example, cf. Dedekind [8], one has $h(-4 \cdot 243) = 9$ with a group $C(3) \times C(3)$. There are four subgroups of order 3, not merely one as would be the case if the group were cyclic. One of these four, comprising $I = (1, 0, 243)$ and the forms $(9, \pm 6, 28)$, represents all primes having 2 as a cubic residue. Another: I and $(4, \pm 2, 61)$ has 3 as a cubic residue; the third: I and $(7, \pm 6, 36)$ has 6; and the last: I and $(13, \pm 4, 19)$ has both $12 = 4 \cdot 3$ and $18 = 2 \cdot 9$ as cubic residues. The discriminant $-4 \cdot 675$, with a group $C(3) \times C(6)$, may be used similarly for $a = 2$, 5, 10, 20, and 50.

It would be digressive to continue this point, but it is not digressive to indicate that the techniques developed above enable us to readily compute the $\mathcal{G}_p$ of (29) for any p, even when Δ is quite large. For example,

$$\Delta = 2^{41} - 1$$

has $39 \mid h(-\Delta)$ by (9). That assists us in computing

$$h(1 - 2^{41}) = 16 \cdot 3 \cdot 13 \cdot 61 \cdot 25. \tag{47}$$

To factor this Δ we would examine the factor 16 in (47) but now we are more interested in the factor 25. It corresponds to the noncyclic subgroup $C(5) \times C(5)$, a structure that is quite rare for quadratic fields. While it took some computation to discover these $C(5)$ factors, it should be noted that their a posteriori *verification* is relatively simple. Thus,

$$F^{\pm 1} = (243235, \pm 136997, 2279474), \qquad F^{\pm 2} = (429619, \mp 305771, 1334042),$$

and

$$G^{\pm 1} = (285283, \pm 160491, 1949626), \qquad G^{\pm 2} = (681100, \pm 418857, 871555),$$

may be shown to all be of order 5 by computing $F^3 = F^{-2}$ and $G^3 = G^{-2}$, and to all have the correct Δ. That suffices to prove that $C(5) \times C(5)$ is contained in this class group.

Recently, there has been much interest in such noncyclic factors and no doubt the foregoing methods will be useful in their investigation. An interesting example substantiating this prediction was discovered shortly after this was written. For the prime $P = 188184253 = 3^6 + 4 \cdot 19^6$, the field $Q(\sqrt{(-3P)})$ has a class group:

$$C(3) \times C(3) \times C(3) \times C(604).$$

Its 3-Sylow subgroup has three factors, and no previously known quadratic field had a p-Sylow subgroup with more than two factors for any $p > 2$. In a forthcoming paper, Peter Weinberger and I prove that for $\Delta = A^6 + 4B^6$ the two fields $Q(\sqrt{\Delta})$ and $Q(\sqrt{(-3\Delta)})$ have 3-Sylow subgroups with an equal number of factors if, and only if, $3 \nmid B$. It follows that the real field $Q(\sqrt{P})$ also has a 3-Sylow subgroup with three factors. In fact, $h(P) = 27$, and one finds the elegant class group $C(3) \times C(3) \times C(3)$.

At the 1962 International Congress, Šaferevič noted that all known quadratic fields having a prime discriminant had cyclic groups, or, at worst, class groups requiring two generators. He raised the question whether there is not a bound on the number of such generators if the discriminant is prime, and subsequently other investigators conjectured (verbally) that there is such a bound, and, in fact, that it equals 2. But $Q(\sqrt{P})$ now shows that this is false.

9. And genera

In §1 we indicated how Lehmer's sieve generated numbers $-\Delta$ having a prescribed quadratic character, and how these discriminants forced us to develop a new method of computing $h(-\Delta)$. Now that this has been done, we find to our pleasure that this new method may also be used to generate all prime solutions Δ themselves. Some composite Δ also appear.

Consider the highly composite product of odd primes:

$$D = \prod_{3}^{61} p. \tag{48}$$

The 17 factors 3, ..., 61 correspond to 17 characters (n/p), and since $D \equiv 3 \pmod 8$ in this example (48), one knows that there are 2^{16} genera in $Q(\sqrt{-D})$ and therefore that $2^{16} \mid h(-D)$. In fact, one finds

$$h(-D) = 2^{16} \cdot 811263 = 53166931968, \tag{49}$$

so there are 811263 classes in each of the 2^{16} genera.

We must digress briefly. Returning to the verification question ($h = h'$) discussed at the close of §§2 and 3, for a highly composite Δ such as (48) the 2-Sylow subgroup $\mathcal{S}$ has many cyclic factors, and for any form F the ratio $m = h/e$ of (13) *must* be large. For example, for (48) we know a priori that $m = 2^{15}M$ for some M since the e forms F^x can contain at most one cyclic factor of $\mathcal{S}$. Previously, we were concerned if m were too large, specifically, if $m > 1000$, but for a Δ such as (48) we need only be concerned if M is large. One can write $(e \cdot 2^{15})M' = h'$ and the multiple $2^{15}e$ is known unequivocally. Only a large M' would cause trouble.

Now consider a new discriminant $-64D = -16 \cdot 4D$ obtained from the D of (48). Dirichlet's Table of Characters, cf. Mathews [9, §130], indicates that the first factor of 4 merely triples the number of *primitive* classes in each genus, and (49) becomes

$$h(-4D) = 2^{16} \cdot 3 \cdot 811263 = 2^{16} \cdot 2433789. \tag{50}$$

But the subsequent factor of 16 introduces two more characters:

$$\chi = (-1)^{(n-1)/2}, \qquad \psi = (-1)^{(n^2-1)/8} \tag{51}$$

and thereby splits each genus G into four genera. It thus splits all odd numbers n represented by G into four sets according as

$$n \equiv 1, 3, 5, \text{ or } 7 \pmod 8).$$

We now have 2^{18} genera and

$$h(-64D) = 2^{18} \cdot 2433789. \tag{52}$$

Now let us introduce a handy notation. If an odd $n \equiv a \pmod 8$ is a quadratic residue for all odd primes 3, 5, ..., p we will say that it is a

$$+aR_p$$

number; if it is a nonresidue for all such p it is a

$$+aN_p$$

number; and if $-n$ is such a residue or nonresidue then n is a

$$-aR_p \quad \text{or} \quad -aN_p$$

number. Since a can be 1, 3, 5, or 7, we therefore have defined 16 different types of numbers for any p. For example,

$$(53) \qquad (-30493/p) = -1 \qquad (p = 3, 5, \ldots, 61)$$

and 30493 is therefore a $-5N_{61}$ number.

From this 30493 we can construct a form

$$(54) \qquad F = (30493, 18442, 30771227851482499357)$$

that represents 30493 and has the discriminant $-64D$. All numbers n represented by F that are prime to $64D$ must also be $-5N_{61}$ numbers and so F represents infinitely many such n. We may also say that F lies in the $-5N_{61}$ genus since this symbol completely defines its character.

Upon examination, we now find that F is of order

$$2 \cdot 2433789.$$

Its even powers F^{2x} all lie in the principal genus, and it may be verified that this is $+1R_{61}$. All n represented by F^{2x} that are prime to $64D$ satisfy

$$(n/p) = +1 \qquad (p = 3, 5, \ldots, 61)$$

and

$$n \equiv 1 \pmod 8.$$

These n are the so-called *pseudosquares*, [4]. On the other hand, all odd powers F^{2x+1} are in the $-5N_{61}$ genus.

Since D has an odd number of prime factors $q \equiv -1 \pmod 4$ and an even number of $q \equiv +1 \pmod 4$, the numbers n in either genus have a Jacobi symbol:

$$(n/D) = +1.$$

It follows that $(-D/n) = +1$, and $-D$ is a quadratic residue for prime n, and also for those composite n where $(-D/p) = +1$ for every prime factor p of n. Therefore, the single $-5N_{61}$ number 30493 leads us to a form F whose powers comprise *all* the forms that represent all the primes $-5N_{61}$, and *all* forms that represent all the primes $+1R_{61}$.

Now $-5N_{61}$ and $+1R_{61}$ are merely 2 out of the 2^{18} genera in (52). Suppose we wished to generate $-3N_{61}$ primes, or $+7R_{61}$ primes, or others of the 16 "pure" types $\pm aR_{61}$ and $\pm aN_{61}$ defined above. Two questions arise. Which of these 16 types are actually genera in (52), and, if they do exist there, how can we locate them among the

$$2^{18} \cdot 2433789 = 638003183616$$

primitive forms of this discriminant? Both questions have very pretty answers, and a "shuttling" process using ambiguous forms gives us an astonishingly efficient solution of the second problem.

The ambiguous form of genus $-5N_{61}$ is

(55) $$F^{2433789} = (78084191920, 78084191920, 12036128094853).$$

By the transformation

(56) $$x = u - v, \qquad y = u + v$$

this form becomes a binomial

(57) $$G(x, y) = 4Ax^2 + By^2$$

with

(58) $$\begin{aligned} A &= 4880261995 &&= 5 \cdot 7 \cdot 17 \cdot 43 \cdot 53 \cdot 59 \cdot 61, \\ B &= 12016607046873 &&= 3 \cdot 11 \cdot 13 \cdot 19 \cdot 23 \cdot 29 \cdot 31 \cdot 37 \cdot 41 \cdot 47. \end{aligned}$$

The partition (58) of the product (48) has a remarkable property. For any odd prime q,

$$q \mid B \text{ implies } (-A/q) = -1, \qquad q \mid A \text{ implies } (-B/q) = -1.$$

It follows that any number $n = G(x, y)$ satisfying $(x, 2B) = (y, 2A) = 1$ will be prime to $64D$ and satisfy

$$(-n/q) = -1 \qquad (q = 3, \ldots, 61).$$

Since $A \equiv 3$ and $B \equiv 1 \pmod 8$, one also verifies that $n \equiv 5 \pmod 8$ and therefore n is a $-5N_{61}$ number. This we already knew since $G = F^{2433789}$.

But now consider the ambiguous forms

(59) $$\begin{gathered} \mathcal{A}_2 = Au^2 + 16Bv^2, \qquad \mathcal{A}_3 = 16Au^2 + Bv^2, \\ \mathcal{A}_4 = A(u - v)^2 + 4B(u + v)^2, \qquad \mathcal{A}_5 = 16u^2 + ABv^2, \\ \mathcal{A}_6 = (u - v)^2 + 4AB(u + v)^2, \qquad \mathcal{A}_7 = 4(u - v)^2 + AB(u + v)^2, \\ I = \mathcal{A}_0 = u^2 + 16ABv^2. \end{gathered}$$

These seven ambiguous forms clearly all have the discriminant $-64D$, and, it may be similarly verified, lie, respectively, in the genera

$$-3N_{61},\ -1N_{61},\ -7N_{61},\ +3R_{61},\ +5R_{61},\ +7R_{61}, \text{ and } +1R_{61}.$$

Since F^2 is of order 2433789 and of the principal genus $+1R_{61}$, and since any of the six $\mathcal{A}_i$, $(i = 2, \ldots, 7)$ of (59) are of order 2, it follows that the composites

(60) $$F_i = F^2\mathcal{A}_i$$

are of order $2 \cdot 2433789$. Therefore, each one generates, respectively, its entire genus by F_i^{2x+1}, and they all have the principal genus F_i^{2x} as their common square.

Note, as in our previous work, that the class number is essential. We use it here to locate the ambiguous form (57). By simply permuting factors of 4, 16, and A in this form we can shuttle over to any other track (60). The single number 30493 therefore leads us to these other genera also.

We have now worked out the details of the case (48). If we had instead

$$(61) \qquad D_1 = \prod_3^{59} p \quad \text{or} \quad D_2 = \prod_2^{59} p$$

we would correspondingly find

$$h(-64D_1) = 2^{17} \cdot 605367, \qquad h(-16D_2) = 2^{17} \cdot 393620.$$

But different D will give us different selections of the 16 pure genera $\pm aR_p$ and $\pm aN_p$ as occurring genera. Let us characterize a particular

$$(61a) \qquad D = \prod_3^{p} q \quad \text{or} \quad D = \prod_2^{p} q$$

by the symbol $x_1x_2x_3$ where x_2 is E or O according as the product comprises an even or odd number of primes $4k + 1$, where x_3 is E or O according to the number of primes $4k - 1$, and where x_1 is O or E according as the factor 2 occurs or not. Then there are eight varieties of D. That of (48) is EEO, D_1 of (61) is EOO, and D_2 is OOO.

One may verify—we omit the details—by using Dirichlet's Table of Characters [9] that each of these eight varieties of D yield eight of the sixteen pure genera. This is shown in Table 1.

Table 1

Type	Genera Occurring
EEE	$\pm aR, \pm aN, \quad (a = 1, 5)$
OEE	$\pm aR, \pm aN, \quad (a = 1, 3)$
EOE	$\pm 1R, \pm 5R, \pm 3N, \pm 7N$
OOE	$\pm 1R, \pm 3R, \pm 5N, \pm 7N$
EEO	$+aR, -aN, \quad$ (all a)
OEO	$+aR, +bN, -bR, -aN, \quad (a = 1, 7; b = 3, 5)$
EOO	$+aR, +aN, \quad$ (all a)
OOO	$+aR, +aN, -bR, -bN, \quad (a = 1, 7; b = 3, 5)$

Thus, for (48) we obtained the 8 genera listed for EEO, and if we wanted the genus $+1N$, say, we would need a different D, say, $\prod_3^{67} p$ of type EEE. Note that the principal genus $+1R$ always occurs, and 14 of the remaining pure types are equidistributed, but (curiously) $-7R$, the so-called "negative squares" [4], never occurs. Not that $-7R$ is without interest; our example (30) is of this character.

The shuttling process cannot be used for every D. For the D_1 of (61), the process works just as before with the partition analogous to (58) being

$$A = 7 \cdot 11 \cdot 13 \cdot 29 \cdot 31 \cdot 37 \cdot 43 \cdot 53,$$
$$B = 3 \cdot \ 5 \cdot 17 \cdot 19 \cdot 23 \cdot 41 \cdot 47 \cdot 59.$$

But for D_2 there are 393620 classes per genus, an even number. Therefore, if F^x

generates all $-3N_{59}$ and $+1R_{59}$ forms, the ambiguous F^{393620} must be in $+1R_{59}$, not $-3N_{59}$. Alternatively, if $-3N_{59}$ has an ambiguous form in its own genus, it cannot have a generator of the entire genus. In fact, one finds that both $-3N_{59}$ and $-5N_{59}$ have the first behavior described: each has a generator and neither an ambiguous form. These generators lead to

$$F^{393620} = G^{393620} = (A, 0, 4B)$$

with

$$A = 3 \cdot 11 \cdot 17 \cdot 23 \cdot 37 \cdot 41 \cdot 47 \cdot 53,$$
$$B = 2 \cdot \ 5 \cdot \ 7 \cdot 13 \cdot 19 \cdot 29 \cdot 31 \cdot 43 \cdot 59.$$

This precludes any shuttling between $-3N_{59}$ and $-5N_{59}$ here, but does yield this second $\mathcal{A}$ in $+1R_{59}$ (besides I) which represents some much smaller pseudosquares than I does.

The $-3N_p$ are of special interest because of their relation to the question of small $h(-\Delta)$, and we note that this genus occurs for exactly one case in (61a) for every p. It occurs for

$$D_3 = \prod_3^7 q \quad \text{and} \quad D_4 = \prod_2^{11} q,$$

and both $-64D_3$ and $-16D_4$ have two classes per genus. In the first case, the $-3N_p$ are both ambiguous: (28, 28, 67) and (43, 26, 43), while in the second neither is: (67, ± 52, 148). In both cases, we generate every prime $-3N_{11}$, including $\Delta = 67$, 163 having $h(-\Delta) = 1$, and $\Delta = 883$, 907 having $h(-\Delta) = 3$. But, of course, the two genera represent different sets of composite $-3N_{11}$, and therefore different sets of Δ having even $h(-\Delta)$: only the first represents $\Delta = 19 \cdot 73$ having $h(-\Delta) = 4$, only the second represents $\Delta = 17 \cdot 131$ having $h(-\Delta) = 6$, while both represent $\Delta = 13 \cdot 271$ having $h(-\Delta) = 6$.

In contrast with Lehmer's DLS 127 we do *not* obtain the primes $-\Delta$ in numerical order, but we do see how they are all algebraically related. As for efficiency, one could easily compute any number of $-3N_{61}$, say, by the form $\mathcal{A}_2$ of (59), and then, with the same program, estimate their $h(-\Delta)$ by (6), and compute the $L(1, \chi_\Delta)$ only of those that are known a priori to be especially small. Further, among so many $-3N_{61}$ there would be some $-3N_{67}$ primes, some $-3N_{71}$, etc., and one could escalate the whole process. Alternatively, one could select starting values from the tables in [4].

10. Acknowledgment

I should like to acknowledge a very long and stimulating correspondence that I carried on with D. H. and Emma Lehmer. All of the foregoing ideas were developed in these letters. A program named SPEEDY for computing the estimate (6) was written by them, as was the first version of CLASNO. Later, with the assistance of Richard Serafin, I wrote a very elaborate, improved CLASNO, including more efficient routines for composition and reduction, the baby step-giant step strategy, and allowing innumerable options for analyzing the p-Sylow subgroups, selecting or ignoring forms (7), compositing as in (46), etc. A very large number of examples were produced by the Lehmers with the DLS 127 and we have discussed these in our joint paper [4].

Appendix 1. Composition and reduction

Since composition has been seldom used in computation recently, and since our version of it here was specifically designed to be most efficient numerically, we include this algorithm for the reader's convenience. The writeup proves commutativity: $F_1F_2 = F_2F_1$ although in the algorithm itself that property is not apparent. Further, we show that if F_1 represents A_1 and F_2 represents A_2 then F_1F_2 represents A_1A_2, the essential multiplicative property.

To compose F_1F_2 from two primitive forms

$$F_1 = (A_1, B_1, C_1), \qquad F_2 = (A_2, B_2, C_2)$$

having

$$B_1^2 - 4A_1C_1 = B_2^2 - 4A_2C_2 \tag{62}$$

set $0 < A_1 \leqq A_2$ without loss of generality. Let $S = \frac{1}{2}(B_1 + B_2)$, $N = B_2 - S$. If $A_1 \mid A_2$, take $Y_1 = 0$, $D = A_1$; otherwise compute $(A_1, A_2) = D$:

$$A_2Y_1 - A_1\mathbf{X}_1 = D. \tag{63}$$

Here, and subsequently, bold quantities such as $\mathbf{X}_1$ are not used in the algorithm and are included only for a symmetric exposition. If $D \mid S$, take $Y_2 = -1$, $X_2 = 0$, $D_1 = D$; otherwise compute $(D, S) = D_1$:

$$SX_2 - DY_2 = D_1. \tag{64}$$

Now

$$(D_1, B_1, A_1C_1/D_1) \quad \text{and} \quad (D_1, B_2, A_2C_2/D_1)$$

are inverses, their product is I, so $F_1F_2 = (V_1, B_1, \mathbf{C}_1\mathbf{D}_1)(V_2, B_2, X)$ where

$$V_1 = A_1/D_1, \qquad V_2 = A_2/D_1, \qquad X = C_2D_1. \tag{65}$$

Let

$$M = Y_1Y_2N - X_2C_2, \qquad \mathbf{M}_0 = X_1Y_2N - X_2C_1.$$

Then, from (62)–(65), $B_2 + 2V_2M = B_1 + 2V_1\mathbf{M}_0$. Let

$$M = QV_1 + R, \tag{66}$$

and set $B_3 = B_2 + 2V_2R = B_1 + 2V_1\mathbf{R}_0$ where $\mathbf{R}_0 = \mathbf{M}_0 - QV_2$. Let

$$Y = X + R(B_2 + V_2R), \qquad \mathbf{Y}_0 = \mathbf{C}_1\mathbf{D}_1 + \mathbf{R}_0(B_1 + V_1\mathbf{R}_0).$$

Then

$$\begin{aligned} &(V_1, B_1, \mathbf{C}_1\mathbf{D}_1) \quad \text{is equivalent to } (V_1, B_3, \mathbf{Y}_0),\\ &(V_2, B_2, X) \qquad \text{is equivalent to } (V_2, B_3, Y), \end{aligned}$$

and since it may be shown by (62)–(64) that $V_1 \mid Y$, say,

$$Y = C_3V_1, \tag{67}$$

the wanted composition is

$$F_1F_2 = (A_3, B_3, C_3) \tag{68}$$

where $A_3 = V_1V_2$. The composite (68) may now require reduction.

Reduction is certainly well known but here, again, certain formulas are more efficient numerically than others. $F = (A, B, C)$ is reduced for *negative* discriminants if $|B| \leqq A \leqq C$. If F is not, compute an equivalent (A_1, B_1, C_1) by

$$B_1 + 2DC = -B, \qquad C_1 = A + \tfrac{1}{2}(B - B_1)D, \qquad A_1 = C, \tag{69}$$

and iterate until the form is reduced.

Appendix 2. Briefly, positive discriminants

For *any* nonsquare positive D, the Gaussian analogue of (2) is

$$\sum_{k=0}^{\infty}\left(\frac{D}{2k+1}\right)\frac{1}{2k+1} = \prod_{p=3}^{\infty}\frac{p}{p-(D/p)} = \frac{h(4D)}{(4D)^{1/2}}\log(T + U\sqrt{D}) \tag{70}$$

where the Jacobi and Legendre symbols are in the sum and product, where $h(4D)$ is the number of equivalence classes of primitive forms having discriminant $4D$, and where $T + U\sqrt{D}$ is the smallest solution >1 of

$$T^2 - DU^2 = +1. \tag{71}$$

We omit, for brevity, the known relations of this $h(4D)$ to $h(D)$ for odd D and for real fields $Q(\sqrt{D})$.

There are two problems that arise in extending §2 to these positive D. First, modifying (6) we now need $\log(T + U\sqrt{D})$ instead of the universal π to make our estimate. For those (very common) D having a small $h(4D)$ and a large $(T + U\sqrt{D})$, the calculation of this latter could easily dominate the entire computation, since, in fact, it could require $O(D^{1/2})$ operations.

Closely related is the other problem. For positive D there may be (very) many reduced forms in every class, in particular in I. How then can we recognize the match (20) when we encounter it? We could form the composite there:

$$F^{H+rg} \cdot F^{-bn}, \tag{72}$$

and if we could recognize that this product is in I we would have it. But such recognition implies a computation essentially equivalent to that of $(T + U\sqrt{D})$.

For many D, then, these problems concerning the fundamental unit would nullify much of the efficiency of our method. But for some D the method can be nearly as efficient as before. Any D having a $(T + U\sqrt{D})$ that is known a priori eliminates the first problem entirely. For example,

$$D = m^2 + 1 \tag{73}$$

satisfies $m^2 - D = -1$ and therefore

$$\log(T + U\sqrt{D})/(4D)^{1/2} = \log(m + \sqrt{D})/\sqrt{D}. \tag{74}$$

Other simple classes of D could be cited. Further, the $(T + U\sqrt{D})$ in these cases is usually relatively small, and therefore I has only a few, easily recognized reduced forms. This largely eliminates the second problem also.

For $D \equiv 1 \pmod 8$ it is known that $h(4D) = h(D) =$ ideal class number for $Q(\sqrt{D})$. An example of (73) is

$$D = 2^{32} + 1, \tag{75}$$

a well-known Fermat number, and one how easily finds by these modifications of §2 that $h(2^{32} + 1) = 4320$ and the class group is $C(32) \times C(3) \times C(9) \times C(5)$. (Gauss indicated [10, p. 359] that he knew of no positive determinant with an odd "exponent of irregularity". But the $C(3) \times C(9)$ above shows that (75) has 3 as its exponent of irregularity. Ironically, Gauss paid special attention [10, p. 447] to this $2^{32} + 1$, since it is the first Fermat number not constructable by ruler and compass.)

The related factorization $F_5 = 2^{32} + 1 = 641 \cdot 6700417$ may be obtained from $h = 4320$ by the method of §5, but again with some modifications. Further, one now has an analogue of (9) in that

$$h(2^{2^n} + 1) \equiv 0 \pmod{2^{n-1} - 1}. \tag{76}$$

In our example (75), $15 \mid h(F_5)$, and we might add that this known factor $C(3)$ gives the aforementioned irregularity $C(3) \times C(9)$ a greater a priori probability.

Another very interesting class of D arc thc

$$D = S_n = (2^n + 3)^2 - 8 \tag{77}$$

that I introduced in [11]. I computed $h(S_{12}) = 27$ by composition there in a method that was the precursor of our present, more highly developed algorithm. For these D the log $(T + U\sqrt{D})$ can be computed by the asymptotic formula [17]:

$$\log (T + U\sqrt{S_n}) = 2n^2 \log 2 + O(n2^{-n}).$$

Further, though there are now many reduced forms in I they all can be recognized instantly [11]. We therefore have no difficulty in computing $h(S_{19}) = 807$ for the first prime S_n following S_{12}. (The prime S_n for $n > 19$ are $n = 27, 28, 32, 36$ (see (5)), 48, 56, 61, ... We will discuss the many remarkable properties of these S_n elsewhere [17].)

To extend our methods to algebraic fields other than quadratic would involve similar difficulties with the unit such as we have encountered here. To date, we have not examined such further extensions seriously.

Appendix 3. A bad estimate

Usually the estimate (6) with $P = 132000$ will be good to 1 part in 1000. Frequently, it is more accurate, but occasionally, somewhat less. If $(-\Delta/p)$ has an extremely biased distribution for the primes p following 132000, our estimate H would be less accurate, our choice of s in (23) would be poor, and the number of compositions needed in (24) would be unduly large. What is the density of the Δ such that an error $\geqq 5\%$ occurs? That is, such that

$$f(\Delta) = \prod_{P}^{\infty} \frac{p}{p - (-\Delta/p)} \tag{78}$$

satisfies

$$f(\Delta) \leqq 1/1.05 \quad \text{or} \quad f(\Delta) \geqq 1.05$$

P. D. T. A. Elliott has kindly computed a bound for me of the density of these (baddies, I think he calls them). He uses probabilistic number theory, specifically, Theorem 1 of his [12], and results contained in his [13]. Suppose

$$P > 2, \qquad 0 < z \leqq 2,$$

and let the density of *all* positive d for which

$$f(d) \leqq \frac{1}{1+z} \quad \text{or} \quad f(d) \geqq 1 + z \tag{79}$$

be designated as $F(P, z)$. Elliott finds that

$$F(P, z) \leqq 2A \exp(-BP[\log(1+z)]^2) \tag{80}$$

for constants A and B of the order of unity [12]. We can now confine ourselves to those $d = \Delta \equiv 0, -1 \pmod 4$ merely by replacing the coefficient 2 in (80) with 4. For $P = 132000$, an error $\geqq 5\%$, or $z = 1/20$, would be very rare indeed.

But suppose there were an even worse Δ having $f(\Delta) \approx 2$ or $f(\Delta) \approx 3$, and therefore suppose that our estimate (6) gave $H \approx \frac{1}{2}h$ or $H \approx \frac{1}{3}h$. The density of such Δ with this P is essentially zero (roughly, e^{-10^5}), and one presumes, but does not know, that the smallest such Δ would be so huge that our method for computing $h(-\Delta)$ would not be feasible. But if there were a sufficiently small Δ of this type a new problem might arise.

Our verification technique for $h = h'$ was to build up the class group by the choice of enough forms (7) until its order h' satisfied $h'/H \approx 1$. Suppose we encountered a case of $f(\Delta) \approx 3$, and suppose (more bad luck) that $3 \mid h$, and suppose (still more bad luck) that in our choice of forms (7) we just happened to hit upon a selection such that the apparent class group had order $h/3$—that is, such that one factor $C(3)$ just happened to elude us so far. Then we would erroneously conclude that this $h/3$ was the class number. (The possibility: $f(\Delta) \approx 2$ and conclusion $h/2$ would be "more" probable, but would be less likely to fool us since the 2-Sylow subgroup ties in so closely with factorization.)

Of course, this is not a practical question. But nonetheless a solution is perhaps available. The use of a constant P in (6) for all Δ is not an efficient choice. For small $h(-\Delta)$, say, $h = 20$, $P = 132000$ is needlessly large, while for Δ the size of (32) a larger P is perhaps already advisable since too little computation is performed in (6) in comparison with that needed for (20). A better balance would be preferable, and the reasonable choice would be to use a $P = O(\Delta^{1/4})$. If this were done, the envisaged counterexample would be even more improbable, and perhaps would be excluded.

Appendix 4. Briefly, techniques

Many improvement techniques, especially devised for use with very large Δ, have been, or could be, investigated. It would be inappropriate to discuss them at length here but they are too pertinent to omit entirely. We mention some of the more interesting very briefly.

a. Whether one uses a fixed P in (6), or a $P = O(\Delta^{1/4})$, the implication above is that one merely uses the final partial product. Actually, the sequence of partial products in this Euler product oscillates randomly around its limit, and presumably a Césaro transform, or some other average, would yield some added accuracy. Further, a Césaro transform would be pertinent for the theoretical problems raised in the previous appendix, and one may anticipate that it would improve the convergence there just as it does in the theory of the Fourier Series.

b. In storing the baby steps (16) one stores only A_n and B_n. They are smaller than C_n, and the latter is redundant anyway. Further, if A_n, B_n exceed one computer word in size, a storing and a matching (20) of only the least significant words of these A_n, B_n would suffice. One then recomputes the exact matching baby form F^{bn} as a check, if necessary.

c. To the present, values of several hundred for s have not been exceeded. Each baby form must be compared with each giant form F^{h+rg}. This means $O(\Delta^{1/4}) \cdot O(\Delta^{1/4}) = O(\Delta^{1/2})$ operations. But for only several hundred steps s this $O(\Delta^{1/2})$ phase of the computation is nonetheless negligible since comparison is so very fast compared with composition. If there were many more steps s, however, one would wish to *sort* the baby forms (according to the size of A_n), sort the giant forms, and *then* compare the two lists. That restores the matching process to $O(\Delta^{1/4+\epsilon})$ operations.

d. Composition is much faster if A_1 is small[4] (see Appendix 1). If $F = (A, B, C)$ has a small A, and if b is also small, the baby steps (16) may be more rapidly computed than the giant steps F^{H+rg}. In that case, one should take s somewhat larger than that given by (23) and allow more baby steps.

e. The Euclid algorithms (63) and (64) consume much of the time in composition. For large Δ, Lehmer's variation, [14] and [15, p. 305], should be tried.

f. The speed of these routines depends markedly on that of the multiprecision routines that are used. Machine code is preferable, and the radical new developments called Toom-Cook arithmetic, cf. [15, p. 258 ff.], should be examined.

g. Many tricks are possible. Suppose we want to factor

$$N = 146527939924199 \tag{81}$$

but do not know yet if it is composite. Since we do not know if $h(-N)$ is even or odd, we must take $b = 1$. Further, since

$$(-N/p) = +1$$

for $p = 2$, 3, and 5, its $h(-N)$ is relatively large. But

$$(-2N/p) = 0, -1, -1$$

for these p, and $h(-8N)$ is actually smaller than $h(-N)$. We would also know that $4 \mid h(-8N)$, and could therefore take $b = 4$. Thus, it is quicker to factor $2N$ than N, and I asked the man who inquired about (81), "Do you mind if I multiply your number by 2 before I factor it?" He said he didn't. Then

$$2N = 2 \cdot 1445599 \cdot 101361401 \tag{82}$$

follows from $h(-8N) = 6578024$ and $\mathcal{S} = C(2) \times C(4)$.

[4]Gauss pointedly emphasized this in [10, p. 393].

h. The $\mathcal{S} = C(2) \times C(4)$ above is isomorphic to Figure 1 with A_3 there the image of $(2, 0, N)$, the parasitic ambiguous form deliberately introduced above. The useful factorizations (82) are the image of A_1 and A_2. What role do G_1 and G_2 and $G_1^2 = G_2^2 = A_3$ play here? Their images have discriminant $-8N$, and dividing these expressions for $-8N$ by -8 we obtain two quadratic partitions:

$$(83)\quad N = 12529693^2 - 2 \cdot 2287495^2, \qquad N = 13297693^2 - 2 \cdot 3892345^2.$$

In effect, (83) gives the four values of $\sqrt{2} \pmod N$. The *Euler-Legendre method* [16], see also [15, p. 351], of factoring an N by combining two quadratic partitions is therefore equivalent to our compositing technique (46).

The difference between our method and theirs is that we create the partitions via the class number, while they presume that the partitions became available in some unspecified way. Such partitions $N = m^2 - an^2$, for any small a, may be wanted in their own right, and the adjunction of the form $(a, 0, N)$, as above, is a means of obtaining them, if they exist.

We may now add to our point C of §6 that just as we make no decision to use Fermat's method: $N = x^2 - y^2$, or not, we make no advanced decision to use our equivalent of Euler-Legendre. We move within $\mathcal{S}$ as our forms (7) direct us, and only after we obtain results such as (34a), (34b) or (83) do we learn which of these classical strategies we were simulating.

i. Aside from a peculiar $2N$ as in point g. above, it would generally be agreed by all, including Gauss [10, sect. 320], that in factoring an integer it is helpful to first remove any of its small divisors, no matter what factorization method is subsequently used. Gauss cites

$$314159265 = 3^2 \cdot 5 \cdot 7 \cdot 997331,$$

and then starts to work on 997331. This initial phase is the trial divisor method. We note that our method really includes this inasmuch as the evaluation of (6) includes the determination of all $p \leqq P$ having $(-\Delta/p) = 0$. The difference between (6) and a pure trial divisor technique is that the latter ignores, and keeps no record of, the distinction: $(-\Delta/p) = \pm 1$, while here we not only record $(-\Delta/p) = 0$, corresponding to *one* form (p, p, c) but we embody in the estimate (6) and list (7) a record of those p having *two* forms $(p, \pm b, c)$, or *none*. Thus, (6) and (7) are a deepening of the trial divisor technique.

We could add this as a fifth characterization in §6. While the trial divisor method retains a basic role in factorization in this viewpoint, its crudity consists of recording only those

$$(84)\qquad -\Delta \equiv r \pmod p$$

having $r = 0$. Actually, by the time one has evaluated (84) for all $p \leqq P$ one has obtained a mass of information concerning the arithmetic nature of Δ, and our method is really based upon the decision not to discard all of this information.

REFERENCES

1. Edmund Landau, *Elementary number theory*, Chelsea, New York, 1958, Def. 38, p. 219. MR **19**, 1159.

2. J. E. Littlewood, *On the class-number of the corpus* $P(\sqrt{-k})$, Proc. London Math. Soc. **28** (1928), 358–372.

3. D. H. Lehmer, *An announcement concerning the Delay Line Sieve* DLS-127, Math. Comp. **20** (1966), 645–646.

4. D. H. Lehmer, Emma Lehmer and Daniel Shanks, *Integer sequences having prescribed quadratic character*, Math. Comp. **24** (1970), 433–451.

5. Daniel Shanks, *Solution of* $x^2 \equiv a \pmod{p}$ *and generalizations* (to appear).

6. ———, *Gauss's ternary form reduction and the 2-Sylow subgroup* (to appear).

7. ———, *Solved and unsolved problems in number theory*. Vol. I, Spartan Books, New York, 1962, Chapter 2. MR **28** #3952.

8. Richard Dedekind, *Über die Anzahl der Idealklassen in reinen kubischen Zahlkörpern*, Crelle's J. **121** (1900), 40–123.

9. G. B. Mathews, *Theory of numbers*, 2nd ed., Chelsea, New York, 1961. MR **23** #A3698.

10. C. F. Gauss, *Untersuchungen über höhere Arithmetik*, Chelsea, New York, 1965. MR **32** #5488.

11. Daniel Shanks, *On Gauss's class number problems*, Math. Comp. **23** (1969), 158–159.

12. P. D. T. A. Elliott, *Some applications of a theorem of Raikov to number theory*, J. Number Theory (to appear).

13. ———, *On the distribution of the quadratic class number*, Litovsk. Mat. Sb. (Russian) (to appear).

14. D. H. Lehmer, *Euclid's algorithm for large numbers*, Amer. Math. Monthly **45** (1938), 370–371.

15. Donald E. Knuth, *Seminumerical algorithms*, The Art of Computer Programming, vol. 2, Addison-Wesley, Reading, Mass., 1969.

16. Adrien-Marie Legendre, *Théorie des nombres*. Tome I, Blanchard reprint, Paris, 1955, Section (236), p. 311.

17. Daniel Shanks, *An interesting sequence:* S_n (to appear).

Applied Mathematics Laboratory
Naval Ship Research and Development Center
Washington, D.C. 20034

AUTHOR INDEX

Italic numbers refer to pages on which a complete reference to a work by the author is given. Roman numbers refer to pages on which a reference is made to a work of the author. For example, under Barsotti would be the page on which a statement like the following occurs: "This theorem was proved earlier by Barsotti [7, Theorem 6] in the following manner. . . ."
Boldface numbers indicate the first page of the articles in this volume.

SUBJECT INDEX